Solutions Guide for
General Chemistry and College Chemistry

EIGHTH EDITIONS
BY HOLTZCLAW AND ROBINSON

John H. Meiser
Ball State University

Frederick K. Ault
Ball State University

D. C. HEATH AND COMPANY
Lexington, Massachusetts   Toronto

Copyright © 1988 by D. C. Heath and Company.

Previous editions copyright © 1984, 1980, and 1976 by D.C. Heath and Company.

All rights reserved. No part of this publication may be reproduced or transmitted in any form or by any means, electronic or mechanical, including photocopy, recording, or any information storage or retrieval system, without permission in writing from the publisher.

Published simultaneously in Canada.

Printed in the United States of America.

International Standard Book Number: 0-669-12866-X

10 9 8 7 6 5 4 3 2 1

Preface

This guide provides students with helpful step-by-step analysis of solutions to specific types of problems. Its present form is expanded from the previous edition to include roughly half the questions at the end of each chapter. The numbers of the exercises selected for inclusion in this manual are printed red in the textbooks. Questions requiring verbal responses are also included. A separate edition, coauthored with Norman Griswold, contains solutions for all questions and is available for instructors who adopt the text.

Each chapter of the guide begins with a short introduction to the particular subject area. The Ready Reference section presents abbreviated definitions and formulas pertinent to the subject area of that chapter. The exercises with their solutions follow. A useful feature to note is that the presentation includes the statement of each problem selected. At least one example of each type of problem in the chapter is chosen for a detailed solution. In most cases a verbal explanation expands the required mathematical manipulations.

The use of this guide should help students become independent learners and gain a better understanding of the quantitative aspects of the sciences. However, students must never merely read over the solutions and think that they have mastered them. The best way to use this guide is for students encountering difficulty to work on the exercise as far as they can. A brief look at the first line or lines of the solution should be used by the student to get started on the problem. If the student becomes "stuck" again, a more detailed look at the solution may be necessary to complete the problem. In the long run, the study of chemistry is not complete without the student's being able to solve problems such as those found in the text.

In producing this manual, we are indebted to Pang Ma and Eugene Wagner, Indiana University Center for Medical Education, and Abbas Pezeshk, Moorhead State University, for their helpful comments and especially to James P. Rybarczyk, Ball State University, for his help in several areas and for his words of encouragement. We also thank our typists, Joanne Beckham and Karen Hall, for the physical body of this manuscript and all the D. C. Heath personnel, especially Mary Le Quesne, our editor; Carmen Wheatcroft, who saw to the overall management of the work; Marret McCorkle, who organized the day-to-day details of producing this work; and Antoinette Schleyer, who coordinated this work with the main texts. Finally, we thank our families for their perseverance with us during the long hours spent in preparing this work.

J. H. M.
F. K. A.

Contents

1. Some Fundamental Concepts   1
2. Symbols, Formulas, and Equations; Elementary Stoichiometry   13
3. Applications of Chemical Stoichiometry   34
4. Thermochemistry   50
5. Structure of the Atom and the Periodic Law   60
6. Chemical Bonding, Part 1: General Concepts   75
7. Chemical Bonding, Part 2: Molecular Orbitals   93
8. Molecular Structure and Hybridization   102
9. Chemical Reactions and the Periodic Table   113
10. The Gaseous State and the Kinetic-Molecular Theory   125
11. Condensed Matter: Liquids and Solids   144
12. Solutions; Colloids   159
13. The Active Metals   179
14. Chemical Kinetics   187
15. An Introduction to Chemical Equilibrium   197
16. Acids and Bases   210
17. Ionic Equilibria of Weak Electrolytes   225
18. The Solubility Product Principle   259
19. Chemical Thermodynamics   280
20. A Survey of the Nonmetals   295
21. Electrochemistry and Oxidation-Reduction   307
22. The Nonmetals, Part 1: Hydrogen, Oxygen, Sulfur, and the Halogens   332
23. The Nonmetals, Part 2: Carbon, Nitrogen, Phosphorus, and the Noble Gases   346

24. Nuclear Chemistry   357
25. The Semi-Metals   364
26. Coordination Compounds   372
27. The Transition Elements   381
28. The Post-Transition Metals   392
29. The Atmosphere and Natural Waters   400
30. Organic Chemistry   404
31. Biochemistry   412

Appendixes

Appendix C: General Physical Constants   419
Appendix D: Solubility Products   420
Appendix E: Formation Constants for Complex Ions   421
Appendix F: Ionization Constants of Weak Acids   422
Appendix G: Ionization Constants of Weak Bases   422
Appendix H: Standard Electrode (Reduction) Potentials   423
Appendix I: Standard Molar Enthalpies of Formation, Standard Molar Free Energies of Formation, and Absolute Standard Entropies (298.15 K (25°C), 1 atm)   424

1

Some Fundamental Concepts

INTRODUCTION

This chapter of the textbook includes a brief introduction to the nature of matter and energy. It also details the units of measurement essential for quantifying and communicating observations of scientific phenomena.

The almost universally accepted system of measurement, the metric system, originally was adopted in France and has undergone continuous improvement since the early nineteenth century. In 1960 the Eleventh Conference on Weights and Measures, as a result of major revisions to the metric system, renamed the system the "International System of Units." The abbreviation of the original French title "le Système International d'Unités," SI, is now the preferred form for identifying the system.

The seven base units of SI are listed in Table 1. Units needed for expressing quantities such as volume, density, velocity, energy, and force are derived from these base units. This manual, as well as the major textbooks, emphasizes SI units.

Table 1 SI Base Units

Quantity	Symbol for Quantity	SI Physical Unit	Symbol
Length	d	Meter	m
Mass	M, or m	Kilogram	kg
Time	t	Second	s
Electric current	I	Ampere	A
Thermodynamic temperature	T	Kelvin	K
Amount of substance	n	Mole	mol
Luminous intensity	I_v	Candela	cd

The problem-solving exercises included in this chapter stress development of skills for each of the following applications: expression of numbers in scientific notation, conversion among SI prefixes, expression of mass-to-volume ratios such as density and specific gravity, conversion of measurements expressed in English units to appropriate SI and derived units, conversion of temperatures from one system to another, and the transfer of heat.

Although scientific work is communicated in terms of SI language, U.S. industry is only gradually moving away from the English system of measurement. As a consequence, scientific workers have to work with two systems.

READY REFERENCE

The SI base units are frequently large with respect to the quantities to be measured. For example, the mass of a dime is approximately 0.0025 kg, and the wavelength of red light is 0.0000007 m. To avoid the cumbersome use of decimals, a system of prefixes specifying a multiplication factor to use with the base units has been developed. The most commonly employed prefixes are shown in Table 2. The SI unit is multiplied by the prefix that permits the numerical value of the measurement to be kept as near to 1 as is practicable.

Table 2 Common Prefixes for Measurement Units

Prefix	Symbol	Factor Times Unit	
kilo	k	1000	$= 10^3$ times
deci	d	1/10	$= 10^{-1}$ times
centi	c	1/100	$= 10^{-2}$ times
milli	m	1/1000	$= 10^{-3}$ times
micro	μ	1/1,000,000	$= 10^{-6}$ times
nano	n	1/1,000,000,000	$= 10^{-9}$ times
pico	p	1/1,000,000,000,000	$= 10^{-12}$ times

(Numbers in parentheses appearing after "Ready Reference" entries refer to sections in the textbook where additional information may be found.)

Density (*D*) (1.13) The density of a substance is the ratio of its mass *M* to its volume *V*. The usual expression for the density of solids and liquids is mass (grams) per cubic centimeter or per milliliter. Based on these units, the densities of elements range from a low of 8.99×10^{-5} g/cm^3 for hydrogen gas to a high of 22.57 g/cm^3 for osmium. Among compounds, water has a maximum density of 1.0 g/cm^3 at 3.98°C. The densities of gases—such as hydrogen or oxygen, for example—are typically reported in units of grams per liter to avoid the excessive use of decimals.

$$\text{Density} = \frac{\text{mass}}{\text{volume}} = \frac{M}{V} = \frac{\text{g}}{\text{mL or cm}^3}$$

Length (meter, m) (1.11) The meter is the standard unit of length, and it is used in conjunction with the prefixes for most linear measurements in science. Another commonly employed non-SI unit is the angstrom (1 Å = 10^{-10} m), which appears in measurements and calculations in spectrographic work.

Mass (kilogram, kg) (1.2) The kilogram is the standard unit of mass and is a measure of the quantity of matter a body possesses. The mass of a body is constant while at rest regardless of geographic position and always is considered constant for chemical studies. Weighings made in the chemical laboratory are balanced against the mass of an object used as a standard; values obtained in this manner are masses. Because the kilogram is a large unit of mass, the more familiar mass in laboratory practice is the gram (g), defined as 1/1000 kg. Stated differently, 1000 g = 1 kg.

Significant figures (1.10) In making measurements it is important to recognize the limitations of the measuring instruments. For example, a typical bathroom scale, although possibly accurate, is calibrated to register weights to the nearest pound and would not be useful for measuring the small quantities used in laboratory experiments; the uncertainty in each measurement would be too great.

Chapter 1 Some Fundamental Concepts

Laboratory measurements require a higher degree of precision or less uncertainty than would be required for weighing oneself on a scale. Common laboratory balances used in general chemistry have a range of uncertainties from 0.01 g to 0.1 g for simple beam balances and from 0.0001 g to 0.001 g for analytical balances.

The reporting of measurements should be made in a way that conveys the maximum information about the measurements to the user. Numbers that are actually read from an instrument are considered to be *significant figures*. Answers to mathematical operations may not contain more significant figures (less uncertainty) than the operant with the smallest number of significant figures (greatest uncertainty). The solved problems in this manual follow the rules on significant figures as these are presented in the textbooks. Many of the solved problems have been analyzed into steps, and in these cases, intermediate answers are written with one more significant figure than is justified in the final answer. *Computations have been made with an electronic calculator. All the numbers that were generated in the calculator from mathematical operations have been carried through to the final answer. The answer is rounded to the proper number of significant figures at the end of the computation.*

Specific gravity (sp gr) (1.13)   The specific gravity of a substance is the ratio of the density of that substance to the density of a substance used as a reference standard. (Both densities are measured at the same temperature.) Water generally serves as the standard for both solids and liquids; air is the standard for gases. Specific gravity may be expressed alternatively as the ratio of the masses of equal volumes of two substances, where one of the substances is arbitrarily designated as the standard. (Again both substances must be at the same temperature.) Notice from the following expression that specific gravity values are unitless:

$$\text{Specific gravity} = \frac{\text{density of unknown}}{\text{density of known}} = \frac{g/cm^3}{g/cm^3} \quad \text{(Units divide out, leaving a dimensionless value.)}$$

Use of prefixes (1.8)   Following is a conversion scheme using some common prefixes and examples.

Length: km $\xrightleftharpoons[1000 \div]{\times 1000}$ m $\xrightleftharpoons[100 \div]{\times 100}$ cm $\xrightleftharpoons[10 \div]{\times 10}$ mm $\xrightleftharpoons[1000 \div]{\times 1000}$ μm $\xrightleftharpoons[1000 \div]{\times 1000}$ nm

Mass: kg $\xrightleftharpoons[1000 \div]{\times 1000}$ g $\xrightleftharpoons[1000 \div]{\times 1000}$ mg $\xrightleftharpoons[1000 \div]{\times 1000}$ μg

Volume: m³ $\xrightleftharpoons[1000 \div]{\times 1000}$ dm³ = L $\xrightleftharpoons[1000 \div]{\times 1000}$ mL $\xrightleftharpoons[10000 \div]{\times 1000}$ μL

Volume (cubic meter, m^3) (1.11)   The cubic meter, the standard unit of volume, is not widely used for measuring laboratory quantities. The volumes of laboratory glassware and other lab ware are more conventionally expressed in cubic centimeters, cm^3, or in terms of liters. Although the liter (L) is not an SI unit, its value is defined to be 1 cubic decimeter (dm^3), or 0.001 m^3. The milliliter (mL) and the cubic centimeter (cm^3) now are defined as equal: 1 mL = 1 cm^3; and 1000 mL = 1000 cm^3 = 1.0 L.

USE OF CALCULATORS

Hand-held calculators are now available to most students. Calculators come in a wide variety of brands and designs, so it would be impossible to attempt to give the details of operation for all of them. However, because Texas Instruments' TI calculators are very popular among students and because most calculators operate according to the same basic principles, this chapter introduces a section in some problems called "Keystrokes," a feature also included in later chapters. "Keystrokes" sections will not appear in each problem but only in those problems in which mathematical operations other than $\times$, $\div$, $+$, and $-$ occur. Thus, when exponentials are introduced, the student will have both the standard mathematical form to observe as well as the actual keystrokes to operate the calculator. It should be noted that when multiplication by 10^x appears by itself in some problems, the number 1 must be pressed before striking the Exp key to enter the exponent. Some "Keystrokes" sections will appear even with the four basic mathematical operations to demonstrate correct use of rounding procedures for expressing the proper number of significant figures.

SOLUTIONS TO CHAPTER 1 EXERCISES

2. What properties distinguish solids from liquids? Solids from gases? Liquids from gases?

   Solution

   The phases of matter are distinguished by physical characteristics or properties. Shape, volume, free surface, color, electrical conductivity, and compressibility are the usual descriptors. The physical state or phase of a substance depends on its temperature. Perhaps the most common example of phase transition with temperature change is the melting of ice (water) to liquid and to steam.

3. Describe a chemical change that illustrates the law of conservation of matter.

   Solution

   The law of conservation of matter means that matter is neither gained nor lost during chemical change. Water, for example, is composed of the elements hydrogen and oxygen in a ratio of two atoms of hydrogen to one atom of oxygen. Atoms of hydrogen and oxygen, however, do not have equal masses. When water molecules are decomposed by electrical current they divide to give:

   $$H_2O \text{ molecule} \xrightarrow{\text{electricity}} 2H \text{ atoms} + 1O \text{ atom}$$

   The decomposition of 18.0 g of water produces 2.0 g of hydrogen plus 16.0 g of oxygen. Conversely, when 2.0 g of hydrogen react with 16.0 g of oxygen, 18.0 g of water is produced. No mass is gained or lost during the reaction process.

Chapter 1 Some Fundamental Concepts   5

5. Describe how a scientist would proceed from the formulation of a question to the establishment of a theory.

Solution

Major scientific theories have typically evolved from attempts to explain observations of events or phenomena that investigators experienced. Questions develop from observations. To answer a question, an organized plan of study designed to gather facts and information about all phases of the question must be developed. We call such an organized plan the *scientific method* of problem solving. Data gathered from experiments are examined for any general relationships that may exist. Tentative explanations, hypotheses, are developed to explain results. If the proposed hypothesis can be used to predict results from further, repeated experiments, it may have enough credibility to justify its status as a theory.

CLASSIFICATION OF MATTER

7. In what ways do heterogeneous mixtures differ from homogeneous mixtures? In what ways are they alike?

Solution

Mixtures are defined as systems containing two or more components that can be separated by physical means. When components mix and blend in a manner such that all regions of the system are the same, the system is homogeneous: soft drinks, sugar water, gasoline, brass, and sterling silver are common examples. Most mixtures in the environment and in our experience are not homogeneous. Rather, deposits of rocks and minerals, river water, blood, and fruit salad are mixtures whose compositions vary throughout and the systems are heterogeneous. They are alike in that physical methods can be used to separate the components.

8. Classify each of the following as a heterogeneous mixture or a homogeneous mixture: blood, ocean water, air, blueberry pancakes, pancake syrup, gasoline, a milkshake, concrete, a bowl of vegetable soup.

Solution

blood	heterogeneous
ocean water	heterogeneous
air	usually considered homogeneous
blueberry pancakes	heterogeneous
blueberry syrup	homogeneous
gasoline	homogeneous
milkshake	heterogeneous
concrete	heterogeneous
vegetable soup	heterogeneous

11. Classify each of the following as a physical or chemical change: condensation of steam, burning of gasoline, souring of milk, dissolving sugar in water,

melting of gold, leaves "turning" in color, explosion of a firecracker, magnetizing a screwdriver, melting of ice.

Solution

condensation of steam	physical
burning of gasoline	chemical
souring of milk	chemical
dissolving of sugar	physical
melting of gold	physical
"turning" of leaves	chemical
explosion of firecrackers	chemical
magnetizing a screwdriver	physical
melting of ice	physical

14. How do molecules of elements and molecules of compounds differ?

Solution

A molecule is the smallest particle of an element or compound that can have a stable, independent existence. Compounds, by definition, are composed of two or more elements. Molecules of compounds — water, for example — are particles that contain different atoms. Elements, such as chlorine (Cl_2), may exist in stable, independent units containing several of the same atoms.

SIGNIFICANT FIGURES

16. Indicate whether each of the following can be determined exactly or must be measured with some degree of uncertainty.

    (a) The number of apples in a bag
    (b) The mass of a watermelon
    (c) The number of gallons of gasoline necessary to fill an automobile gas tank
    (d) The number of meters in exactly 5 km
    (e) The mass of this textbook
    (f) The number of minutes in one year
    (g) The time required to drive from Dallas to Chicago at an average speed of 53 miles per hour

    Solution

    (a) Counted exactly. Truckloads would not be counted.
    (b) With uncertainty
    (c) With uncertainty—probably to the nearest one-tenth of a gallon
    (d) Exact conversion
    (e) Uncertainty—perhaps to 0.1 g
    (f) Exact—defined in terms of common units
    (g) With considerable uncertainty

18. How many significant figures are contained in each of the following numbers: 113; 207.033; 0.0820; 0.04109; 3.2×10^{-8}; 9.74150×10^{-4}; 17.0?

    Solution

    113, 3 s.f.; 207.033, 6 s.f.; 0.0820, 3 s.f.; 0.04109, 4 s.f.; 3.2×10^{-8}, 2 s.f.; 9.74150×10^{-4}, 6 s.f.; 17.0, 3 s.f.

Chapter 1 Some Fundamental Concepts   7

20. Express the following numbers in exponential notation: 711.0; 0.239; 90743; 134.2; 0.05499; 10000.0; 0.000000738592.

    Solution

    711.0 (4 s.f.) to 7.110×10^2
    0.239 (3 s.f.) to 2.39×10^{-1}
    90743 (5 s.f.) to 9.0743×10^4
    134.2 (4 s.f.) to 1.342×10^2
    0.05499 (4 s.f.) to 5.499×10^{-2}
    10000.0 (6 s.f.) to 1.00000×10^4
    0.000000738592 (6 s.f.) to 7.38592×10^{-7}

21. Perform the following calculations and give the answer with the correct number of significant figures.

    (a) 1228×3442
    (b) $5.6 \times 10^{-2} \times 7.41 \times 10^3$
    (e) $1.4 + 7.340 + 4.7593$

    Solution

    (a) $1228 \times 3442 = ?$ Calculator shows 4226776. Each multiplier has four significant figures. Therefore, the product must be rounded to four figures as 4227000 or 4.227×10^6.
    (b) $5.6 \times 10^2 \times 7.41 \times 10^3 = ?$ The multiplier 5.6×10^2 has two significant figures and limits the product to two figures: 4.1×10^6.

    KEYSTROKES: 5.6 $\boxed{\text{EE}}$ 2 × 7.41 $\boxed{\text{EE}}$ 3 = 4.1496×10^6

    (e) $1.4 + 7.340 + 4.7593 = ?$ When the numbers are keyed in the calculator and summed, the answer reads 13.4993. Since 1.4 has only one position right of the decimal, the sum is rounded to 13.5.

22. It is necessary to determine the density of a liquid to four significant figures. The volume of solution can be measured to the nearest 0.01 cm³.

    (a) What is the minimum volume of sample that can be used for the measurement?

    Solution

    (a) Density = mass/volume. For density to have four significant figures, the minimum number of significant figures for either determinant must be four. Since volume measurement accuracy is limited to 0.01 cm³, to get four figures requires at least 10.00 cm³.

METRIC SYSTEM: SI UNITS

26. Indicate the SI base units appropriate to express the following.

    (a) the length of a 500-mile race
    (b) the mass of an elephant
    (c) the volume of a swimming pool
    (d) the speed of an automobile
    (e) the density of the metal platinum

(f) the area of a football field
(g) the maximum temperature at the South Pole on April 1, 1913

Solution

(a) kilometer
(b) kilogram
(c) cubic meter
(d) kilometers per hour
(e) grams per cubic centimeter
(f) square meters
(g) kelvin

30. Complete the following conversions.

(a) 13 g = _____ kg
(c) 3451 mg = _____ g
(e) 4.18 g = _____ mg
(h) 17.38 km = _____ cm

Solution

Making a conversion between measurement units requires us first to identify an appropriate conversion factor, or identity, that relates the two units. The actual conversion operation involves multiplying the given unit times the fraction created from the identity, which causes the common unit to divide out leaving the desired unit.

(a) 13 g = _____ kg

Identity: 1 kg = 1000 g

$$\text{Mass (kg)} = 13 \text{ g} \times \frac{1 \text{ kg}}{1000 \text{ g}} = 0.013 \text{ kg}$$

(c) 3451 mg = _____ g

Identity: 1 g = 1000 mg

$$\text{Mass (g)} = 3451 \text{ mg} \times \frac{1 \text{ g}}{1000 \text{ mg}} = 3.451 \text{ g}$$

(e) 4.18 g = _____ mg

Identity: 1 g = 1000 mg

$$\text{Mass (mg)} = 4.18 \text{ g} \times \frac{1000 \text{ mg}}{1 \text{ g}} = 4180 \text{ mg} = 4.18 \times 10^3 \text{ mg}$$

(h) 17.38 km = _____ cm

Two identities are involved:

  1 km = 1000 m
  1 m = 100 cm

Make the conversion in two steps: first to meters, then to centimeters.

1. $\text{Distance (m)} = 17.38 \text{ km} \times \dfrac{1000 \text{ m}}{\text{km}} = 17{,}380 \text{ m}$

2. $\text{Distance (cm)} = 17{,}380 \text{ m} \times \dfrac{100 \text{ cm}}{\text{m}} = 1{,}738{,}000 \text{ cm} = 1.738 \times 10^6 \text{ cm}$

Chapter 1 Some Fundamental Concepts   9

CONVERSION OF UNITS

33. Soccer is played with a round ball between 69 and 71 cm in circumference and weighing between 400 and 450 g (two significant figures). What are these specifications in inches and ounces?

   Solution

   Identities: 1 in = 2.54 cm

   1 oz = 28.35 g

   Circumference:

   1. Distance (in) = 69 cm $\times \dfrac{1 \text{ in}}{2.54 \text{ cm}}$ = 27 in (2 s.f.)

   2. Distance (in) 71 cm $\times \dfrac{1 \text{ in}}{2.54 \text{ cm}}$ = 28 in (2 s.f.)

   Mass:

   1. Mass (oz) = 400 g $\times \dfrac{1 \text{ oz}}{28.35 \text{ g}}$ = 14 oz

   2. Mass (oz) = 450 g $\times \dfrac{1 \text{ oz}}{28.35 \text{ g}}$ = 16 oz

36. A 170-cm tall track-and-field athlete weighs 49.8 kg. Is she more likely to be a distance runner or a shot putter?

   Solution

   Convert the athlete's height and weight to English units.

   From Appendix B: 1 inch = 2.54 cm

   1 kg = 2.2046 lb

   Height = 170 cm $\times \dfrac{1 \text{ in}}{2.54 \text{ cm}} \times \dfrac{1 \text{ ft}}{12 \text{ in}}$ = 5.577 ft (or 5 ft 7 in)

   Weight = 49.8 kg $\times \dfrac{2.2046 \text{ lb}}{\text{kg}}$ = 110 lb

   Her height and weight fit those of a distance runner.

40. Gasoline is sometimes sold by the liter. How many liters are required to fill a 10.0-gallon (liquid, U.S.) gas tank?

   Solution

   Identities: 1 L = 1.057 qt

   1 gal = 4 qt

   V (L) = 10.0 gal $\times \dfrac{4 \text{ qt}}{\text{gal}} \times \dfrac{1 \text{ L}}{1.057 \text{ qt}}$ = 37.8 L

46. The distance between the centers of the two oxygen atoms in an oxygen molecule is 1.21 Å. What is this distance in centimeters and in inches?

Solution

Identities: $1 \text{ Å} = 1 \times 10^{-8}$ cm

$1 \text{ in} = 2.54$ cm

Distance (cm) = $1.21 \text{ Å} \times \dfrac{1 \times 10^{-8} \text{ cm}}{1 \text{ Å}} = 1.21 \times 10^{-8}$ cm

Distance (in) = $1.21 \times 10^{-8} \text{ cm} \times \dfrac{1 \text{ in}}{2.54 \text{ cm}} = 4.76 \times 10^{-9}$ in

49. Calculate the density of aluminum if 13.8 cm³ has a mass of 37.3 g.

Solution

Definition: Density = mass/volume

$$D = \dfrac{37.3 \text{ g}}{13.8 \text{ cm}^3} = 2.70 \, \dfrac{\text{g}}{\text{cm}^3}$$

51. What is the mass of each of the following?

(a) 6.00 cm³ of bromine; density = 2.928 g/cm³
(b) 1000 mL of octane; density = 0.702 g/cm³

Solution

(a) Definition: Density = mass/volume. Rearrange the definition to solve for M in terms of V and D; $M = D \times V$.

$$M \text{ (g)} = 6.00 \text{ cm}^3 \times 2.928 \, \dfrac{\text{g}}{\text{cm}^3} = 17.6 \text{ g (3 s.f.)}$$

(b) From $D = M/V$, $M = D \times V$ and 1 mL = 1 cm³.

$$M \text{ (g)} = 1000 \text{ mL} \times 0.702 \, \dfrac{\text{g}}{\text{cm}^3} = 702 \text{ g}$$

52. What is the volume of each of the following?

(a) 25 g of iodine, density = 4.93 g/cm³

Solution

(a) Definition: $D = M/V$. Rearrange the definition to solve for V in terms of M and D: $V = M/D$.

$$V \text{ (cm}^3 \text{ or mL)} = \dfrac{25 \text{ g}}{4.93 \text{ g/mL}} = \dfrac{25 \text{ g}}{4.93} \times \dfrac{\text{mL}}{\text{g}} = 5.1 \text{ mL}$$

TEMPERATURE

55. Normal body temperature is 98.6°F. What is the temperature in degrees Celsius and in kelvins?

Chapter 1 Some Fundamental Concepts 11

Solution

Identities: $°C = \frac{5}{9}(°F - 32)$

$K = °C + 273.15$

$°C = \frac{5}{9}(98.6 - 32) = \frac{5}{9}(66.6) = 37.0°C$

$K = 37.0 + 273.15 = 310.2 \text{ K}$

56. The Voyager I flyby of Saturn revealed the surface temperature of the moon Titan to be 93 K. What is the surface temperature in degrees Celsius and degrees Fahrenheit?

Solution

Identities: $°C = K - 273.15$

$°F = \frac{9}{5}°C + 32$

$°C = 93 - 273.15 = -180°C$

$°F = \frac{9}{5}(-180) + 32 = -292°F$

ADDITIONAL EXERCISES

59. How do the densities of most substances change as a result of an increase in temperature?

Solution

Most substances undergo a volume increase with an increase in temperature. From the definition of density, $D = M/V$, when the volume is increased while maintaining a constant mass, the fraction decreases, yielding a lower density.

65. Solids will float in a liquid with a density greater than the solid. A pound of butter forms a block about $6.5 \times 6.5 \times 12.0$ cm. Will this butter float in water (density 1.0 g/cm^3)?

Solution

Butter floats if its density is less than that of water. Is it?

Identities: 1 lb = 453.6 g

$$D = \frac{M}{V}$$

$V(\text{cm}^3) = 6.5 \text{ cm} \times 6.5 \text{ cm} \times 12.0 \text{ cm} = 507 \text{ cm}^3$

$$D = \frac{453.6 \text{ g}}{507 \text{ cm}^3} = 0.89 \frac{\text{g}}{\text{cm}^3} \text{ (2 s.f.)}$$

Therefore, it will float.

71. What mass in kilograms of concentrated hydrochloric acid is contained in a standard 5.0-pint container? The specific gravity of concentrated hydrochloric acid is 1.21.

Solution

Definition:

$$\text{Specific gravity} = \frac{D \text{ acid solution}}{D \text{ water}} = \frac{D\,(\text{HCl})}{D\,(\text{H}_2\text{O})}$$

Rearrange to obtain $D\,(\text{HCl})$:

$$D\,(\text{HCl}) = D\,(\text{H}_2\text{O}) \times \text{Sp gr} = 1.00\,\frac{\text{g}}{\text{mL}} \times 1.21 = 1.21\,\text{g/mL}$$

Identities: $1.0\,\text{L} = 1.057\,\text{qt}$ or $0.946\,\text{L} = 1\,\text{qt}$

and $0.473\,\text{L} = 1\,\text{pt} = 473\,\text{mL}$

$$V\,(\text{mL}) = 5.0\,\text{pint} \times 473\,\frac{\text{mL}}{\text{pt}} = 2365\,\text{mL}$$

$$M\,(\text{kg}) = 2365\,\text{mL} \times 1.21\,\frac{\text{g}}{\text{mL}} \times \frac{1\,\text{kg}}{1000\,\text{g}} = 2.9\,\text{kg (2 s.f.)}$$

2

Symbols, Formulas, and Equations; Elementary Stoichiometry

INTRODUCTION

The interactions of matter that occur both in nature and in the laboratory are described in the language of chemistry, which includes symbols, formulas, and equations. Symbols consisting of a single capital letter or a capital letter and one or two small letters, such as O, N, Cl, Na, and Unh, are used as shorthand abbreviations for elements. Combinations of symbols are used to write formulas of compounds; for example, methane, CH_4; vitamin C (ascorbic acid), $C_6H_8O_6$; and table sugar (sucrose), $C_{11}H_{22}O_{11}$. Elements and compounds enter into chemical reactions and are described in the form of an equation using symbols and formulas.

Chemical equations are similar to household recipes in that both provide information for preparing products from particular quantities of reactants. Unlike household recipes, however, chemical equations usually do not detail reaction conditions or laboratory procedures other than noting that energy or a specific catalyst must be used in conjunction with the reactants. The purpose of a chemical equation is to convey quickly to the reader pertinent information about the reaction that includes the substances involved and their quantitative ratio.

In this chapter you have the opportunity to study quantitative aspects of reactions. By working with the symbols, formulas, and equations introduced here, you soon will become comfortable with the basic language tools of chemistry.

READY REFERENCE

Atomic weight (2.2)   The atomic weight of an element is the average mass of all atoms composing a normal sample of the element. For example, consider a sample of naturally occurring carbon atoms. Most of the carbon atoms (98.89%) would contain six protons and six neutrons, ^{12}C; some (1.11%) would contain six protons and seven neutrons, ^{13}C; and a trace (1 atom in 10^{12} atoms) would contain six protons and eight neutrons, ^{14}C. Carbon-12, -13, and -14 are called isotopes of carbon and differ only in the number of neutrons in their nuclei.

Carbon-12 has been arbitrarily assigned a mass of exactly 12.00 and is used as the reference standard for determining the masses of other atoms. Based on this standard the atomic weight of carbon is 12.011 and represents the average mass of the atoms composing a statistical sample of carbon atoms.

Empirical formula (2.1)   This is a formula that gives the simplest whole number ratio of atoms.

Formula weight (FW) (2.3)   The formula weight of a substance is the sum of the atomic weights of all of the atoms appearing in a formula. For example, the formula weight of copper(II) sulfate, $CuSO_4$, is calculated in the following way:

$$\text{Number of atoms:} \quad FW = \overset{1\ Cu}{1(63.546)} + \overset{1\ S}{1(32.06)} + \overset{4\ O}{4(15.9994)} = 159.60$$

Formula weights are conventionally written as unitless entities, with atomic mass units implied or understood.

Gram-atomic weight (g-at. wt) (2.2)   This is the atomic weight of an element expressed in grams.

Gram-formula weight (g-form. wt) (2.3)   Gram-formula weight is the formula weight of a substance expressed in grams.

Gram-molecular weight (g-mol. wt) (2.3)   This is the molecular weight of a substance expressed in grams.

Limiting reagent (2.9)   The reagent completely consumed by a reaction is called the limiting reagent. From a balanced chemical equation representing a reaction, the proper mole ratio for reactants is found. If the reaction mixture contains one reactant in excess of the proper mole ratio, the other reactant(s) will be completely consumed. The amount of product produced by the reaction, therefore, is limited by the amount of reagent available for reaction. The excess amount of the one reactant will remain unreacted. As an analogy, a chemical reaction may be compared to the assembly of bicycles begun with three bicycle frames and eight tires. Three complete bicycles can obviously be built, with two tires in excess. The available bicycle frames limit (like the limiting reagents) the number of bicycles that can be built when tires are in excess.

Mole (mol) (2.2)   The mole is defined as a number equal to the number of atoms in exactly 12 g of carbon-12. This number, called Avogadro's number, 6.022×10^{23}, has wide application in chemistry. For the purposes of this chapter, 1 mol can be interpreted in the following ways:

> The number of atoms in a gram-atomic weight of an elemental substance. For example, the number of atoms in 22.9898 g of sodium (Na) is 6.022×10^{23}.
>
> The number of formula units in a gram-formula weight of an ionic substance. For example, the number of formula units in 58.443 g of sodium chloride (NaCl) is 6.022×10^{23}.
>
> The number of molecules in a gram-molecular weight of a molecular substance. For example, the number of molecules in 44.010 g of carbon dioxide (CO_2) is 6.022×10^{23}.

Identities used for computations involving the mole and quantities of substances should be interpreted as equivalencies and not equalities as in mathematics. As cited above, 1 mole of sodium atoms (g-at. wt) has a mass of 22.9898 g, and the equivalence is written

1 mol Na ≙ g-at. wt Na = 22.9898 g ≙ 6.022×10^{23} atoms

where the symbol ≙ is read "is equivalent to" or "corresponds to." This equivalence can be used to calculate the amount of a substance (in moles) contained in a specified amount of that substance in the following ways:

$$\text{Moles} = \text{mass of substance} \times \frac{1 \text{ mol}}{\text{g-at. wt, g-form. wt, or g-mol. wt of substance}}$$

or

$$= \text{number of particles (atoms, formula units, or molecules)} \times \frac{1 \text{ mol}}{6.022 \times 10^{23} \text{ particles}}$$

Molecular weight (mol. wt) (2.3)   Molecular weight is the formula weight of a substance that exists as discrete molecules.

Percent yield (2.10)   The percent yield for a reaction is the yield fraction: actual yield divided by theoretical yield, times 100%.

$$\% \text{ yield} = \frac{\text{actual yield}}{\text{theoretical yield}} \times 100\%$$

Theoretical yield (2.10)   The theoretical yield is the potential amount of product that can be produced in a reaction. The theoretical yield is determined by the mole ratio in the balanced equation and calculated from the amounts of available reactant(s).

Yield (2.10)   The actual amount of product produced in a chemical reaction is the yield.

SOLUTIONS TO CHAPTER 2 EXERCISES

Note: Students sometimes get answers that differ in the last decimal place from those given. This generally reflects different ways of rounding off intermediate steps in the calculation. The answers given have been determined by rounding off to the correct number of significant figures at the end of the calculation. Molecular weights were calculated using all significant figures in the atomic weights of the constituent atoms.

SYMBOLS, FORMULAS, AND CHEMICAL EQUATIONS

2. Explain why the symbol for the element sulfur and the formula for a molecule of sulfur differ.

   Solution

   The symbol for the element sulfur is S. Sulfur in its natural state consists of molecules that contain eight sulfur atoms.

3. Determine the empirical formulas of the following compounds:
   (a) dinitrogen tetraoxide, N_2O_4;
   (b) hydrazine, N_2H_4;
   (c) phosphoric acid, H_3PO_4;
   (d) the anesthetic cyclopropane, C_3H_6;
   (e) fructose, $C_6H_{12}O_6$;

Solution

(a) The simplest whole number ratio of nitrogen atoms to oxygen atoms is obtained by dividing both subscripts by 2.

$$N_2O_4 \div 2 = N_1O_2 \text{ or } NO_2$$

(b) $N_2H_4 \div 2 = NH_2$
(c) The subscripts 3, 1, and 4 are in their lowest factor state. Therefore, the molecular formula, H_3PO_4, is also the empirical formula.
(d) $C_3H_6 \div 3 = CH_2$
(e) $C_6H_{12}O_6 \div 6 = CH_2O$

4. Write a balanced equation that describes each of the following chemical reactions:

   (a) Acetylene gas, C_2H_2, burns in air forming gaseous carbon dioxide, CO_2, and water.
   (b) When heated, pure aluminum reacts with air to give Al_2O_3.
   (c) Gypsum, $CaSO_4 \cdot 2H_2O$, decomposes when heated, giving calcium sulfate, $CaSO_4$, and water.
   (e) During photosynthesis in plants, carbon dioxide and water are converted into glucose, $C_6H_{12}O_6$, and oxygen, O_2.
   (g) Water vapor reacts with sodium metal to produce gaseous hydrogen, H_2, and solid sodium hydroxide, NaOH.

Solution

(a) In setting up problems of this type, first set up the equation describing the reaction. Write the starting materials as reactants with a plus sign between them followed by an arrow and then the products.

$$C_2H_2 + O_2 \text{ (only oxygen gas reacts in the air)} \longrightarrow CO_2 + H_2O$$

Now make the number of atoms on both sides equal. Balance the number of carbons by placing the coefficient 2 before CO_2.

$$C_2H_2 + \underline{\hspace{1cm}} O_2 \longrightarrow 2CO_2 + H_2O$$

Hydrogen is already balanced. Since there are 5 oxygen atoms on the right and only 2 on the left (count them now), multiply O_2 by $\frac{5}{2}$, giving

$$C_2H_2 + \tfrac{5}{2}O_2 \longrightarrow 2CO_2 + H_2O$$

Fractional exponents are not normally used. Remove the fraction by multiplying the entire equation by 2. The final result is

$$2C_2H_5 + 5O_2 \longrightarrow 4CO_2 + 2H_2O$$

(b) $4Al + 3O_2 \longrightarrow 2Al_2O_3$

Chapter 2 Symbols, Formulas, and Equations; Elementary Stoichiometry 17

(c) $CaSO_4 \cdot 2H_2O \xrightarrow{\Delta} CaSO_4 + 2H_2O$
(e) $6CO_2 + 6H_2O \rightarrow C_6H_{12}O_6 + 6O_2$
(g) $2Na + 2H_2O \rightarrow 2NaOH + H_2$

5. Balance the following equations:

   (a) $Pt + Cl_2 \rightarrow PtCl_4$
   (d) $Sc_2O_3 + SO_3 \rightarrow Sc_2(SO_4)_3$
   (e) $Al + H_2SO_4 \rightarrow Al_2(SO_4)_3 + H_2$
   (g) $CuCO_3 + HCl \rightarrow CuCl_2 + H_2O + CO_2$

   Solution

   (a) $Pt + 2Cl_2 \rightarrow PtCl_4$
   (d) $Sc_2O_3 + 3SO_3 \rightarrow Sc_2(SO_4)_3$
   (e) $2Al + 3H_2SO_4 \rightarrow Al_2(SO_4)_3 + 3H_2$
   (g) $CuCO_3 + 2HCl \rightarrow CuCl_2 + H_2O + CO_2$

7. Determine both the formula and the number of moles of the missing substances in the following equations:

   (a) $Zn + CuSO_4 \rightarrow \underline{} + ZnSO_4$
   (b) $N_2 + \underline{} \rightarrow 2NH_3$
   (c) $2K + 2H_2O \rightarrow 2KOH + \underline{}$
   (d) $2C_6H_6 + \underline{} \rightarrow 12CO_2 + 6H_2O$
   (e) $2MnO_2 + 4KOH + \underline{} \rightarrow 2H_2O + 2K_2MnO_4$
   (f) $2AgNO_3 + 2\underline{} \rightarrow Ag_2O + H_2O + 2NaNO_3$

   Solution

   (a) $Zn + CuSO_4 \rightarrow Cu + ZnSO_4$
   (b) $N_2 + 3H_2 \rightarrow 2NH_3$
   (c) $2K + 2H_2O \rightarrow 2KOH + H_2$
   (d) $2C_6H_6 + 15O_2 \rightarrow 12CO_2 + 6H_2O$
   (e) $2MnO_2 + 4KOH + O_2 \rightarrow 2H_2O + 2K_2MnO_4$
   (f) $2AgNO_3 + 2NaOH \rightarrow Ag_2O + H_2O + 2NaNO_3$

ATOMIC AND MOLECULAR WEIGHTS, MOLES

8. If the mass of a ^{12}C atom were redefined to be exactly 18 nmu (new mass units), what would the mass of an average oxygen atom be in nmu?

   Solution

   Atomic weights are determined by using ^{12}C as the weight standard with this isotope having a mass of exactly 12.0 amu (atomic mass units). Any change in the definition of the standard will produce a proportional change in all other atomic weights. Therefore, the new mass of oxygen is

   ^{12}C (nmu) $= \frac{18}{12} = 1.5$ times larger

   Mass (O) (nmu) $= 1.5 (15.9994) = 23.9991$ nmu

9. A sample of CsCl contains 0.55418 g of Cs and 0.14783 g of Cl. From this information, calculate the atomic weight of Cs using 35.453 as the atomic weight of Cl.

Solution

The substance, CsCl, contains one Cs atom and one Cl atom. Their combining masses are proportional to their atomic weights. The atomic weight of Cs is greater than that of Cl by the fraction

$$\frac{\text{Mass (Cs)}}{\text{Mass (Cl)}} = \frac{0.55418 \text{ g}}{0.14783 \text{ g}} = 3.7487655$$

Multiplying this factor by the known atomic weight, we have

$$\text{Atomic weight (Cs)} = 3.7487655 \times 35.453 = 132.90$$

11. Determine the molecular weight of each of the following compounds:

    (a) Hydrogen bromide, HBr
    (c) Sulfuric acid, H_2SO_4
    (g) Aspirin, $C_6H_4(CO_2H)(CO_2CH_3)$

Solution

(a) HBr: $1.0079 + 79.904 = 80.912$

(c) H_2SO_4: $\quad 2 \times 1.0079 = \quad 2.0158$
$\qquad\qquad 1 \times 32.06 \;\; = \quad 32.06$
$\qquad\qquad 4 \times 15.9994 = \quad 63.9976$
$\qquad\qquad\qquad\qquad\qquad\;\; 98.0734 \to 98.07$

(g) Aspirin: $9C + 8H + 4O$
$\qquad\qquad 9 \times 12.011 \quad = 108.099$
$\qquad\qquad 8 \times 1.0079 \quad = \quad 8.0632$
$\qquad\qquad 4 \times 15.9994 \;\, = \quad 63.9976$
$\qquad\qquad\qquad\qquad\qquad\;\; 180.1598 \to 180.160$

12. Calculate the mass, in grams, of one mole of each of the following compounds:

    (d) Silver sulfate, Ag_2SO_4
    (e) Acetic acid, CH_3CO_2H
    (f) Boric acid, $B(OH)_3$

Solution

(d) Ag_2SO_4: $\quad 2 \times 107.868 \; = 215.736$
$\qquad\qquad\;\; 1 \times 32.06 \quad\;\; = \quad 32.06$
$\qquad\qquad\;\; 4 \times 15.9994 \; = \quad 63.9976$
$\qquad\qquad\qquad\qquad\qquad\;\; 311.7936 \text{ g} \to 311.79 \text{ g}$

(e) CH_3CO_2H: $\quad 2 \times 12.011 \; = 24.022$
$\qquad\qquad\qquad 4 \times 1.0079 = \quad 4.0316$
$\qquad\qquad\qquad 2 \times 15.9994 = 31.9988$
$\qquad\qquad\qquad\qquad\qquad\;\; 60.0524 \text{ g} \to 60.052 \text{ g}$

(f) $B(OH)_3$: $\quad 1 \times 10.81 \quad = 10.81$
$\qquad\qquad\;\; 3 \times 15.9994 = 47.9982$
$\qquad\qquad\;\; 3 \times 1.0079 \;\; = \quad 3.0237$
$\qquad\qquad\qquad\qquad\qquad\;\; 61.8319 \text{ g} \to 61.83 \text{ g}$

Chapter 2 Symbols, Formulas, and Equations; Elementary Stoichiometry

13. Determine the number of moles in each of the following:

   (a) 4.26 g of ammonia, NH_3
   (c) 30.0 g of ethylene, C_2H_4
   (d) 1.95×10^{-2} g of the amino acid glycine, $CH_2(NH_2)CO_2H$

   Solution

   (a) Given a specified mass of a substance, the number of moles in that mass is determined by dividing the mass by the gram-molecular weight of the substance.

   $$\text{g-mol. wt } (NH_3) = 14.0067 \text{ g} + 3(1.0079 \text{ g}) = 17.0304 \text{ g}$$

   $$\text{Moles } (NH_3) = 4.26 \text{ g} \times \frac{1 \text{ mol}}{17.0304 \text{ g}} = 0.250 \text{ mol}$$

   (c) $\text{g-mol. wt } (C_2H_4) = 2(12.011 \text{ g}) + 4(1.0079 \text{ g}) = 28.0536 \text{ g}$

   $$\text{Moles } (C_2H_2) = 30.0 \text{ g} \times \frac{1 \text{ mol}}{28.0536 \text{ g}} = 1.07 \text{ mol}$$

   (d) $\text{g-mol. wt (glycine)} = 2(12.011 \text{ g}) + 5(1.0079 \text{ g}) + 2(15.9994 \text{ g}) + 14.0067 \text{ g}$
   $= 75.067 \text{ g}$

   $$\text{Moles (glycine)} = 1.95 \times 10^{-2} \text{ g} \times \frac{1 \text{ mol}}{75.067 \text{ g}} = 2.60 \times 10^{-4} \text{ mol}$$

14. Determine the mass, in grams, of the following:

   (a) 1.78×10^2 mol of potassium bromide, KBr
   (c) 1.518×10^{-3} mol of phosphoric acid, H_3PO_4
   (e) 0.22930 mol of aluminum sulfate, $Al_2(SO_4)_3$

   Solution

   (a) Given a quantity of substance in moles, its mass is determined by multiplying the number of moles times the gram-molecular weight of the substance.

   $$\text{g-mol. wt } (KBr) = 39.0983 \text{ g} + 79.904 \text{ g} = 119.0023 \text{ g}$$

   $$\text{Mass} = 1.78 \times 10^2 \text{ mol} \times \frac{119.0023 \text{ g}}{1 \text{ mol}} = 2.12 \times 10^4 \text{ g}$$

   (c) $\text{g-mol. wt } (H_3PO_4) = 3(1.0079 \text{ g}) + 30.97376 \text{ g} + 4(15.9994 \text{ g})$
   $= 97.99236 \text{ g}$

   $$\text{Mass} = 1.518 \times 10^{-3} \text{ mol} \times \frac{97.99236 \text{ g}}{1 \text{ mol}} = 1.488 \times 10^{-1} \text{ g}$$

   (e) $\text{g-mol. wt } (Al_2(SO_4)_3) = 2(26.98154 \text{ g}) + 3(32.06 \text{ g}) + 12(15.9994 \text{ g})$
   $= 342.14 \text{ g}$

   $$\text{Mass} = 0.22930 \text{ mol} \times \frac{342.14 \text{ g}}{1 \text{ mol}} = 78.453 \text{ g}$$

17. The herbicide Treflan, $C_{13}H_{16}N_2O_4F_3$, is applied at the rate of 450 g (two significant figures) per acre to control weeds in corn, cantaloupes, cotton, and

other crops. What is this application in moles per acre? In molecules per cm^2? (1 acre = 4.047×10^7 cm^2)

Solution

$$\text{g-mol. wt} = 13(12.011 \text{ g}) + 16(1.0079 \text{ g}) + 2(14.0067 \text{ g}) + 4(15.9994 \text{ g}) + 3(18.998403 \text{ g}) = 321.276 \text{ g}$$

$$\text{Moles} = 450 \text{ g} \times \frac{1 \text{ mol}}{321.276 \text{ g}} = 1.4 \text{ mol on one acre}$$

$$\text{Number of molecules in 1.4 mol} = 1.4 \text{ mol} \times \frac{6.022 \times 10^{23} \text{ molecules}}{1 \text{ mol}}$$
$$= 8.43 \times 10^{23} \text{ molecules}$$

$$\frac{\text{Number of molecules}}{1 \text{ cm}^2} = \frac{8.43 \times 10^{23} \text{ molecules}}{1 \text{ acre}} \times \frac{1 \text{ acre}}{4.047 \times 10^7 \text{ cm}^2}$$
$$= 2.1 \times 10^{16} \text{ molecules/cm}^2$$

20. An average 55 kg woman contains 7.5×10^{-3} mole of hemoglobin (molecular weight: 64,456) in her blood. What is this quantity in grams? In pounds?

Solution

$$7.5 \times 10^{-3} \text{ mol} \times \frac{64,456 \text{ g}}{1 \text{ mol}} = 480 \text{ g}$$

$$480 \text{ g} \times \frac{1 \text{ lb}}{454 \text{ g}} = 1.1 \text{ lb}$$

21. Determine the moles of hydrogen atoms, the moles of uranium atoms, the moles of water molecules, and the moles of oxygen atoms in 0.100 mol of the mineral carnotite, $K_2(UO_2)_2(VO_4)_2 \cdot 3H_2O$.

Solution

$$\text{Mol (H atoms)} = \frac{6 \text{ mol H}}{\text{mol mineral}} \times 0.100 \text{ mol} = 0.600 \text{ mol}$$

$$\text{Mol (U atoms)} = \frac{2 \text{ mol U}}{\text{mol mineral}} \times 0.100 \text{ mol} = 0.200 \text{ mol}$$

$$\text{Mol (H}_2\text{O molecules)} = \frac{3 \text{ mol H}_2\text{O}}{\text{mol mineral}} \times 0.100 \text{ mol} = 0.300 \text{ mol}$$

$$\text{Mol (O atoms)} = \frac{15 \text{ mol O}}{\text{mol mineral}} \times 0.100 \text{ mol} = 1.50 \text{ mol}$$

22. Determine the grams of zirconium and of oxygen in 0.3384 mol of the mineral zircon, $ZrSiO_4$.

Solution

0.3384 mol $ZrSiO_4$ contains 0.3384 mol Zr and 4×0.3384, or 1.354 mol oxygen atoms.

Chapter 2 Symbols, Formulas, and Equations; Elementary Stoichiometry   21

$$0.3384 \text{ mol} \times \frac{91.22 \text{ g}}{1 \text{ mol}} = 30.87 \text{ g Zr}$$

$$1.354 \text{ mol} \times \frac{16.00 \text{ g}}{1 \text{ mol}} = 21.66 \text{ g O}$$

23. Determine the number of

   (a) grams of sulfur in 0.16 mol SO_3
   (b) grams of magnesium in 7.52 mol MgS
   (c) grams of carbon in 0.01008 mol of novocain, $C_{13}H_{21}N_2O_2Cl$
   (d) moles of hydrochloric acid in 785.4 g of hydrochloric acid, HCl
   (e) moles of oxygen atoms in 51.7 g of gypsum, $MgSO_4 \cdot 2H_2O$
   (f) moles of carbon atoms in 0.6163 g of niacin, $C_6H_5NO_2$

Solution

(a) 0.16 mol SO_3 contains 0.16 mol S.

$$0.16 \text{ mol} \times \frac{32.1 \text{ g}}{1 \text{ mol}} = 5.1 \text{ g S}$$

(b) 7.52 mol MgS contains 7.52 mol Mg.

$$7.52 \text{ mol} \times \frac{24.30 \text{ g}}{1 \text{ mol}} = 183 \text{ g Mg}$$

(c) 0.01008 mol novocain contains 13×0.01008 (or 0.1310) mol C.

$$0.1310 \text{ mol} \times \frac{12.01 \text{ g}}{1 \text{ mol}} = 1.574 \text{ g C}$$

(d) $785.4 \text{ g} \times \dfrac{1 \text{ mol}}{36.46 \text{ g}} = 21.54 \text{ mol HCl}$

(e) $51.7 \text{ g} \times \dfrac{1 \text{ mol}}{156.4 \text{ g}} = 0.3306 \text{ mol } MgSO_4 \cdot 2H_2O$

   0.3306 mol $MgSO_4 \cdot 2H_2O$ contains 6×0.3306 (or 1.98) mol O.

(f) $0.6163 \text{ g} \times \dfrac{1 \text{ mol}}{123.11 \text{ g}} = 5.006 \times 10^{-3} \text{ mol } C_6H_5NO_2$

   5.006×10^{-3} mol $C_6H_5NO_2$ contains $6 \times 5.006 \times 10^{-3}$ (or 3.004×10^{-2}) mol C.

24. Determine which of the following contains the greatest mass of hydrogen: 1 mol of CH_4; 0.4 mol of C_3H_8; 1.5 mol of H_2.

Solution

Determine which quantity contains the greatest number of moles of hydrogen atoms.

$$\text{Mol (H) in } CH_4 = \frac{4 \text{ mol H}}{\text{mol } CH_4} \times 1 \text{ mol } CH_4 = 4 \text{ mol H}$$

$$\text{Mol (H) in } C_3H_8 = \frac{8 \text{ mol H}}{\text{mol } C_3H_8} \times 0.4 \text{ mol } C_3H_8 = 3.2 \text{ mol H}$$

22   Chapter 2   Symbols, Formulas, and Equations; Elementary Stoichiometry

$$\text{Mol (H) in H}_2 = \frac{2 \text{ mol H}}{\text{mol H}_2} \times 1.5 \text{ mol H}_2 = 3.0 \text{ mol H}$$

Answer: 1 mol CH_4

28. Determine which of the following contains the greatest total number of atoms: 15.0 g of aluminum metal; 1.25×10^{23} molecules of carbon monoxide, CO; 0.10 mol of carbon tetrachloride, CCl_4; 5 g of sulfur dioxide, SO_2.

Solution

Each quantity must be reduced to the number of atoms composing the substance.

Al: Number of atoms = $15.0 \text{ g} \times \dfrac{1 \text{ mol}}{26.98154 \text{ g}} \times \dfrac{6.022 \times 10^{23} \text{ atoms}}{\text{mol}}$

$= 3.35 \times 10^{23}$ atoms

CO: Number of atoms = 1.25×10^{23} molecules $\times \dfrac{2 \text{ atoms}}{\text{molecule}}$

$= 2.50 \times 10^{23}$ atoms

CCl_4: Number of atoms = $0.10 \text{ mol} \times \dfrac{6.022 \times 10^{23} \text{ molecules}}{\text{mol}} \times \dfrac{5 \text{ atoms}}{\text{molecule}} = 3.01 \times 10^{23}$

SO_2: Number of atoms = $5 \text{ g} \times \dfrac{1 \text{ mol}}{64.06 \text{ g}} \times \dfrac{6.022 \times 10^{23} \text{ molecules}}{\text{mol}} \times \dfrac{3 \text{ atoms}}{\text{molecule}}$

$= 1 \times 10^{23}$

Answer: 15.0 g Al.

EMPIRICAL FORMULAS AND PERCENT COMPOSITION

29. Calculate the percent composition of each of the following compounds to three significant figures.

   (a) Potassium chloride, KCl
   (c) Chromium(III) oxide, Cr_2O_3

Solution

(a) Based on the formula of a substance, the percent composition of each element is its formula mass (atomic weight times the number of times it appears in the formula) divided by the formula weight of the substance, then multiplied by 100 to convert the fraction to percentage. Be sure to include *all* the mass of the desired element.

Formula weight (KCl) = $39.0983 + 35.453 = 74.551$

$$\% \text{ K} = \frac{39.0983}{74.551} \times 100 = 52.4\%$$

$$\% \text{ Cl} = \frac{35.453}{74.551} \times 100 = 47.6\%$$

Chapter 2 Symbols, Formulas, and Equations; Elementary Stoichiometry

(c) Formula weight $(Cr_2O_3) = 2(51.996) + 3(15.9994) = 151.990$

$$\% \ Cr = \frac{2(51.996)}{151.990} \times 100 = 68.4\%$$

$$\% \ O = \frac{3(15.9994)}{151.990} \times 100 = 31.6\%$$

31. Which of the following fertilizers contains the highest percentage of nitrogen: urea, N_2H_4CO; ammonium nitrate, NH_4NO_3; ammonium sulfate, $(NH_4)_2SO_4$?

Solution

Use the formula weight of each compound to calculate the percentage of nitrogen.

Urea: Formula weight $(N_2H_4CO): 2(14.0067) + 4(1.0079) + 12.011 + 15.9994 = 60.0554$

$$\% \ N = \frac{2(14.0067)}{60.0554} \times 100 = 46.64\%$$

Ammonium nitrate: Formula weight $(NH_4NO_3) = 2(14.0067) + 4(1.0079) + 3(15.9994) = 80.0432$

$$\% \ N = \frac{2(14.0067)}{80.0432} \times 100 = 35.00\%$$

Ammonium sulfate: Formula weight $((NH_4)_2SO_4) = 2(14.0067) + 8(1.0079) + 32.06 + 4(15.9994) = 132.13$

$$\% \ N = \frac{2(14.0067)}{132.13} \times 100 = 21.2\%$$

Therefore, urea has highest nitrogen content.

33. The most common flavoring agent added to food contains 39.3% Na and 60.7% Cl. Determine the empirical formula of this compound.

Solution

In order to determine the empirical formula, a relationship between percent composition and atom composition must be established. The percent composition of each element in a compound can be found either by dividing its mass by the total mass or by dividing the total formula weight of that element as it appears in the formula (atomic weight times the number of times the element appears in the formula) by the formula weight of the compound. From the latter perspective the percent composition of an element needs only to be divided by its respective atomic weight to find the relative ratio of atoms in the formula. From these numbers the whole number ratio of elements in the compound can be found by dividing each ratio by the number representing the smallest ratio. This process can be broken into two simple steps:

Step 1: Divide each element's percentage by its atomic weight.

Na: $\dfrac{39.3}{22.98977} = 1.71$

Cl: $\dfrac{60.7}{35.453} = 1.71$

This operation establishes the relative ratio of sodium to chlorine in the formula.

Step 2: To establish a whole number ratio of Na to Cl, divide each factor by the smallest factor. In this case, both factors are equal, meaning that the ratio of atoms is 1 to 1.

Na: $\dfrac{1.71}{1.71} = 1$

Cl: $\dfrac{1.71}{1.71} = 1$

The empirical formula of this substance is NaCl.

36. Determine the empirical formula of sodium nitrite, a compound containing 33.32% Na, 20.30% N, and 46.38% O, which is used to preserve bacon and other meat.

Solution

Assume 100.0 g of compound.

Na: $33.32 \text{ g} \times \dfrac{1 \text{ mol}}{22.9898 \text{ g}} = 1.449 \text{ mol}$ $\dfrac{1.449}{1.449} = 1.000$

N: $20.30 \text{ g} \times \dfrac{1 \text{ mol}}{14.0067 \text{ g}} = 1.449 \text{ mol}$ $\dfrac{1.449}{1.449} = 1.000$

O: $46.38 \text{ g} \times \dfrac{1 \text{ mol}}{15.9994 \text{ g}} = 2.899 \text{ mol}$ $\dfrac{2.899}{1.449} = 2.001$

Empirical formula: $NaNO_2$

38. Most polymers are very large molecules composed of simple units repeated many times. Thus they often have relatively simple empirical formulas. Determine the empirical formulas of the following polymers:

   (b) Polystyrene; 92.3% C, 7.7% H
   (c) Saran; 24.8% C, 2.0% H, 73.1% Cl
   (d) Orlon; 67.9% C, 5.70% H, 26.4% N

Solution

(b) C: $92.3 \text{ g} \times \dfrac{1 \text{ mol}}{12.011 \text{ g}} = 7.68 \text{ mol}$

   H: $7.7 \text{ g} \times \dfrac{1 \text{ mol}}{1.0079 \text{ g}} = 7.64 \text{ mol}$

   Empirical formula: CH

(c) C: $24.8 \text{ g} \times \dfrac{1 \text{ mol}}{12.011 \text{ g}} = 2.07 \text{ mol}$

   H: $2.0 \text{ g} \times \dfrac{1 \text{ mol}}{1.0079 \text{ g}} = 2.0 \text{ mol}$

   Cl: $73.1 \text{ g} \times \dfrac{1 \text{ mol}}{35.453 \text{ g}} = 2.06 \text{ mol}$

   Empirical formula: CHCl

(d) Step 1:

C: $\dfrac{67.9}{12.011} = 5.65$

H: $\dfrac{5.70}{1.0079} = 5.66$

N: $\dfrac{26.4}{14.0067} = 1.88$

Step 2:

C: $\dfrac{5.65}{1.88} = 3$

H: $\dfrac{5.66}{1.88} = 3$

N: $\dfrac{1.88}{1.88} = 1$

Empirical formula: C_3H_3N

40. Dichloroethane, a compound containing carbon, hydrogen, and chlorine, is often used for dry cleaning. The molecular weight of dichloroethane is 99. A sample of dichloroethane is found to contain 24.3% carbon and 71.6% chlorine. What are its empirical and molecular formulas?

Solution

The molecular formula is a whole number multiple of the empirical formula. And the molecular weight is a whole number multiple of the empirical weight. The solution sequence is to determine the empirical formula, its formula weight, and then the molecular formula.

Step 1: C: $\dfrac{24.3}{12.011} = 2.02$

H: $\dfrac{4.1}{1.0079} = 4.07$

Cl: $\dfrac{71.6}{35.453} = 2.02$

Step 2: $\dfrac{2.02}{2.02} = 1$

$\dfrac{4.07}{2.02} = 2$

$\dfrac{2.02}{2.02} = 1$

Empirical formula = CH_2Cl = 49.5
Mol. wt = empirical formula weight × number of formula units
99 = 49.5 × number of formula units

Solve for the number of formula units: $\dfrac{99}{49.5} = 2$

Molecular formula = $2(CH_2Cl) = C_2H_4Cl_2$

46. A 1.728-g magnesium carbonate sample decomposes upon heating, giving gaseous CO_2 and a residue of 0.821 g of MgO. Determine the empirical formula of this compound.

Solution

To determine the empirical formula, the mass of each element in the sample must be known. The mass of MgO is known from which the mass of Mg can be calculated. The mass of O in MgO can be calculated but must be added to the mass of O in CO_2. The mass of C can be determined from the mass of CO_2.

$$\text{Mass }(CO_2) = (1.728 - 0.821)\text{ g} = 0.907\text{ g}$$

$$\%\text{ C in }CO_2 = \frac{12.011}{44.010} \times 100 = 27.29\%$$

$$\text{Mass (C)} = 0.907\text{ g} \times 0.2729 = 0.2475\text{ g}$$

$$\text{Mass (O)} = 0.907 - 0.2475\text{ g} = 0.6595\text{ g}$$

$$\%\text{ Mg in MgO} = \frac{24.305}{40.3044} \times 100 = 60.30\%$$

$$\text{Mass (Mg)} = 0.821\text{ g} \times 0.6030 = 0.4951\text{ g}$$

$$\text{Mass (O)} = 0.821 - 0.4951 = 0.326\text{ g}$$

$$\text{Mass (O) total} = 0.6595 + 0.326 = 0.986\text{ g}$$

Finally, the empirical formula:

Step 1:   Mg: $\dfrac{0.4951}{24.305} = 0.0204$

C: $\dfrac{0.2475}{12.011} = 0.0206$

O: $\dfrac{0.986}{15.9994} = 0.0616$

Step 2: $\dfrac{0.0204}{0.0204} = 1$

$\dfrac{0.0206}{0.0204} = 1$

$\dfrac{0.0616}{0.0204} = 3$

Empirical formula: $MgCO_3$

CHEMICAL CALCULATIONS INVOLVING EQUATIONS

50. If 28.1 g of Si reacts with N_2 giving 0.333 mole of Si_3N_4 according to the equation

$$3Si + 2N_2 \longrightarrow Si_3N_4$$

which of the following can be determined by using only the information given in this exercise: (a) the moles of Si reacting, (b) the moles of N_2 reacting, (c)

Chapter 2 Symbols, Formulas, and Equations; Elementary Stoichiometry   27

the atomic weight of N, (d) the atomic weight of Si, (e) the mass of Si_3N_4 produced?

Solution

(a) From the equation, 0.333 mol of Si_3N_4 requires 3×0.333 mol of Si atoms or 1 mol of Si.

(b) The number of moles of N_2 reacting is one-half the number of moles of N atoms produced in Si_3N_4:

$$\text{mol (N) atoms} = \frac{4 \text{ mol N}}{1 \text{ mol } Si_3N_4} \times 0.333 \text{ mol } Si_3N_4 = 1.332$$

$$\text{mol (N}_2) = \frac{1.332}{2} = 0.666 \text{ mol}$$

(c) This can be done only if the total mass is known independently. If the mass of Si_3N_4 produced is known, subtract the mass of Si used to obtain the mass of N in the product. Then divide by 1.332 obtained above to find the mass of 1 mol of nitrogen atoms.

(d) Since 1 mol of Si was used and since all of the starting Si is consumed, the molar mass is 28.1 g.

(e) The mass produced can be calculated only if atomic weights are known.

$$\text{Mass} = \frac{140.283 \text{ g } Si_3N_4}{\text{mol}} \times 0.333 \text{ mol} = 46.7 \text{ g } Si_3N_4$$

52. (a) How many moles of H_2 are produced by the reaction of 1.24 mol of H_3PO_4 in the following reaction?

$$2Cr + 2H_3PO_4 \longrightarrow 3H_2 + 2CrPO_4$$

(b) What mass of H_2, in grams, is produced?

Solution

(a) According to the equation, 2 mol H_3PO_4 produces 3 mol H_2. The amount of H_2 is:

$$\text{mol (H}_2) = 1.24 \text{ mol } H_3PO_4 \times \frac{3 \text{ mol } H_2}{2 \text{ mol } H_3PO_4} = 1.86 \text{ mol } H_2$$

(b) $\quad 2 \text{ Cr} + \underset{2 \text{ mol}}{\overset{1.24 \text{ mol}}{2H_3PO_4}} \longrightarrow \underset{3 \text{ mol}}{\overset{\text{mass?}}{3H_2}} + 2CrPO_4$

$$\text{Mass (H}_2) = 1.24 \text{ mol } H_3PO_4 \times \frac{3 \text{ mol } H_2}{2 \text{ mol } H_3PO_4} \times \frac{2.0158 \text{ g}}{\text{mol } H_2} = 3.75 \text{ g}$$

55. Silicon carbide, SiC, a very hard material used as an abrasive on sandpaper and in other applications, is prepared by the reaction of pure sand, SiO_2, with carbon at high temperature. Carbon monoxide, CO, is the other product of this reaction. Write the balanced equation for the reaction and calculate how much SiO_2 is required to produce 1.00 kg of SiC.

Solution

$$SiO_2 + 3C \longrightarrow SiC + 2CO$$

28 Chapter 2 Symbols, Formulas, and Equations; Elementary Stoichiometry

From the balanced equation, 1 mol of SiO_2 produces 1 mol of SiC. The unknown is the mass of SiO_2 required to produce 1.00 kg (1000 g) of SiC. To calculate the mass of SiO_2 required, calculate the formula weights for SiO_2 and SiC, set up the equation relating the unknown mass to the known amount of SiC, then write the mole relationship of SiO_2 to SiC.

$$\underset{\text{1 mol}}{\underset{\text{Mass?}}{SiO_2}} + 3C \longrightarrow \underset{\text{1 mol}}{\underset{\text{1000 g}}{SiC}} + 2CO$$

Formula weights: $(SiO_2) = 60.0843$ g; $(SiC) = 40.0955$ g

$$\text{Mass } (SiO_2) = 1000 \text{ g SiC} \times \frac{1 \text{ mol SiC}}{40.0955 \text{ g}} \times \frac{1 \text{ mol } SiO_2}{1 \text{ mol SiC}} \times \frac{60.0843 \text{ g } SiO_2}{1 \text{ mol}}$$

$$= 24.940 \text{ mol SiC} \times \frac{1 \text{ mol } SiO_2}{1 \text{ mol SiC}} \times \frac{60.0843 \text{ g } SiO_2}{1 \text{ mol } SiO_2} = 1.50 \text{ kg}$$

57. Tooth enamel consists of hydroxyapatite, $Ca_5(PO_4)_3(OH)$. This substance is converted to the more decay-resistant fluorapatite, $Ca_5(PO_4)_3F$, by treatment with tin(II) fluoride, SnF_2 (commonly referred to as stannous fluoride). Products of this reaction are SnO and water. What mass of hydroxyapatite can be converted to fluorapatite by reaction with 0.100 g of SnF_2?

Solution

The balanced equation is

$$2Ca_5(PO_4)_3(OH) + SnF_2 \longrightarrow 2Ca_5(PO_4)_5F + SnO + H_2O$$

The unknown is the amount of hydroxyapatite required to react with 0.100 g of SnF_2. Determine formula weights of the reactants, set up mole ratios from the balanced equation and calculate the unknown.

Formula weights: $(SnF_2) = 118.69 + 2(18.998) = 156.686$ g

$(Ca_5(PO_4)_3(OH)) = 5(40.08) + 3(30.97376) + 13(15.9994) + 1.0079$

$= 502.32$ g

From the balanced equation,

$$\underset{\text{2 mol requires 1 mol}}{\underset{\text{Mass?}}{2Ca_5(PO_4)_3(OH)}} + \underset{}{\underset{\text{0.100 g}}{SnF_2}}$$

$$\text{Mass} = 0.100 \text{ g } SnF_2 \times \frac{1 \text{ mol}}{156.686 \text{ g } SnF_2} \times \frac{2 \text{ mol } Ca_5(PO_4)_3(OH)}{1 \text{ mol } SnF_2}$$

$$\times \frac{502.32 \text{ g } Ca_5(PO_4)_3(OH)}{1 \text{ mol } Ca_5(PO_4)_3(OH)} = 0.641 \text{ g } Ca_5(PO_4)_3(OH)$$

LIMITING REAGENTS

60. What is the limiting reagent when 0.25 mol of P_4 and 0.25 mol of O_2 react according to the following equation?

$$P_4 + 5O_2 \longrightarrow P_4O_{10}$$

Solution

A balanced equation establishes the ratio of reactants to each other and the reactants' ratio to the products. In this case, 1 mol of P_4 requires 5 mol of O_2.

Chapter 2 Symbols, Formulas, and Equations; Elementary Stoichiometry   29

$$P_4 + 5O_2 \longrightarrow P_4O_{10}$$
$$\text{1 mol} : \text{5 mol}$$

Given that 0.25 mol each of P_4 and O_2 are available, which reactant will limit the production of P_4O_{10}? Stated differently, do these quantities represent the proper ratio? If not, one reactant will be in excess: some will remain unreacted.

Since this question deals only with the reactants, begin by assuming that all the available P_4 is to be reacted. This being the case, how much O_2 is required?

$$\text{Mol }(O_2) \text{ required} = 0.25 \text{ mol } P_4 \times \frac{5 \text{ mol } O_2}{1 \text{ mol } P_4} = 1.25 \text{ mol}$$

Only 0.25 mol of O_2 is available; therefore, O_2 is the limiting reagent.

Suppose we had chosen O_2 to be totally reacted. Now, how much P_4 is required?

$$\text{Mol }(P_4) \text{ required} = 0.25 \text{ mol } O_2 \times \frac{1 \text{ mol } P_4}{5 \text{ mol } O_2} = 0.05 \text{ mol}$$

Only 0.05 mol P_4 can be used by the available O_2, therefore leaving unreacted P_4. The limiting reagent is O_2.

62. What is the limiting reagent when 1.00 g of Si and 1.00 g of N_2 combine according to the following reaction?

$$3Si + 2N_2 \longrightarrow Si_3N_4$$

Solution

As in exercise 60, this question deals only with reactants. Given that 1.00 g each of Si and N_2 are available, which one limits the reaction? Arbitrarily assume that all of the available Si is reacted. How much N_2 is required?

$$\begin{array}{cc} 1.00 \text{ g} & 1.00 \text{ g} \\ 3Si + 2N_2 & \longrightarrow Si_3N_4 \\ 3 \text{ mol} : 2 \text{ mol} \end{array}$$

Formula weights: Si = 28.0855 g
                   N_2 = 28.0134 g

$$\text{Mass }(N_2) \text{ required} = 1.00 \text{ g Si} \times \frac{1 \text{ mol Si}}{28.0855 \text{ g Si}} \times \frac{2 \text{ mol } N_2}{3 \text{ mol Si}} \times \frac{28.0134 \text{ g } N_2}{1 \text{ mol } N_2}$$
$$= 0.665 \text{ g}$$

Since 0.655 g of N_2 is required and 1.00 g is available, N_2 is in excess and Si is the limiting reagent.

PERCENT YIELD

66. A sample of calcium oxide, CaO, weighing 0.69 g (ACTUAL) was prepared by heating 1.31 g of calcium carbonate. What was the percent yield of the reaction?

$$CaCO_3(s) \xrightarrow{\Delta} CaO(s) + CO_2(g)$$

Solution

Begin by calculating the mass of CaO that can be produced from 1.31 g of $CaCO_3$, assuming that all CaO can be recovered.

30   Chapter 2   Symbols, Formulas, and Equations; Elementary Stoichiometry

$$\underset{1\text{ mol}}{\underset{1.31\text{ g}}{CaCO_3}} \longrightarrow \underset{1\text{ mol}}{\overset{Mass?}{CaO}} + CO_2$$

Formula weights: $CaCO_3 = 100.09$
$CaO = 56.05$

For complete conversion,

$$\text{Mass (CaO)} = 1.31\text{ g CaCO}_3 \times \frac{1\text{ mol CaCO}_3}{100.09\text{ g CaCO}_3} \times \frac{1\text{ mol CaO}}{1\text{ mol CaCO}_3} \times \frac{56.08\text{ g CaO}}{1\text{ mol CaO}}$$

$$= 0.734\text{ g CaO (theoretical yield)}$$

Actual recovery $= 0.69$ g

$$\text{Percent yield} = \frac{\text{actual yield}}{\text{theoretical yield}} \times 100 = \frac{0.69\text{ g}}{0.734\text{ g}} \times 100 = 94\%$$

68. Freon-12, CCl_2F_2, is prepared from CCl_4 by reaction with HF. The other product of this reaction is HCl. What is the percent yield of a reaction that produces 12.5 g of CCl_2F_2 from 32.9 g of CCl_4?

Solution

Write and balance the equation for the reaction.

$$\underset{1\text{ mol}}{\overset{32.9\text{ g}}{CCl_4}} + 2HF \longrightarrow \underset{1\text{ mol}}{\overset{\text{Theoretical mass?}}{CCl_2F_2}} + 2HCl$$

Formula weights: $CCl_4 = 153.82$
$CCl_2F_2 = 120.89$

For complete conversion;

$$\text{Mass (CCl}_2\text{F}_2) = 32.9\text{ g CCl}_4 \times \frac{1\text{ mol CCl}_4}{153.82\text{ g CCl}_4} \times \frac{1\text{ mol CCl}_2\text{F}_2}{1\text{ mol CCl}_4} \times \frac{120.89\text{ g CCl}_2\text{F}_2}{1\text{ mol CCl}_2\text{F}_2}$$

$$= 25.86\text{ g CCl}_2\text{F}_2$$

$$\text{Percent yield} = \frac{12.5\text{ g}}{25.86\text{ g}} \times 100 = 48.3\%$$

70. Citric acid, $C_6H_8O_7$, a component of jams, jellies, and fruity soft drinks, is prepared industrially by fermentation of sucrose by the mold *Aspergillus niger*. The overall reaction is

$$C_{12}H_{22}O_{11} + H_2O + 3O_2 \longrightarrow 2C_6H_8O_7 + 4H_2O$$

What is the amount, in kilograms, of citric acid produced from 1 metric ton $(1.000 \times 10^3$ kg) of sucrose in a reaction if the yield is 92.3%?

Solution

Calculate the mass of citric acid by assuming 100% yield. The expected yield is 92.3% of this amount.

Chapter 2 Symbols, Formulas, and Equations; Elementary Stoichiometry

$$\underset{1 \text{ mol}}{C_{12}H_{22}O_{11}} \overset{1.000 \times 10^3 \text{ kg}}{+} H_2O + 3O_2 \overset{100\% \text{ yield?}}{\longrightarrow} \underset{2 \text{ mol}}{2C_6H_8O_7} + 4H_2O$$

Formula weights: $C_{12}H_{22}O_{11} = 342.30$; $C_6H_8O_7 = 192.12$

Once the ratio of masses is established, the formula weights are independent of units. This means that either mass in grams or kilograms can be used as long as they are used consistently. One mole of $C_{12}H_{22}O_{11}$, 342.30 g, produces two moles of $C_6H_8O_7$, 2×192.12 g = 384.24 g. Using kilogram values, 342.30 kg of $C_{12}H_{22}O_{11}$ produces 2×192.12 kg of citric acid, or 384.24 kg.

For 100% yield,

$$\text{Mass } (C_6H_8O_7) = 1.000 \times 10^3 \text{ kg sucrose} \times \frac{1 \text{ mol-kg sucrose}}{342.30 \text{ kg sucrose}}$$

$$\times \frac{2 \text{ mol-kg citric acid}}{1 \text{ mol-kg sucrose}} \times \frac{192.12 \text{ kg citric acid}}{\text{mol-kg citric acid}}$$

$$= 1122.5 \text{ kg citric acid}$$

At 92.3% yield,

$$1122.5 \text{ kg} \times 0.923 = 1036 \text{ kg} = 1.04 \times 10^3 \text{ kg}$$

74. Addition of 0.403 g of sodium oxalate, $Na_2C_2O_4$, to a solution containing 1.48 g of uranyl nitrate, $UO_2(NO_3)_2$, yields 1.073 g of solid $UO_2(C_2O_4) \cdot 3H_2O$.

$$UO_2(NO_3)_2(aq) + Na_2C_2O_4(aq) + 3H_2O \longrightarrow UO_2(C_2O_4) \cdot 3H_2O(s) + 2NaNO_3(aq)$$

Determine the limiting reagent and percent yield of this reaction.

Solution

Using the balanced equation, determine which reactant quantity produces the smallest theoretical yield. This quantity represents the largest amount of product that can be produced. Then calculate the percent yield.

$$\underset{1 \text{ mol}}{UO_2(NO_3)_2} \overset{1.48 \text{ g}}{+} \underset{1 \text{ mol}}{Na_2C_2O_4} \overset{0.403 \text{ g}}{+} 3H_2O \longrightarrow \underset{1 \text{ mol}}{UO_2(C_2O_4) \cdot 3H_2O} \overset{\text{Theor. yield?}}{+} 2NaNO_3$$

Formula weights:

$UO_2(NO_3)_2 = 394.04$

$Na_2C_2O_4 = 134.00$

$UO_2(C_2O_4) \cdot 3H_2O = 412.09$

Reaction 1:

$$\text{Mass (Product)} = 1.48 \text{ g } UO_2(NO_3)_2 \times \frac{1 \text{ mol } UO_2(NO_3)_2}{394.04 \text{ g } UO_2(NO_3)_2}$$

$$\times \frac{1 \text{ mol } UO_2(C_2O_4) \cdot 3H_2O}{1 \text{ mol } UO_2(NO_3)_2} \times \frac{412.0 \text{ g } UO_2(C_2O_4) \cdot 3H_2O}{1 \text{ mol } UO_2(C_2O_4) \cdot 3H_2O}$$

$$= 1.55 \text{ g } UO_2(C_2O_4) \cdot 3H_2O$$

Reaction 2:

$$\text{Mass (Product)} = 0.403 \text{ g Na}_2\text{C}_2\text{O}_4 \times \frac{1 \text{ mol Na}_2\text{C}_2\text{O}_4}{134.00 \text{ g Na}_2\text{C}_2\text{O}_4} \times \frac{1 \text{ mol UO}_2(\text{C}_2\text{O}_4)\cdot 3\text{H}_2\text{O}}{1 \text{ mol Na}_2\text{C}_2\text{O}_4}$$

$$\times \frac{412.09 \text{ g UO}_2(\text{C}_2\text{O}_4)\cdot 3\text{H}_2\text{O}}{1 \text{ mol UO}_2(\text{C}_2\text{O}_4)\cdot 3\text{H}_2\text{O}}$$

$$= 1.239 \text{ g UO}_2(\text{C}_2\text{O}_4)\cdot 3\text{H}_2\text{O}$$

$Na_2C_2O_4$ is the limiting reagent. An amount of $UO_2(NO_3)_2$ is left unreacted.

$$\text{Percent yield} = \frac{1.073 \text{ g}}{1.239 \text{ g}} \times 100 = 86.6\%$$

ADDITIONAL EXERCISES

80. Sulfur may be removed from coal by washing powdered coal with NaOH(*aq*). The reaction may be represented as

$$R_2S(s) + 2\text{NaOH}(aq) \longrightarrow R_2O(s) + \text{Na}_2S(aq) + H_2O(l)$$

where R_2S represents the organic sulfur present in the coal. What mass of NaOH(*s*), in kilograms, is required to react with the sulfur in 1.00 metric ton (1000 kg) of coal that contains 1.6% S by mass?

Solution

The equation indicates that 1 mol of S reacts with 2 mol of NaOH: R is not important to the calculation.

Active S = 0.016 × 1000 kg = 16 kg

$$\underset{1 \text{ mol}}{\underset{16 \text{ kg}}{R_2S}} + \underset{2 \text{ mol}}{\underset{\text{Mass?}}{2\text{NaOH}}} \longrightarrow R_2O + \text{Na}_2S + H_2O$$

Formula weights: S = 32.06

NaOH = 40.00

$$\text{Mass (NaOH)} = 16 \text{ kg S} \times \frac{1 \text{ mol-kg S}}{32.06 \text{ kg S}} \times \frac{2 \text{ mol-kg NaOH}}{1 \text{ mol-kg S}} \times \frac{40.00 \text{ kg NaOH}}{\text{mol-kg NaOH}}$$

$$= 40 \text{ kg NaOH}$$

81. Concentrated sulfuric acid with a density of 1.84 g/cm³ contains 97.5% H_2SO_4 by mass. How many liters of this acid are required to manufacture 1000 kg of phosphate fertilizer, $Ca(H_2PO_4)_2$, according to the reaction

$$2\text{Ca}_5(\text{PO}_4)_3\text{F} + 7\text{H}_2\text{SO}_4 + 14\text{H}_2\text{O} \longrightarrow 2\text{Ca}(\text{H}_2\text{PO}_4)_2 + 7(\text{CaSO}_4 \cdot 2\text{H}_2\text{O}) + 2\text{HF}$$

Solution

First, determine the mass of H_2SO_4 required to produce 1000 kg of phosphate, assuming 100% yield. Second, use this value to calculate the volume of the acid concentrate required. From the equation,

Chapter 2 Symbols, Formulas, and Equations; Elementary Stoichiometry

$$\underset{\text{7 mol-kg}}{\underset{\text{Mass?}}{7H_2SO_4}} \longrightarrow \underset{\text{2 mol-kg}}{\underset{1000 \text{ kg}}{2Ca(H_2PO_4)_2}}$$

Formula weights: $H_2SO_4 = 98.073$

$Ca(H_2PO_4)_2 = 234.054$

$$\text{Mass } (H_2SO_4) \text{ required} = 1000 \text{ kg } Ca(H_2PO_4)_2 \times \frac{1 \text{ mol } Ca(H_2PO_4)_2}{234.054 \text{ kg } Ca(H_2PO_4)_2}$$

$$\times \frac{7 \text{ mol } H_2SO_4}{2 \text{ mol } Ca(H_2PO_4)_2} \times \frac{98.073 \text{ kg } H_2SO_4}{\text{mol-kg } H_2SO_4}$$

$$= 1466.565 \text{ kg } H_2SO_4$$

$$\frac{\text{Mass } H_2SO_4}{\text{Liter concentrate}} = \frac{1.84 \text{ g}}{\text{cm}^3} \times 1000 \frac{\text{cm}^3}{\text{L}} \times 0.975 = 1.794 \times 10^3 \frac{\text{g}}{\text{L}}$$

$$= 1.794 \text{ kg/L}$$

$$\text{Volume (L)} = 1466.565 \text{ kg} \times \frac{1 \text{ L}}{1.794 \text{ kg}} = 817 \text{ L}$$

3
Applications of Chemical Stoichiometry

INTRODUCTION

We now extend the presentation of Chapter 2 to include broader classes of reactions and equations and to bring out their utility in the study of chemistry. Various examples are used to emphasize systematic procedures for analyzing a problem into its essential components.

Although the seeming simplicity of chemical equations tends to convey the idea that reactions occur in a single step, most reactions require completion of several steps before the desired product is formed. To illustrate the complexity of what appears to be a simple reaction, consider the production of table salt, $NaCl(s)$, from its elements according to the equation

$$Na(s) + \tfrac{1}{2}Cl_2(g) \longrightarrow NaCl(s)$$

The production of $NaCl(s)$ is exothermic (that is, proceeds with the liberation of heat) and occurs spontaneously in the presence of sunlight or heat. Without going into great detail, the overall reaction process can be broken down into the following five steps:

1. $Na(s) + 108.8 \text{ kJ} \longrightarrow Na(g)$ (endothermic, requires heat to proceed)
2. $Na(g) + 496.0 \text{ kJ} \longrightarrow Na^+ + e^-$ (endothermic)
3. $\tfrac{1}{2}Cl_2(g) + 119.7 \text{ kJ} \longrightarrow Cl(g)$ (endothermic)
4. $Cl(g) + e^- \longrightarrow Cl^-(g) + 364.8 \text{ kJ}$ (exothermic, heat is given off)
5. $Na^+(g) + Cl^-(g) \longrightarrow NaCl(g) \longrightarrow NaCl(s) + 770.7 \text{ kJ}$ (exothermic)

The total heat released in producing 1 mol of $NaCl(s)$ is 411.0 kJ, obtained by subtracting the sum of the endothermic values from the sum of the exothermic values. Much of the work in chemical research lies in the elucidation of similar reaction processes and their attendant energy changes. That study begins in Chapter 4.

Most of the reactions studied in general chemistry involve substances in solution. The solution phase facilitates and/or provides the medium for reactions to occur. When substances are in solution, the solution must be described in terms of how much substance (solute) is contained in a specified quantity (volume or mass) of solvent. Although several units of concentration are commonly used in chemistry, molarity, M, is presented in this chapter to enable you to begin work with solutions in the laboratory as soon as possible.

The solutions to many problems involve several steps that can generally be simplified into a convenient format for a systematic solution. To develop such a

Chapter 3 Applications of Chemical Stoichiometry

format, read each problem carefully in order to identify what is known and unknown. Then write the equation for the reaction and balance it as necessary. The next step is to develop an algebraic equation that relates the known quantities to those that are unknown, and then solve for the unknown quantity(ies) accordingly. After studying examples and solving a few problems, you will begin to develop a personal method for organizing data and solving problems. Do not become concerned that your method may be slightly different from the way the text approaches the problem—after all, solving problems is a human endeavor, and probably no two people solve problems in exactly the same way.

There are many possible methods in chemistry for solving problems that involve reactants and products. The problem-solving format used in the text is basically the same as that used in this manual, but you occasionally will notice slightly different organizational practices that should enable you to see still other ways of approaching problems.

READY REFERENCE

Molarity (M) (3.1) Molarity is a unit of concentration defined as the moles of solute present in 1 L of solution.

$$\text{Molarity} = \frac{\text{moles of solute}}{\text{volume of solution (liters)}} = \frac{\text{mol}}{\text{L}}$$

This definition can be rearranged to express the amount of solute (in moles) contained in a given volume of solution:

$$\text{Amount of solute} = \text{molarity}\left(\frac{\text{mol}}{\text{L}}\right) \times \text{vol (L)} = \text{mol}$$

The volume of a solution can be expressed in terms of molarity and the amount of solute:

$$\text{Volume in liters} = \frac{\text{moles of solute}}{\text{molarity}}$$

The definition of molarity is readily extended to calculate the amount of solute required to prepare a specified volume of solution.

$$\text{Molarity} = \frac{\text{mol}}{\text{vol (L)}} = \frac{\text{mass (g)} \times \dfrac{1 \text{ mol}}{\text{g-form. wt}}}{\text{vol (L)}}$$

$$\text{Mass} = \text{molarity} \times \text{vol (L)} \times \text{g-form. wt} = \frac{\text{mol}}{\text{L}} \times \text{L} \times \frac{\text{g}}{\text{mol}} = \text{g}$$

We shall use the following abbreviations to express common terms throughout the remainder of this manual:

Atomic weight: at. wt
Formula weight (grams): g-form. wt
Molecular weight (grams): g-mol. wt
Molecular weight: mol. wt
Molarity = mol/L = M

SOLUTIONS TO CHAPTER 3 EXERCISES

MOLARITY AND SOLUTIONS

1. Calculate the concentration in moles per liter for each of the following solutions:

   (a) 98.1 g of sulfuric acid, H_2SO_4, in 1.00 L of solution
   (c) 90.0 g of acetic acid, CH_3CO_2H, in 0.750 L of solution
   (e) 0.1374 g of copper sulfate, $CuSO_4$, in 13 mL of solution

   Solution

   (a) $\text{Molarity} = \dfrac{\text{mol } H_2SO_4}{\text{volume of solution in liters}}$

   First convert mass in grams to moles, and then substitute the proper terms into the definition.

   g-mol. wt $(H_2SO_4) = 98.07$ g

   $\text{mol } (H_2SO_4) = 98.1 \text{ g} \times \dfrac{1 \text{ mol}}{98.07 \text{ g}} = 1.00 \text{ mol}$

   $M = \dfrac{1.00 \text{ mol}}{1.00 \text{ L}} = 1.00 \text{ M}$

   (c) g-mol. wt $(CH_3CO_2H) = 60.05$ g

   $\text{mol } (CH_3CO_2H) = 90.0 \text{ g} \times \dfrac{1 \text{ mol}}{60.05 \text{ g}} = 1.50 \text{ mol}$

   $M = \dfrac{1.50 \text{ mol}}{0.750 \text{ L}} = 2.00 \text{ M}$

   (e) g-mol. wt $(CuSO_4) = 159.60$ g

   $\text{mol } (CuSO_4) = 0.1374 \text{ g} \times \dfrac{1 \text{ mol}}{159.60 \text{ g}} = 0.0008609 \text{ mol}$

   $V = 13 \text{ mL} \times \dfrac{1 \text{ L}}{1000 \text{ mL}} = 0.013 \text{ L}$

   $M = \dfrac{0.0008609 \text{ mol}}{0.013 \text{ L}} = 0.066 \text{ M} = 6.6 \times 10^{-2} \text{ M}$

2. Determine the mass of solute present in each of the following solutions:

   (a) 0.450 L of 1.00 M NaCl solution
   (c) 455 mL of 3.75 M HCl solution
   (e) 25.38 mL of 9.721×10^{-2} M KOH solution

   Solution

   (a) Using the definition of molarity, rearrange the equation to yield moles in terms of volume and molarity. Finally, convert moles to mass in grams.

   $M = \dfrac{\text{mol}}{\text{L}}$ and $\text{mol} = M \times \text{L}$

Chapter 3 Applications of Chemical Stoichiometry   37

$$\text{g-form. wt (NaCl)} = 58.443 \text{ g}$$

$$\text{mol (NaCl)} = M \times L = 1.00 \, \frac{\text{mol}}{\text{L}} \times 0.450 \text{ L} = 0.450 \text{ mol}$$

$$\text{Mass (NaCl)} = 0.450 \text{ mol} \times 58.443 \, \frac{\text{g}}{\text{mol}} = 26.3 \text{ g}$$

(c) g-mol. wt (HCl) = 36.46 g

$$\text{mol (HCl)} = 3.75 \text{ M} \times 455 \text{ mL} \times \frac{1.0 \text{ L}}{1000 \text{ mL}} = 1.706 \text{ mol}$$

$$\text{Mass (HCl)} = 1.706 \text{ mol} \times 36.46 \, \frac{\text{g}}{\text{mol}} = 62.2 \text{ g}$$

(e) g-mol. wt (KOH) = 56.1056 g

$$\text{mol (KOH)} = M \times L = 9.721 \times 10^{-2} \text{ M} \times 25.38 \text{ mL} \times \frac{1 \text{ L}}{1000 \text{ mL}}$$
$$= 2.467 \times 10^{-3} \text{ mol}$$

$$\text{Mass (KOH)} = 2.467 \times 10^{-3} \text{ mol} \times 56.1056 \, \frac{\text{g}}{\text{mol}} = 1.384 \times 10^{-1} \text{ g}$$

3. Determine the moles of solute required to make the indicated amount of solution:

   (a) 1.00 L of 1.00 M LiNO$_3$ solution
   (c) 275 mL of 0.5151 M KClO$_4$ solution
   (e) 1856 mL of 0.1475 M H$_3$PO$_4$ solution

   Solution

   (a) Rearrange the definition of molarity to solve for moles in terms of volume and molarity.

   $$M = \frac{\text{mol}}{\text{L}} \quad \text{so} \quad \text{mol} = M \times L$$

   $$\text{mol (LiNO}_3) = 1.00 \, \frac{\text{mol}}{\text{L}} \times 1.00 \text{ L} = 1.00 \text{ mol}$$

   (c) Convert mL to L and solve as in (a).

   $$V(\text{L}) = 275 \text{ mL} \times \frac{1 \text{ L}}{1000 \text{ mL}} = 0.275 \text{ L}$$

   $$\text{mol (KClO}_4) = 0.5151 \text{ M} \times 0.275 \text{ L} = 0.142 \text{ mol}$$

   (e) $\text{mol (H}_3\text{PO}_4) = 0.1475 \text{ M} \times 1856 \text{ mL} \times \frac{1 \text{ L}}{1000 \text{ mL}} = 0.2738 \text{ mole}$

5. Cow's milk contains an average of 4.5 g of the sugar lactose, $C_{12}H_{22}O_{11}$, per 0.100 L of milk. What is the molarity of lactose in this milk?

   Solution

   The question asks for molarity. Determine the known factors, then use the definition of molarity to solve for molarity. Known factors: Lactose ($C_{12}H_{22}O_{11}$) is the solute. Mass = 4.5 g; V = 0.100 L.

$$M = \frac{\text{mol (Lactose)}}{L}$$

g-mol. wt (Lactose) = 342.30 g

$$\text{mol (Lactose)} = 4.5 \text{ g} \times \frac{1 \text{ mol}}{342.30 \text{ g}} = 0.0131 \text{ mol}$$

Substitute in the formula:

$$M = \frac{0.0131 \text{ mol}}{0.100 \text{ L}} = 0.13 \text{ M}$$

7. An iron content of 0.1 mg/L in drinking water can usually be detected by taste. Will the iron in a water sample with an $FeCO_3$ concentration of 5.25×10^{-7} M be detected by taste?

Solution

The allowable concentration is in terms of Fe^{3+} present, not $FeCO_3$ present. Convert the molar concentration of $FeCO_3$ to mg $FeCO_3$ per liter, then use the percentage of Fe in $FeCO_3$ to calculate the mass of Fe^{3+} present. Convert:

$$\frac{\text{mol } FeCO_3}{L} \rightarrow \frac{\text{mg } FeCO_3}{L} \rightarrow \frac{\text{mg } Fe^{3+}}{L}$$

g-mol. wt $(FeCO_3) = 115.86$; % Fe $= \frac{55.847}{115.86} \times 100 = 48.20\%$

$$\text{Mass } (FeCO_3) = 5.25 \times 10^{-7} \frac{\text{mol}}{L} \times 115.86 \frac{g}{\text{mol}} \times \frac{1000 \text{ mg}}{g} = 6.08 \times 10^{-2} \text{ mg/L}$$

$$\text{Mass } (Fe^{3+}) = 6.08 \times 10^{-2} \frac{\text{mg}}{L} \times 0.4820 = 0.0293 \frac{\text{mg}}{L}$$

$$= 2.93 \times 10^{-2} \text{ mg Fe/L}$$

This mass is less than that needed for detection.

8. How many moles of hydrochloric acid, HCl, are required to react with the sodium hydroxide in 25.0 mL of a 0.100 M NaOH solution.

Solution

According to the balanced equation, the number of moles of HCl required equals the number of moles of NaOH used.

$NaOH(aq) + HCl(aq) \rightarrow NaCl(aq) + H_2O(l)$

1 mol NaOH : 1 mol HCl

mol (NaOH) $= M \times L = 0.100$ M $\times 0.0250$ L $= 0.00250$ mol

mol (HCl) needed $= 0.00250$ mol $= 2.50 \times 10^{-3}$ mol

11. What mass of $PbCrO_4$, the pigment "chrome yellow," often used by artists, can be produced by addition of excess Na_2CrO_4 to 1.00 L of a 0.493 M solution of $Pb(NO_3)_2$?

$Na_2CrO_4(aq) + Pb(NO_3)_2(aq) \rightarrow PbCrO_4(s) + 2NaNO_3(aq)$

Chapter 3   Applications of Chemical Stoichiometry   39

Solution

The equation shows that PbCrO$_4$ is produced from Pb(NO$_3$)$_2$ on a one mol to one mol basis. Since Na$_2$CrO$_4$ is used in excess, the mass of PbCrO$_4$ depends only on the amount of Pb(NO$_3$)$_2$ used.

mol (Pb(NO$_3$)$_2$) used = mol (PbCrO$_4$) produced

mol (Pb(NO$_3$)$_2$) used = 0.493 M × 1.00 L = 0.493 mol

g-mol. wt (PbCrO$_4$) = 323.19 g

Mass (PbCrO$_4$) = 0.493 mol × $\dfrac{323.19 \text{ g}}{\text{mol}}$ = 159 g

13. An excess of silver nitrate, AgNO$_3$, reacted with 25.00 mL of a solution of CaCl$_2$, producing Ca(NO$_3$)$_2$ and 4.498 g of AgCl. What is the molarity of the CaCl$_2$ solution?

Solution

Write a balanced equation for the reaction, then compare moles of AgCl produced versus moles of CaCl$_2$ required for this amount of product.

CaCl$_2$(*aq*) + 2AgNO$_3$(*aq*) ⟶ 2AgCl(*aq*) + Ca(NO$_3$)$_2$(*aq*)

1 mol CaCl$_2$ produces 2 mol AgCl. In steps:

mol AgCl produced ⟶ mol CaCl$_2$ required ⟶ *M* CaCl$_2$

g-form. wt AgCl = 143.32 g

mol (AgCl) = 4.498 g × $\dfrac{1 \text{ mol}}{143.32 \text{ g}}$ = 0.03138 mol

mol (CaCl$_2$) required = 0.03138 mol AgCl × $\dfrac{1 \text{ mol CaCl}_2}{2 \text{ mol AgCl}}$ = 0.01569 mol

$M = \dfrac{0.01569 \text{ mol CaCl}_2}{25.00 \text{ mL} \times \dfrac{1 \text{ L}}{1000 \text{ mL}}}$ = 0.6277 M

16. The lead nitrate, Pb(NO$_3$)$_2$, in 47.29 mL of a 0.1001 M solution reacts with all the aluminum sulfate, Al$_2$(SO$_4$)$_3$, in 25.00 mL of solution. What is the molar concentration of the Al$_2$(SO$_4$)$_3$ in the original Al$_2$(SO$_4$)$_3$ solution?

3Pb(NO$_3$)$_2$(*aq*) + Al$_2$(SO$_4$)$_3$(*aq*) ⟶ 3PbSO$_4$(*s*) + 2Al(NO$_3$)$_3$(*aq*)

Solution

Given the balanced equation, the number of moles of Al$_2$(SO$_4$)$_3$ in 25.00 mL of solution must equal one-third the number of moles of Pb(NO$_3$)$_2$ in the reaction volume. In steps:

mol Pb(NO$_3$)$_2$ present ⟶ mol Al$_2$(SO$_4$)$_3$ ⟶ *M* Al$_2$(SO$_4$)$_3$

mol (Pb(NO$_3$)$_2$) = 0.1001 M × 47.29 mL × $\dfrac{1 \text{ L}}{1000 \text{ mL}}$ = 0.0047337 mol

$$\text{mol (Al}_2(\text{SO}_4)_3) = 0.0047337 \text{ mol} \times \frac{1 \text{ mol Al}_2(\text{SO}_4)_3}{3 \text{ mol (Pb(NO}_3)_2)} = 0.0015779 \text{ mol}$$

$$M \text{ (Al}_2(\text{SO}_4)_3) = \frac{0.0015779 \text{ mol}}{25.00 \text{ mL} \times \frac{1 \text{ L}}{1000 \text{ mL}}} = 0.06312 \text{ M} = 6.312 \times 10^{-2} \text{ M}$$

TITRATION

17. What is the concentration of NaCl in a solution if titration of 10.00 mL of the solution with 0.250 M $AgNO_3$ requires 17.05 mL of the $AgNO_3$ solution to reach the end point?

$$AgNO_3(aq) + NaCl(aq) \longrightarrow AgCl(s) + NaNO_3(aq)$$

Solution

The end point in the reaction occurs when all of the available $AgNO_3$ has been reacted with the NaCl added. From the balanced equation, $AgNO_3$ and NaCl react on a one mol to one mol basis. Therefore

mol (NaCl) in 10.00 mL = mol ($AgNO_3$) present

$$\text{mol (AgNO}_3) = 0.250 \text{ M} \times 17.05 \text{ mL} \times \frac{1 \text{ L}}{1000 \text{ mL}} = 0.0042625 \text{ mol}$$

mol (NaCl) in 10.00 mL = 0.0042625 mol

$$M \text{ (NaCl)} = \frac{0.0042625 \text{ mol}}{0.01000 \text{ L}} = 0.426 \text{ M}$$

19. What is the H_2SO_4 concentration in a solution of H_2SO_4 that requires 31.91 mL to titrate 2.474 g of K_2CO_3?

$$H_2SO_4(aq) + K_2CO_3(s) \longrightarrow K_2SO_4(aq) + H_2O(l) + CO_2(g)$$

Solution

According to the equation, H_2SO_4 and K_2CO_3 react on a 1 mol to 1 mol basis. The number of moles of H_2SO_4 in 31.91 mL equals the number of moles in 2.474 g of K_2CO_3.

mol K_2CO_3 $\longrightarrow$ mol H_2SO_4 $\longrightarrow$ M H_2SO_4

g-form. wt (K_2CO_3) = 138.206 g

$$\text{mol (K}_2\text{CO}_3) = 2.474 \text{ g} \times \frac{1 \text{ mol}}{138.206 \text{ g}} = 0.01790 \text{ mol}$$

mol (H_2SO_4) present = 0.01790 mol

$$M \text{ (H}_2\text{SO}_4) = \frac{0.01790 \text{ mol}}{0.03191 \text{ L}} = 0.5610 \text{ M}$$

22. Titration of a 20.0 mL sample of a particularly acidic rain required 1.7 mL of 0.0811 M NaOH to reach the end point. If we assume that the acidity of the rain is due to the presence of sulfuric acid, H_2SO_4, in the rain water, what was the concentration of this sulfuric acid solution?

$$H_2SO_4(aq) + 2NaOH(aq) \longrightarrow Na_2SO_4(aq) + 2H_2O(l)$$

Chapter 3 Applications of Chemical Stoichiometry 41

Solution

The balanced equation indicates a ratio of 2 mol NaOH to 1 mol H_2SO_4. Therefore, the number of moles of H_2SO_4 in 20.0 mL of rain equals one-half the number of moles of NaOH in the titrant volume.

mol NaOH used $\longrightarrow$ mol H_2SO_4 present $\longrightarrow M\ H_2SO_4$

$$\text{mol (NaOH)} = 0.0811\ M \times 1.7\ \text{mL} \times \frac{1\ L}{1000\ \text{mL}} = 0.000138\ \text{mol}$$

$$\text{mol (H}_2SO_4) = 0.000138\ \text{mol NaOH} \times \frac{1\ \text{mol H}_2SO_4}{2\ \text{mol NaOH}} = 0.0000690\ \text{mol}$$

$$M\ (H_2SO_4) = \frac{0.0000690\ \text{mol}}{0.0200\ L} = 0.0034\ M = 3.4 \times 10^{-3}\ M$$

24. Crystalline potassium hydrogen phthalate, $KHC_8H_4O_4$, is often used as a standard acid for standardizing basic solutions because it is easy to purify and to weigh. If 1.5428 g of this salt is titrated with a solution of $Ba(OH)_2$, the reaction is complete when 22.51 mL of the solution has been added. What is the concentration of the $Ba(OH)_2$ solution?

$$2KHC_8H_4O_4 + Ba(OH)_2 \longrightarrow BaK_2(C_8H_4O_4)_2 + 2H_2O$$

Solution

The molar ratio in the balanced equation is 2 mol $KHC_8H_4O_4$ to 1 mol $Ba(OH)_2$. Therefore, the number of moles of $Ba(OH)_2$ in 22.51 mL of solution equals one-half the number of moles of $KHC_8H_4O_4$ in 1.5428 g.

mol $KHC_8H_4O_4$ $\longrightarrow$ mol $Ba(OH)_2$ $\longrightarrow M\ Ba(OH)_2$

g-mol. wt $(KHC_8H_4O_4) = 204.2234$ g

$$\text{mol (KHC}_8H_4O_4) = 1.5428\ \text{g} \times \frac{1\ \text{mol}}{204.2234\ \text{g}} = 0.0075545\ \text{mol}$$

$$\text{mol (Ba(OH)}_2) = 0.0075545\ \text{mol} \times \frac{1\ \text{mol Ba(OH)}_2}{2\ \text{mol KHC}_8H_4O_4} = 0.003777\ \text{mol}$$

$$M\ (Ba(OH)_2) = \frac{0.003777\ \text{mol}}{0.02251\ L} = 0.1678\ M$$

26. (a) A 5.00-mL sample of vinegar, a solution of acetic acid (CH_3CO_2H), was titrated with a 0.240 M NaOH solution. If 16.96 mL was required to reach the end point, what was the molar concentration of the acetic acid in the vinegar?

$$CH_3CO_2H + NaOH \longrightarrow CH_3CO_2Na + H_2O$$

(b) If the density of the vinegar is 1.005 g/cm^3, what is the percent of acetic acid by mass in the vinegar?

Solution

(a) On a 1 mol to 1 mol reaction basis, the number of moles CH_3CO_2H in 5.00 mL of vinegar equals the number of moles of NaOH in 16.96 mL of 0.240 M solution.

42   Chapter 3   Applications of Chemical Stoichiometry

$$\text{mol NaOH used} \rightarrow \text{mol CH}_3\text{CO}_2\text{H} \rightarrow M \text{ CH}_3\text{CO}_2\text{H}$$

$$\text{mol (NaOH)} = 0.240 \text{ M} \times 16.96 \text{ mL} \times \frac{1 \text{ L}}{1000 \text{ mL}} = 0.004070 \text{ mol}$$

$$M \text{ (vinegar)} = \frac{0.004070 \text{ mol}}{0.00500 \text{ L}} = 0.814 \text{ M}$$

(b) Calculate the mass of CH_3CO_2H per liter of vinegar. Then, given the density of vinegar, calculate the mass of one liter of vinegar. Divide mass of CH_3CO_2H by the solution mass to get percentage by mass.

$$\frac{\text{mol (CH}_3\text{CO}_2\text{H})}{\text{L}} \rightarrow \frac{\text{mass (CH}_3\text{CO}_2\text{H})}{\text{L}} \rightarrow \frac{\text{mass (CH}_3\text{CO}_2\text{H}) \times 100}{\text{mass of 1 L of vinegar}} = \%$$

g-mol. wt $(CH_3CO_2H) = 60.0524$ g

$$\text{Mass (CH}_3\text{CO}_2\text{H}) = 0.814 \frac{\text{mol}}{\text{L}} \times 60.0524 \frac{\text{g}}{\text{mol}} = 48.883 \frac{\text{g}}{\text{L}}$$

$$\text{Mass of 1.0 L of vinegar} = 1.005 \frac{\text{g}}{\text{cm}^3} \times 1000 \frac{\text{cm}^3}{\text{L}} = 1005 \frac{\text{g}}{\text{L}}$$

$$\% \text{ CH}_3\text{CO}_2\text{H} = \frac{48.883 \text{ g}}{1005 \text{ g}} \times 100 = 4.86\%$$

CONVERSION BETWEEN QUANTITIES

27. Begin with the expression for each of the following conversions, and derive the expression for the reverse conversion:

    (a) Grams A to moles A
    (c) Moles of solute to molar concentration (given the volume of solution)
    (d) Percent A and mass of A to mass of a sample containing A

    Solution

    (a) mol A = grams A $\times$ 1 mol/g-form. wt A, so grams A = g-form. wt A $\times$ mol A
    (c) M = mol solute/V (L) so mol solute = V (L) $\times$ M
    (d) We need to determine what mass of sample with % A contains the stated mass.

    $$\frac{\% \text{ A}}{100\%} \times \text{sample mass} = \text{mass of A}$$

    $$\text{Sample mass} = \frac{\text{mass of A} \times 100\%}{\% \text{ A}}$$

29. Determine the moles of $Ca_3(PO_4)_2$ and the mass of Ca contained in a 374-g sample of phosphate rock that contains 89.2% $Ca_3(PO_4)_2$.

    Solution

    Given the rock mass and the percent $Ca_3(PO_4)_2$, calculate the mass of $Ca_3(PO_4)_2$ in 374 g, then convert to moles. Calculate percent Ca in $Ca_3(PO_4)_2$, then multiply the mass of the rock by the percent of Ca to obtain the mass of Ca.

Chapter 3 Applications of Chemical Stoichiometry   43

g-form. wt $(Ca_3(PO_4)_2) = 310.18$ g; g-at. wt $(Ca) = 40.08$ g

Mass $(Ca_3(PO_4)_2)$ in sample $= 0.892 \times 374$ g $= 333.6$ g

$$\text{mol }(Ca_3(PO_4)_2) = 333.6 \text{ g} \times \frac{1 \text{ mol}}{310.18 \text{ g}} = 1.08 \text{ mol}$$

$$\% \text{ Ca in } Ca_3(PO_4)_2 = \frac{40.08 \times 3}{210.18} \times 100 = 38.76\%$$

Mass (Ca) in sample $= 0.3876 \times 333.6$ g $= 129$ g

31. What is the molar concentration of isotonic saline used for intravenous injections, which has a density of 1.007 g/mL and contains 0.95% by mass of NaCl?

Solution

Convert density in g/mL to g/L, then calculate the mass of NaCl at 0.95% in one liter of solution. Use the definition of molarity to calculate the molarity of NaCl.

$$\frac{\text{g solution}}{\text{mL}} \rightarrow \frac{\text{g solution}}{\text{L}} \rightarrow \frac{\text{g NaCl}}{\text{L}} \rightarrow \frac{\text{mol NaCl}}{\text{L}}$$

$$\frac{\text{Mass solution}}{\text{L}} = 1.007 \frac{\text{g}}{\text{mL}} \times 1000 \frac{\text{mL}}{\text{L}} = 1007 \text{ g/L}$$

$$\frac{\text{Mass NaCl}}{\text{L}} = 1007 \frac{\text{g}}{\text{L}} \times 0.0095 = 9.57 \text{ g/L}$$

$$M \text{ (NaCl)} = \frac{9.57 \text{ g}}{\text{L}} \times \frac{1 \text{ mol}}{58.44 \text{ g}} = 0.16 \text{ M}$$

34. What volume of 0.250 M potassium hydroxide solution can be prepared from 174 g of KOH that contains 12.8% water by mass?

Solution

Calculate the mass of KOH in the 174 g sample, then use the definition of molarity to calculate the volume of 0.250 M solution that can be prepared from this mass of KOH.

Mass of KOH $\rightarrow$ mol KOH $\rightarrow$ liters of solution

Mass (KOH) $= 174$ g $\times 0.872 = 151.7$ g

$$\text{mol (KOH)} = 151.7 \text{ g} \times \frac{1 \text{ mol}}{56.106 \text{ g}} = 2.704 \text{ mol}$$

$$M = \frac{\text{mol}}{\text{L}} \text{ so L} = \frac{\text{mol}}{M} = \frac{2.704 \text{ mol}}{0.250 \text{ M}} = 10.8 \text{ L}$$

37. Concentrated sulfuric acid as produced industrially often contains 98.0% H_2SO_4 by mass and has a density of 1.92 g/mL. What is the concentration of H_2SO_4 in this acid solution?

Solution

Use the density to calculate the mass of one liter of solution, then multiply by the percent H_2SO_4 to get the mass of H_2SO_4 per liter of solution. Determine molarity from the mass of H_2SO_4 per liter.

$$\frac{\text{Mass of solution}}{\text{L}} = 1.92\ \frac{\text{g}}{\text{mL}} \times 1000\ \frac{\text{mL}}{\text{L}} = 1.92 \times 10^3\ \text{g/L}$$

$$\frac{\text{Mass (H}_2\text{SO}_4)}{\text{L}} = \frac{1.92 \times 10^3\ \text{g}}{\text{L}} \times 0.980 = 1.882 \times 10^3\ \text{g/L}$$

$$M = \frac{\text{mol H}_2\text{SO}_4}{\text{L}} = \frac{1.882 \times 10^3\ \text{g}}{\text{L}} \times \frac{1\ \text{mol}}{98.07\ \text{g}} = 19.2\ M$$

41. What volume of gaseous carbon dioxide (density 1.964 g/L) is required to carbonate 12 ounces (355 mL) of a beverage with a density of 1.067 g/cm³ that contains 0.66% CO_2 by mass?

Solution

Calculate the mass of CO_2 required in the solution and based on its density, calculate the volume needed.

Mass of CO_2 required ⟶ volume of CO_2 required

$$\text{Mass (CO}_2) = 355\ \text{mL} \times 1.067\ \frac{\text{g}}{\text{cm}^3} \times 0.0066\ CO_2 = 2.50\ \text{g CO}_2$$

$$D = \frac{\text{mass}}{\text{volume}} \quad \text{so} \quad \text{volume} = \frac{\text{mass}}{D}$$

$$V\ (\text{L}) = \frac{2.50\ \text{g}}{1.964\ \text{g/L}} = 1.3\ \text{L}$$

44. How many grams of an insect-repellent solution containing 17.0% of N,N-dimethyl-metatoluamide, $C_{10}H_{13}NO$, can be prepared from 1.00 mol of $C_{10}H_{13}NO$?

Solution

Determine the mass of 1.00 mol of $C_{10}H_{13}NO$. Then find the mass of solution whose composition is 17.0% $C_{10}H_{13}NO$.

$$0.170 \times \text{mass of solution} = 163.22\ \text{g}$$

$$\text{Mass of solution} = \frac{163.22\ \text{g}}{0.170} = 960\ \text{g}$$

CHEMICAL STOICHIOMETRY

45. How many grams of CaO are required for reaction with the HCl in 27.5 mL of a 0.523 M HCl solution? The equation for the reaction is

$$CaO + 2HCl \longrightarrow CaCl_2 + H_2O$$

Solution

Given the balanced equation, one mole of CaO is required for each two moles of HCl contained in the solution.

mol HCl ⟶ mol CaO needed ⟶ mass of CaO

mol (HCl) = 0.523 M × 0.0275 L = 0.01438 mol

Chapter 3 Applications of Chemical Stoichiometry 45

$$\text{mol (CaO)} = 0.01438 \text{ mol HCl} \times \frac{1 \text{ mol CaO}}{2 \text{ mol HCl}} = 0.00719 \text{ mol}$$

$$\text{Mass (CaO)} = 0.00719 \text{ mol} \times \frac{56.08 \text{ g}}{\text{mol}} = 0.403 \text{ g}$$

47. Aspirin, $C_6H_4(CO_2H)(CO_2CH_3)$, can be prepared in the chemistry laboratory by the reaction of salicylic acid, $C_6H_4(CO_2H)(OH)$, with acetic anhydride, $(CH_3CO)_2O$.

$$2C_6H_4(CO_2H)(OH) + (CH_3CO)_2O \longrightarrow 2C_6H_4(CO_2H)(CO_2CH_3) + H_2O$$

What volume of acetic anhydride (density, 1.0820 g/cm³) is required to produce 1.00 kg of aspirin, assuming a 100% yield?

Solution

The volume of acetic anhydride required can be calculated from its mass determined from the stoichiometry of the reaction.

Mass of $(CH_3CO)_2O$ required $\longrightarrow$ volume of solution $(D = 1.0820 \text{ g/cm}^3)$

In the equation, one mol of acetic anhydride produces two mol of aspirin.
g-mol.wts:

AA = 102.09 g; Asp = 180.150 g

$$\text{Mass (AA)} = 1.00 \times 10^3 \text{ g Asp} \times \frac{1 \text{ mol}}{180.160 \text{ g}} \times \frac{1 \text{ mol AA}}{2 \text{ mol Asp}} \times \frac{102.09 \text{ g}}{\text{mol AA}} = 283.33 \text{ g}$$

$$D = \frac{M}{V} \longrightarrow V = \frac{M}{D} = \frac{283.33 \text{ g}}{1.0820 \text{ g/cm}^3} = 262 \text{ cm}^3 = 262 \text{ mL}$$

50. What volume in L of air (density of 1.20 g/L and 21.0% O_2 by mass) is required to burn 1.00 gallon of gasoline (density, 0.780 g/cm³), which can be represented by C_8H_{16}? Assume a 100% yield and complete consumption of the O_2.

$$C_8H_{16}(l) + 12O_2(g) \longrightarrow 8CO_2(g) + 8H_2O(g)$$

Solution

Use the balanced equation to calculate the mass of oxygen required. Given the mass, use the gas density to calculate the volume required.

$V\ C_8H_{16} \longrightarrow$ mol $C_8H_{16} \longrightarrow$ mass O_2 needed $\longrightarrow V$ air needed

$$\text{Mass }(C_8H_{16}) = 1.00 \text{ gal} \times \frac{3785 \text{ cm}^3}{\text{gal}} \times 0.780 \frac{\text{g}}{\text{cm}^3} = 2952 \text{ g}$$

$$\text{mol }(C_8H_{16}) = 2952 \text{ g} \times \frac{1 \text{ mol}}{112.21 \text{ g}} = 26.31 \text{ mol}$$

$$\text{mol }(O_2) \text{ required} = 26.31 \text{ mol } C_8H_{16} \times \frac{12 \text{ mol } O_2}{1 \text{ mol } C_8H_{16}} = 315.70 \text{ mol}$$

$$\text{Mass }(O_2) = 315.70 \text{ mol} \times \frac{32.0 \text{ g}}{\text{mol}} = 1.010 \times 10^4 \text{ g}$$

$$\frac{\text{Mass (O}_2)}{\text{L air}} = 1.20 \ \frac{g}{L} \times 0.21 = \frac{0.252 \ g \ O_2}{L}$$

$$V \text{(air)} = \frac{1.010 \times 10^4 \ g \ O_2}{0.252 \ g \ O_2/L} = 4.01 \times 10^4 \ L$$

52. Bronzes used in bearings are often alloys (solid solutions) of copper and aluminum. A 1.953-g sample of one such bronze was analyzed for its copper content by the sequence of reactions given below. The aluminum also dissolves, but we need not consider it because aluminum sulfate, $Al_2(SO_4)_3$, does not react with KI.

$$Cu + 2H_2SO_4 \longrightarrow CuSO_4 + 2H_2O + SO_2$$

$$2CuSO_4 + 4KI \longrightarrow 2CuI + I_2 + 2K_2SO_4$$

$$I_2 + 2Na_2S_2O_3 \longrightarrow Na_2S_4O_6 + 2NaI$$

If 35.06 mL of 0.837 M $Na_2S_2O_3$ is required to react with the I_2 formed, what is the percent Cu in the sample?

Solution

Careful analysis of the three balanced equations shows the following reaction ratio: 2 mol Cu : 2 mol $CuSO_4$: 1 mol I_2, which requires 2 mol $Na_2S_2O_3$. This ratio leads a person to the conclusion that one mole of $Na_2S_2O_3$ reflects the consumption of one mole of copper in the sample. Stepwise:

$$\text{mol } Na_2S_2O_3 \text{ used} \longrightarrow \text{mol Cu in sample} \longrightarrow \text{percent Cu}$$

$$\text{mol (Na}_2S_2O_3) \text{ used} = 0.837 \ M \times 0.03506 \ L = 0.02934 \ \text{mol}$$

$$\text{mol (Cu) in sample} = 0.02934 \ \text{mol } Na_2S_2O_3 \times \frac{2 \ \text{mol Cu}}{2 \ \text{mol } Na_2S_2O_3} = 0.02934 \ \text{mol}$$

$$\text{Mass (Cu)} = 0.02934 \ \text{mol} \times 63.546 \ \frac{g}{\text{mol}} = 1.8644 \ g$$

$$\% \ Cu = \frac{1.8644 \ g}{1.953 \ g} \times 100 = 95.5\%$$

54. Glauber's salt, $Na_2SO_4 \cdot 10H_2O$, is an important industrial chemical that is isolated from naturally occurring brines in New Mexico. A 0.3440-g sample of this material was allowed to react with an excess of $Ba(NO_3)_2$, and 0.2398 g of $BaSO_4$ was isolated. What is the percent of $Na_2SO_4 \cdot 10H_2O$ in the sample analyzed?

$$Na_2SO_4 \cdot 10H_2O(aq) + Ba(NO_3)_2(aq) \longrightarrow BaSO_4(s) + 2NaNO_3(aq) + 10H_2O$$

Solution

One mole of Glauber's salt produces one mole of $BaSO_4$. How much salt is required to produce 0.2398 g of $BaSO_4$?

$$\text{Mass } BaSO_4 \longrightarrow \text{mol } BaSO_4 \longrightarrow \text{mol G. salt} \longrightarrow \text{mass G. salt} \longrightarrow \% \ G. \ salt$$

g-form.wts: G. salt = 322.17; $BaSO_4$ = 233.39

$$\text{mol (BaSO}_4) = 0.2398 \ g \times \frac{1 \ \text{mol}}{233.39 \ g} = 0.0010275 \ \text{mol}$$

Chapter 3 Applications of Chemical Stoichiometry 47

mol (G. salt used) = 0.0010275

$$\text{Mass (G. salt)} = 0.0010275 \times 322.17 \ \frac{\text{g}}{\text{mol}} = 0.33102 \text{ g}$$

$$\% \ Na_2SO_4 \cdot 10 \ H_2O = \frac{0.33102 \text{ g}}{0.3440} \times 100 = 96.23\%$$

ADDITIONAL EXERCISES

59. The average plasma volume in an adult is 39 mL per kilogram of weight. The average concentration of sodium ion, Na^+, is 0.142 M. What is the mass of sodium ion present in the serum of a 75-kg (165-lb) adult? What mass of NaCl, which is composed of Na^+ and Cl^- ions, would be required to provide this much sodium?

Solution

Total plasma volume ⟶ mol Na^+ ⟶ mass Na^+ ⟶ mass NaCl

$$V \text{(Plasma)} = \frac{39 \text{ mL}}{\text{kg}} \times 75 \text{ kg} = 2925 \text{ mL} = 2.925 \text{ L}$$

$$\text{mol } (Na^+) = M \times L = 0.142 \text{ M} \times 2.925 \text{ L} = 0.42 \text{ mol}$$

$$\text{Mass } (Na^+) = 0.42 \text{ mol} \times 23.0 \ \frac{\text{g}}{\text{mol}} = 9.5 \text{ g } Na^+$$

$$\text{mol (NaCl)} = \text{mol } Na^+ = 0.42 \text{ mol}$$

$$\text{Mass (NaCl)} = 0.42 \text{ mol} \times \frac{1 \text{ mol}}{58.5 \text{ g}} = 25 \text{ g NaCl}$$

61. A copper ion, Cu^{2+}, content of 1.0 mg/L causes a bitter taste in drinking water. What mass of $CuSO_4$ will give a copper content of 1.0 mg/L and what is the $CuSO_4$ concentration in mol/L?

Solution

Determine the mass of $CuSO_4$ required by calculating the percent Cu in $CuSO_4$ and then relating the percentage to the mass of Cu needed to affect taste. g-form. wts:

$CuSO_4 = 159.60$, $Cu = 63.546$

$$\% \ Cu = \frac{63.546}{159.60} \times 100 = 39.81\%$$

$0.3981 \ (\text{Mass } CuSO_4) = 1.0 \text{ mg}$

Mass = 2.5 mg

$$M \text{ (}CuSO_4\text{)} = \frac{2.5 \text{ mg} \times \dfrac{1 \text{ g}}{1000 \text{ mg}} \times \dfrac{1 \text{ mol}}{159.60 \text{ g}}}{1.0 \text{ L}} = 1.6 \times 10^{-5} \text{ M}$$

65. The amounts of active ingredients per tablet of several antacids, and equations for how these ingredients react with stomach acid (HCl), are given below. Assuming that the stomach acid is undiluted (0.155 M), calculate what volume of acid in mL will react with a pure sample of the active ingredients of each tablet.

(a) Phillip's Tablets, 0.311 g Mg(OH)$_2$

$$Mg(OH)_2 + 2HCl \longrightarrow MgCl_2 + 2H_2O$$

Solution

(a) Mass Mg(OH)$_2$ $\longrightarrow$ mol Mg(OH)$_2$ $\longrightarrow$ mol HCl needed $\longrightarrow$ V HCl needed

1 mol Mg(OH)$_2$ requires 2 mol HCl

g-form. wt (Mg(OH)$_2$) = 58.3196 g

$$\text{mol (Mg(OH)}_2\text{)} = 0.311 \text{ g} \times \frac{1 \text{ mol}}{58.3196 \text{ g}} = 0.005333 \text{ mol}$$

$$\text{mol (HCl) required} = 0.005333 \text{ mol} \times \frac{2 \text{ mol HCl}}{1 \text{ mol Mg(OH)}_2} = 0.01066 \text{ mol}$$

$$M = \frac{\text{mol}}{V} \longrightarrow V = \frac{\text{mol}}{M} = \frac{0.01066 \text{ mol}}{0.155 \text{ M}} = 0.0688 \text{ L} = 68.8 \text{ mL}$$

68. On average, protein contains 15.5% nitrogen by mass. This nitrogen can be converted to gaseous ammonia, NH$_3$(g), and analyzed by the Kjeldahl technique. The ammonia reacts with an excess of boric acid, H$_3$BO$_3$

$$NH_3(g) + H_3BO_3(aq) \longrightarrow (NH_4)H_2BO_3(aq)$$

and the product titrated with standardized hydrochloric acid.

$$(NH_4)H_2BO_3 + HCl \longrightarrow NH_4Cl + H_3BO_3$$

What is the percent by mass of protein in a sample of lobster meat if titration of the (NH$_4$)H$_2$BO$_3$ formed in a Kjeldahl analysis of a 4.9-g meat sample requires 34.05 mL of 0.2011 M HCl solution?

Solution

The reaction summary is

1 mol N $\longrightarrow$ 1 mol NH$_3$ $\longrightarrow$ 1 mol (NH$_4$)H$_2$BO$_3$ $\longrightarrow$ 1 mol HCl

mol (N) = mol (HCl) used = 0.2011 M $\times$ 0.03405 L = 0.006848 mol

$$\text{Mass (N) in sample} = 0.006848 \text{ mol} \times \frac{14.0067 \text{ g N}}{\text{mol}} = 0.0959 \text{ g}$$

Since the N mass is about 15.5% of the protein mass, the protein mass is

0.155 (protein mass) = 0.0959 g

Protein = 0.619 g

$$\% \text{ protein in meat} = \frac{0.619 \text{ g}}{4.95 \text{ g}} \times 100 = 12.5\%$$

Chapter 3 Applications of Chemical Stoichiometry   49

72. What is the limiting reagent when 5.0×10^{-2} moles of HNO_3 react with 225 mL of 0.10 M $Ca(OH)_2$ solution?

$$Ca(OH)_2(aq) + 2HNO_3(aq) \longrightarrow Ca(NO_3)_2(aq) + 2H_2O(l)$$

Solution

Given the two reactant quantities, calculate the amount of one of the products (water for simplicity) which each quantity would produce assuming a sufficient amount of the other is available. The smallest amount of product indicates the limiting reagent.

1. $\text{mol } H_2O = 5.0 \times 10^{-2} \text{ mol } HNO_3 \times \dfrac{2 \text{ mol } H_2O}{2 \text{ mol } HNO_3} = 5.0 \times 10^{-2} \text{ mol}$

2. $\text{mol } Ca(OH)_2 \text{ available} = 0.10 \text{ M} \times 0.225 \text{ L} = 0.0225 \text{ mol}$

   $\text{mol } H_2O = 0.0225 \text{ mol} \times \dfrac{2 \text{ mol } H_2O}{1 \text{ mol } Ca(OH)_2} = 0.045 \text{ mol}$

Available $Ca(OH)_2$ produces 0.045 mol versus 0.05 mol from HNO_3, so $Ca(OH)_2$ is the limiting factor.

4

Thermochemistry

INTRODUCTION

The study of the heat absorbed or evolved in chemical reactions and phase changes is the major focus of this chapter. We begin by learning how heat is transferred and about its measurement or calorimetry through the expression $q = cm\,\Delta T$, where q is the heat, c is the specific heat, M is the mass in grams, and ΔT is the temperature change.

Under constant-pressure conditions, we find that the heat change is a measure of the enthalpy of a substance. The standard molar enthalpy of formation and the enthalpy of combustion are two enthalpies produced under set conditions that provide a useful reference for further calculations based on Hess's law. This law allows the manipulation of chemical equations and their associated enthalpies as though they were mathematical expressions. The fuller study of thermodynamics, of which thermochemistry is but one part, is reserved for a later chapter.

READY REFERENCE

Calorie (cal) (4.1)   The calorie is an older unit of energy and is still very much a part of our language. It is widely used for work in nutrition, physiology, and the health sciences in general. For purposes of conversion, 1 joule equals 0.239 calories or 1 calorie (cal) equals 4.184 J. The calorie is liberally defined as the amount of heat required to raise the temperature of 1.0 g of water 1.0°C. The nutrition calorie (Cal) equals 1.0 kcal or 4184 J (4.184 kJ).

Endothermic reaction (4.4)   A reaction that occurs with the loss or removal of heat from the surroundings.

Enthalphy (H) (4.4)   The heat content or enthalphy of the system. The change in enthalphy ΔH is the quantity of heat absorbed or liberated by the system when a reaction takes place at constant pressure; therefore, $\Delta H = q$. If a reaction is endothermic, q is positive; if exothermic, q is negative. By definition, $\Delta H = \Delta E + \Delta(PV)$, or $\Delta H = \Delta E + P\,\Delta V$ for a constant-pressure process.

Exothermic reaction (4.1)   A reaction that occurs with the production of heat.

Heat (q) (4.1)   Heat is a form of energy that can be measured only as it passes from one boundary to another. A positive sign for heat indicates the heat is gained by the body. A negative sign associated with heat represents a heat loss.

Chapter 4 Thermochemistry

Heat of combustion (4.4) The enthalpy change for the combustion (combination with oxygen) of 1 mole of a substance under standard conditions.

Hess's law (4.5) For any process that can be considered the sum of several stepwise processes, the enthalpy change for the total process must equal the sum of the enthalpy changes for the various steps.

Joule (J) (4.1) A joule is the amount of heat or other energy equal to the kinetic energy ($\frac{1}{2} Mv^2$) of an object having mass M of exactly 2 kilograms moving with a velocity v of exactly 1 meter per second.

Specific heat capacity (c) (4.1) Energy is intangible in that only changes in energy can be measured. Although the joule serves as the standard unit of energy, changes in heat must be determined by measuring temperature changes of specific substances. The specific heat capacity (formerly known as the specific heat) of a substance is the amount of heat required to change the temperature of 1 gram of that substance by 1°C; the specific heat capacity for a given substance is unique to that substance. An algebraic interpretation of this definition applied to any substance is

Heat = sp. heat × mass × temp. change

$$q = c \times M \times \Delta T$$

The transfer of heat between two bodies (heat flow) always occurs from the hotter body to the colder body. In the process of heat flow both bodies eventually reach an equilibrium temperature. Because heat is conserved, the amount of heat lost by the hotter body (a minus quantity) equals the amount of heat gained by the colder body (a positive quantity).

$$q_{\text{gained}} = -q_{\text{lost}}$$

or $q_{\text{lost}} + q_{\text{gained}} = 0$

Standard molar enthalpy of formation (4.4) This is the change in enthalpy when one mole of a pure substance is formed from the free elements in their most stable state under standard conditions. For any free element in its most stable form, the value of the standard molar enthalpy is zero.

Standard state (4.4) An agreed-upon specific set of conditions designed to facilitate the handling of data. The standard state of a pure substance is taken as 25°C (298.15 K) and 1 atm pressure.

State (4.3) The condition of the system defined by n, P, V, and T.

State of function (4.3) A function that depends only on the particular state of a system and not on how the system got to the particular state.

System (4.3) That part of the universe on which we focus our attention and with whose properties we are concerned.

Thermochemistry (Introduction) The determination and study of the heat absorbed or evolved in chemical reactions and in changes of phase.

SOLUTIONS TO CHAPTER 4 EXERCISES

HEAT

3. (a) If 14.5 kJ of heat was added to the 800 g of water in Fig. 4.2, how much would its temperature increase?

(b) If 29 kcal of heat was added to the 800 g of water in Fig. 4.2, how much would its temperature increase?

Solution

(a) $q = cM\,\Delta T$ Rearrange and solve for ΔT; be sure to use joules consistently.

$$\Delta T = \frac{q}{cM} = \frac{14{,}500 \text{ J}}{800 \text{ g} \times 4.18 \text{ J/(g °C)}} = 4.34°C$$

(b) $q = cM\,\Delta T$ Use c expressed in cal/g °C and express the heat in cal.

$$\Delta T = \frac{q}{cM} = \frac{29{,}000 \text{ cal}}{800 \text{ g} \times 1.00 \text{ cal/(g °C)}} = 36°C$$

4. Explain the difference between *heat capacity* and *specific heat* of a substance.

Solution

The term *heat capacity* refers to the quantity of heat required to increase a substance's temperature by 1 degree Celsius.

Heat capacity = specific heat (J/g °C) × mass (g)

Heat capacity depends upon the amount of mass present. The *specific heat* is the quantity of heat required to raise the temperature of 1 gram of a substance 1 degree C. Specific heat is independent of the amount of mass.

5. How much heat, in joules and in calories, must be added to a 75.0-g iron block with a specific heat of 0.451 J/g °C to increase its temperature from 25°C to its melting temperature of 1535°C?

Solution

$q = cM\,\Delta T$

$\Delta T = (T_f - T_i) = (1535 - 25)°C = 1510°C$

$q\text{ (J)} = 0.451\dfrac{\text{J}}{\text{g °C}} \times 75.0 \text{ g} \times 1510°C = 51{,}100 \text{ J} = 5.11 \times 10^4 \text{ J}$

$q\text{ (cal)} = 51{,}100 \text{ J} \times \dfrac{1.00 \text{ cal/(g °C)}}{4.18 \text{ J/(g °C)}} = 12{,}200 \text{ cal} = 1.22 \times 10^4 \text{ cal}$

7. Calculate the heat capacity, in joules and in calories, of the following

(a) 28.4 g (1 ounce) of water (specific heat 4.184 J/g °C)

(c) 45.8 g of nitrogen gas (specific heat 1.04 J/g °C)

Solution

(a) Heat capacity = specific heat × mass = 4.184 J/g °C × 28.4 g
= 119 J/°C. In calories,

$$119 \frac{J}{°C} \times \frac{1 \text{ cal}}{4.184 \text{ J}} = 28.4 \text{ cal}/°C$$

(c) Heat capacity = 1.04 J/g °C × 45.8 g = 47.6 J/°C; $\frac{47.6 \text{ J}/°C}{4.184 \text{ J}/\text{cal}} = 11.4 \text{ cal}/°C$

8. Calculate the specific heat of water in the units Calories/lb °F (1 Calorie is the nutritional calorie = 1000 calories). The specific heat of water is 4.184 J/g °C.

Solution

Use unit conversions to make the conversion.

$$c = 4.184 \frac{J}{g\ °C} \times \frac{1°C}{\frac{9}{5}°F} \times \frac{453.6 \text{ g}}{1.00 \text{ lb}} \times \frac{1 \text{ cal}}{4.184 \text{ J}} \times \frac{1 \text{ Cal}}{1000 \text{ cal}} = 0.2520 \text{ Cal}/(\text{lb °F})$$

10. How much will the temperature of 180 mL of coffee at 95°C be reduced when a 45-g silver spoon (specific heat 0.24 J/g °C) at 25°C is placed in the coffee and the two are allowed to reach the same temperature? Assume the coffee has the same density and specific heat as water.

Solution

Because of the law of conservation of energy we write

$$q_{spoon} + q_{coffee} = 0 \quad q_{spoon} = -q_{coffee};$$

$$c_{spoon} \times M \times \Delta T = c_{coffee} \times \text{mL} \times \frac{g}{\text{mL}} \times \Delta T$$

$$0.24 \frac{J}{g\ °C} \times 45 \text{ g} \times (T_f - 25°C) = -4.18 \frac{J}{g\ °C} \times 180 \text{ mL} \times \frac{1.00 \text{ g}}{1.00 \text{ mL}} \times (T_f - 95°C)$$

$$10.8\ T_f - 270 = -752.4\ T_f + 71478$$

$$763.2\ T_f = 71748$$

$$T_f = 94°C$$

$$T_i - T_f = 95° - 94° = 1°C$$

CALORIMETRY

13. A 0.500-g sample of KCl is added to 50.0 g of water in a coffee cup calorimeter (Fig. 4.3). If the temperature decreases by 1.05°C, what amount of heat is involved in the dissolution of the KCl assuming the heat capacity of the resulting solution is 4.18 J/g °C?

Solution

Assume that both the KCl and water are at the same temperature. Then the entire temperature change can be attributed to the heat of dissolution.

$$q = cM\ \Delta T = 4.18 \frac{J}{g\ °C} \times (50.0 \text{ g} + 0.500 \text{ g}) \times (-1.05°C) = -222 \text{ cal}$$

This means that 222 calories has been absorbed from the water and the KCl in order for the KCl to dissolve.

16. A sample of 0.562 g of carbon in the form of graphite is burned in oxygen in a bomb calorimeter (Fig. 4.4). The temperature of the calorimeter increases from 26.74°C to 27.63°C. If the specific heat of the calorimeter and its contents is 20.7 kJ/°C, how much heat was released by this reaction?

Solution

$$q = \text{heat capacity} \times (T_f - T_i)$$

$$q = 20.7 \frac{\text{kJ}}{°\text{C}} \times (27.63 - 26.74)°\text{C} = 20.7 \times 0.89 \text{ kJ} = 18 \text{ kJ}$$

The mass of the sample is not needed to solve this problem; but see exercise 20.

ENTHALPY CHANGES

17. Which of the heats of combustion in Table 4.1 are also molar heats of formation?

Solution

Heats of formation are always written: elements in standard states and proper stoichiometric ratios forming one mole of product. From Table 4.1 all entries under carbon, magnesium, nitrogen, silver, sulfur as well as

$$H_2(g) + \tfrac{1}{2}O_2(g) \longrightarrow H_2O(l)$$

under hydrogen, are heats of formation. The other entry under hydrogen,

$$H_2(g) + \tfrac{1}{2}O_2(g) \longrightarrow H_2O(g)$$

is not a heat of formation, because $H_2O(g)$ is not its standard state. See exercise 18.

18. Does $\Delta H°_{f_{H_2O(g)}}$ differ from $\Delta H°_{298}$ for the reaction $2H_2(g) + O_2(g) \longrightarrow 2H_2O(g)$? If so, how?

Solution

Heats of formation, such as $\Delta H°_{f_{H_2O(g)}}$, refer to the formation of one mole of the substance with each reactant and product in its standard state. The value of $\Delta H°_{298}$ refers to the reaction as written; in this case, the reaction is the formation of 2 mol of $H_2O(g)$, not 1 mole of $H_2O(g)$ as would be required for its value to be the same as $\Delta H°_{f_{H_2O(g)}}$.

20. When 2.50 g of methane burns in oxygen, 125 kJ of heat is produced. What is the molar heat of combustion of methane under these conditions?

Solution

Methane has a molecular weight of 16.0432. The 2.50 g burned represents 2.50 g/16.0432 g mol^{-1} or 0.1558 mol. Since 0.1558 mol produced -125 kJ (an exothermic quantity), -125 kJ/0.1558 mol $= -802$ kJ mol^{-1}.

Chapter 4   Thermochemistry

21. In 1774 oxygen was prepared by Joseph Priestly by heating red mercury(II) oxide with sunlight focused through a lens. How much heat is required to decompose 1 mol of HgO(s) to Hg(l) and $O_2(g)$ under standard conditions?

Solution

The reaction $HgO(s) \rightarrow Hg(l) + \frac{1}{2}O_2(g)$ under standard state conditions is the reverse of the reaction which forms one mole of HgO(s) from the elements in their most stable states under standard state conditions. Thus for

$$HgO(s) \rightarrow Hg(l) + \tfrac{1}{2}O_2(g) \qquad \Delta H^\circ_{298} = -\Delta H^\circ_{f_{HgO(s)}}$$

$$= -(-90.83 \text{ kJ}) = 90.83 \text{ kJ}$$

23. The reaction of graphite with oxygen is described by the equation

$$C(s, graphite) + O_2(g) \rightarrow CO_2(g)$$

Assume the reaction was run at constant pressure and calculate the molar enthalpy of formation of $CO_2(g)$ from the data in exercise 16. What conditions would be required for this to be a standard molar enthalpy of formation?

Solution

Under constant pressure conditions the heat evolved is the enthalpy change, in this case, −18,000 J. Since only 0.562 g of carbon is used, convert the amount of heat actually evolved to the amount that would be evolved if one mole of carbon were actually burned. (From the stoichiometry of the equation, 1 mole C → 1 mole CO_2.)

$$\Delta H = \frac{-18,000 \text{ J}}{0.562 \text{ g C}} \times \frac{12.011 \text{ g C}}{\text{mol C}} \times \frac{1 \text{ mol C}}{1 \text{ mol CO}_2} = -390,000 \text{ J} = -3.9 \times 10^2 \text{ kJ mol}^{-1}$$

25. Assume that the solutions have densities of 1.00 g/mL and calculate the approximate value of ΔH for the reaction

$$NaCl(aq) + AgNO_3(aq) \rightarrow AgCl(s) + NaNO_3(aq)$$

using the data in exercise 14. (The result is only approximate because the density of the solutions is assumed to be 1.00 g/mL).

Solution

The solution has a volume of 50.0 g × 1.00 mL/g = 50.0 mL = 0.0500 L. The amount of NaCl and $AgNO_3$ present is 0.050 L × 0.200 M = 0.0100 mol.

$$\Delta H = -q = -cM\,\Delta T = -4.2 \frac{\text{J}}{\text{g °C}} \times 100.0 \text{ g} \times (25.67 - 24.10)°C = -659 \text{ J}$$

$$\Delta H^\circ_{rx} = \frac{-659 \text{ J}}{0.0100 \text{ mol}} = -65,900 \text{ J mol}^{-1} = -65.9 \text{ kJ mol}^{-1}$$

HESS'S LAW

28. Calculate ΔH for the process

$$Os(s) + 2O_2(g) \rightarrow OsO_4(g)$$

from the following information:

$$Os(s) + 2O_2(g) \longrightarrow OsO_4(s) \quad \Delta H = -391 \text{ kJ}$$
$$OsO_4(s) \longrightarrow OsO_4(g) \quad \Delta H = 56.4 \text{ kJ}$$

Solution

Addition of the two reaction equations as though they were mathematical expressions gives

$$Os(s) + 2O_2(g) \longrightarrow OsO_4(g) \quad \Delta H° = -391 \text{ kJ} + 56.4 \text{ kJ} = -335 \text{ kJ}$$

30. Calculate $\Delta H°_{298}$ for the process

$$Zn(s) + S(s) + 2O_2(g) \longrightarrow ZnSO_4(s)$$

from the following information:

$$Zn(s) + S(s) \longrightarrow ZnS(s) \quad \Delta H° = -206.0 \text{ kJ}$$
$$ZnS(s) + 2O_2(g) \longrightarrow ZnSO_4(s) \quad \Delta H° = -776.8 \text{ kJ}$$

Solution

Addition of the two reaction equations as though they were mathematical equations gives

$$Zn(s) + S(s) + 2O_2(g) \longrightarrow ZnSO_4(s)$$
$$\Delta H° = -206.0 \text{ kJ} + (-776.8 \text{ kJ}) = -982.8 \text{ kJ}$$

32. Calculate $\Delta H°_{298}$ for the process

$$3Co(s) + 2O_2(g) \longrightarrow Co_3O_4(s)$$

from the following information:

$$Co(s) + \tfrac{1}{2} O_2(g) \longrightarrow CoO(s) \quad \Delta H°_{298} = -237.9 \text{ kJ}$$
$$3CoO(s) + \tfrac{1}{2} O_2(g) \longrightarrow Co_3O_4(s) \quad \Delta H°_{298} = -177.5 \text{ kJ}$$

Solution

Simple addition is not sufficient to give the desired expression. Multiply the first expression throughout by 3.

$$3Co(s) + \tfrac{3}{2}O_2(g) \longrightarrow 3CoO(s) \quad \Delta H° = 3 \times (-237.9 \text{ kJ})$$

Addition of the second equation

$$3CoO(s) + \tfrac{1}{2}O_2(g) \longrightarrow Co_3O_4(s) \quad \Delta H° = -177.5 \text{ kJ}$$

gives

$$3Co(s) + 2O_2(g) \longrightarrow Co_3O_4(s) \quad \Delta H° = -713.7 \text{ kJ} + (-177.5 \text{ kJ})$$
$$= -891.2 \text{ kJ}$$

33. Calculate the standard molar enthalpy of formation of NO(g) from the following data:

$$N_2(g) + 2O_2(g) \longrightarrow 2NO_2(g) \quad \Delta H°_{298} = 66.4 \text{ kJ}$$
$$2NO(g) + O_2(g) \longrightarrow 2NO_2(g) \quad \Delta H°_{298} = -114.1 \text{ kJ}$$

Chapter 4 Thermochemistry

Solution

Hess's law can be applied to the two equations above by reversing the sense of the second equation. The first equation is a formation reaction and is so indicated by writing $\Delta H^\circ_{f_{298}}$.

$$N_2(g) + 2O_2(g) \longrightarrow 2NO_2(g) \qquad \Delta H^\circ_{f_{298}} = 66.4 \text{ kJ}$$

$$2NO_2(g) \longrightarrow 2NO(g) + O_2(g) \qquad \Delta H^\circ = 114.1 \text{ kJ}$$

Adding gives

$$N_2(g) + O_2(g) \longrightarrow 2NO(g) \qquad \Delta H^\circ = 180.5 \text{ kJ}$$

This is the heat of formation of 2 moles of NO. For 1 mole,

$$\Delta H^\circ_{f_{298}} = \frac{180.5 \text{ kJ}}{2} = 90.3 \text{ kJ mol}^{-1} \text{ of NO}$$

36. (a) Calculate the molar heat of combustion of propane, $C_3H_8(g)$ from the formation of $H_2O(l)$ and $CO_2(g)$. The enthalpy of formation of propane is -104 kJ/mol.
(b) Calculate the molar heat of combustion of butane, $C_4H_{10}(g)$ from the formation of $H_2O(l)$ and $CO_2(g)$. The enthalpy of formation of butane is -126 kJ/mol.
(c) Both propane and butane are used as gaseous fuels. Which one has the higher heat of combustion per gram?

Solution

(a) The heat of combustion is given by the reaction

$$C_3H_8(g) + 5O_2(g) \longrightarrow 3CO_2(g) + 4H_2O(l)$$

and can be found by adding the appropriate equations:

$$3[C(s) + O_2(g) \longrightarrow CO_2(g)] \qquad 3[\Delta H^\circ_f = -393.5 \text{ kJ mol}^{-1}]$$

$$4[H_2(g) + \tfrac{1}{2}O_2(g) \longrightarrow H_2O(l)] \qquad 4[\Delta H^\circ_f = -285.8 \text{ kJ mol}^{-1}]$$

$$C_3H_8(g) \longrightarrow 3C(s) + 4H_2(g) \qquad -[\Delta H^\circ_f = -104 \text{ kJ mol}^{-1}]$$

$$\Delta H_{comb} = 3(-393.5 \text{ kJ mol}^{-1}) + 4(-285.8 \text{ kJ mol}^{-1}) - (-104 \text{ kJ mol}^{-1})$$
$$= -2219.2 = -2.22 \times 10^3 \text{ kJ mol}^{-1}$$

(b) The heat of combustion for butane is given by the reaction

$$C_4H_{10}(g) + \tfrac{13}{2}O_2(g) \longrightarrow 4CO_2(g) + 5H_2O(l)$$

Adding the appropriate equations gives the heat of combustion.

$$4[C(s) + O_2(g) \longrightarrow CO_2(g)] \qquad 4[\Delta H^\circ_f = -393.5 \text{ kJ mol}^{-1}]$$

$$5[H_2(g) + \tfrac{1}{2}O_2(g) \longrightarrow H_2O(l)] \qquad 5[\Delta H^\circ_f = -285.8 \text{ kJ mol}^{-1}]$$

$$C_4H_{10}(g) \longrightarrow 4C(s) + 5H_2(g) \qquad -[\Delta H^\circ_f = -126 \text{ kJ mol}^{-1}]$$

$$\Delta H_{comb} = 4(-393.5 \text{ kJ mol}^{-1}) + 5(-285.8 \text{ kJ mol}^{-1}) + 126 \text{ kJ mol}^{-1}$$
$$= -2.88 \times 10^3 \text{ kJ mol}^{-1}$$

(c) For propane: $-2.22 \times 10^3 \text{ kJ mol}^{-1}/44.09 \text{ g mol}^{-1} = 50.4 \text{ kJ g}^{-1}$
For butane: $-2.88 \times 10^3 \text{ kJ mol}^{-1}/58.12 \text{ g mol}^{-1} = 49.6 \text{ kJ g}^{-1}$

Propane has a slightly higher heat per gram.

FUEL AND FOOD

39. Ethanol, C_2H_5OH, is used as a fuel for motor vehicles, particularly in Brazil. (a) Write the balanced equation for the combustion of ethanol to $CO_2(g)$ and $H_2O(g)$, and, using the data in Appendix I, calculate the heat of the combustion of 1 mol of ethanol. (b) The density of ethanol is 0.7893 g mL^{-1}. Calculate the heat of combustion of exactly 1 mL of ethanol. (c) Assuming that the mileage an automobile gets is directly proportional to the heat of combustion of the fuel, calculate how many times farther an automobile could be expected to go on 1 gal of gasoline than on 1 gal of ethanol. Assume that gasoline has the heat of combustion and the density of n-octane, C_8H_{18} ($\Delta H_f^\circ = -208.4$ kJ mol^{-1}; density = 0.7025 g mL^{-1}).

Solution

(a) $C_2H_5OH(l) + 3O_2(g) \longrightarrow 3H_2O(g) + 2CO_2(g)$

The heat of combustion is found by combining the appropriate equations.

$C_2H_5OH(l) \longrightarrow 2C(s) + 3H_2(g) + \frac{1}{2}O_2(g)$ $-[\Delta H_f^\circ = -277.7]$

$2[C(s) + O_2(g) \longrightarrow CO_2(g)]$ $2[\Delta H_f^\circ = -393.5]$

$3[H_2(g) + \frac{1}{2}O_2(g) \longrightarrow H_2O(g)]$ $3[\Delta H_f^\circ = -241.82]$

$\Delta H_{comb} = 277.7$ kJ mol^{-1} + 2(−393.5 kJ mol^{-1}) + 3(−241.82 kJ mol^{-1})

$= -1234.8$ kJ mol^{-1}

(b) The molecular weight of ethanol is 46.07 g mol^{-1}. Then

46.07 g mol^{-1}/0.7893 g mL^{-1} = 58.37 mL mol^{-1}.

$\Delta H_{comb} = -1234.8$ kJ mol^{-1}/58.37 mL mol^{-1} = -23.41 kJ mL^{-1}

(c) The heat of combustion expression is

$C_8H_{18}(l) + \frac{25}{2}O_2(g) \longrightarrow 8CO_2(g) + 9H_2O(g)$

From the heats of formation we have

$C_8H_{18}(l) \longrightarrow 8C(s) + 9H_2(g)$ $-[\Delta H_f^\circ = -208.4$ kJ mol$^{-1}]$

$8[C(s) + O_2(g) \longrightarrow CO_2(g)]$ $8[\Delta H_f^\circ = -393.5$ kJ mol$^{-1}]$

$9[H_2(g) + \frac{1}{2}O_2(g) \longrightarrow H_2O(g)]$ $9[\Delta H_f^\circ = -241.82$ kJ mol$^{-1}]$

$\Delta H_{comb} = 208.4$ kJ mol^{-1} + 8 (−393.5 kJ mol^{-1}) + 9(−241.82 kJ mol^{-1})

$= 5116.0$ kJ mol^{-1}

114.15 g mol^{-1}/0.7025 g mL^{-1} = 162.49 mL mol^{-1}

$\Delta H_{comb} = \dfrac{-5116.0 \text{ kJ mol}^{-1}}{162.49 \text{ mL mol}^{-1}} = -31.48$ kJ mL

$\dfrac{\Delta H_{comb} \text{ (octane)}}{\Delta H_{comb} \text{ (ethanol)}} = \dfrac{-31.48 \text{ kJ/mL}}{-21.16 \text{ kJ/mL}} = 1.488$ times farther

43. The oxidation of the sugar glucose, $C_6H_{12}O_6$, is described by the following equation.

$C_6H_{12}O_6(s) + 6O_2(g) \longrightarrow 6CO_2(g) + 6H_2O(l)$ $\Delta H = -2816$ kJ

Chapter 4 Thermochemistry

Metabolism of glucose gives the same products, although the glucose reacts with oxygen in a series of steps in the body. (a) How much heat in kilojoules is produced by the metabolism of 1.0 g of glucose? (b) How many nutritional Calories (1 cal = 4.184 J; 1 nutritional Cal = 1000 cal) are produced by the metabolism of 1.0 g of glucose?

Solution

(a) The molecular weight of glucose is 180.1572 g mol^{-1}.

$$\frac{\Delta H_{comb}}{g} = \frac{-2816 \text{ kJ mol}^{-1}}{180.16 \text{ g mol}^{-1}} = -15.6 \text{ kJ/g or } -16 \text{ kJ/g}$$

to the correct number of significant figures.

(b) $\dfrac{-15.6 \text{ kJ}}{g} \times \dfrac{1000 \text{ J}}{1 \text{ kJ}} \times \dfrac{\text{Cal}}{1000 \text{ cal}} = 3.7$ Calories

ADDITIONAL EXERCISES

44. A sample of WO$_2$(s) with a mass of 0.9745 g was "burned" in oxygen in a bomb calorimeter like that shown in Fig 4.4, and the temperature increased by 1.310°C.

    (a) Calculate the heat released if the heat capacity of the calorimeter and product is 872.3 J/°C.
    (b) The product of the reaction is WO$_3$(s). Assume that the heat produced is proportional to the enthalpy change of the reaction in the calorimeter and calculate the enthalpy change of the reaction per mole of product.
    (c) The enthalpy of formation of WO$_3$(s) under these condition is −842.91 kJ mol^{-1}. From these data, calculate the enthalpy of formation of WO$_2$(s).

Solution

(a) 1.310°C × 872.3 J/°C = −1142.7 = −1143 J   (heat released)
(b) The molecular weight of WO$_2$(s) is 215.849 g mol^{-1}. Since 0.9745 g liberated 1.143 kJ, then

$$\Delta H_{comb} = \frac{215.849 \text{ g mol}^{-1}}{0.9745 \text{ g}} \times -1.1427 \text{ kJ} = -253.1 \text{ kJ mol}^{-1}$$

(c)   WO$_2$(s) + $\frac{1}{2}$O$_2$(g) ⟶ WO$_3$(s) $\Delta H_{comb} = -253.1$ kJ mol^{-1}

   WO$_3$(s) ⟶ W(s) + $\frac{3}{2}$O$_2$(g) $-[\Delta H_f^\circ = -842.9$ kJ mol$^{-1}]$

Adding the first two equations gives the reverse of the desired equation.

   WO$_2$(s) ⟶ W(s) + O$_2$(g) $-\Delta H^\circ = 589.8$ kg mol^{-1}

Then

   W(s) + O$_2$(g) ⟶ WO$_2$(s) $\Delta H_f^\circ = -589.8$ kJ mol^{-1}

5

Structure of the Atom and the Periodic Law

INTRODUCTION

Ancient records point to the fact that human beings have long pondered the basic structure of matter in the hope of better understanding themselves and their surroundings. The early writings of the physician and philosopher Empedocles of Agrigentum in Sicily (ca. 490–435 B.C.) suggested that all visible objects are composed of four unchanging elements — air, earth, water, and fire. The Greek philosophers Leucippos and his pupil, Demokritos of Abdera (ca. 460–370 B.C.), postulated that matter is composed of small particles of these four elements and that these particles are in motion. This was the first evidence of an atomic and kinetic theory. Building upon these ideas, Aristotle moved away from the atomistic concept and popularized a theory to explain the transformation of one "element" to another. His ideas, including that of the void, or "ether," dominated Western scientific thought for centuries. Unfortunately, Aristotle's theory did not lend itself to experimental verification, so no one tried to test his theories. Consequently, very little experimental work took place for the purpose of testing theories until about 1800.

Recently, the development of sophisticated instruments has enabled scientists who study atoms to probe into the atoms' structure. We know now that the nature and composition of atoms depend on the interaction of the many smaller particles that provide the building blocks of all atoms. Several theoretical atomic models were proposed during the early 1900s to explain atomic phenomena and to predict atomic and molecular behavior.

This chapter treats the discovery of atomic particles, the unparalleled experimentation during the evolution of modern atomic theory, the particulars of the quantum mechanical model of the atom, and the relationship of atomic structure to the Periodic Table. Remember as you proceed that the development of the models and theories represents the attempt by scientists to advance our knowledge of matter and that the latest models generally will require further refinement in the future.

READY REFERENCE

Aufbau Process (5.12)   The aufbau process is the way in which electronic structures of complex atoms are formed by assigning one electron at a time to the lowest energy sublevel available based on Hund's rule and the Pauli exclusion principle.

Bohr model energies (5.7)

$$E_n = -2.179 \times 10^{-18} \frac{Z^2}{n^2} \text{ J} \quad \text{or} \quad -2.179 \times 10^{-18}/n^2 \text{ J}$$

where Z is the atomic number. For a hydrogen atom $Z = 1$.

The energy calculated from this equation is the energy associated with a specific energy level in a hydrogenlike atom or ion. By convention all the values calculated using this equation have a negative sign, and when $n = 1$ the value is -2.179×10^{-18} J. As n increases (higher energy levels), the values of E increase and approach zero at $n = \infty$. The energy required to raise an electron from $n = 1$ to $n = \infty$ is the ionization energy for a hydrogenlike atom or ion in its ground state.

Constants and equations (5.8)

Planck's constant $(h) = 6.626 \times 10^{-34}$ J s

$\qquad\qquad\qquad\quad = 6.626 \times 10^{-27}$ erg s

$\qquad\qquad\qquad\quad = 1.584 \times 10^{-37}$ kcal s

Speed of light $(c) = 2.998 \times 10^8$ m s^{-1} (vacuum)

Rydberg equation $E = 2.179 \times 10^{-18} \left(\dfrac{1}{n_1^2} - \dfrac{1}{n_2^2} \right)$ J, where $n_1 < n_2$

The Rydberg equation is empirically derived from spectral data resulting from studies on hydrogen. The energy value calculated from the equation is the energy associated with a photon produced by an electron transition from a higher energy level (n_2) to a lower energy level (n_1) in a hydrogen atom.

Electromagnetic wave characteristics (5.8) Visible light and other types of radiant energy described in this chapter are propagated as waves in the form of electromagnetic radiation consisting of two components; one with an electric field, the other with a magnetic field. These components are coupled at right angles to each other and travel as a sine wave as shown in the following figure, where E is the electric field and H is the magnetic field.

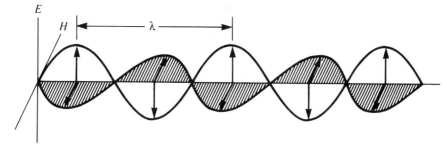

Electron affinity (5.15) The electron affinity (EA) of an atom is a measure of the attraction a neutral gaseous atom in its ground state has for an additional electron. In general, atoms that have half-filled orbitals or incomplete valence levels will attract additional electrons and release energy upon ionization. Some examples defining the electron affinity for several elements along with their numerical values are

General case: $X(g) + e^- \longrightarrow X^-(g)$ + energy

$Cl(g) + e^- \longrightarrow Cl^-(g) + 3.48 \times 10^5$ J/mol $\qquad$ EA = 3.48×10^5 J/mol

$O(g) + e^- \longrightarrow O^-(g) + 2.25 \times 10^5$ J/mol $\qquad$ EA = 2.25×10^5 J/mol

$Mg(g) + e^- + 2.89 \times 10^4$ J/mol $\longrightarrow Mg^-(g)$ $\qquad$ EA = -2.89×10^4 J/mol

Energy-wavelength-frequency relations (5.8)   Velocity, frequency, and wavelength are related through the equation $\lambda\nu = c$. The velocity of all electromagnetic radiation passing through a continuous medium (air, water, etc.) is a constant and varies only with the medium. Here, the assumption will be made that the velocity of radiation is constant and that the medium of transmission is a vacuum in which the velocity = 2.998×10^8 m/s. Because the value of c is constant, λ and ν are inversely related to each other; that is,

λ(short)ν(high) = constant

λ(long)ν(low) = constant

The energy of a photon is related to frequency as

$E = h\nu$ (h = Planck's constant)

in which energy is directly proportional to frequency:

High energy ⟷ high frequency

Low energy ⟷ low frequency

where ⟷ may be read as "is associated with." The relationship between energy and wavelength is derived by solving for ν in the equation $c = \lambda\nu$ and substituting $E = h\nu$:

$\lambda\nu = c$ and $\nu = c/\lambda$

$E = h\nu = hc/\lambda$

This is an inverse relationship of energy to wavelength; that is,

High energy ⟷ short wavelength

Low energy ⟷ long wavelength

These relationships can be extended to the frequency of light in the following way:

High energy ⟷ high frequency ⟷ short wavelength

Low energy ⟷ low frequency ⟷ long wavelength

The proper units for use with energy relationships are

E in joules = $h\nu$ = $(6.626 \times 10^{-34}$ J s$)$ $(\nu$ in 1/s or s$^{-1})$

E in joules = $\dfrac{hc}{\lambda} = \dfrac{(6.626 \times 10^{-34} \text{ J s})\,(c \text{ in m/s})}{\lambda \text{ in m}}$

Frequency (ν, nu) (5.8)   Frequency is the number of wavelengths passing a reference point in unit time. The unit for frequency is per second and is usually written as 1/s or s^{-1}. The words *cycle* and *wave* are implied but not written.

Hund's Rule (5.11)   Atomic energy levels are filled by electrons in such a way as to achieve the lowest possible total energy state for the atom. Hund's rule states that each orbital in a subshell or sublevel must be singly filled before any one orbital is completely (doubly) filled. In addition, all electrons in singly filled orbitals within a sublevel must have parallel spins; that is, their spin orientations are the same (aligned).

Ionization potential (5.15)   The ionization potential (IP) of an atom, frequently called *ionization energy*, is the amount of energy required to completely remove the most loosely bound electron from a gaseous atom in its ground state. Values for ionization potentials are usually listed in terms of mole quantities; some examples are

General case: $X(g) + \text{energy} \longrightarrow X^+(g) + e^-$

$Li(g) \longrightarrow Li^+(g) + e^-$ $IP = 5.19 \times 10^5$ J/mol

$N(g) \longrightarrow N^+(g) + e^-$ $IP = 1.40 \times 10^6$ J/mol

$He(g) \longrightarrow He^+(g) + e^-$ $IP = 2.37 \times 10^6$ J/mol

Pauli Exclusion Principle (5.11)   The rigorous solution of the wave equation representing electrons orbiting an atomic nucleus involves four components. Four quantum numbers describe the behavior of these components and represent the effective volume of space in which an electron moves, the shape of the orbital in space, the orientation of the electronic charge cloud in space, and the direction of the spin of each electron on its own axis. The Pauli exclusion principle states that each electron has a unique set of four quantum numbers; that is, no two electrons in the same atom can have the same four quantum numbers.

Representative elements (5.14)   The representative elements are the A-group elements, IA through VIIA and Group IIB, as listed in the Periodic Table. These elements have valence electrons either in s or s and p sublevels. Their valence electrons are in the same primary level (n level) as their periodic location. For example, the three valence electrons in gallium, $Z = 31$, are in the fourth n level, that is $4s^2$ and $4p^1$ electrons.

Transition elements (5.14)   The transition elements are metals and are located in the B groups of the Periodic Table (except Group IIB). These elements have outer electrons that include the $(n - 1)d$ sublevels, where n is the period location of the elements. For example, titanium, Ti, $Z = 22$, is in period 4, but its outermost electrons are in the $3d$ orbitals. The period 4 transition metals are relatively abundant and are extremely important to the world steel industry. With a few exceptions the transition elements in periods 5 and 6 are in limited supply as far as commercially exploitable deposits are concerned. Several are highly valued for use in jewelry, coinage, and electronic systems.

Valence electrons (5.14)   Valence electrons are the electrons that occupy the highest energy level of an atom. The electrons in the outermost sublevel when that sublevel is not in the highest energy level (for example, $3d$ electrons when the $4s$ level is occupied) are also valence electrons. The valence electrons are largely responsible for the chemical behavior of the atom. All elements within a periodic group or family have the same outer electron configuration, so they also have the same number of valence electrons. For the representative elements, the number of valence electrons in an atom equals its periodic group number.

Velocity (c) (5.8)   The velocity of a wave is the distance traveled by the wave per unit time. For electromagnetic radiation in a vacuum, $c = 2.998 \times 10^8$ m s^{-1}.

Wavelength (λ, lambda) (5.8)   Wavelength is the distance between equivalent positions on adjacent waves, for example, from one peak to the next.

SOLUTIONS TO CHAPTER 5 EXERCISES

ATOMIC STRUCTURE

1. What is the evidence that the cathode rays in a discharge tube are negatively charged particles? What is their source?

    Solution

    Cathode rays are deflected by magnetic and electric fields, and the direction of deflection is the same as that shown by negatively charged particles in similar fields. The source of cathode rays is the cathode (negatively charged electrode).

2. How are cathode rays and β particles similar?

    Solution

    Cathode rays and β particles are both shown to be composed of electrons, the cathode rays being generated from a Crookes tube, the β-particles being generated from nuclear decay processes.

5. Must the numbers of protons, electrons, and neutrons be the same or different in each of the following:

    (a) two different isotopes of the same element,
    (b) two atoms of the same element with the same mass,
    (c) two atoms of different elements with the same mass number?

    Solution

    (a) Same number of protons and electrons in neutral atom, different numbers of neutrons possible.
    (b) Same number of protons, electrons, and neutrons.
    (c) Different numbers of protons, electrons and neutrons, but the sum of protons and neutrons must be equal.

6. In what way is the lightest hydrogen nucleus unique?

    Solution

    Hydrogen (protium) is the only element with no neutron in the nucleus.

7. Using the data in Table 5.2, sketch the mass spectrum expected for a sample of neon atoms.

    Solution

    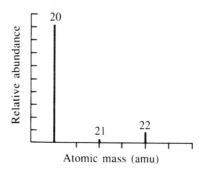

Chapter 5 Structure of the Atom and the Periodic Law   65

For neon the relative abundances and mass of the isotopes are

$^{20}_{10}$Ne   90.92%   19.9924 amu
$^{21}_{10}$Ne   0.257%   20.9940 amu
$^{22}_{10}$Ne   8.82%   21.9914 amu

8. From the data given in Table 5.2, calculate the atomic weight of naturally occurring neon.

Solution

The contribution to the total mass of neon from each isotope is in direct proportion to that isotope's percentage in the natural abundance of neon.

Weighted average mass = 0.9092(19.9924 amu) + 0.00257(20.9940 amu)
   + 0.0882(21.9914 amu) = 18.177 + 0.05395 + 1.9396 = 20.17

10. Determine the number of protons and neutrons in the nuclei of each of the following elements.

(a) $^{7}_{3}$Li,

(c) $^{27}_{13}$Al,

Solution

(a) Mass number Z equals 7 and is the sum of protons and neutrons in the nucleus. The lower left subscript indicates 3 protons in the nucleus. Therefore, 7 − 3 = 4 neutrons are in the nucleus.

(c) Nucleus: $^{27}_{13}$Al
    Protons: 13
    Neutrons: 14

11. Identify the following elements:

(a) $^{9}_{4}$X,

(d) $^{182}_{74}$X,

Solution

(a) This is an element with 4 protons in the nucleus; to establish its identity, merely look for the element whose atomic number is 4. This isotope corresponds to the most abundant isotope of beryllium, Be.

(d) Atomic number 74: X = W (tungsten).

THE BOHR MODEL OF THE ATOM

13. What does it mean to say that the energies of the electrons in hydrogen atoms are quantized?

Solution

"Quantized" means that only certain values are permitted. The statement means that the electron is permitted to have only certain values of energy; no electron may have an energy between those values.

14. Figure 5.13 gives the energies of an electron in hydrogen atoms for values of n from 1 to 5. What is the energy, in eV, of an electron with $n = 6$ in a hydrogen atom?

    Solution

    The equation giving the energy of a specific energy level is

    $$E_n = \frac{-kZ^2}{n^2}.$$

    For hydrogen, Z is equal to 1; n is the energy level, 6 in this case. The value of k is 2.179×10^{-18} J. Therefore, $E_6 = -2.179 \times 10^{-18}$ J/$(6)^2 = -6.053 \times 10^{-20}$ J. Since 1 eV = 1.602×10^{-19} J,

    $$E_6 = -6.053 \times 10^{-20} \text{J} \times 1 \text{ eV}/1.602 \times 10^{-19} \text{ J} = -0.3778 \text{ eV}$$

    KEYSTROKES: 2.179 $\boxed{+/-}$ $\boxed{\text{exp}}$ $\boxed{+/-}$ 18 ÷ 6 = ÷ 6 = ÷ 1.602 $\boxed{\text{exp}}$ $\boxed{+/-}$ 19 = -0.3778263

16. Excited H atoms with electrons in very high-energy levels have been detected. What is the radius (in centimeters) of a H atom with an electron characterized by an n value of 110? How many times larger is that than the radius of the H atom in its ground state?

    Solution

    The radius of the orbit is equal to $n^2 a_0/Z$ where $a_0 = 0.529$ Å. For hydrogen, the charge on the nucleus, Z, equals 1. For $n = 110$,

    $$r = (110)^2(0.529 \times 10^{-8} \text{ cm})/1 = 6.40 \times 10^{-5} \text{ cm}$$

    This is 6.40×10^{-5} cm/0.529×10^{-8} cm = 12,100 times larger than the Bohr radius. Such atoms are called *Rydberg* atoms.

17. Calculate the radius of a Li^{2+} ion using the Bohr model of the electronic structure of this ion.

    Solution

    The radius is given by $r = n^2 a_0/Z$. The electron in Li^{2+} must be in the $n = 1$ level, the lowest possible in Li. The charge on the nucleus is $+3$ and, therefore, $Z = 3$.

    $$r = 1^2(0.529 \text{ Å})/3 = 0.176 \text{ Å}$$

ORBITAL ENERGIES AND SPECTRA

20. (a) From the data given in Fig. 5.13, calculate the energy, in eV, required to ionize a hydrogen atom. Calculate the energy in J per atom.
    (b) Using the Bohr model, calculate the energy, in eV, required to ionize a hydrogen atom. Calculate the energy in J per atom.

    Solution

    (a) The energy required to ionize the electron from hydrogen is equal to the energy absorbed to raise the electron from the $n = 1$ level to the $n = \infty$ level, that is, 13.595 eV, as seen from the figure.

Chapter 5 Structure of the Atom and the Periodic Law 67

$$13.595 \text{ eV} \times 1.6022 \times 10^{-19} \text{ J/eV} = 2.1782 \times 10^{-18} \text{ J}$$

Electron volts is an energy unit on a per atom basis. Therefore, the energy required per atom is 2.1782×10^{-18} J.

(b) From the Bohr model

$$E = E_{n_1} - E_{n_2} = 2.179 \times 10^{-18} \left(\frac{1}{n_1^2} - \frac{1}{n_2^2} \right) \text{ J}$$

$$= 2.179 \times 10^{-18} \left(\frac{1}{1^2} - \frac{1}{\infty^2} \right) = 2.179 \times 10^{-18} \text{ J/atom}$$

In electron volts,

$$\frac{2.179 \times 10^{-18} \text{ J}}{1.6022 \times 10^{-19} \text{ J/eV}} = 13.600 \text{ eV}$$

21. When heated, lithium atoms emit photons of red light with a wavelength of 6708 Å. What is the energy in J and the frequency of this light?

Solution

From

$$E = h\nu = h\frac{c}{\lambda} = \frac{6.626 \times 10^{-34} \text{ J s } (2.998 \times 10^{8} \text{ m s}^{-1})}{6708 \times 10^{-10} \text{ m}}$$

$$= 2.961 \times 10^{-19} \text{ J}$$

$$\nu = \frac{c}{\lambda} = \frac{2.998 \times 10^{10} \text{ cm s}^{-1}}{6708 \times 10^{-8} \text{cm}} = 4.469 \times 10^{14} \text{ s}^{-1}$$

23. Using the Bohr model, calculate the energy, in eV and J, of the light emitted by a transition from the $n = 2$ (L) to the $n = 1$ (K) shell in He$^+$.

Solution

$$E = -\frac{kZ^2}{n^2} = -2.179 \times 10^{-18} \text{ J}(Z)^2 \left(\frac{1}{n_1^2} - \frac{1}{n_2^2} \right)$$

$$= -2.179 \times 10^{-18} \text{ J } (2)^2 \left(\frac{1}{1^2} - \frac{1}{2^2} \right) = -2.179 \times 10^{-18} (2)^2(\tfrac{3}{4})$$

$$= -6.537 \times 10^{-18} \text{ J}$$

The energy of the light emitted is the absolute value of this quantity, or 6.537×10^{-18} J.

In eV,

$$E = |-13.595 \text{ eV } (2)^2(\tfrac{3}{4})| = 40.785 \text{ eV}$$

27. Light that looks blue has a wavelength of 4800 Å. What is the frequency of this light? What is the energy of a photon of this blue light, in joules?

Solution

$$c = \lambda\nu \quad \text{or} \quad \nu = \frac{c}{\lambda} = \frac{2.998 \times 10^{10} \text{ cm s}^{-1}}{4800 \times 10^{-8} \text{ cm}}$$

$$= 6.246 \times 10^{14} \text{s}^{-1}$$

$$E = h\nu = h\frac{c}{\lambda} = \frac{6.626 \times 10^{-34} \text{ J s } (2.998 \times 10^8 \text{ m s}^{-1})}{4800 \times 10^{-10} \text{ m}}$$
$$= 4.138 \times 10^{-19} \text{ J}$$

28. Does a photon of the blue light described in exercise 27 have enough energy to excite the electron in a hydrogen atom from the $n = 1$ shell to the shell with $n = 2$?

Solution

The transition from $n = 1$ to $n = 2$ requires $(13.595 - 3.399)$ eV from Figure 5.13. This is 10.196 eV $\times$ 1.6022 $\times$ 10^{-19} J (eV)$^{-1}$ or 1.6336 $\times$ 10^{-18} J. This is more energy than contained in the light of exercise 27; therefore the answer is no.

29. X rays are produced when the electron stream in an X-ray tube knocks an electron out of a low-lying shell of an atom in the target, and an electron from a higher shell falls into the lower-lying shell. The X ray is the photon given off as the electron falls into the lower shell. The most intense X rays produced by an X-ray tube with a copper target have wavelengths of 1.542 Å and 1.392 Å. These X rays are produced when an electron from the L or M shell falls into the K shell of a copper atom. Calculate the energy separation in eV of the K, L, and M shells in copper.

Solution

$$E = h\nu = \frac{hc}{\lambda}$$

For 1.542 Å: (L,K)

$$E = \frac{6.626 \times 10^{-34} \text{ J s} \times 2.998 \times 10^{10} \text{ cm s}^{-1}}{1.542 \times 10^{-8} \text{ cm}} = 1.288 \times 10^{-15} \text{ J or 8040 eV}$$

For 1.392 Å : (M,K)

$$E = \frac{6.626 \times 10^{-34} \text{ J s} \times 2.998 \times 10^{10} \text{ cm s}^{-1}}{1.392 \times 10^{-8} \text{ cm}} = 1.427 \times 10^{-15} \text{ J or 8908 eV}$$

The difference between these levels (M, L) is $8908 - 8040 = 868$ eV

THE QUANTUM MECHANICAL MODEL OF THE ATOM

31. How do electron shells, subshells, and orbitals differ?

Solution

The electron shell is a concept from Bohr theory that describes an exact path that an electron must reside in with an exact amount of energy.

Subshells relate to a modified Bohr theory in which they refer to different paths of about the same energy. The electrons that reside in these have different characteristics as determined by their spectra. This term has been carried over into quantum theory and refers to the *s, p, d,* and *f* orbitals.

Orbitals have specific shape and specifically refer to the quantum mechanical picture of the atom, in which probability plays a direct role.

33. What information about an electron in an atom is available from the wave function that describes the electron?

Solution

The wave function gives all of the information available about an electron: the probability of finding the electron in a particular region of space, its energy, and the extent of its travel.

35. Which of the following orbitals are degenerate in an atom with two or more electrons: $3d_{xy}$, $4s$, $4p_z$, $4d_{xy}$, $4p_x$?

Solution

Degenerate are the $4p_z$ and $4p_x$ orbitals.

36. Consider the atomic orbitals, (i), (ii), and (iii) shown below in outline.

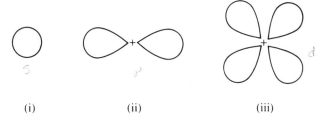

(i)   (ii)   (iii)

(a) What is the maximum number of electrons that can be contained in atomic orbital (iii)?
(b) How many orbitals with the same value of l as orbital (i) can be found in the shell $n = 4$? How many as orbital (ii)? How many as orbital (iii)?
(c) What is the smallest n value possible for an electron in an orbital of type (iii)? of type (ii)? of type (i)?
(d) What are the l values that characterize each of these three orbitals?
(e) Arrange these orbitals in order of increasing energy in the M shell. Is this order different in other shells?

Solution

(a) No orbital can contain more than 2 electrons. Therefore, the maximum number is 2 not only for (iii) but also for any other orbital.
(b) Orbital (i) represents an s orbital; s is never degenerate and, therefore, only one can be found in shell $n = 4$. Orbital (ii) represents a p orbital. There are 3 degenerate orbitals of this type in $n = 4$. Orbital (iii) represents a d orbital. Five degenerate d orbitals can be held in $n = 4$.
(c) The lowest value of n for which a d-type orbital is permitted is 3; for a p-type, 2; for an s-type, 1.
(d) For (i), $l = 0$; for (ii), $l = 1$; for (iii), $l = 2$.
(e) i < ii < iii; no.

38. What type of orbital is occupied by an electron with the quantum numbers $n = 3$, $l = 2$? How many degenerate orbitals of this type are found in a multielectron atom?

Solution

No matter what the n value (as long as it is 3 or greater), an electron with an l value of 2 occupies a d orbital. In this case it is a $3d$ electron. There are five degenerate orbitals with this same energy.

70 Chapter 5 Structure of the Atom and the Periodic Law

39. Write the quantum numbers of the six electrons in a carbon atom. For example, the quantum numbers for one of the 2s electrons will be $n = 2, l = 0, m = 0, s = +\frac{1}{2}$.

Solution

n	l	m	s
1	0	0	$+\frac{1}{2}$
1	0	0	$-\frac{1}{2}$
2	0	0	$+\frac{1}{2}$
2	0	0	$-\frac{1}{2}$
2	1	$+1$	$+\frac{1}{2}$
2	1	-1	$+\frac{1}{2}$

ELECTRONIC CONFIGURATIONS

41. Using complete subshell notation ($1s^2 2s^2 3p^6$, etc.), predict the electronic configuration of each of the following atoms:

   (a) $^{16}_{8}O$,

   (d) $^{40}_{18}Ar$,

   (g) $^{55}_{25}Mn$,

   (j) $^{175}_{71}Lu$,

Solution

   (a) $1s^2 2s^2 2p^4$
   (d) $1s^2 2s^2 2p^6 3s^2 3p^6$
   (g) $1s^2 2s^2 2p^6 3s^2 3p^6 3d^5 4s^2$
   (j) $1s^2 2s^2 2p^6 3s^2 3p^6 3d^{10} 4s^2 4p^6 4d^{10} 4f^{14} 5s^2 5p^6 5d^1 6s^2$

42. Using complete subshell notation ($1s^2 2s^2 2p^6$, etc.), predict the electronic configuration of each of the following ions:

   (a) N^{3-},
   (c) Al^{3+},
   (f) Pb^{2+},
   (g) Ce^{4+},

Solution

   (a) $1s^2 2s^2 2p^6$
   (c) $1s^2 2s^2 2p^6$
   (f) $1s^2 2s^2 2p^6 3s^2 3p^6 3d^{10} 4s^2 4p^6 4d^{10} 4f^{14} 5s^2 5p^6 5d^{10} 6s^2$
   (g) $1s^2 2s^2 2p^6 3s^2 3p^6 3d^{10} 4s^2 4p^6 4d^{10} 5s^2 5p^6$

43. Identify the atoms whose electronic configurations are given below.

   (a) $1s^2 2s^2 2p^6$
   (c) $[Ar]4s^2 3d^7$
   (e) $[Kr]5s^2 4d^{10} 5p^2$

Chapter 5 Structure of the Atom and the Periodic Law 71

Solution

(a) Ne
(c) Co
(e) Sn

44. The 4s orbitals fill before the 3d orbitals when building up electronic structures by the aufbau process. However, electrons are lost from the 4s orbital before they are lost from the 3d orbitals when transition elements are ionized. The ionization of copper ([Ar]$3d^{10}4s^1$) gives Cu$^+$ ([Ar]$3d^{10}$), for example. The electronic structures of a number of transition metal ions are given below; identify the transition metals.

   (a) M^{3+}, [Ar]$3d^3$
   (c) M^{3+}, [Ar]$3d^4$
   (e) M^{3+}, [Kr]$4d^6$

Solution

(a) Cr^{3+}
(c) Mn^{3+}
(e) Rh^{3+}

46. Which of the following electronic configurations describe an atom in its ground state and which describe an atom in an excited state?

   (a) [He]$2s^1$
   (b) [Ar]$5s^14d^3$
   (c) [Ar]$4s^23d^{10}4p^4$
   (d) [Ar]$4s^23d^94p^3$

Solution

To be in a ground state the normal order of filling must be observed.

(a) ground state
(b) excited state; ground state is [Ar]$4s^23d^2$
(c) ground state
(d) excited state; ground state is [Ar]$4s^23d^{10}4p^2$

47. Which of the following sets of quantum numbers describes the most easily removed electron in a boron atom in its ground state? Which of the electrons described is most difficult to remove?

   (a) $n = 1, l = 0, m = 0, s = -\frac{1}{2}$
   (b) $n = 2, l = 1, m = 0, s = -\frac{1}{2}$
   (c) $n = 2, l = 0, m = 0, s = \frac{1}{2}$
   (d) $n = 3, l = 1, m = 1, s = -\frac{1}{2}$
   (e) $n = 4, l = 1, m = 1, s = \frac{1}{2}$

Solution

An electron in the highest level for boron would be the most easily removed because it has the most energy. This corresponds to condition (b). The electron with the least energy (most stable) is the most difficult to remove, (a).

THE PERIODIC TABLE AND PERIODIC PROPERTIES

51. In terms of electronic structure, explain why there are 2 elements in the first period of the Periodic Table, 8 elements in the second and third periods, 18 elements in the fourth and fifth periods, and 32 elements in the sixth period.

Solution

In the first period, only 2 electrons are allowed in a $1s$ orbital corresponding to the possibility of IA and IIA elements. In the second and third periods, only 8 electrons in $2s$ and $2p$ orbitals and in $3s$ and $3p$ orbitals are allowed corresponding to one element for families IA to VIII. In like manner only 18 electrons in $4s4p4d$ and $5s5p5d$ orbitals can be accommodated in the fourth and fifth periods, again corresponding to one new element for each electron. The same holds true for the 32 electrons allowed in the sixth period, with the f orbitals now filling to accommodate 14 more electrons, each corresponding to a new element.

52. Which groups have the following electronic structures in their valence shells? (n represents the principal quantum number.)

Solution

(a) ns^2
(c) ns^2np^3
(e) $ns^2(n-1)d^2$

Solution

(a) Alkaline earth metals, group IIA
(c) Group VA
(e) Group IVB

55. Most representative metals that form positive ions in chemical compounds do so by the loss of all of their valence electrons. Which group in the Periodic Table forms tripositive ions, M^{3+}?

Solution

Group IIIA

57. Why is the radius of an atom larger than the radius of a positive ion formed from it?

Solution

When an atom loses an electron to become a positive ion, the positive charge on the nucleus attracts the remaining electrons more strongly, thus pulling the electrons close to the nucleus and making the radius of the positive ion smaller than the atom from which it was formed.

58. Why is the radius of an atom smaller than the radius of a negative ion formed from it?

Chapter 5 Structure of the Atom and the Periodic Law

Solution

When an atom gains an electron to become a negatively charged ion, the nucleus cannot hold the electron as strongly as the others and the ion becomes larger than the atom from which it was formed.

59. Explain the decreasing radius of the isoelectronic ions Na^+, Mg^{2+}, Al^{3+}.

Solution

Na^+ has 11 positive charges in the nucleus; Mg^{2+} has 12 positive charges in the nucleus; Al^{3+} has 13 positive charges in the nucleus. In each case the same number of electrons are held by an increasing positive charge which tends to shrink the ions as listed, Na^+ being largest with Al^{3+} being the smallest.

63. Arrange each of the atoms or ions in the following groups in order of increasing size, based on their location in the Periodic Table.

(a) Cs, K, Li, Na, Rb
(c) Br^-, Ca^{2+}, K^+, Se^{2-}
(e) As^{3-}, Ca^{2+}, Cl^-, K^+

Solution

(a) Li, Na, K, Rb, Cs
(c) Ca^{2+}, K^+, Br^-, Se^{2-}
(e) Ca^{2+}, K^+, Cl^-, As^{3-}

ADDITIONAL EXERCISES

64. Consider one of the isotopes of uranium used in nuclear fission, $^{233}_{92}U$.

(a) How many protons, neutrons, and electrons are contained in this atom?
(b) How many protons, neutrons, and electrons are contained in a $^{233}_{92}U$ ion with a charge of +3?

Solution

(a) In the neutral atom there are 92 protons, 92 electrons and 141 neutrons.
(b) In the +3 ion, there are 92 protons, 89 electrons, and 141 neutrons.

72. Figure 5.15 shows the radial probability density of three of the lowest-energy orbitals in a hydrogen atom. Which other orbitals are degenerate with each of these three?

Solution

In Figure 5.15, the 1s, 2p, and 3d radial probability densities are shown. There are 2 additional p orbitals and 4 additional d orbitals with exactly the same distributions. When 2s, 3p, or 4d densities are plotted, there are differences in the plots.

76. The energy required to remove an electron from a Si^{3+} ion is 45.08 eV. To remove an electron from an Al^{3+} ion requires 120.2 eV. Explain the large difference.

Solution

Removal of one more electron from Al^{3+} would require that a particularly stable octet of electrons be disturbed. This need accounts for the high energy required. For Si^{3+}, one more electron remains before the octet is reached. Consequently a lower ionization energy is expected.

77. Assume that, in another universe, the values of the quantum number m are limited to zero or positive integers up to the value of l for a particular subshell; thus for $l = 2$, m could be 0, 1, or 2. If, with this restriction, all other quantum numbers behaved as described in this chapter, describe the electronic configuration of nitrogen (at. no. 7) and the configuration of fluorine (at. no. 9). How many unpaired electrons would these atoms contain?

Solution

In the new universe, N: $1s^2 2s^2 2p^3$; F: $1s^2 2s^2 2p^4 3s^1$, where the maximum number of electrons in the p orbitals is 4. Each element would have one unpaired electron.

80. The gaseous ion X^{2+} has no unpaired electrons in its most stable state. If X is a representative element, it could be a member of one of two possible groups. Which ones are they?

Solution

X^{2+} is a gaseous ion and has no unpaired electrons. Possibilities with no unpaired electrons can come from IIA and IVA because all other p containing electron groups would have an unpaired electron.

85. Each of the following electron configurations is found in a negative ion. Arrange them in increasing order of their size. The atomic number is given for each.

(a) $(Z = 7)$ [Ne]
(b) $(Z = 8)$ [Ne]
(c) $(Z = 9)$ [Ne]
(d) $(Z = 34)$ [Kr]
(e) $(Z = 35)$ [Kr]

Solution

With an electron arrangement of Ne, the size increases $Z = 9$, $Z = 8$, $Z = 7$; the same trend is noted with $Z = 35$, $Z = 34$. The complete order is c, b, a, e, d.

88. The value of one of the quantum numbers describing the electron that is lost when a silver atom forms an Ag^+ ion cannot be predicted. Which quantum number is it?

Solution

The spin quantum number may be either $+\frac{1}{2}$ or $-\frac{1}{2}$; its value is completely random as electrons are lost when the ions are formed.

6

Chemical Bonding, Part 1: General Concepts

INTRODUCTION

This chapter specifically treats the types of bonds formed when atoms interact to form molecules or ion pairs. These bonds are either interatomic or intramolecular (within the same molecule), as would exist between hydrogen and oxygen in a water molecule or between sodium and chlorine in a sodium chloride ion pair. Discussed later in the text are the interactions between neighboring molecules or ion pairs (intermolecular bonds) that provide the bases for understanding the properties and structures of liquids and solids.

The chemical bonds between atoms in molecules vary greatly in terms of their strength, bond length or distance between the atoms, and directional character. Some bonds are relatively weak. For example, the intermolecular bonds in napthalene holding napthalene molecules together are very weak. A slight amount of heat is sufficient to rupture the bonds and allow the napthalene molecules to escape the structure. Other bonds such as those holding calcium, oxygen, and silicon in cement are very strong by comparison. These atoms rarely escape the cement structure.

In this chapter the discussion of bonding is limited to ionic and covalent bonding. The characteristics of elements leading to these types of bonds and compounds produced through these bonds are emphasized. Bond strength as well as methods used for naming compounds are also discussed.

READY REFERENCE

Anion (6.1)   An atom or group of atoms that possesses a greater number of electrons than protons is called an *anion*. The anion has a negative charge equal to the charge difference. Elemental anions are usually formed from nonmetals; metallic elements rarely form negative ions. The dissociation of compounds often produces two ions — one anion and one cation.

Bond energy (6.9)   The enthalpy change required to break a particular bond in a mole of molecules is known as the bond energy and is a measure of the strength of a covalent bond.

Covalent bond (6.3)   Two atoms may unite to form a covalent bond in which one bonding electron from each atom is shared between the two atoms. In general, these bonds form between atoms of elements that have little tendency to give up electrons. Indeed, covalent bonds may form between atoms of the same element to form diatomic molecules, including H_2, N_2, O_2 and F_2. Reactions involving different nonmetallic elements form compounds through covalent bonding; these compounds include SO_2, N_2O_4, HCl, and CH_4. Carbon is particularly noteworthy because it can readily form bonds with itself and with other elements. Generally speaking, convalent bonds form between nonmetallic elements close to each other in the Periodic Table. These elements will generally have electronegativity differences less than 1.7.

Electronegativity (6.5)   Electronegativity is based on a numerical scale representing the tendencies of an atom to attract electrons in a covalent bond. These values range from a low of 0.9 for cesium, representing little attraction for electrons, to a high of 4.1 for fluorine. In general, metallic elements have lower electronegativities than nonmetallic elements; metallic elements tend to give up electrons in chemical reactions, while nonmetallic elements tend to gain electrons.

Electronegativities can be used to predict the nature of bonding between two elements. Elements that form primarily ionic bonds have electronegativities that differ by more than 1.7, while predominantly covalent bonds are formed between elements having differences less than 1.7. In real chemical bonds, regardless of electronegativity differences, both ionic and covalent characters occur to some extent. For example, the bond between sodium and chlorine in NaCl is mostly ionic, but some sharing of electrons does exist; in H_2, although the hydrogen atoms have little tendency to give up electrons, the bond is still about 5% ionic. Standard definitions of ionic and covalent bonding are inadequate, because real bonds show characteristics of both types of bonding.

Ionic bond (6.1)   Ionic bonds result from electrostatic attractions between two oppositely charged particles called ions. These mutual attractions produce bonds between ions to form ionic substances, such as sodium chloride, magnesium fluoride, and molten salts. The formation of ionic bonds when two elements react involves the transfer of electrons from one element to the other. For example, in the formation of sodium chloride, NaCl, the single valence electron of sodium is removed and transferred to complete the valence level of chlorine. This transfer creates the ion pair shown:

$$Na^x + :\ddot{Cl}\cdot \longrightarrow Na^+ + :\ddot{Cl}:_x^-$$

Isoelectronic (6.3)   This term refers to two molecules or ions that have the same arrangement of lone pairs and bonding pairs of electrons.

Lewis formula and structure (6.1 and 6.3)   The Lewis formula of an element is a means of showing the valence electrons available for bonding. The valence electrons are designed by symbols, such as $\cdot$, $\circ$, $\times$, as shown in the following:

$$\circ Mg \circ \qquad \times \overset{\times}{Al} \times \qquad :\ddot{S}\cdot \qquad :\ddot{I}\cdot$$

Compound formation is shown through the use of the Lewis formula of individual atoms by bringing the symbols next to each other and arranging the valence electrons so that eight electrons (where possible) occur about each symbol. For example:

$$Na\!:\!\ddot{Br}\!: \qquad \text{or} \qquad :\ddot{Cl}\!:\!\overset{:\ddot{Cl}:}{\underset{}{Al}}\!\!\!\times\!\!\ddot{Cl}\!:$$

Note that although different symbols may be used for convenience to designate electrons, all electrons are the same regardless of origin.

Chapter 6 Chemical Bonding, Part 1: General Concepts

Oxidation number (6.10) The oxidation number or oxidation state is a bookkeeping method to keep track of the electrons during chemical reactions. A few of the rules are:

1. The oxidation number of an atom of any element in its elemental form is zero;
2. For a monatomic ion the oxidation number is equal to the charge of the ion;
3. Oxidation number of fluorine in a compound is always -1;
4. The elements of Group IA in compounds are $+1$;
5. The elements of Group IIA are $+2$;
6. The elements of Group VIIA have a value of -1 when combined with less electronegative elements;
7. Oxygen is generally -2; it is -1 in peroxides; $+2$ with F;
8. Hydrogen is -1 in compounds with less electronegative elements, and $+1$ in compounds with more electronegative elements;
9. The sum of the oxidation numbers of all atoms in a neutral compound is zero, whereas in an ion it is equal to the charge on the ion.

Polar bonds (6.8) Atoms do not usually share electrons equally in covalent bonds. This is quite evident for covalently bonded compounds where the electronegativity difference is rather large; the more electronegative element has the strongest attraction for the shared electrons. This unequal attraction causes a shift of the shared electrons toward the more electronegative element, resulting in the electrons being nearer the more electronegative element most of the time. Covalent bonds in which the electrons are shared unequally are called *polar bonds*. For example, in hydrogen chloride, HCl, chlorine is more electronegative than hydrogen, 3.0 vs. 2.1. The electron pair in the bond is shifted toward chlorine, making the electron cloud region about chlorine relatively negative in charge with respect to the region about hydrogen. The partial charges are indicated by the symbol δ with the appropriate charge written to the upper right. Thus,

$$\overset{\delta^+}{H} \overset{x}{\underset{\cdot}{\rightarrow}} \overset{\delta^-}{\underset{xx}{\overset{xx}{Cl_x^x}}}$$

Resonance (6.7) The concept of resonance is introduced because our drawings are inadequate to accurately describe what is really taking place in bonding. When two or more Lewis structures with the same arrangement (placement) of atoms can be written for a molecule or ion, the *actual* distribution of electrons is a resonance *hybrid* or an average of that shown by the various Lewis structures. Each structure is called a *resonance form*.

SOLUTIONS TO CHAPTER 6 EXERCISES

IONS AND IONIC BONDING

1. Why does a cation have a positive charge?

   Solution

   Atoms or groups of atoms that have fewer electrons than protons have a net positive charge and are called cations. Monatomic cations are formed mostly from metallic elements; some examples indicating possible ion charges are

| | | Protons + electrons in ions |
Atom	Ion	= overall charge
Na − 1e⁻ →	Na⁺	11p⁺ + 10e⁻ = +1
Ca − 2e⁻ →	Ca²⁺	20p⁺ + 18e⁻ = +2
Al − 3e⁻ →	Al³⁺	13p⁺ + 10e⁻ = +3
Pb − 2e⁻ →	Pb²⁺	82p⁺ + 80e⁻ = +2

The reason for the positive charges relates to the ion tending to achieve a noble gas configuration in most cases.

2. Why does an anion have a negative charge?

Solution

Atoms or groups of atoms that have more electrons than protons have a net negative charge. Monatomic anions are formed mostly from nonmetallic elements; some examples indicating the charge on each ion are

| | | Protons + electrons in ions |
Atom	Ion	= overall charge
Cl + 1e⁻ → Cl⁻		17p⁺ + 18e⁻ = −1
O + 2e⁻ → O²⁻		8p⁺ + 10e⁻ = −2
N + 3e⁻ → N³⁻		7p⁺ + 10e⁻ = −3

The addition of one or more electrons to form these anions relates to the ions tending toward completing a shell of electrons of noble gas configuration.

4. Predict the charge on the monatomic ions formed from the following elements in ionic compounds containing these elements:

(a) As,
(c) Ca,
(e) F,
(f) Ga,
(h) S,
(j) Mg,
(l) K

Solution

The charge on monatomic ions can be predicted from the oxidation number rules. In the formation of ionic compounds, metallic elements lose electrons and become positive ions, whereas nonmetallic elements gain electrons and become negative ions. Furthermore, elements tend to gain or lose sufficient electrons to achieve a noble gas electron configuration. The electrons gained or lost by the above elements in the formation of ionic compounds are

(a) $As(Ar3d^{10}4s^24p^3) \rightarrow As^{3+}(Ar3d^{10}4s^2) + 3e^-$
(c) $Ca(Ar4s^2) \rightarrow Ca^{2+}(Ar) + 2e^-$
(e) $F(He2s^22p^5) + e^- \rightarrow F^-(Ne)$
(f) $Ga(Ar3d^{10}4s^24p) \rightarrow Ga^{3+}(Ar3d^{10}) + 3e^-$
(h) $S(Ne3s^23p^4) + 2e^- \rightarrow S^{2-}(Ne3s^23p^6)$ or $S^{2-}(Ar)$
(j) $Mg(Ne3s^2) \rightarrow Mg^{2+}(Ne) + 2e^-$
(l) $K(Ar4s^1) \rightarrow K^+(Ar) + e^-$

Chapter 6 Chemical Bonding, Part 1: General Concepts

6. The elements that form the greatest concentration of monatomic ions in sea water are Cl, Na, Mg, Ca, K, Br, Sr, and F. Write the Lewis symbols for the ions formed from these elements.

    Solution

    $:\ddot{\underset{..}{Cl}}:^-$, Na^+, Mg^{2+}, Ca^{2+}, K^+, $:\ddot{\underset{..}{Br}}:^-$, Sr^{2+}, $:\ddot{\underset{..}{F}}:^-$.

8. Write the formulas of each of the following ionic compounds using Lewis symbols:

    (a) NaBr,
    (b) MgCl$_2$,
    (c) Li$_2$O,

    Solution

    (a) $Na^+ \left[:\ddot{\underset{..}{Br}}: \right]^-$

    (b) $\left[:\ddot{\underset{..}{Cl}}: \right]^- Mg^{2+} \left[:\ddot{\underset{..}{Cl}}: \right]^-$

    (c) $Li^+ \left[:\ddot{\underset{..}{O}}: \right]^{2-} Li^+$

10. Why is it incorrect to speak of the molecular weight of NaCl?

    Solution

    Sodium chloride, NaCl, is formed from chloride ions each of which is surrounded by six sodium ions. Each sodium ion in turn is surrounded by six chloride ions. Consequently a sodium ion does not "belong" to any one chloride ion and thus no molecule exists.

COVALENT BONDING

12. What are the characteristics of atoms that will form a covalent bond?

    Solution

    Covalent bonds are favored when the atoms have little tendency to give or accept electrons. Such bonds generally form between nonmetallic elements close to each other in the Periodic Table. These elements will generally have electronegativity differences less than 1.7.

14. Predict which of the following compounds are ionic and which are covalent, based on the location of their constituent elements in the Periodic Table.

    (a) SO$_2$
    (c) ClF$_3$
    (e) MgO
    (g) H$_2$O
    (j) Cl$_2$CO

    Solution

    (a) covalent
    (c) covalent

(e) ionic
(g) covalent
(j) covalent

15. How are single, double, and triple bonds similar? How do they differ?

Solution

When possible, the formation of bonds between atoms results in completing the valence levels of the elements involved to achieve noble gas electronic structures. The terms single, double, and triple bonds refer to covalent bonds. In ionic bonds one atom has lost one electron to another atom to form an ion pair. In contrast, both bonding atoms in a single covalent bond share a pair of electrons. Furthermore, single bonds can form between two atoms through direct atomic orbital overlap without mixing (hybridization) of orbitals.

Double and triple bonds are covalent bonds in which elements must share four and six electrons, respectively. In contrast to single bonds, the formation of these bonds requires that some or all of the atomic orbitals of the bonding atoms be mixed, that is, hybridized, to create orbitals having compatible geometries for bonding.

Note that there is a gradual increase in the bond energy and decrease in the bond distance. For example, consider the carbon-carbon bonds in the following organic molecules:

Ethane H_3C-CH_3 distance = 1.50 Å, E = 368 kJ/mol
Ethene $H_2C=CH_2$ distance = 1.34 Å, E = 682 kJ/mol
Ethyne $HC\equiv CH$ distance = 1.20 Å, E = 828 kJ/mol

Note also that singly bonded atoms as in ethane, C—C, are free to rotate about each other to form different conformational geometries. On the other hand, doubly or triply bonded atoms as in ethene, —C=C—, or ethyne, —C≡C—, are fixed in a planar geometry that does not permit free rotation but does allow a degree of twisting.

17. Write Lewis structures for the following:

(a) O_2
(b) CO_2
(c) H_2CO
(d) FNO
(e) H_2CCH_2

Solution

(a) 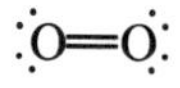 :O=O:

Note that in this case the Lewis structure is inadequate to depict the fact that experimental studies have shown two unpaired electrons in each oxygen molecule, a condition known as *paramagnetism*.

(b) 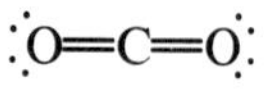 :O=C=O:

(c) H₂C=O (with lone pairs on O)

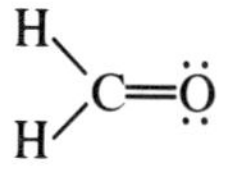

(d) 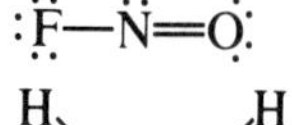 :F—N=O:

(e) 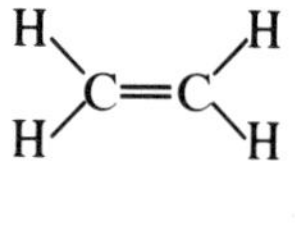 H₂C=CH₂

Chapter 6 Chemical Bonding, Part 1: General Concepts   81

21. The skeleton structures of a number of biologically active molecules are given below. Complete the Lewis structure of these molecules.

   (a) The amino acid alanine
   (b) Urea
   (d) Carbonic acid
   (f) Vitamin C (ascorbic acid)

Solution

(a) 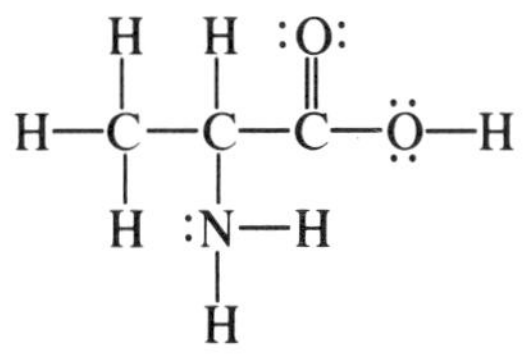
```
        H   H   :Ö:
        |   |   ||
   H — C — C — C — Ö — H
        |   |
        H   :N — H
            |
            H
```

(b) 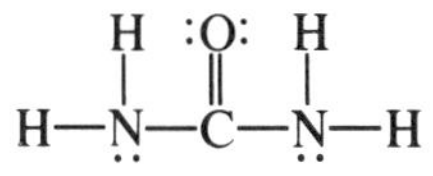
```
        H   :Ö:   H
        |   ||    |
   H — N — C — N — H
       ··        ··
```

(d) 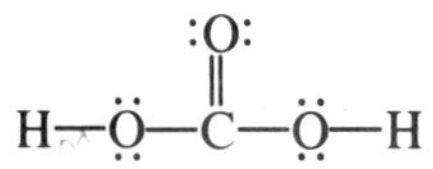
```
        :Ö:
        ||
   H — Ö — C — Ö — H
```

(f) 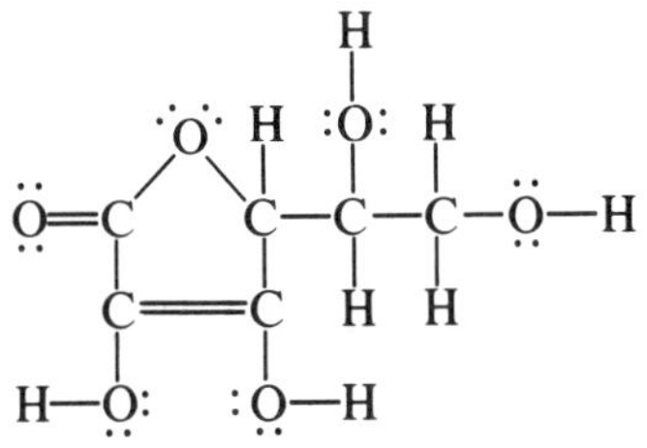
```
              H
              |
        :Ö·  H  :Ö:  H
        /    |   |   |
   Ö = C    C — C — C — Ö — H
   ··   \\   |   |
         C = C   H   H
        /     \\
   H — Ö:   :Ö — H
```

24. The following species do not satisfy the noble gas structure rule. Write the Lewis structure for each.

   (a) BCl₃,
   (c) AsF₅,
   (d) ICl₄⁻,
   (f) S₂F₁₀ (contains an S—S bond),
   (g) PCl₆⁻.

Solution

(a) 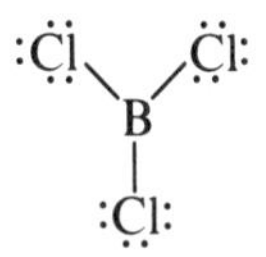
```
   :Cl̈    :Cl̈:
      \\  /
       B
       |
      :Cl̈:
```

(c) 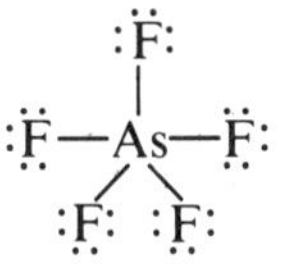
```
        :F̈:
         |
   :F̈ — As — F̈:
        / \\
      :F̈:  :F̈:
```

(d) 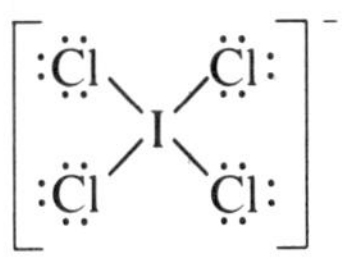
```
   ⎡ :Cl̈    :Cl̈: ⎤⁻
   ⎢    \\  /     ⎥
   ⎢     I        ⎥
   ⎢    /  \\     ⎥
   ⎣ :Cl̈    :Cl̈: ⎦
```

82   Chapter 6 Chemical Bonding, Part 1: General Concepts

(f) 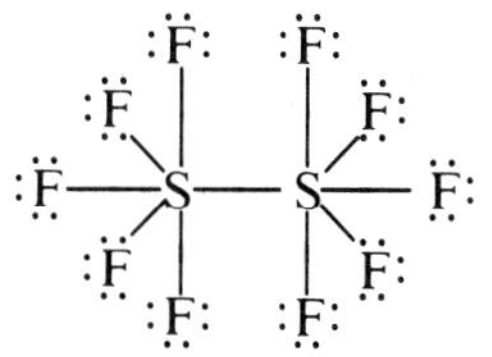 Structure showing two S atoms bonded together, each with four F atoms attached (S$_2$F$_{10}$-like structure with all fluorines showing lone pairs).

(g) 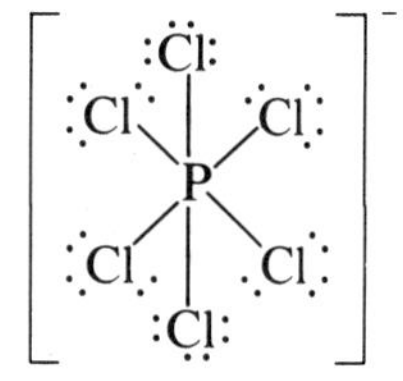 $[PCl_6]^-$ structure showing P center bonded to six Cl atoms, each with lone pairs, enclosed in brackets with a negative charge.

RESONANCE

26. Draw the resonance forms for the following:

    (a) selenium dioxide, OSeO
    (b) the nitrite ion, NO_2^-
    (c) the carbonate ion, CO_3^{2-}
    (d) the acetate ion,

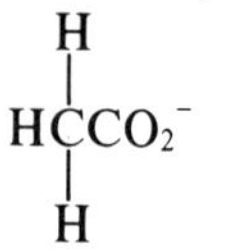

    $$HC(H)(H)CO_2^-$$

Solution

A more systematic way of distributing electrons is needed particularly when resonance is possible. The best way is to follow these simple steps:

1. Add the valence electrons of all atoms participating in the molecule.
2. Arrange the atoms of the compound to correspond to its known structure. (If the structure is not known, place the most distinctive atom in the center and arrange the other atoms symmetrically around it.)
3. Place two electrons between each atom, then complete the octet of electrons around each of the outer atoms.
4. Any electrons of the original valence electrons that are not utilized by the preceding operation are placed on the central atom.
5. Rearrange electrons, shifting electron pairs from an outer atom to form double bonds where necessary to complete the octet on the central atom.

The steps are illustrated only for

(a) selenium dioxide, OSeO.

   Step 1. Selenium is in family VIA so that it has 6 valence electrons as does each oxygen. There are therefore a total of 18 electrons to be accommodated.

   Step 2. A symmetrical distribution gives

   O   Se   O

   Step 3. O:Se:O, then :Ö—Se—Ö:

Chapter 6 Chemical Bonding, Part 1: General Concepts

Step 4. 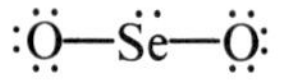 :Ö—Se—Ö:

Step 5.  :Ö—Se—Ö: → :O=Se—Ö:

An equally likely arrangement is possible leading to a resonance picture:

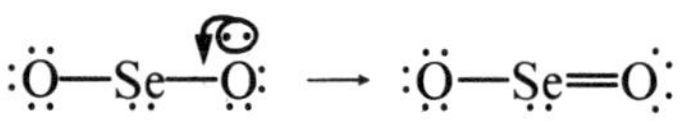

:Ö—Se—Ö: → :Ö—Se=O:

or

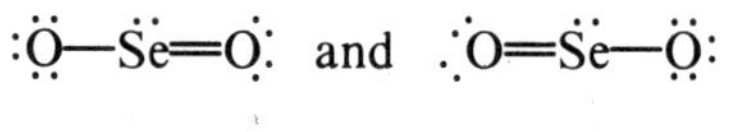

:Ö—Se=O: and :O=Se—Ö:

(b) 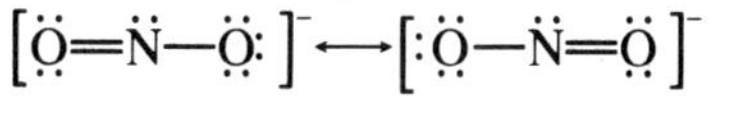 $[\ddot{O}=N-\ddot{O}:]^- \longleftrightarrow [:\ddot{O}-N=\ddot{O}]^-$

(c) Three resonance structures of CO_3^{2-} shown.

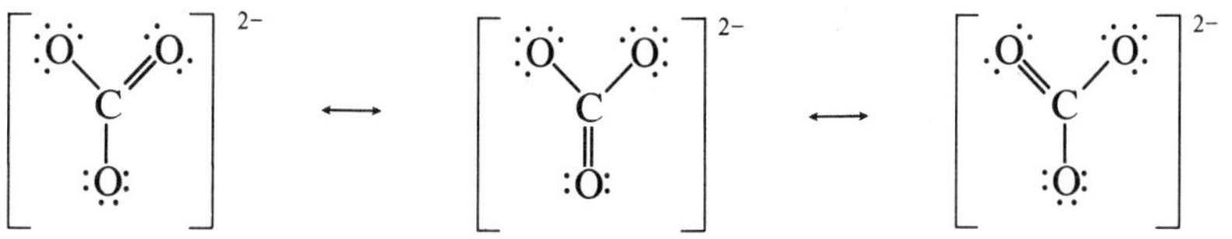

(d) Two resonance structures shown for acetate-like ion with H-C(H)(H)-C(=O)(O⁻).

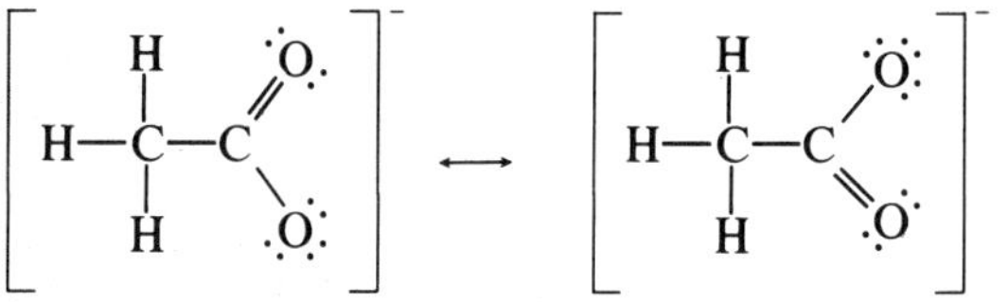

29. Write the resonance forms for the nitrate ion, NO_3^-, and show that this ion is isoelectronic with the carbonate ion, CO_3^{2-}.

Solution

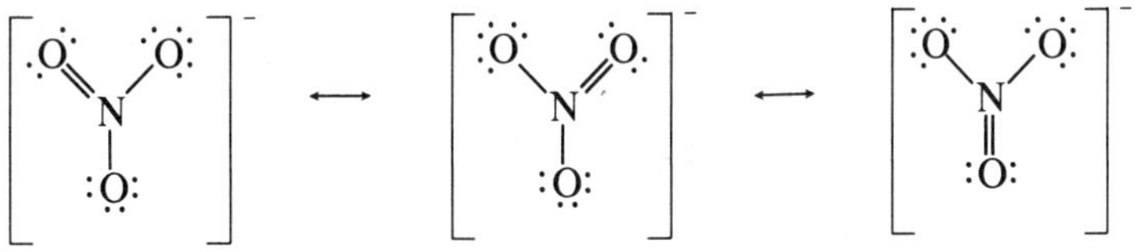

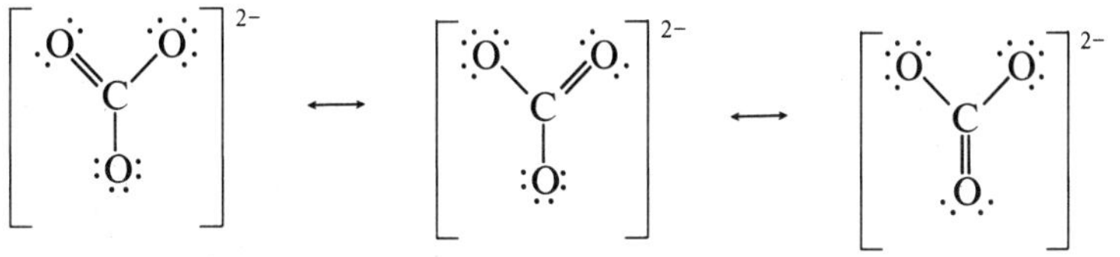

Three resonance structures of NO_3^- shown.

Three resonance structures of CO_3^{2-} shown.

31. The skeleton structure of benzene, C_6H_6, is given below. Write the resonance forms of benzene

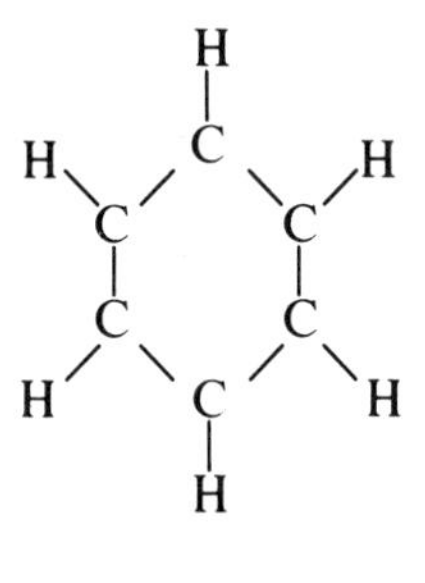

Solution

[Two Kekulé structures of benzene shown side by side, with alternating double bonds in opposite positions]

ELECTRONEGATIVITY AND POLAR BONDS

34. How does the electronegativity difference of the atoms in a covalent bond affect the polarity of the bond?

    Solution

    The larger the electronegativity difference of the atoms in a covalent bond, the greater the polarity of the bond. The more electronegative atom is partially negative, and the other atom becomes partially positive.

36. Explain why some molecules that contain polar bonds are not polar.

    Solution

    A polar bond involves the unequal sharing of electrons by the bonding atoms. The bonding electrons shift toward the more electronegative atom, resulting in a net imbalance in electronic charge, measured by the dipole moment, in that direction. Molecules may contain several polar bonds and as a whole may be nonpolar if there are other bonds in the molecule that offset or balance the imbalance in electronic distribution. For example, in carbon dioxide both carbon-oxygen double bonds are polar and both dipole moments are in the direction of oxygen. The molecule is linear and the two dipole moments are equal but opposite in direction and hence cancel.

    $$O^- = {}^+C^+ = {}^-O$$

38. Which of the following molecules or ions contain polar bonds? O_3, P_4, S_2^{2-}, NO_2^-, ClO_3^-, H_2S, BH_4^-, HCN.

    Solution

    NO_2^-, ClO_3^-, H_2S, BH_4^-, HCN all contain polar bonds because of the difference in electronegativities between different atoms.

40. One of the most important polar molecules is water, H_2O, a bent molecule with an angle of 105° between the two O—H bonds. Which end of the O—H bond in water is positive and which is negative? Sketch a figure of the H_2O molecule and indicate the positive and negative ends of the molecule.

Chapter 6 Chemical Bonding, Part 1: General Concepts

Solution

Oxygen is more electronegative than hydrogen. Therefore, the oxygen end of the O—H bond is negative and the other end is positive.

$$\delta^+ \, H \underset{105°}{\overset{\overset{\delta^-}{O}}{\diagdown\diagup}} H \, \delta^+$$

BOND ENERGIES

41. Using the data in Appendix I, calculate the bond energies of F_2, Cl_2, and FCl. All are gases in their most stable form at standard state conditions.

Solution

The bond energy D for a diatomic molecule equals the change in standard enthalpy for the dissociation of the molecule. In general, the reaction can be expressed as

$$XY(g) \longrightarrow X(g) + Y(g) \qquad \Delta H° \text{ (reaction)} = D$$

For F—F:

$$F_2(g) \longrightarrow F(g) + F(g)$$

$$D = \Delta H° = 2\Delta H°_{f_{F(g)}} - \Delta H°_{f_{F_2(g)}} = 2(78.99 \text{ kJ}) - 0 \text{ kJ}$$
$$= 158.0 \text{ kJ per mole of bonds}$$

For Cl—Cl:

$$Cl_2(g) \longrightarrow Cl(g) + Cl(g)$$

$$D = \Delta H° = 2\Delta H°_{f_{Cl(g)}} - \Delta H°_{f_{Cl_2(g)}} = 2(121.68 \text{ kJ}) - 0 \text{ kJ}$$
$$= 243.36 \text{ kJ per mole of bonds}$$

For F—Cl:

$$FCl(g) \longrightarrow F(g) + Cl(g)$$

$$D = \Delta H° = \Delta H°_{f_{F(g)}} + \Delta H°_{f_{Cl(g)}} - \Delta H°_{f_{FCl(g)}}$$
$$= 78.99 + 121.68 - (-54.48) \text{ (units of kJ per mole of bonds)}$$
$$= 255.15 \text{ kJ per mole of bonds}$$

43. Using the data in Appendix I, calculate the Ti—Cl single bond energy in $TiCl_4$.

Solution

The energy of four Ti—Cl single bonds, $4D_{Ti-Cl}$, is equal to the standard enthalpy of the reaction

$$TiCl_4 \longrightarrow Ti(g) + 4Cl(g)$$

Using Hess's law,

$$4D_{Ti-Cl} = \Delta H° = \Delta H°_{f_{Ti(g)}} + 4\Delta H°_{f_{Cl(g)}} - \Delta H°_{f_{TiCl_4(g)}}$$
$$= 469.9 \text{ kJ} + 4(121.68 \text{ kJ}) - (-763.2 \text{ kJ})$$
$$= 1719.8 \text{ kJ}$$

The average Ti—Cl bond energy, D_{Ti-Cl}, is $1719.8/4 = 430.0$ kJ per mole of Ti—Cl bonds.

45. The enthalpy of formation of $AsF_5(g)$ has been determined to be -16.46 kJ g^{-1} of arsenic using the reaction $2As(s) + 5F_2(g) \rightarrow 2AsF_5(g)$. Using this information and the data in Appendix I, calculate the As—F single bond energy in AsF_5.

Solution

The enthalpy of formation is given on a per gram basis. First convert the enthalpy to a per mole basis multiplying by the atomic weight of As, 74.92 g mol^{-1}. Thus,

$$-16.46 \text{ kJ g}^{-1} \times 74.92 \text{ g mol}^{-1} = -1233.2 \text{ kJ mol}^{-1}$$

Then,

$$5D_{As-F} = \Delta H° = \Delta H°_{f_{As(g)}} + 5\, \Delta H°_{f_{F(g)}} - \Delta H°_{f_{AsF_5(g)}}$$
$$= 303 \text{ kJ} + 5(78.99 \text{ kJ}) - (-1233.18) = 1931 \text{ kJ}$$

$$D_{As-F} = \frac{1931 \text{ kJ}}{5} = 386 \text{ kJ}$$

46. Using the bond energies in Table 6.3 calculate the approximate enthalpy change for each of the following reactions.

(a) $H_2(g) + F_2(g) \rightarrow 2HF(g)$
(c) $CH_4(g) + Cl_2(g) \rightarrow CH_3Cl(g) + HCl(g)$
(e) $CH_3CH_2CH_3(g) \rightarrow CH_3CH=CH_2(g) + H_2(g)$

Solution

(a) Heats of reaction are calculated from bond energies by using the same procedure that is used with enthalpies:

Δ bond energy = bond energy of products − bond energy of reactants

Each bond in every compound involved in the reaction must be considered. First write the structural formula of the compounds to better determine the bonds involved.

$$H-H + F-F \rightarrow H-F + H-F$$

Then write the number of bond types.

Reactants	Products
1 H—H	2 H—F
1 F—F	

The change in bond energies is obtained by taking the value of the bond energies of the products minus the values of those of the reactants.

$$-2\Delta H_f = \Delta \text{ bond energy} = \underset{\text{Products}}{2(569)} - \underset{\text{Reactants}}{(436 + 160)}$$
$$= +542 \text{ kJ mol}^{-1}$$

Because the change in bond energies is positive, more energy is released when the bonds in HF are formed than is required to break the bonds in

H₂ and F₂. Consequently, the heat of the reaction has the same numerical value but is negative.

$$\Delta H = \frac{-542}{2} = 271 \text{ kJ mol}^{-1}$$

(c) $\text{CH}_4(g) + \text{Cl}_2(g) \longrightarrow \text{CH}_3\text{Cl}(g) + \text{HCl}(g)$

Products		Reactants	
4 C—H	1 Cl—Cl	3 C—H	1 H—Cl
		1 C—Cl	

$-\Delta H = \Delta$ bond energy = 3(415) + 330 + 432 − 4(415) − 243
$\qquad\qquad\qquad\qquad\;\; = 104 \text{ kJ mol}^{-1}$

$\Delta H = -104 \text{ kJ mol}^{-1}$

(e) $\text{CH}_3\text{CH}_2\text{CH}_3(g) \longrightarrow \text{CH}_3\text{CH}=\text{CH}_2(g) + \text{H}_2(g)$

8 C—H $\;D = 415$ kJ mol⁻¹ $\quad$ 6 C—H $\;D = 415$ kJ mol⁻¹ $\quad$ 1 H—H
2 C—C $\;D = 345$ kJ mol⁻¹ $\quad$ 1 C—C $\;D = 345$ kJ mol⁻¹ $\quad D = 436$ kJ mol⁻¹
$\qquad\qquad\qquad\qquad\qquad\;$ 1 C=C $\;D = 611$ kJ mol⁻¹

$-\Delta H = \Delta$ bond energy = products − reactants
$\qquad\;\; = 6(415) + 345 + 611 + 435 − 8(415) − 2(345)$
$\qquad\;\; = -2(415) − 345 + 611 + 435 = -129 \text{ kJ mol}^{-1}$

$\Delta H = 129 \text{ kJ}$

OXIDATION NUMBERS

47. What are the oxidation numbers of the following elements in their compounds?

    (a) F,
    (b) Mg,
    (c) Rb,
    (d) Na,
    (e) K,
    (f) Ca.

Solution

In each of these cases we use the fact that the oxidation number can be determined by the ease with which the outermost electron(s) is removed (gained) from (by) the neutral atom as indicated by its position in the periodic table.

(a) Belongs to family VIIA. Lacks one electron of complete octet. Easily gains one electron to become negatively charged. Oxidation state −1.
(b) Family IIA, oxidation state +2.
(c) Family IA, +1.
(d) Family IA, +1.
(e) Family IA, +1.
(f) Family IIA, +2.

50. Determine the oxidation number of each of the following elements in the compound indicated:

   (a) B in B_2O_3
   (b) C in CF_4
   (d) P in P_4O_{10}
   (e) Co in $LiCoO_2$
   (g) S in SO_2Cl_2

Solution

(a) Since oxygen is always -2 except in peroxides, its oxidation number of $3 \times -2 = -6$ must be balanced by the charge on B. There are two boron atoms, so each must have an oxidation number of $+6/2 = +3$.
(b) F is -1, so the oxidation number of carbon must balance $4 \times -1 = -4$. Therefore its oxidation number is $+4$.
(d) Total charge for $O = 10 \times -2 = -20$. Oxidation number $P = +20/4 = +5$.
(e) $Li = +1$, $O = -2$. Assignment in $LiCoO_2$ is $+1 + ? + (2 \times -2) = 0 = +1 + Co - 4$. To give a neutral species: $Co = +3$.
(g) $O = -2$, $Cl = 1$. $S + 2(-2) + 2(-1) = S - 6 = 0$. To balance this requires $S = +6$.

53. Sodium thiosulfate, $Na_2S_2O_3$, is used as a fixing agent in photography. What is the oxidation number of each of the elements in $Na_2S_2O_3$?

Solution

O is -2 except in peroxide (—O—O—). Na is always $+1$. To balance $+2$ and -6 or a net -4 on the other elements, each S must have an oxidation state $+2$.

NAMES OF COMPOUNDS

55. Do the suffixes -*ous* and -*ic* correspond to any particular oxidation state? How are they used?

Solution

Suffixes -*ous* and -*ic* do not correspond to a particular oxidation state. The suffix -*ic* refers to the higher of two oxidation states possessed by an element, whereas the suffix -*ous* refers to the lower oxidation state.

56. Name the following binary compounds:

   (a) NaF,
   (b) CuS,
   (e) KH,
   (g) $MgCl_2$,
   (i) Na_2O,
   (k) Mg_3N_2.

Solution

(a) sodium fluoride
(b) copper(II) sulfide or cupric sulfide

Chapter 6 Chemical Bonding, Part 1: General Concepts

- (e) potassium hydride
- (g) magnesium chloride
- (i) sodium oxide
- (k) magnesium nitride or trimagnesium dinitride

59. Name the following compounds. (Since the metal atom in each can exhibit two or more oxidation states, it will be necessary to identify the oxidation number of the metal as you name the compound.)

- (a) $FeCl_3$,
- (b) Cu_2S,
- (c) $MnBr_2$,
- (d) $Pb(OH)_2$,
- (e) $TlNO_3$
- (f) $Cr(ClO_4)_3$, —
- (g) $SnCl_4$,
- (h) $CuSO_4$.

Solution

- (a) iron(III) chloride or ferric chloride
- (b) copper(I) sulfide or cuprous sulfide
- (c) manganese(II) bromide
- (d) lead(II) hydroxide
- (e) thallium(I) nitrate
- (f) chromium(III) perchlorate
- (g) tin(IV) chloride or stannic chloride
- (h) copper(II) sulfate or cupric sulfate

60. Name the following compounds, using Greek prefixes for each element as needed:

- (a) NO_2,
- (c) N_2O_4,
- (e) Cl_2O_7,

Solution

- (a) nitrogen dioxide
- (c) dinitrogen tetraoxide
- (e) dichlorine heptaoxide

FORMULAS OF COMPOUNDS

61. Why must a sample of calcium chloride, $CaCl_2$, contain two chloride ions for each calcium ion?

Solution

Calcium atoms easily lose 2 electrons to become positively charged (+2). Each chlorine atom gains 1 electron to complete a shell and becomes negatively charged (−1). In order to maintain charge neutrality, there must be two chloride ions for each calcium ion.

63. Using the ionic charges as a guide, write the formula of each of the following compounds:

   (a) magnesium perchlorate
   (b) sodium nitrate
   (c) calcium sulfate
   (d) potassium hydroxide
   (e) ammonium phosphate
   (f) aluminum nitrate
   (g) sodium cyanide

   Solution

   (a) $Mg(ClO_4)_2$
   (b) $NaNO_3$
   (c) $CaSO_4$
   (d) KOH
   (e) $(NH_4)_3PO_4$
   (f) $Al(NO_3)_3$
   (g) $NaCN$

64. Using the indicated oxidation state as a guide, write the formula of each of the following compounds:

   (a) cobalt(II) bromide
   (b) lead(II) sulfide
   (c) mercury(II) oxide
   (d) iron(II) nitrate
   (e) copper(I) sulfide
   (f) chromium(II) sulfate
   (g) tin(II) perchlorate

   Solution

   (a) $CoBr_2$
   (b) PbS
   (c) HgO
   (d) $Fe(NO_3)_2$
   (e) Cu_2S
   (f) $CrSO_4$
   (g) $Sn(ClO_4)_2$

66. The following compounds have been approved by the FDA as food additives and may be used in foods as neutralizing agents, sequestering agents, sources of minerals, etc. What is the chemical formula of each compound?

   (a) potassium iodide
   (b) sulfur dioxide
   (c) calcium oxide
   (d) calcium chloride
   (e) hydrogen chloride
   (f) sodium hydroxide
   (g) magnesium carbonate √
   (h) ammonium carbonate
   (i) carbon dioxide
   (j) copper(I) iodide

Solution

(a) KI
(b) SO$_2$
(c) CaO
(d) CaCl$_2$
(e) HCl
(f) NaOH
(g) MgCO$_3$
(h) (NH$_4$)$_2$CO$_3$
(i) CO$_2$
(j) CuI

ADDITIONAL EXERCISES

69. X may indicate a different representative element in each of the following Lewis formulas. To which group does X belong in each case?

    (a) 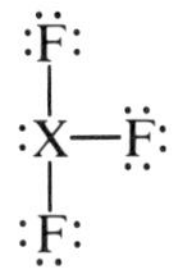
    $$\begin{array}{c} :\ddot{F}: \\ | \\ :\ddot{X}-\ddot{F}: \\ | \\ :\ddot{F}: \end{array}$$

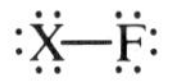

    (b) $:\ddot{X}-\ddot{F}:$

    (c)
    $$\begin{array}{c} :\ddot{Cl}: \\ | \\ :\ddot{Cl}-X-\ddot{Cl}: \\ | \\ :\ddot{Cl}: \end{array}$$

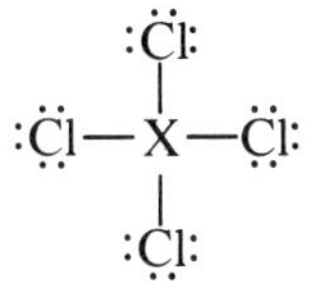

    (d) 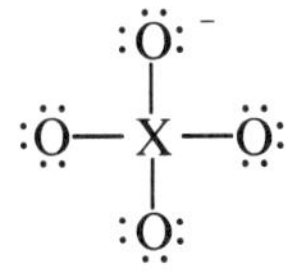
    $$\begin{array}{c} :\ddot{O}:^- \\ | \\ :\ddot{O}-X-\ddot{O}: \\ | \\ :\ddot{O}: \end{array}$$

Solution

Count the number of valence electrons associated with X in each formula; the group number equals the number of valence electrons.

(a) Number of valence electrons shared with F = 3, plus unshared pair = 5; therefore, X belongs to Group V.
(b) Number of valence electrons shared with F = 1, plus 3 unshared pairs = 7; therefore, X belongs to Group VII.
(c) Number of valence electrons shared with Cl atoms = 4. Therefore, X belongs to Group IV.
(d) Eight e$^-$ are shared coordinate covalently with O. One of these electrons is added from an outside source to give an anion with 7 valence electrons; therefore, X belongs to Group VII.

70. The number of covalent bonds between two atoms is called bond order. Bond order may be an integer, as with :N≡N: (bond order = 3), or a fraction, as with SO$_2$ (bond order = 1$\frac{1}{2}$ due to the resonance O=S—O ⟷ O—S=O). Arrange the molecules and ions in each of the following groups in increasing

bond order for the bond indicated. Unshared pairs have been omitted for clarity.

(a) C—O bond order: C≡O; O=C=O; CH₃—O—O—CH₃; HCO₂⁻

$$\left(H-C\begin{smallmatrix}\nearrow O \\ \searrow O^-\end{smallmatrix} \longleftrightarrow H-C\begin{smallmatrix}\nearrow O^- \\ \searrow O\end{smallmatrix} \right)$$

(b) N—N bond order: CH₃—N=N—CH₃; N₂; H₂N—NH₂;
N₂O (N=N=O ⟷ N≡N—O).

(c) C—N bond order: H₃C—NH₂; CH₃CONH⁻

$$\left(CH_3-C\begin{smallmatrix}\nearrow O \\ \searrow NH^-\end{smallmatrix} \longleftrightarrow CH_3-C\begin{smallmatrix}\nearrow O^- \\ \searrow NH\end{smallmatrix} \right)$$

H₂C=NH; H₃C—C≡N.

(d) C—C bond order: H₃C—CH₃; HC≡CH; H₂C=CH₂; C₆H₆

(benzene resonance structures)

Solution

Increasing bond order:

(a) CH₃—O—O—CH₃, (1); HCO₂⁻, (1½); O=C=O, (2); C≡O, (3).
(b) H₂N—NH₂, (1); CH₃—N=N—CH₃, (2); N₂O, (2½); N₂, (3).
(c) H₃C—NH₃, (1); CH₃CONH⁻, (1½); H₂C=NH, (2); H₃C—C≡N, (3).
(d) H₃C—CH₃, (1); C₆H₆, (1½); H₂C=CH₂, (2); HC≡CH, (3).

7

Chemical Bonding, Part 2: Molecular Orbitals

INTRODUCTION

In this chapter the focus is on how electrons are distributed throughout a molecule to accomplish the bonding studied in Chapter 6. The main theory presently used for this explanation is the molecular orbital theory, which has proved very useful for our understanding of chemistry. The individual atoms of a molecule are visualized at their normal interatomic distance but without their electrons. Orbitals much like the s and p orbitals with which we are familiar are constructed encompassing the entire molecule. The electrons which otherwise would be on the individual atoms are filled into these molecular orbitals, always filling the lowest energy molecular orbital first. The conclusions drawn from the completed molecular orbital diagram give us insight as to how bonding takes place and give us the ability to make predictions about the chemical behavior of the molecule.

READY REFERENCE

Antibonding orbitals (7.1)   Orbitals of higher energy that tend to destabilize a molecule.

Bonding orbitals (7.1)   Orbitals of lower energy that cause bond formation.

Bond order (7.3)   The bond order refers to the strength of a bond. The higher the number, the stronger the bond. It is the net number of bonding pairs of electrons between atoms.

Degenerate orbitals (7.1)   These are orbitals with the same energy, such as a set of three p-type orbitals.

Molecular orbital (7.1)   A wave description of the bonding within a molecule.

Overlap (7.1)   The degree to which two orbitals interact so that electron(s) may be shared between the atoms on which the orbitals are based.

Pi (π) ***orbital*** (7.1)   A molecular orbital in which the maximum charge density of the electrons lies away from the line connecting the centers of the bonded atoms.

Sigma (σ) orbital (7.1) A molecular orbital in which the maximum charge density of the electrons lies along the line connecting the two atoms involved in the bond.

SOLUTIONS TO CHAPTER 7 EXERCISES

MOLECULAR ORBITALS

1. Why does an electron in a bonding molecular orbital formed from two atomic orbitals have a lower energy than in either of the individual atomic orbitals?

   Solution

   In a molecular orbital the electrons are attracted by two nuclei rather than only one. The greater attraction that results reduces the otherwise repulsive forces between nuclei and, therefore, the bonding molecular orbital is stabilized (i.e., has a lower energy) relative to the individual atoms.

5. Draw diagrams showing the molecular orbitals that can be formed by combining two s orbitals, by combining two p orbitals, and by combining an s and a p orbital.

   Solution

   For homonuclear molecules the energy of the s orbitals on each atom is the same. The p orbitals have higher energy than the s orbitals. And, again for homonuclear molecules, the energy of the p orbitals on each atom is the same.

   Combining two s orbitals:

   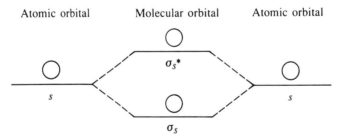

   The lower molecular orbital will be occupied first.

Combining two p orbitals (p's aligned toward each other are indicated σ, those parallel are indicated π):

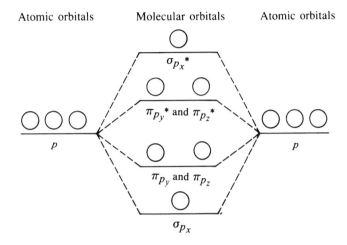

Combining an s and a p orbital: The energy of the s on one atom must be similar to the energy of the p orbital on the other atom. Electrons are shown for convenience.

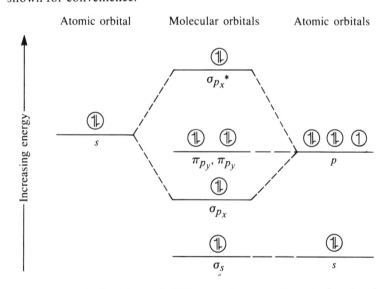

6. Describe the similarities and differences between the set of molecular orbitals formed by end-to-end overlap of p orbitals and the set of molecular orbitals formed by side-to-side overlap of p orbitals.

Solution

In both cases bonding occurs and the energy of the molecule is lowered. Generally for end-to-end bonding (formation of a σ orbital), the overlap is greater than for side-to-side bonding (formation of a π orbital); greater overlap means a stronger bond.

HOMONUCLEAR DIATOMIC MOLECULES

8. Using molecular orbital energy diagrams and the electron occupancy in the molecular orbitals, compare the stability of Li_2, Be_2^+, and Be_2.

Solution

The molecular orbital diagrams for Li_2 and for Be_2 are shown in Figures 7.8 and 7.9, respectively. The MO diagram for Be_2^+ is

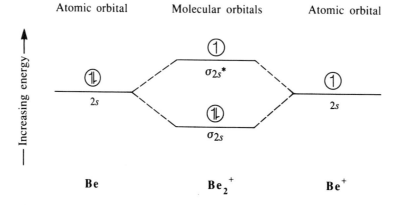

Li_2 has two electrons in a bonding orbital, thus giving it a bond order of 1. In Be_2^+ the electron in the antibonding orbital cancels the effect of one of the electrons in the bonding orbital, giving Be_2^+ a bond order of $\frac{1}{2}$. Be_2 is not stable and has a bond order of zero.

10. The acetylide ion, C_2^{2-}, is a component of calcium acetylide. Draw the molecular orbital energy diagram of this ion and write the molecular orbital electron configuration.

Solution

Lower-energy electrons in the $1s$ orbital of each atom are not shown. See diagram on page 97.

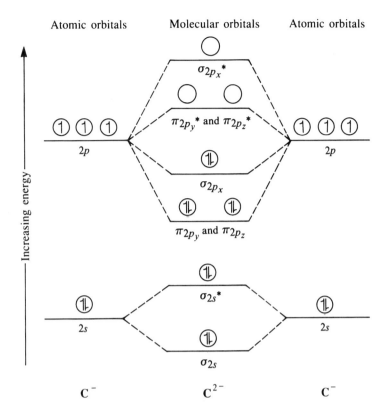

The electron configuration is $KK(\sigma_{2s})^2 (\sigma_{2s}*)^2 (\pi_{2p_y}, \pi_{2p_z})^4 (\sigma_{2p_x})^2$.

12. Draw a molecular orbital energy diagram for the ion S_2^{2-}, using only the orbitals in the valence shell of sulfur that are occupied in the free atom. What is the electron configuration of S_2^{2-}?

Solution

The $1s$, $2s$, and $2p$ orbitals are not drawn since they are much lower in energy and do not participate in bonding. Assume that the order of filling sulfur orbitals is the same as in oxygen. See diagram on page 98.

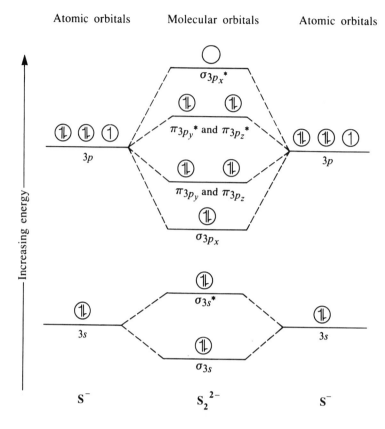

The electron configuration is KKLL $(\sigma_{3s})^2 (\sigma_{3s}*)^2 (\sigma_{3p_x})^2 (\pi_{3p_y}, \pi_{3p_z})^4 (\pi_{3p_y}*, \pi_{3p_z}*)^4$.

HETERONUCLEAR DIATOMIC MOLECULES

13. How does the molecular orbital energy diagram for a diatomic molecule involving atoms of two different elements differ from that for a diatomic molecule made up of two atoms of the same element?

    Solution

    For two atoms of the same type, the corresponding orbitals have the same energy. This facilitates overlap of orbitals and bond formation. For diatomic molecules with differing atoms, the corresponding orbitals of the more electronegative element will lie at lower energy. If the energies of the orbitals on the two different atoms are much different, hybridization occurs so that better overlap is achieved between the new hybrid orbitals on the different atoms.

15. What does the presence of a polar bond in a molecule imply about the shapes of its molecular orbitals?

    Solution

    Presence of a polar bond will generally mean that the orbital energies used to form the molecular orbital are different, thus leading to unsymmetrical molecular orbitals.

16. Draw the molecular orbital energy diagrams for CO and NO⁺. Show that CO and NO⁺ are isoelectronic and have the same bond order. How does the molecular orbital electron configuration of these species compare with that for NO?

Solution

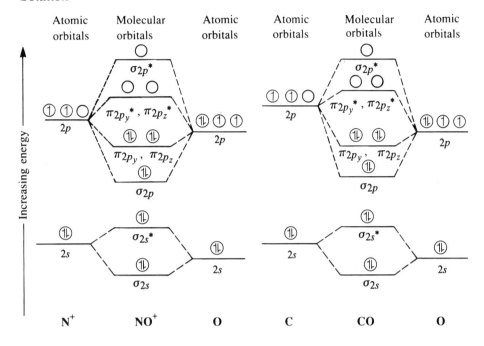

As seen here, both CO and NO⁺ have the same electron arrangement, making them isoelectric. Both have a bond order of three. The molecular orbital diagram of NO has one more electron and it is in a $\pi_{2p_y}*$ orbital. The electron thus reduces the bond order of NO to 2.5 and, of course, NO is no longer isoelectronic with CO and NO⁺.

BOND ORDER

20. Determine the bond orders of the homonuclear diatomic molecules formed by the first ten elements of the Periodic Table.

Solution

Element	Bond Order	Element	Bond Order
H_2	1	C_2	2
He_2	0	N_2	3
Li_2	1	O_2	2
Be_2	0	F_2	1
B_2	1	Ne_2	0

22. Arrange the following in increasing order of bond energy: F_2, F_2^+, F_2^-.

Solution

The increasing order of bond energy is indicated by increasing bond order. The electron configuration for fluorine is given in Section 7.11. The bond order for F_2 is 1; with one less electron, F_2^+ has a bond order of 1.5; with one electron more the bond order is 0.5 for F_2^-. Increasing order is: F_2^-, F_2, F_2^+.

23. The Lewis structure of the nitric oxide anion (N=O⁻) indicates a bond order of 2 for this ion. Beginning with the molecular orbital energy diagram for NO, work out the bond order for NO⁻, using molecular orbital theory.

Solution

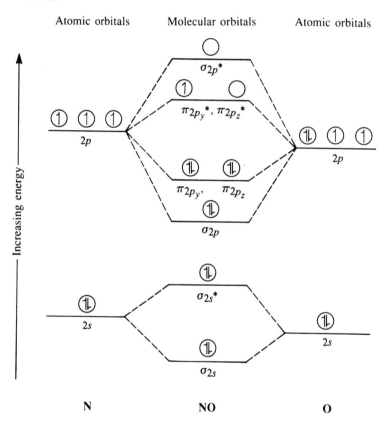

The additional electron of NO⁻ would occupy the empty $\pi_{2p_z}^*$ orbital. Subtraction of the two antibonding electrons from the total of six bonding electrons leaves four bonding electrons, or a bond order of 2.

ADDITIONAL EXERCISES

27. Like the ionization energy of an atom (Section 5.15), the ionization energy of a molecule is the energy necessary to remove the least tightly bound electron from the molecule. Arrange the molecules or ions in each of the following groups in increasing order of first ionization energy:

(a) N_2, N_2^{2+}, N_2^-
(b) NO, NO⁻, NO⁺

Chapter 7 Chemical Bonding, Part 2: Molecular Orbitals

Solution

(a) increasing first ionization energy ⟶ $N_2^- < N_2 < N_2^{2+}$

(b) The bond strength of the least tightly bound electron is reduced by its being in an antibonding orbital. From the figures in exercises 16 and 23, we can see that two antibonding orbitals are occupied in NO^-, one in NO, and none in NO^+. The increasing order of ionization energy is NO^-, NO, NO^+.

28. Predict whether each of the following atoms, ions, and molecules contains only paired electrons or contains some unpaired electrons.

(a) Li
(b) Be^+
(c) Be_2^{2+}

Solution

(a) unpaired
(b) unpaired
(c) paired

30. Identify the homonuclear diatomic molecules or ions that have the following electron configurations:

(a) X_2^+: $KK(\sigma_{2s})^2(\sigma_{2s}^*)^1$
(c) X_2^-: $KK(\sigma_{2s})^2(\sigma_{2s}^*)^2(\pi_{2p_y}, \pi_{2p_z})^1$

Solution

(a) This lies in the second energy level and has only 3 electrons. If this species is not charged, there would be 4 electrons in the $n = 2$ energy level of the molecule or only two electrons per atom. This corresponds to Be and Be_2^+.

(c) Again this is in the second energy level. Without the extra electron, each atom had to contribute 2 electrons from its $2s$ level, thus corresponding to two atoms of beryllium forming Be_2^-.

8

Molecular Structure and Hybridization

INTRODUCTION

Early scientists had no knowledge of the three-dimensional arrangement of atoms in molecules and ions, but as chemists sought to rationalize the course of chemical reactions, it became obvious that making reliable predictions requires knowledge of many factors, including the structure of molecules and ions. It is fortunate that a simple and reliable system exists today to predict geometry of chemical entities. The valence-shell electron-pair repulsion theory (VSEPR theory) is outstanding in its ability to predict the shapes of molecules and ions. The VSEPR theory coupled with hybridization concepts gives a clear understanding of atom interactions leading to particular shapes.

READY REFERENCE

Bond angle (8.1) Starting from a common atom, this angle is measured from one bond joining that atom to the next bond joining that atom.

Bond length (8.1) The distance along a straight line joining the nuclei of two bonded atoms.

Double bond (8.8) A stronger bond than a single bond, consisting of one sigma bond and one pi bond.

Hybridization (8.3) The mixing of atomic orbitals to make the same number of new orbitals with the same energy, having partial character of each of the original orbitals.

Pi (π) bond (8.7) A bond in which the electron density lies off the line adjoining the two nuclei.

Sigma (σ) bond (8.3) A bond in which the maximum electron density lies along the line adjoining the two nuclei.

SOLUTIONS TO CHAPTER 8 EXERCISES

MOLECULAR STRUCTURE

1. How many regions of high electron density are required around an atom to form a linear arrangement of these regions about the atom? A trigonal planar arrangement? A tetrahedral arrangement? A trigonal bipyramidal arrangement? An octahedral arrangement? What are the angles between the regions in each of these arrangements?

   Solution

   Consider x in the following drawing to represent a central atom. In order to achieve bonding, each line must be a region of high electron density.

   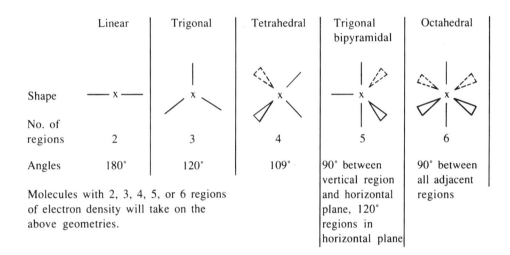

	Linear	Trigonal	Tetrahedral	Trigonal bipyramidal	Octahedral
Shape	—x—				
No. of regions	2	3	4	5	6
Angles	180°	120°	109°	90° between vertical region and horizontal plane, 120° regions in horizontal plane	90° between all adjacent regions

   Molecules with 2, 3, 4, 5, or 6 regions of electron density will take on the above geometries.

2. Predict the structure of each of the following molecules:

   (a) BH_3
   (b) BeH_2
   (c) PCl_5
   (d) SeF_6
   (e) SiH_4

   Solution

   The procedure is simple.

   1. Count the number of valence electrons.
   2. Distribute the atoms symmetrically about the unique or central atom.
   3a. Place two electrons between each atom.
   3b. Complete the octet of electrons on outermost atoms.
   4. If there are extra electrons still available, they are placed on the central atom.

   The following examples illustrate these steps.

   (a) Step 1: No. of valence electrons: B = 3, H = 1 each, total 6.

   Step 2:   H

                B

           H   H

104   Chapter 8   Molecular Structure and Hybridization

Step 3:

```
      H
      |
      B
    / \
   H   H
```

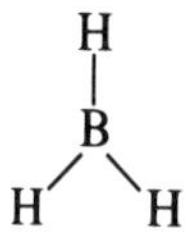

Molecule is trigonal.

(b) Step 1: No. of valence electrons: Be = 2, H = 1 each, total 4.

Step 2:   H   Be   H

Step 3:   H:Be:H

Molecule is linear.

(c) Step 1: No of valence electrons: P = 5, Cl = 7 each, total 40.

Steps 2 and 3:

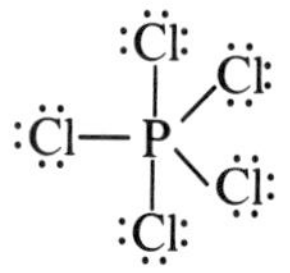

Total number of electrons is 40; molecule is a trigonal bypyramid.

(d) Step 1: No. of valence electrons: Se = 6, F = 7 each, total 48.

Steps 2 and 3a:

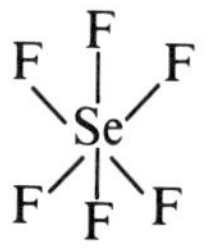

```
     F
   F | F
    \|/
    Se
    /|\
   F F F
```

Step 3b:

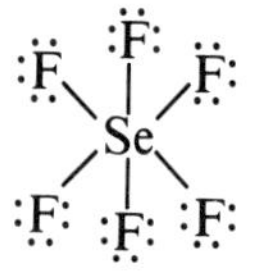

Total number of electrons used is 48; six bonds mean this is an octahedron.

(e) Step 1: No. of valence electrons: Si = 4, H = 1 each; total 8.

Steps 2 and 3:

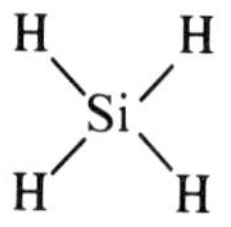

```
   H     H
    \   /
     Si
    /   \
   H     H
```

All eight electrons are used; molecule is a tetrahedron.

3. Predict the structure of each of the following ions:

(a) XeF_6^{2+}
(b) BF_4^-
(c) CH_3^+
(d) SF_5^+
(e) BH_2^+

Solution

(a) Step 1: No. of valence electrons: Xe = 8, F = 7 each; total would be 50, but because there is a 2+ charge on the ion, this indicates two electrons are missing; therefore, there are 50 − 2 = 48 electrons to account for.

Chapter 8 Molecular Structure and Hybridization

Steps 2 and 3a:
$$\left[\begin{array}{c} F \quad F \\ F\diagdown | \diagup F \\ Xe \\ F \diagup | \diagdown F \\ F \quad F \end{array}\right]^{2+}$$

Step 3b:
$$\left[\begin{array}{c} :\!\ddot{F}\!: \quad :\!\ddot{F}\!: \quad :\!\ddot{F}\!: \\ :\!\ddot{F}\!\diagdown | \diagup\!\ddot{F}\!: \\ Xe \\ :\!\ddot{F}\!\diagup | \diagdown\!\ddot{F}\!: \\ :\!\ddot{F}\!: \quad :\!\ddot{F}\!: \quad :\!\ddot{F}\!: \end{array}\right]^{2+}$$

All electrons are accounted for; ion is octahedral.

(b) Step 1: No. of valence electrons; B = 5, F = 7 each; total 33, but because of the negative charge there are 33 + 1 = 34 electrons to account for.

Steps 2 and 3a:
$$\left[\begin{array}{c} F \quad F \\ \diagdown \diagup \\ B \\ \diagup \diagdown \\ F \quad F \end{array}\right]^{-}$$

Step 3b:
$$\left[\begin{array}{c} :\!\ddot{F}\!: \quad :\!\ddot{F}\!: \\ \diagdown \diagup \\ B \\ \diagup \diagdown \\ :\!\ddot{F}\!: \quad :\!\ddot{F}\!: \end{array}\right]^{-}$$

At this stage 32 electrons have been placed, leaving 2 unaccounted. By Step 4 these are placed on the central atom.

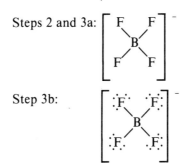

Since there are 5 electron density regions, the ion will be trigonal bipyramidal. The electrons will take a trigonal plane position since the repulsion will be less there: two 90° and two 120° repulsions are less than three 90° repulsions.

(c) Step 1: Total electrons, 6.
Steps 2 and 3:

$$\left[\begin{array}{c} H \quad H \\ \diagdown \diagup \\ C \\ | \\ H \end{array}\right]^{+}$$

Ion is trigonal.

(d) Step 1: Total electrons: 40.
Steps 2 and 3:

$$\left[\begin{array}{c}\ddot{\mathrm{F}}:\phantom{\ddot{\mathrm{F}}:}\ddot{\mathrm{F}}:\\ \ddot{\mathrm{F}}-\mathrm{S}\diagdown\ddot{\mathrm{F}}:\\ :\ddot{\mathrm{F}}:\phantom{\ddot{\mathrm{F}}:}\end{array}\right]^+$$

Ion is trigonal bipyramidal.

(e) Step 1: Total electrons: 4
Steps 2 and 3: $[\mathrm{H-B-H}]^+$
Ion is linear.

6. Predict the structure of each of the following molecules:

   (a) CO_2
   (b) IOF_5
   (c) F_2CO
   (d) SO_2F_2

Solution

(a) By Steps 1 to 3 we have 16 electrons to account for:

$:\ddot{\mathrm{O}}-\mathrm{C}-\ddot{\mathrm{O}}:$

In Lewis dot structure, electron pairs must be brought from the outer oxygens to form double-bonded carbon. In determining structure, the important thing is the number of regions of electron density. Existence of two regions implies a linear molecule.

(b) Step 1: No. of valence electrons: I = 7, O = 6, F = 7 each, total 48.

Steps 2 and 3a:

$$\begin{array}{c}F\\ F\diagdown|\diagup O\\ I\\ F\diagup|\diagdown F\\ F\end{array}$$

Without finishing Step 3b, it can be seen that all 36 electrons are used to complete the octets. Since there are six regions of electron density, the molecule is octahedral.

(c) Step 1: No. of valence electrons: total = 24.

Steps 2 and 3a:

$$\begin{array}{c}F\diagdown\\ \quad C-O\\ F\diagup\end{array}$$

Step 3 b:

$$\begin{array}{c}:\ddot{F}\diagdown\\ \phantom{:\ddot{F}}\quad C-\ddot{O}:\\ :\ddot{F}\diagup\end{array}$$

All electrons are used. To complete the octet, an electron pair can shift from the oxygen to the carbon; but three regions of high electron density are apparent, so that the structure is trigonal.

Chapter 8 Molecular Structure and Hybridization   107

(d) Step 1:   No. valence electrons: total = 32

Steps 2 and 3:

$$\begin{array}{c} :\ddot{\text{F}}: \\ | \\ :\ddot{\text{O}}-\text{S}-\ddot{\text{O}}: \\ | \\ :\ddot{\text{F}}: \end{array}$$

Molecule has four regions of electron density and is therefore tetrahedral.

8. Predict the geometry around the indicated atom or atoms:

   (a) The nitrogen atom in nitrous acid, HNO_2 (HONO).
   (b) The nitrogen atom and the two carbon atoms in glycine, the simplest amino acid. The skeleton structure of glycine is

   $$\begin{array}{c} \text{H} \quad\quad \ddot{\text{O}}. \\ | \quad\quad \nearrow \\ \text{H}-\text{C}-\text{C} \\ | \quad\quad \searrow \\ \text{NH}_2 \quad \ddot{\text{O}}-\text{H} \end{array}$$

   (e) The central oxygen atom in the ozone molecule, O_3.
   (h) The carbon atom in the carbonate ion, CO_3^{2-}; the bicarbonate ion, HCO_3^- ($HOCO_2^-$); and carbonic acid, H_2CO_3 [$(HO)_2CO$].

Solution

(a) Using the procedure from exercise 2, we count 18 electrons and assume that nitrogen is the central atom. Then the formula

$$\text{H}-\ddot{\text{O}}-\text{N}-\ddot{\text{O}}:$$

uses 16 electrons, leaving 2 to be placed on the central nitrogen. The resulting three regions of high electron density about the nitrogen means that the geometry is trigonal.

(b) The Lewis structure is

$$\begin{array}{c} \text{H} \quad\quad \ddot{\text{O}}. \\ | \quad\quad \nearrow \\ \text{H}-\text{C}-\text{C} \\ | \quad\quad \searrow \\ \text{H}-\ddot{\text{N}}: \quad \ddot{\text{O}}-\text{H} \\ | \\ \text{H} \end{array}$$

This structure uses all 30 electrons. There are three regions of high electron density around the carbon attached to O, so this carbon has trigonal symmetry. The other C and N each have four regions of high electron density about them and have tetrahedral geometry.

(e) The Lewis structure is

$$\overset{\ddot{\text{O}}}{\underset{:\ddot{\text{O}}\quad\ddot{\text{O}}:}{\diagup\,\diagdown}} \quad \longleftrightarrow \quad \overset{\ddot{\text{O}}}{\underset{:\ddot{\text{O}}\quad\ddot{\text{O}}:}{\diagup\,\diagdown}}$$

Three regions of high electron density about the central oxygen indicate a geometry that is trigonal.

(h) The Lewis structure for CO_3^{2-} is

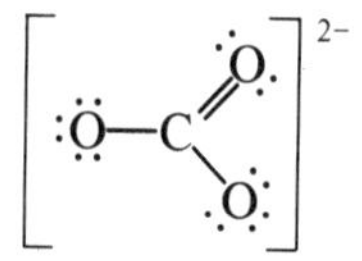

$$\left[\ddot{\mathrm{\ddot{O}}}\mathrm{-C}\begin{array}{c}\mathscr{O}\\ \ddot{\mathrm{\ddot{O}}}\mathrm{:}\end{array}\right]^{2-} \quad \text{(plus other resonance forms)}$$

Formation of bicarbonate ion occurs by addition of one proton, H^+, to an oxygen atom in carbonate ion. There is no change in the number of electrons and the geometry of the central atom remains unchanged. Carbonic acid is carbonate ion plus two protons attached to different oxygen atoms. Again, the number of electrons is unchanged and the geometry remains the same. All species have trigonal geometry around carbon.

HYBRID ORBITALS

11. What are the angles between the hybrid orbitals in each of the following sets: sp, sp^2, sp^3, sp^3d, sp^3d^2?

    Solution

sp	180°
sp^2	120°
sp^3	109°
sp^3d	120° and 180°
sp^3d^2	90°

12. Identify the hybridization of each carbon atom in each of the following molecules:

    (a) vinyl chloride, the compound used to make the plastic PVC (polyvinyl chloride), $H_2C\text{=}CHCl$

    (c) acetic acid, the compound that gives vinegar its acidic taste

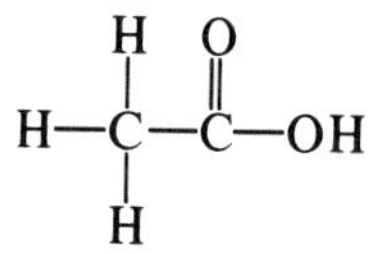

    $$\begin{array}{c} H \quad O \\ | \quad \| \\ H-C-C-OH \\ | \\ H \end{array}$$

    Solution

    (a) Both carbons are attached to three other atoms. Hybridization is sp^2.
    (c) Carbon bonded to O is sp^2, the other carbon is sp^3.

13. Identify the hybridization of the indicated atom in each of the following molecules:

    (a) S in SF_4
    (b) Sb in $SbCl_5$
    (c) B in BH_3
    (d) N in NO_2^+
    (e) S in SO_4^{2-}

    Solution

    (a) sp^3d
    (b) sp^3d
    (c) sp^2

(d) *sp*

(e) *sp*³

15. Show the distribution of electrons in the valence orbitals of the central atom in each of the following molecules or ions (1) when it is the free atom in the ground state, (2) following electron promotion, and (3) following bond formation:

   (a) SiH₄
   (c) AlH₃
   (e) IF₆⁺ (consider this ion as being formed from I⁺ and 6 F atoms)

Solution

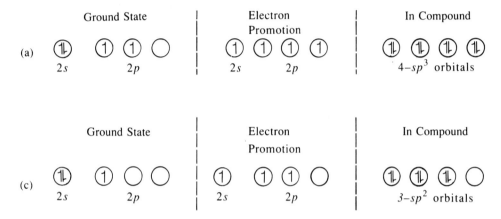

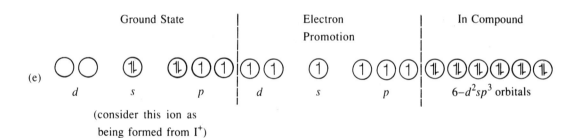

(consider this ion as being formed from I⁺)

18. Write Lewis structures for the molecules and ions listed below. Utilizing both the valence shell electron-pair repulsion theory and the concept of hybridization, and remembering that single bonds are σ bonds, double bonds are a σ plus a π bond, and triple bonds are a σ plus two π bonds, indicate the type of hybridization of each central atom in the listed molecules and ions.

   (a) BO_3^{3-}
   (c) ClNO

(e) SO_4^{2-}
(h) N_2O_4 (contains an N—N bond)
(j) BrCN

Solution

(a)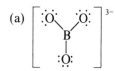

sp^2 hybrid orbitals based on B. (Double or π bonds do not form with boron compounds.)

(c) :Cl̈—N̈=Ö:

sp^2 hybrid orbitals on N but a p orbital is also used on N.

(e) 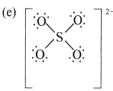 sp^3 hybridization

(h)

All bonds are sp^2 hybrids on N with one π bond associated with each N.

Other resonance forms are possible.

(j) :B̈r—C≡N: sp hybridization

19. Some forms of solid phosphorus(V) chloride contain PCl_4^+ and PCl_6^- ions. Upon heating, gaseous molecules of PCl_5 are formed. Describe the hybridization of the phosphorus atom in PCl_4^+, PCl_6^-, and PCl_5.

Solution

PCl_4^+, sp^3; PCl_6^-, sp^3d^2; PCl_5, sp^3d

20. Which of the following carbon-carbon bonds would be the strongest: (a) bond between two sp^3 hybridized carbon atoms, (b) a bond between two sp^2 hybridized carbon atoms, (c) a bond between two sp hybridized carbon atoms?

Solution

A bond between two sp hybridized carbon atoms would be the strongest, because it consists of a σ bond and two π bonds.

ADDITIONAL EXERCISES

23. XF_3 has a trigonal planar molecular structure. To which main group of the Periodic Table does X belong?

Solution

Since three bonds are formed and the structure is trigonal planar, there are no extra electrons on the central atom which would demand higher hybridization. Therefore, there must be only three valence electrons; this corresponds to an atom in Group IIIA.

25. If XCl_3 exhibits a dipole moment, X could be a member of either of two groups in the Period Table. Which ones?

Solution

If XCl_3 has a dipole moment, X cannot be in Group IIIA, because it would be symmetrical without a dipole moment. If X belonged to Group VA, the molecule would have tetrahedral hybridization but would be pyramidal. Therefore, it would have a dipole moment. Again, if X were from VIIA, the hybridization would be dsp^3 and the shape of the molecule is T-shaped. Therefore, it would have a dipole moment.

26. Elemental sulfur consists of covalently bonded, puckered rings of eight sulfur atoms, S_8. The S—S—S angles in the ring are 107.9°. The production of sulfuric acid, $SO_2(OH)_2$, from sulfur involves oxidation of sulfur to SO_2, then to SO_3, and finally reaction with water to give sulfuric acid. Trace the changes in the hybridization of the sulfur atoms during this sequence of reactions.

Solution

In S—S—S the hybridization is sp^3.
In SO_2 the hybridization is sp^2.
In SO_3 the hybridization is sp^2.
In H_2SO_4 the hybridization is sp^3.

28. One-half mol of a compound contains 0.5 mol of C atoms and 1.0 mol of S atoms. Identify the molecular structure of the compound and the hybridization of the carbon atom.

Solution

One mol of compound contains 1 mol of C atoms and 2 mol of S atoms. For CS_2,

$$\ddot{\underset{..}{S}}=C=\ddot{\underset{..}{S}}$$

the bonding hybridization is sp and the molecule is linear.

30. In 1986, 3.7 million tons of a compound with a molecular weight of 32 and containing 37.5% C, 12.5% H, and 50.0% O were prepared in the United States for use as a solvent and to manufacture other chemicals. Write the Lewis structure of this compound, describe the geometry around the C and O atoms in the molecule, and identify the hybridization of the C and O atoms.

Solution

Molar composition is

C: $\frac{375}{12.0} = 3.125$ H: $\frac{12.5}{1.0} = 12.5$ O: $\frac{50.0}{16.0} = 3.125$

Ratio is $C_1H_4O_1$; in this ratio the molecular weight is 32. The compound is CH_3OH.

$$\begin{array}{c} H \\ | \\ H-C-\ddot{O}-H \\ | \\ H \end{array}$$

There is sp^3 hybridization on C and on O. There is a tetrahedral arrangement about C, and the C—O—H bond is angular (109°).

31. A compound with a molecular weight of 30 contains 40% C, 6.7% H, and 53.3% O. Determine the molecular formula of the compound, write its Lewis structure, and identify the hybridization of the C atom.

Solution

C: $\dfrac{40}{12.0} = 3.33$ H: $\dfrac{6.7}{1.0} = 6.7$ O: $\dfrac{53.3}{16.0} = 3.33$

Ratio is $C_1H_2O_1$. With this as the lowest number of atoms in the compound, the molecular weight is 30, which corresponds to the compound formaldehyde,

$$\begin{array}{c} H \\ \diagdown \\ C=\ddot{O}\!: \\ \diagup \\ H \end{array}$$

The hybridization of C is sp^2.

9
Chemical Reactions and the Periodic Table

INTRODUCTION

For the previous eight chapters, your text has concentrated on stoichiometry and the development of bonding concepts based upon the nature of the electron arrangement in atoms. In this chapter we learn to make predictions of chemical reactivity and even of the compounds to be formed based upon the reactants' position within the periodic table, its characteristics with respect to oxidation-reduction powers, or behavior as an acid or base.

We now study classifications of chemical reactions to help in the prediction of reactions we meet for the first time. As an outgrowth of the study of the variation in properties within the periodic table, we investigate the variation of metallic and nonmetallic behavior of the representative elements. In a sense, once this chapter is mastered, most of the inorganic chemical reactions important to industry can be explained within a relatively simple framework.

READY REFERENCE

Brønsted acid and base (9.2) Brønsted acid is a compound that donates a hydrogen ion or proton (H^+) to another compound. A Brønsted base is a compound that accepts a hydrogen ion.

Catalyst (9.8) A catalyst is a substance that increases the rate of both forward and reverse reactions in a reversible reaction. It itself is unchanged in the reaction.

Combination reaction (9.3) A combination or addition reaction occurs when two or more substances combine to give another substance.

Decomposition reaction (9.3) A decomposition reaction occurs when one compound breaks down (decomposes) into two or more substances.

Electrolyte (9.2) An electrolyte is a compound that dissolves in water to produce solutions that contain ions.

Metathetical reactions (9.3) A metathetical reaction is a reaction in which two compounds exchange parts.

Nonelectrolytes (9.2) Compounds that do not ionize in aqueous solution are nonelectrolytes.

Oxidation (9.3)   A process in which a substance loses one or more electrons. Originally oxidation meant the union of a substance with oxygen.

Oxidation-reduction reaction (9.3)   In this type of reaction, simultaneous oxidation and reduction occur.

Reduction (9.4)   A process in which the substance gains one or more electrons. Originally reduction meant removal of oxygen.

Reversible reaction (9.2, 9.3)   A reversible reaction is a reaction that can proceed in either direction.

Salt (9.2)   The result (other than water) of an acid-base reaction; a salt is an ionic compound composed of positive ions (cations) and negative ions (anions).

Strong acids and bases (9.2)   A strong acid or base dissolves in water to give essentially 100% yield of protons or hydroxide ions respectively.

Strong electrolyte (9.2)   Strong electrolytes are ionic compounds and those covalent compounds such as HCl which give essentially a 100% yield of ions in aqueous solution.

Weak acid and base (9.2)   A weak acid or base ionizes in dilute solution giving a small percent of hydronium ion or hydroxide ion respectively.

Weak electrolytes (9.2)   Those compounds, such as weak acids and ammonia which give a low percentage yield of ions in water, are called weak electrolytes.

SOLUTIONS TO CHAPTER 9 EXERCISES

TYPES OF COMPOUNDS

2. Identify the cations and anions in the following compounds:

   (a) LiCl
   (d) Ga_2O_3
   (e) K_2SO_4
   (g) $Ca(H_2PO_4)_2$
   (i) $[PCl_4][PCl_6]$

   Solution

   Cations are positively charged ions; anions are negatively charged ions. Formulas are written with the positively charged species placed first in the formulas. The cations in the following compounds are listed.

	Cation	Anion
(a)	Li^+	Cl^-
(d)	Ga^{3+}	O^{2-}
(e)	K^+	SO_4^{2-}
(g)	Ca^{2+}	$H_2PO_4^-$
(i)	PCl_4^+	PCl_6^-

3. When the following compounds dissolve in water, do they give solutions of acids, bases, or salts?

   (a) KI
   (b) P$_4$O$_{10}$
   (c) H$_3$PO$_4$
   (e) NH$_4$Cl
   (g) NaClO$_4$

Solution

Solutions derived from the cations of strong bases and of strong acids form neutral solutions. If the anion derived from a weak acid is coupled with a cation from a strong base, the solution will be basic. Conversely, if the cation derived from a weak base is coupled with an anion from a strong acid the solution will be acidic.

(a) This is derived from a strong acid and a strong base respectively. The material is a salt and the solution is neutral.
(b) Nonmetallic oxides normally form acids in water. This forms H$_3$PO$_4$.
(c) This is a weak acid in water.
(e) This is a salt derived from the cation of a weak base and anion of a strong acid. It is acidic.
(g) This forms a neutral salt since its ions are derived from strong acids and bases.

4. Explain why sulfuric acid, H$_2$SO$_4$, a covalent molecule, dissolves in water and behaves as an electrolyte.

Solution

The bond between H and O in H$_2$SO$_4$ is polarized because of the difference in electronegativity. In water, the dipole interaction of water with the hydrogen on sulfuric acid facilitates the breaking of the O—H bond putting H$^+$ into solution. Sulfuric acid, therefore, reacts with water to form H$_3$O$^+$ and HSO$_4^-$.

5. Indicate whether each of the following compounds and elements dissolves in water to give a solution of a strong electrolyte, a weak electrolyte, or a non-electrolyte:

   (a) H$_2$S
   (b) In(NO$_3$)$_3$
   (c) Ba(OH)$_2$
   (f) NH$_4$NO$_3$
   (l) NH$_3$
   (m) Xe

Solution

(a) H$_2$S: weak electrolyte
(b) In(NO$_3$)$_3$: strong electrolyte
(c) Ba(OH)$_2$: strong electrolyte
(f) NH$_4$NO$_3$: strong electrolyte
(l) NH$_3$: in aqueous solution, a weak base forms a weak electrolyte.
(m) Xe: a nonelectrolyte.

6. Sodium hydrogen carbonate, also called sodium bicarbonate or baking soda, $NaHCO_3$, is sometimes kept in chemical laboratories because it can neutralize either acid or base spills. Explain how it can act either as a base or as an acid.

Solution

Hydrogen carbonate reacts in the following way in the presence of acid or base:

With acid: $NaHCO_3 + H^+ = Na^+ + H_2CO_3$
With base: $NaHCO_3 + OH^- = Na^+ + CO_3^{2-} + H_2O$

The effect is to neutralize either the acid or the base.

TYPES OF CHEMICAL REACTIONS

9. Classify the following as acid-base reactions or oxidation-reduction reactions:
   (a) $Na_2S + 2HCl \longrightarrow 2NaCl + H_2S$
   (b) $2Na + 2HCl \longrightarrow 2NaCl + H_2$
   (e) $K_3P + 2O_2 \longrightarrow K_3PO_4$

Solution

(a) acid-base
(b) oxidation-reduction
(e) oxidation-reduction

10. Identify the atoms that are oxidized and reduced, the change in oxidation number for each, and the oxidizing and reducing agents in each of the following equations:
    (a) $Mg + NiCl_2 \longrightarrow MgCl_2 + Ni$
    (c) $C_2H_4 + 3O_2 \longrightarrow 2CO_2 + 2H_2O$
    (e) $2K_2S_2O_3 + I_2 \longrightarrow K_2S_4O_6 + 2KI$

Solution

(a) Oxidation states—Mg: 0 to +2; Ni: +2 to 0. Mg is oxidized; Ni^{2+} is reduced. $NiCl_2$ is oxidizing reagent; Mg is reducing agent.
(c) Oxidation states—C: −2 to +4; O: 0 to −2. Carbon is oxidized; oxygen is reduced. C_2H_4 is the reducing agent; O_2 is the oxidizing agent.
(e) Oxidation states—S: +2 to $+\frac{5}{2}$; I_2: 0 to −1. Sulfur is oxidized; I_2 is reduced. $K_2S_2O_3$ is the reducing agent; I_2 is the oxidizing agent.

METALS AND NONMETALS

11. From the positions of their components in the Periodic Table, predict which member in each of the following pairs will:
    (a) conduct electricity: Cl_2 or Al
    (c) react as a base: $P(OH)_3$ or $Al(OH)_3$
    (e) reduce Sn^{2+} to Sn: Mg or Cl_2
    (h) react with water to give a solution of an acid: In_2O_3 or P_4O_6

Solution

(a) Cl_2 is a nonmetal and does not conduct. Al is metallic and will conduct.

(c) Metal hydroxides function as bases. Al is a metal and forms the base.

(e) Nonmetals, such as Cl_2, oxidize metals. Mg is a more active metal than Sn, so Mg should reduce Sn^{2+} to Sn.

(h) Nonmetallic oxides form acids when placed in water. P is a nonmetal so P_4O_6 should form an acid. In is a metal as seen from its position in the Periodic Table.

12. From the positions of the elements in the Periodic Table, predict which member in each of the following pairs will:

    (a) reduce S_8: Cl_2 or Mg
    (c) oxidize Si: Ca or Cl_2
    (e) reduce SeO_2: Ca or Cl_2

    ## Solution

    (a) Mg: Metals are good reducing agents and Mg is a metal.
    (c) Cl_2: Nonmetallic elements are oxidizing agents, and Cl_2 is nonmetallic.
    (e) Ca: The active metal will reduce the nonmetallic oxide.

13. From the positions of the elements in the Periodic Table, predict which member of each of the following will be:

    (a) more basic: CsOH or BrOH
    (c) more acidic: H_2SO_3 or $HClO_3$
    (e) more easily reduced: F_2 or I_2
    (g) more easily oxidized: Ca or S

    ## Solution

    (a) CsOH: Hydroxides of metals are basic; hydroxides of nonmetals are acidic.
    (c) $HClO_3$: Other things being equal, the more nonmetallic element, Cl in this case, will form the stronger acid.
    (e) F_2: The stronger nonmetal is more easily reduced.
    (g) Ca: Metals are more easily oxidized than nonmetals.

15. A sample of an element that conducts electricity is broken into small pieces by cracking it with a hammer. These pieces react with chlorine to form a volatile liquid chloride. Could the element be aluminum? Silicon? Iodine? Explain your answers.

    ## Solution

    Iodine normally would not conduct electricity since it is a nonmetal. Aluminum and silicon can conduct. Aluminum has a low electronegativity and would tend to form an ionic compound which would not volatilize. Silicon, with a higher electronegativity, would tend to form a covalent compound which could volatilize. Therefore, silicon would be the element present.

COMMON OXIDATION NUMBERS

18. Explain why the most negative oxidation number of an atom of Group VIA is equal to the number of electrons required to fill its valence shell.

    Solution

    In Group VIA there are six valence electrons. Only two more electrons can be accommodated in the valence shell, and two is the maximum negative oxidation state. No more orbitals are available for occupancy.

19. Which of the following elements will exhibit a positive oxidation number when combined with phosphorus?

    (a) Mg
    (b) F
    (c) Al
    (d) O

    Solution

    (a) Mg, positive
    (b) F, negative
    (c) Al, positive
    (d) O, negative

23. Without reference to Table 9.6, identify those representative metals that may display two or more common oxidation numbers in their compounds.

    Solution

    Representative metals with two or more oxidation numbers are: Sn: +4, +2; Sb: +5, +3, −3; Hg: +2, +1; Tl: +3, +1; Pb: +4, +2; Bi: +5, +3

THE ACTIVITY SERIES

25. One of the elements in the activity series (Table 9.7) will not produce hydrogen when it reacts with steam, but it will form hydrogen when it reacts with hydrochloric acid. Reaction of this element with a solution of $Ni(NO_3)_2$ produces $Ni(s)$. What is this element?

    Solution

    Problems of this type are best solved with reference to Table 9.7. There we find Co, Ni, Sn, Pb, which can react with HCl to liberate hydrogen but cannot react with steam to liberate hydrogen. For a reaction to occur, the substance must lie above the metal atom of the material reacting. Therefore, the element must be Co.

27. With the aid of the activity (electromotive) series (Table 9.7), predict whether or not the following reactions will take place:

    (b) $Al_2O_3 + 3H_2 \longrightarrow 2Al + 3H_2O$
    (c) $2Au + Fe^{2+} \longrightarrow 2Au^+ + Fe$
    (f) $K_2O + H_2 \longrightarrow 2K + H_2O$
    (h) $3Pd + 2Au^{3+} \longrightarrow 2Au + 3Pd^{2+}$

Chapter 9 Chemical Reactions and the Periodic Table 119

Solution

(b) No. Al lies in a group of elements whose oxides are not reduced by hydrogen.

(c) No. Au is less active than Fe, as indicated by its position at the bottom of the activity series. Consequently we would not expect the reaction to occur.

(f) No. Potassium, like aluminum, lies in a group of elements whose oxides are not reduced by hydrogen.

(h) Yes. Palladium (Pd) lies above gold in the activity series so that Pd will displace Au ions from solution. In other words, any metal in the series will displace any other below it from a dilute aqueous solution of a soluble compound.

29. What mass of copper can be recovered from a solution of copper(II) sulfate by addition of 2.11 kg of aluminum?

Solution

Aluminum is the limiting reagent. The reaction is

$$2Al + 3CuSO_4 \longrightarrow 3Cu + Al_2(SO_4)_3$$

From the stoichiometry, 2 mol Al $\longrightarrow$ 3 mol Cu.

$$g\ Cu = 2110\ g\ Al \times \frac{1\ mol\ Al}{26.98154\ g} \times \frac{3\ mol\ Cu}{2\ mol\ Al} \times \frac{63.546\ g}{1\ mol\ Cu}$$

$$= 7450\ g = 7.45\ kg$$

INDUSTRIAL CHEMICALS

30. The following reactions are all similar to those of the industrial chemicals described in Section 9.9. Complete and balance the equations for these reactions.

(a) reaction of a weak base and a strong acid

$$NH_3 + HClO_4 \longrightarrow$$

(c) preparation of a soluble source of calcium and phosphorus used in animal feeds

$$Ca_3(PO_4)_2 + H_3PO_4 \longrightarrow$$

(e) preparation of a soluble silver salt for silver plating

$$Ag_2CO_3 + HNO_3 \longrightarrow$$

(g) hardening of plaster containing slaked lime

$$Ca(OH)_2 + CO_2 \longrightarrow$$

(j) the reaction of baking powder that produces carbon dioxide gas and causes bread to rise

$$NaHCO_3 + NaH_2PO_4 \longrightarrow$$

(l) preparation of strontium hydroxide by electrolysis of a solution of strontium chloride

$$SrCl_2(aq) + H_2O(l) \xrightarrow{\text{Electrolysis}}$$

120   Chapter 9   Chemical Reactions and the Periodic Table

Solution

(a) $NH_3 + HClO_4 \longrightarrow NH_4^+ + ClO_4^-$

(c) $Ca_3(PO_4)_2 + 4H_3PO_4 \longrightarrow 3Ca(H_2PO_4)_2$. Calcium phosphate is too insoluble to provide soluble phosphorus to plants and animals. Calcium dihydrogen phosphate is soluble.

(e) $Ag_2CO_3 + 2HNO_3 \longrightarrow 2Ag^+ + 2NO_3^- + H_2O + CO_2$

(g) $Ca(OH)_2 + CO_2 \longrightarrow CaCO_3 + H_2O$

(j) $NaHCO_3 + NaH_2PO_4 \longrightarrow 2Na^+ + HPO_4^{2-} + CO_2 + H_2O$. The underlying equations are

$$H_2PO_4^- + H_2O \longrightarrow H_3O^+ + HPO_4^{2-}$$

$$HCO_3^- + H_3O^+ \longrightarrow 2H_2O + CO_2(g)$$

(l) $SrCl_2(aq) + 2H_2O(l) \xrightarrow{\text{Electrolysis}} Sr(OH)_2(aq) + Cl_2(g) + H_2(g)$

PREDICTION OF REACTION PRODUCTS

31. Complete and balance the equations for the following oxidation-reduction reactions. In some cases there may be more than one correct answer depending on the amounts of reactant used, as illustrated in Examples 9.5 and 9.6.

    (a) $Mg + O_2 \longrightarrow$

    (c) $K + S_8 \longrightarrow$

    (e) $P_4 + O_2 \longrightarrow$

    (g) $P_4O_6 + O_2 \longrightarrow$

    (k) $Sr(s) + CH_3CO_2H(aq) \longrightarrow$

    (m) $Mg(s) + PbO(s) \xrightarrow{\Delta}$

    (o) $AlP + O_2 \xrightarrow{\Delta}$

Solution

(a) From the fact that Mg is a metal in Group IIA, it tends to lose electrons and form Mg^{2+}; O is a nonmetal in Group VIA and tends to gain two electrons.

$$2Mg + O_2 \longrightarrow 2MgO$$

(c) The metal K forms K^+ ions and the nonmetal S is in Group VIA and will form S^{2-} ions in oxidation-reduction. Ionic charge balance in $(K^+)_2S^{2-}$ gives

$$16K + S_8 \longrightarrow 8K_2S$$

(e) Nonmetals oxidize less electronegative nonmetals. $P_4 + 3O_2 \longrightarrow P_4O_6$ or $P_4 + 5O_2 \longrightarrow P_4O_{10}$

(g) Nonmetals oxidize less electronegative nonmetals. $P_4O_6 + 2O_2 \longrightarrow P_4O_{10}$

(k) Metals react with acids, forming H_2 and salts. Strontium is in Group IIA, and it loses electrons to become Sr^{2+}. The hydrogen ion gains an electron to form hydrogen gas

$$Sr(s) + 2CH_3CO_2H(aq) \longrightarrow Sr^{2+}(aq) + 2CH_3CO_2^-(aq) + H_2(g)$$

(m) This is another oxidation-reduction reaction that can be understood from the activity series.

$$Mg(s) + PbO(s) \longrightarrow MgO(s) + Pb(s)$$

Chapter 9 Chemical Reactions and the Periodic Table

(o) Nonmetals oxidize less electronegative nonmetals. $4AlP + 6O_2 \longrightarrow 2Al_2O_3 + P_4O_6$

33. Military lasers use the very intense light produced when fluorine combines explosively with hydrogen. What is the balanced equation for this reaction?

Solution

$$H_2(g) + F_2(g) \longrightarrow 2HF(g)$$

36. Dow Chemical Company for many years prepared bromine from the sodium bromide, NaBr, in Michigan brine by treating the brine with chlorine gas. Write a balanced equation for the process.

Solution

$$2NaBr(aq) + Cl_2(g) \longrightarrow 2Na^+(aq) + 2Cl^-(aq) + Br_2(aq)$$

37. Metallic copper is produced by roasting ore containing Cu_2S in air, followed by the reduction of the product with excess carbon. The gas escaping from the reduction process is burned to provide part of the heat for the roasting process. Write the balanced equation for each of these three steps.

Solution

$$2Cu_2S(s) + 3O_2(g) \longrightarrow 2Cu_2O(s) + 2SO_2(g)$$

$$Cu_2O(s) + C(s) \longrightarrow 2Cu(s) + CO(g)$$

$$2CO(g) + O_2 \longrightarrow 2CO_2(g)$$

38. Complete and balance the equations for the following acid-base reactions. If the reactions are run in water as a solvent, write the reactants and products as solvated ions. In some cases there may be more than one correct answer, depending on the amounts of reactants used.

(a) $HBr(g) + In_2O_3(s) \longrightarrow$
(c) $Al(OH)_3(s) + HNO_3(aq) \longrightarrow$
(e) $Li_2O(s) + CH_3CO_2H(l) \longrightarrow$

Solution

(a) $6HBr(g) + In_2O_3(s) \longrightarrow 2InBr_3(s) + 3H_2O(l)$
(c) $Al(OH)_3(s) + 3HNO_3(l) \longrightarrow Al^{3+}(aq) + 3NO_3^-(aq) + 3H_2O(l)$
(e) $Li_2O(s) + 2CH_3CO_2H(l) \longrightarrow 2Li^+(aq) + 2CH_3CO_2^-(aq) + H_2O(l)$

39. The following salts contain anions of weak acids. Write a balanced equation for the reaction of a solution of each with an excess of the indicated acid.

(a) $LiCH_3CO_2 + H_3PO_4 \longrightarrow$
(c) $NaCN + HCl \longrightarrow$
(e) $CaHPO_4 + HClO_4 \longrightarrow$
(g) $KNO_2 + HBr \longrightarrow$

Solution

In each of the following, the principle is that the stronger acid will release the weaker acid from its salt.

(a) $LiCH_3CO_2 + H_3PO_4 \longrightarrow LiH_2PO_4 + CH_3CO_2H$
(c) $NaCN + HCl \longrightarrow HCN + NaCl$
(e) $CaHPO_4 + 2HClO_4 \longrightarrow H_3PO_4 + Ca(ClO_4)_2$
(g) $KNO_2 + HBr \longrightarrow HNO_2 + KBr$

40. Complete and balance the equations of the following reactions, each of which is used to remove hydrogen sulfide from natural gas:

   (a) $Ca(OH)_2(s) + H_2S(g) \longrightarrow$
   (c) $Na_2CO_3(aq) + H_2S(g) \longrightarrow$

   Solution

   (a) $Ca(OH)_2(s) + H_2S(g) \longrightarrow CaS(s) + 2H_2O(l)$
   (c) $Na_2CO_3(aq) + H_2S(g) \longrightarrow Na_2S(s) + H_2CO_3(aq)$

41. The medical laboratory test for cyanide ion, CN^-, involves separation of the cyanide ion from a blood, urine, or tissue sample by the addition of sulfuric acid. The gaseous product is absorbed in a sodium hydroxide solution and then analyzed. Write balanced equations that describe the separation and absorption.

   Solution

   Separation: $CN^-(aq) + H_2SO_4(aq) \longrightarrow HCN(g) + HSO_4^-(aq)$

   Absorption: $HCN(g) + NaOH(aq) \longrightarrow NaCN(aq) + H_2O(aq)$

42. Calcium propionate is sometimes added to bread to retard spoilage. This compound can be prepared by the reaction of calcium carbonate, $CaCO_3$, with propionic acid, $C_2H_5CO_2H$, which has properties similar to those of acetic acid. Write the balanced equation for the formation of calcium propionate.

   Solution

   $CaCO_3 + 2C_2H_5CO_2H \longrightarrow Ca(C_2H_5CO_2)_2 + CO_2 + H_2O$

ADDITIONAL EXERCISES

45. Complete and balance the following equations. If the reactions are run in water as a solvent, write the reactants and products as solvated ions. In some cases there may be more than one correct answer, depending on the amounts of reactants used.

   (a) $Ca(OH)_2(s) + HCl(g) \longrightarrow$
   (c) $Ca + S_8 \longrightarrow$
   (e) $MgH_2 + H_2O \longrightarrow$
   (g) $Si + F_2 \longrightarrow$
   (i) $Ca(CH_3CO_2)_2 + H_2SO_4 \xrightarrow{H_2O}$
   (k) $H_2 + Br_2 \longrightarrow$
   (m) $NH_3(g) + H_2S(g) \longrightarrow$
   (o) $Al(s) + HClO_4(aq) \longrightarrow$
   (q) $NaF + HNO_3 \xrightarrow{H_2O}$

Solution

(a) $Ca(OH)_2(s) + 2HCl(g) \longrightarrow CaCl_2(aq) + 2H_2O(l)$

(c) $8Ca + S_8 \longrightarrow 8CaS$

(e) The hydride ion, H^-, is very strong base and picks up hydrogen ion from water.

$$MgH_2 + 2H_2O \longrightarrow Mg(OH)_2(aq) + 2H_2(g)$$

(g) $Si + 2F_2 \longrightarrow SiF_4$

(i) $Ca(CH_3CO_2)_2 + H_2SO_4 \xrightarrow{H_2O} CaSO_4 + 2CH_3CO_2H$

(k) $H_2(g) + Br_2(g) \longrightarrow 2HBr(g)$

(m) $2NH_3(g) + H_2S(g) \longrightarrow (NH_4)_2S(s)$

(o) $2Al(s) + 6HClO_4(aq) \longrightarrow 2Al(ClO_4)_3(aq) + 3H_2(g)$

(q) $Na^+(aq) + F^-(aq) + H^+(aq) + NO_3^-(aq) \longrightarrow Na^+(aq) + NO_3^-(aq) + HF(aq)$

46. Hydrogen sulfide, H_2S, is removed from natural gas by passing the raw gas through a solution of ethanolamine, $HOCH_2CH_2NH_2$, whose behavior is similar to that of NH_3 (which can be viewed as HNH_2) in its acid-base reactions. After the solution is saturated, the H_2S can be recovered by heating the solution to reverse the reaction. Part of the recovered H_2S is burned with air, and the product is allowed to react with the remaining H_2S to produce sulfur, S_8, which is easier to ship than H_2S. Write the reaction of ethanolamine with hydrogen sulfide, showing the Lewis structures of the reactants and products. Write the balanced equations for the reactions that lead to the formation of sulfur from hydrogen sulfide.

Solution

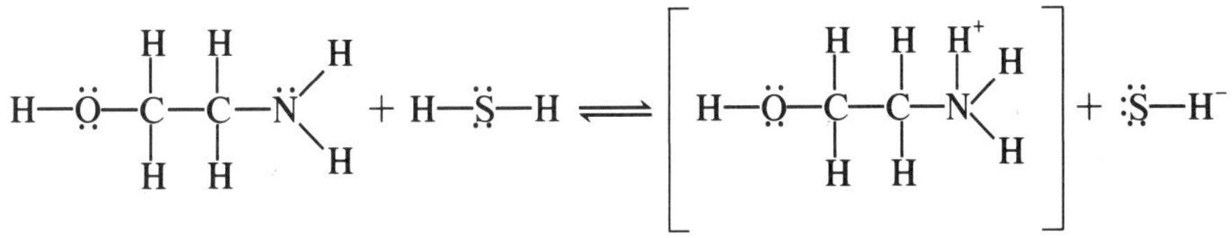

[Lewis structure: $H-\ddot{O}-C(H)(H)-C(H)(H)-N(H)(H)H + H-\ddot{S}-H \rightleftharpoons [H-\ddot{O}-C(H)(H)-C(H)(H)-N(H)(H)(H)]^+ + \ddot{S}-H^-$]

$2H_2S(g) + 3O_2(g) \longrightarrow 2H_2O(g) + 2SO_2(g)$

$SO_2(g) + 2H_2S(g) \xrightarrow{\Delta} 3S + 2H_2O$

51. Indicate whether water acts as an acid, a base, an oxidizing agent, or a reducing agent in each of the following reactions. Write an equation for each reaction.

(a) Na is added to water
(c) Na_2O is added to water
(e) NaF is added to water

Solution

(a) Water acts as an oxidizing agent. H_2 and NaOH are formed.

$$2Na + 2H_2O \longrightarrow 2NaOH + H_2$$

(c) Water acts as an acid. NaOH is formed.

$$Na_2O + H_2O \longrightarrow 2NaOH$$

(e) Water acts as an acid. HF is formed.

$$F^-(aq) + H_2O(l) \longrightarrow HF(aq) + OH^-(aq)$$

52. What is the percent by mass of phosphorus in the compound formed by the reaction of phosphorus with magnesium?

Solution

The reaction is

$$6Mg + P_4 \longrightarrow 2Mg_3P_2$$

$$\text{percent P} = \frac{2 \times \text{at. wt P}}{\text{mol. wt Mg}_3\text{P}_2} \times 100 = \frac{2 \times 30.97376 \text{ g} \times 100}{3 \times 24.305 \text{ g} + 2 \times 30.97376 \text{ g}}$$

$$= \frac{61.94752}{134.86252} \times 100 = 45.93\%$$

54. What is the hydroxide ion concentration in a solution formed by dissolution of 1.00 g of calcium metal in enough water to form 250 mL of the solution?

Solution

The reaction is

$$Ca(s) + H_2O(l) \longrightarrow Ca(OH)_2(aq) + H_2(g)$$

$$\text{mol OH}^- = 1.00 \text{ g Ca} \times \frac{1 \text{ mol Ca}}{40.08 \text{ g Ca}} \times \frac{2 \text{ mol OH}^-}{1 \text{ mol Ca}} = 0.0499 \text{ mol}$$

$$\text{Molarity of OH}^- = [\text{OH}^-] = \frac{0.0499 \text{ mol Ca}}{0.250 \text{ L}} = 0.200 \text{ M}$$

58. How does the hybridization of the sulfur atom change when SO_2 is dissolved in water?

Solution

The reaction is $SO_2 + H_2O \longrightarrow H_2SO_3$. In SO_2 the hybridization is sp^2; in H_2SO_3 the hybridization is sp^3.

10
The Gaseous State and the Kinetic-Molecular Theory

INTRODUCTION

The study of gases has occupied a very important position in human thought and imagination from the earliest recorded times. For example, the Greek Anaximenes (d. ca. 528 B.C.) thought that air constituted the primary matter from which all other substances came into being. In this context, water could be formed by the condensation of air. Although today's perspective is different, the concept then was one of different states of matter, the gaseous state being the most important.

Although the Greeks did not develop the scientific method, experimentation with gases did occur before the seventeenth century. Hero of Alexandria (ca. A.D. 62–150) described many mechanisms operated by steam, even a steam engine; the pressure of gases was the motive power in all these systems. Hero, therefore, had a very clear idea of the nature of gases, and in many respects anticipated the kinetic theory of gases. His ideas on combustion were very close to those of Lavoisier, who lived at the beginning of modern chemistry. However, scientists such as Hero did not anticipate the discoveries of the eighteenth century because they tended to employ experiments only to demonstrate preconceived hypotheses.

The ability to study gases *scientifically* became possible when, in 1643, the Italian physicist Evangelista Torricelli invented the mercury barometer, which was the first means available to measure pressure. Although the common unit of measurement of pressure is the millimeter of mercury (mm Hg), or *torr,* named after Torricelli; the SI unit is the pascal.

The British chemist Robert Boyle was quick to exploit this advance and in 1662 showed that the pressure of a gas and its volume were inversely related. A French physicist, Amontons, discovered the relation between the temperature and the volume of gas almost immediately after Boyle's work. But it was a century later before Charles, and still much later before the French chemist Joseph Gay-Lussac, independently showed the same temperature-volume dependence.

A theoretical explanation of the behavior of gases was developed in 1738 by the Swiss mathematician Daniel Bernoulli who laid the foundation for the modern kinetic-molecular theory of gases.

READY REFERENCE

Avogadro's law (10.7)   Equal volumes of all gases, measured under the same conditions of temperature and pressure, contain the same number of molecules. For 0°C and 1.00 atm (STP, standard temperature and pressure), the volume of one mole of an ideal gas is 22.4 liters.

Boyle's law (10.3, 10.15)   The volume of a given mass of gas held at constant temperature is inversely proportional to its pressure. Mathematically this can be expressed as

$$V = \text{constant} \times \frac{1}{P} = k \times \frac{1}{P} \quad \text{or} \quad PV = \text{constant} = k$$

Charles's law (10.4, 10.15)   The volume of a given mass of gas is directly proportional to its Kelvin temperature, holding pressure constant. This may be stated in equation form:

$$V = \text{constant} \times T = k \times T \quad \text{or} \quad \frac{V}{T} = \text{constant} = k$$

where k is a constant different from the one in Boyle's law.

Dalton's law of partial pressure (10.1, 10.15)   The total pressure of a mixture of gases equals the sum of the partial pressures of the component gases. If the pressures of the individual gases A, B, C, and so forth are designated P_A, P_B, P_C, and so forth, then the total pressure P_T may be written as

$$P_T = P_A + P_B + P_C + \cdots$$

This law is useful in calculating the pressure of a gas collected over water. The vapor pressure of the water is fixed for a particular temperature. Consequently, if the vapor pressure of water at the temperature of the vessel is subtracted from the total pressure, the pressure of the gas is determined.

Gay-Lussac's law of combining volumes (10.6, 10.7)   After studying a large number of reactions, Gay-Lussac generalized that in reactions involving gases at constant temperature and pressure, the volumes of the gases can be expressed as a ratio of small whole numbers.

Graham's law of diffusion of gases (10.13, 10.15)   This law states that the rates of diffusion of different gases are inversely proportional to the square roots of their densities or molecular weights. Mathematically,

$$\frac{\text{Rate of diffusion of gas A}}{\text{Rate of diffusion of gas B}} = \frac{\sqrt{\text{density B}}}{\sqrt{\text{density A}}} = \frac{\sqrt{\text{mol. wt B}}}{\sqrt{\text{mol. wt A}}}$$

Note that the time required for the diffusion of a gas is inversely proportional to its rate. We can write, assuming equal moles of A and B,

$$\frac{\text{Time for diffusion of gas A}}{\text{Time for diffusion of gas B}} = \frac{\sqrt{\text{density A}}}{\sqrt{\text{density B}}} = \frac{\sqrt{\text{mol. wt A}}}{\sqrt{\text{mol. wt B}}}$$

Ideal gas equation (10.8, 10.17)   This equation relates the variables P, V, T of a gas to the number of moles, n, of gas present. The equation is

$$PV = nRT$$

Chapter 10 The Gaseous State and the Kinetic-Molecular Theory

where R is the universal gas constant and can be expressed in several different units. Expressions showing common units are

$$P\,(\text{atm}) \times V\,(\text{L}) = n\,(\text{mol}) \times R\left(\frac{\text{L atm}}{\text{mol K}}\right) \times T\,(\text{K})$$

$$P\,(\text{Pa}) \times V\,(\text{L}) = n\,(\text{mol}) \times R\left(\frac{\text{L Pa}}{\text{mol K}}\right) \times T\,(\text{K})$$

where R has the values

0.08206 L atm/mol K   or   8314 L Pa/mol K

Kinetic-molecular theory (10.14)   In this theory an ideal gas consists of separate particles, either atoms or molecules. The volume occupied by the individual particles is small compared to the total volume of the gas. As a result, the particles in the gaseous state are relatively far apart and have no attraction for one another.

A second premise concerning the ideal gas is that the gas particles are in continuous motion and travel in straight lines with various speeds. As particles collide with each other, no net loss in average kinetic energy occurs; that is, collisions among particles occur as though the particles are perfectly elastic bodies.

Finally, the average kinetic energy ($\frac{1}{2}mu^2$) of all molecules, even of different gases, is the same at the same temperature.

These premises explain the gas laws in a qualitative way and can be shown to predict the ideal gas law, $PV = nRT$, a mathematical interpretation of the behavior or characteristics of gases.

Kelvin temperature (10.5)   A more fundamental temperature scale than the Celsius or Fahrenheit scale, the Kelvin scale (K) is independent of the working fluid. Its relation to the Celsius scale is

$$K = {}^\circ C + 273.15$$

The interval between degrees is the same on the Kelvin and Celsius scales. All temperatures involving the gas laws should be expressed on the Kelvin scale before an attempt to work numerical problems involving multiplication and division.

Pressure (10.2)   Gas pressure provides the most easily measured gas property. Pressure is the force exerted upon a unit area of surface. Common units for pressure are dynes per cm^2 and millimeters of mercury (mm Hg) or torr (= 1 mm Hg). Pressure is also expressed in terms of the standard atmosphere (atm), which equals 760 mm Hg. The SI pressure unit, which will gain more recognition with time, is the pascal (Pa). The pascal is expressed in terms of kg s^{-2} m^{-1}. The relations necessary for interconversion from one unit to another are

$$1 \text{ atm} = 760 \text{ mm Hg} = 760 \text{ torr} = 101{,}325 \text{ Pa}$$

SOLUTIONS TO CHAPTER 10 EXERCISES

PRESSURE

1. A typical barometric pressure in Denver, Colorado, is 615 torr. What is this pressure in atmospheres and kilopascals?

    Solution

    Convert 615 torr to atmospheres:

Identity: 760 torr = 1 atm = 101.3 kPa

$$P\,(\text{atm}) = 615\,\text{torr} \times \frac{1\,\text{atm}}{760\,\text{torr}} = 0.809\,\text{atm}$$

Convert 615 torr to kilopascals:

$$P\,(\text{kPa}) = 615\,\text{torr} \times \frac{101.3\,\text{kPa}}{760\,\text{torr}} = 82.0\,\text{kPa}$$

3. A medical laboratory catalog describes the pressure in a cylinder of a gas as 14.82 MPa (1 MPa = 10^6 Pa). What is the pressure of this gas in atmospheres and torr?

Solution

Convert 14.82 MPa to atm.

$$P\,(\text{atm}) = 14.82\,\text{MPa} \times 1000\,\frac{\text{kPa}}{\text{MPa}} \times \frac{1\,\text{atm}}{101.3\,\text{kPa}} = 146.3\,\text{atm}$$

$$P\,(\text{torr}) = 146.3\,\text{atm} \times \frac{760\,\text{torr}}{1\,\text{atm}} = 1.112 \times 10^5\,\text{torr}$$

5. A biochemist adds carbon dioxide, CO_2, to an evacuated bulb like the one shown in the figure at 19.8°C and stops when the difference in the heights of the mercury columns, h, is 4.75 cm. What is the pressure of CO_2 in the bulb in atmospheres, torr, and kilopascals?

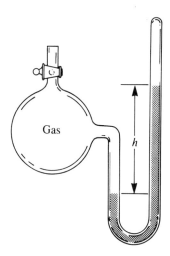

Solution

The pressure exerted by CO_2 equals 4.75 cm Hg or 47.5 mm Hg. Converting 47.5 mm Hg to the desired units follows.

Identities: 1 atm = 760 torr = 760 mm Hg = 101.3 kPa

$$P\,(\text{atm}) = 47.5\,\text{mm Hg} \times \frac{1\,\text{atm}}{760\,\text{mm Hg}} = 0.0625\,\text{atm}$$

$$P\,(\text{torr}) = 47.5\,\text{mm Hg} = 47.5\,\text{torr}$$

$$P\,(\text{kPa}) = 47.5\,\text{mm Hg} \times \frac{101.3\,\text{kPa}}{760\,\text{mm Hg}} = 6.33\,\text{kPa}$$

Chapter 10 The Gaseous State and the Kinetic-Molecular Theory   129

THE GAS LAWS

9. The volume of a sample of carbon monoxide, CO, is 1.40 L at 2.25 atm and 467 K. What volume will it occupy at 4.50 atm and 467 K?

   Solution

   Begin by listing P, V, and T data in tabular form. This procedure helps to identify all factors and to quickly determine the unknown.

   V_1: 1.40 L P_1: 2.25 atm T_1: 467 K
   V_2: ? P_2: 4.50 atm T_2: 467 K

   Since $T_1 = T_2$, the change in pressure will produce a change in volume according to Boyle's law.

   $$P_1V_1 = k = P_2V_2$$

   or

   $$V_2 = \frac{P_1V_1}{P_2} = \frac{1.40 \text{ L} \times 2.25 \text{ atm}}{4.50 \text{ atm}} = 0.700 \text{ L}$$

12. A sample of nitrogen gas, N_2, occupies a volume of 1.94 L at a pressure of 98.74 kPa. What volume will it occupy at the same temperature if the pressure is 452 torr?

    Solution

    List all the V and P data. We can disregard T, which remains constant.

    V_1: 1.94 L P_1: 98.74 kPa
    V_2: ? P_2: 452 torr

    Convert P_1 and P_2 to equivalent units:

    101.3 kPa = 760 torr

    $$P_1 \text{ (torr)} = 98.74 \text{ kPa} \times \frac{760 \text{ torr}}{101.3 \text{ kPa}} = 740.8 \text{ torr}$$

    The increase in pressure will reduce the volume according to Boyle's law.

    $$V_1P_1 = k = V_2P_2 \quad \text{so} \quad V_2 = \frac{V_1P_1}{P_2} = \frac{1.94 \text{ L} \times 740.8 \text{ torr}}{452 \text{ torr}} = 3.18 \text{ L}$$

14. A cylinder of oxygen for medical use contains 35.4 L of oxygen at a pressure of 150 atm. What is the volume of this oxygen, in liters, at 0.98 atm of pressure and the same temperature?

    Solution

    List all V and P data. We can disregard T, which remains constant.

    V_1: 35.4 L P_1: 150 atm
    V_2: ? P_2: 0.98 atm

    $$V_1P_1 = k = V_2P_2$$

    so

    $$V_2 = 35.4 \text{ L} \times \frac{150 \text{ atm}}{0.98 \text{ atm}} = 5.4 \times 10^3 \text{ L}$$

17. What is the volume of a sample of ethane, C_2H_6, at 467 K and 2.25 atm if it occupies 1.405 L at 300 K and 2.25 atm?

Solution

List all V, P, and T data.

V_1: 1.405 L P_1: 2.25 atm T_1: 300 K
V_2: ? P_2: 2.25 atm T_2: 467 K

Since there is no change in pressure, the increase in temperature will produce an increase in volume according to Charles's law.

$$\frac{V_1}{T_1} = k = \frac{V_2}{T_2}$$

so

$$V_2 = \frac{V_1 T_2}{T_1} = \frac{1.405 \text{ L} \times 467 \text{ K}}{300 \text{ K}} = 2.19 \text{ L}$$

19. The gas in a 1.00-L bottle at 25°C can be put into a 1-qt (0.946-L) bottle at the same pressure if the temperature is reduced. What temperature is required?

Solution

List all V and T data. We can disregard P, which remains constant.

V_1: 1.00 L T_1: 25°C = 298 K
V_2: 0.946 L T_2: ?

To maintain a constant pressure while reducing the volume requires a reduction in temperature. Converting T_1 to kelvins and applying Charles's law gives:

$$\frac{V_1}{T_1} = k = \frac{V_2}{T_2}$$

or

$$T_2 = \frac{V_2 T_1}{V_1} = \frac{0.9462 \times 298 \text{ K}}{1.00 \text{ L}} = 282 \text{ K} = 9°C$$

21. A gas occupies 275 mL at 100°C and 380 Pa. What final temperature is required to decrease the pressure to 305 Pa if the volume is held constant?

Solution

List all V, P, and T data.

V_1: 275 mL P_1: 380 Pa T_1: 100°C = 373 K
V_2: 275 mL P_2: 305 Pa T_2: ?

Convert T_1 to kelvins and apply the general gas law equation, realizing that $V_1 = V_2$.

$$\frac{V_1 P_1}{T_1} = \frac{V_2 P_2}{T_2} \quad \text{and} \quad \frac{P_1}{T_1} = \frac{P_2}{T_2} \quad \text{and} \quad T_2 = \frac{T_1 P_2}{P_1}$$

$$T_2 = \frac{373 \text{ K} \times 305 \text{ Pa}}{380 \text{ Pa}} = 299 \text{ K} = 26°C$$

Chapter 10 The Gaseous State and the Kinetic-Molecular Theory   131

23. A gas occupies a volume of 12 L at 685.4 torr and 85.6°C. What would be its volume at 98.7 kPa and 64.8°C?

Solution

List all V, P, and T data.

V_1: 12 L P_1: 685.4 torr T_1: 85.6°C = 358.6 K
V_2: ? P_2: 98.7 kPa T_2: 64.8°C = 337.8 K

Convert P_1 and P_2 to equivalent pressure units.

760 torr = 101.3 kPa

$$P_1 \text{ (kPa)} = 685.4 \text{ torr} \times \frac{101.3 \text{ kPa}}{760 \text{ torr}} = 91.357 \text{ kPa}$$

Convert T_1 and T_2 to kelvins and then apply the general gas law equation to solve for V_2.

$$\frac{V_1 P_1}{T_1} = \frac{V_2 P_2}{T_2}$$

$$V_2 = \frac{V_1 P_1 T_2}{P_2 T_1} = \frac{12 \text{ L} \times 91.357 \text{ kPa} \times 337.8 \text{ L}}{98.7 \text{ kPa} \times 358.6 \text{ K}} = 10 \text{ L}$$

25. A high-altitude balloon is filled with 1.41×10^4 L of hydrogen at a temperature of 21°C and a pressure of 745 torr. What is the volume of the balloon at a height of 20 km, where the temperature is −48°C and the pressure is 63.1 torr?

Solution

List all V, P, and T data.

V_1: 1.41×10^4 L P_1: 745 torr T_1: 21°C = 294 K
V_2: ? P_2: 63.1 torr T_2: −48°C = 225 K

Use the general gas law equation to calculate V_2 given that both pressure and temperature change.

$$\frac{V_1 P_1}{T_1} = \frac{V_2 P_2}{T_2}$$

$$V_2 = \frac{V_1 P_1 T_2}{P_2 T_1} = \frac{1.41 \times 10^4 \text{ L} \times 745 \text{ torr} \times 225 \text{ K}}{63.1 \text{ torr} \times 294 \text{ K}}$$

$$= 1.27 \times 10^5 \text{ L}$$

27. A spray can is used until it is empty, except for the propellant gas, which has a pressure of 1344 torr at 23°C. If the can is thrown into a fire ($T = 475°C$), what will be the pressure (in atm) in the hot can?

Solution

List all P and T data. We can disregard V, which remains constant.

P_1: 1344 torr T_1: 23°C = 295 K
P_2: ? T_2: 475°C = 748 K

The volume of the can is fixed and the pressure will increase dramatically with the large increase in temperature. Convert T_1 and T_2 to kelvins and apply the general gas law equation to solve for P_2 while keeping V constant.

$$\frac{V_1 P_1}{T_1} = \frac{V_2 P_2}{T_2}$$

$$P_2 = \frac{P_1 T_2}{T_1} = \frac{1344 \text{ torr} \times 748 \text{ K}}{296 \text{ K}} = 3.4 \times 10^3 \text{ torr}$$

28. How many moles of carbon monoxide, CO, are contained in a 327.2 mL bulb at 48.1°C if the pressure is 149.3 kPa?

Solution

The quantity of gas (moles) is unknown. Given V, P, and T, the amount can be calculated by using the ideal gas equation: $PV = nRT$.

$$V = 327.2 \text{ mL}, \quad P = 149.3 \text{ kPa}, \quad T = 48.1°C$$

To use these data with the equation, we must convert volume to liters, temperature to kelvins and use R as 8.314 [(L × kPa)/(mol × K)] Rearrange $PV = nRT$ to solve for n.

$$n = \frac{PV}{RT} = \frac{(149.3 \text{ kPa})(0.3272 \text{ L})}{(8.314 \text{ L kPa/mol K})(321.15 \text{ K})} = 0.0183 \text{ mol}$$

29. How many moles of chlorine gas, Cl_2, are contained in a 10.3-L tank at 21.2°C if the pressure is 63.3 atm?

Solution

V, P, and T values are stated, and the amount of gas (moles) is unknown. Rearrange the ideal gas equation to solve for n.

$$V = 10.3 \text{ L}, \quad P = 63.3 \text{ atm}, \quad T = 21.2°C = 294.35 \text{ K}$$

$$R = 0.08206 \text{ L atm/mol K}$$

$$n = \frac{PV}{RT} = \frac{(63.3 \text{ atm})(10.3 \text{ L})}{(0.08206 \text{ L atm/mol K})(294.35 \text{ K})} = 27.0 \text{ mol}$$

32. What is the temperature of a 0.274-g sample of methane, CH_4, confined in a 300.0-mL bulb at a pressure of 198.7 kPa?

Solution

The ideal gas equation is used here because only one set of conditions is given. The amount (n) must be calculated from the mass and the molecular weight of CH_4.

$$PV = nRT \quad \text{and} \quad T = \frac{PV}{nR}$$

$$T = \frac{(198.7 \text{ kPa})(0.3000 \text{ L})}{\dfrac{0.274 \text{ g}}{16.0426 \text{ g/mol}} \times 8.314 \dfrac{\text{L kPa}}{\text{mol K}}} = 420 \text{ K} \quad \text{or} \quad 147°C$$

37. (a) What mass of laughing gas, N_2O, occupies a volume of 0.250 L at a temperature of 325 K and a pressure of 113.0 kPa?
    (b) What is the density of N_2O under these conditions?

Chapter 10 The Gaseous State and the Kinetic-Molecular Theory

Solution

(a) Calculate the mass from $PV = nRT$.

$$n = \frac{PV}{RT} = \frac{(113.0 \text{ kPA}) (0.250 \text{ L})}{(8.314 \text{ L kPa/mol K}) (325 \text{ K})} = 0.010455$$

$$\text{Mass (N}_2\text{O)} = 0.010455 \text{ mol} \times 44.0128 \frac{\text{g}}{\text{mol}} = 0.460 \text{ g}$$

(b) $D = \dfrac{\text{mass}}{\text{volume}} = \dfrac{0.460 \text{ g}}{0.250 \text{ L}} = 1.84 \text{ g/L}$

40. Calculate the density of nitrogen, N_2, at a temperature of 373 K and a pressure of 108.3 kPa.

Solution

The density is calculated from the volume of one mole of nitrogen measured at the stated conditions. Rearranging $PV = nRT$ and solving for V gives

$$P = 108.3 \text{ kPa}, \quad T = 373 \text{ K}, \quad n = 1.0 \text{ mole}$$

$$V = \frac{nRT}{P} = \frac{(1.0 \text{ mol}) (8.314 \text{ L kPA/mol K}) (373 \text{ K})}{108.3 \text{ kPA}} = 28.634 \text{ L}$$

$$D = \frac{\text{mass}}{\text{volume}} = \frac{28.0134 \text{ g}}{28.634 \text{ L}} = 0.978 \text{ g/L}$$

43. A cylinder of a standard gas for calibration of blood gas analyzers contains 3.5% CO_2, 10.0% O_2, and the remainder N_2 at a total pressure of 145 atm. What is the partial pressure of each component of this gas? (The percentages given indicate the percent of the total pressure due to each component.)

Solution

The total pressure of a system, P_T, equals the sum of the pressures exerted by the individual gases. The total pressure is a function of the total number of gas particles present and is proportional to the individual percentages.

$$P_T = P_{CO_2} + P_{O_2} + P_{N_2}$$

$$P_{CO_2} = 3.5\% \text{ of } 145 \text{ atm} = 0.035 \times 145 \text{ atm} = 5.075 = 5.1 \text{ atm}$$

$$P_{O_2} = (0.10)(145 \text{ atm}) = 14.5 \text{ atm}$$

$$P_{N_2} = P_T - P_{CO_2} - P_{O_2} = 145 \text{ atm} - 5.1 \text{ atm} - 14.5 \text{ atm} = 125 \text{ atm}$$

44. Most mixtures of hydrogen gas with oxygen gas are explosive. However, a mixture that contains less than 3.0% O_2 is not. If enough O_2 is added to a cylinder of H_2 at 33.2 atm to bring the total pressure to 33.5 atm, is the mixture explosive?

Solution

The partial pressure of O_2 is calculated from the total pressure by subtracting the partial pressure of H_2.

$$P_T = P_{O_2} + P_{H_2} \quad \text{and} \quad P_{O_2} = P_T - P_{H_2}$$

$$P_{O_2} = 33.5 \text{ atm} - 33.2 \text{ atm} = 0.3 \text{ atm}$$

$$\text{Percent } O_2 = \frac{0.3 \text{ atm}}{33.5 \text{ atm}} \times 100 = 0.9\% \text{ of the total}$$

The mixture is not explosive.

46. A mixture of 0.200 g He, 1.00 g of O_2, and 0.820 g Ne is contained in a closed container at STP. What is the volume of the container, assuming that the gases exhibit ideal behavior?

Solution

The volume of STP is a function of the total number of moles of gas in the container. Using $P_T V = n_T RT$, calculate n_T and rearrange the equation to solve for V.

$$n_{He} = 0.200 \text{ g} \times \frac{1 \text{ mol}}{4.0026} = 0.04997$$

$$n_{O_2} = 1.00 \text{ g} \times \frac{1 \text{ mol}}{31.9988 \text{ g}} = 0.03125$$

$$n_{Ne} = 0.820 \text{ g} \times \frac{1 \text{ mol}}{20.179 \text{ g}} = 0.04064$$

$$n_T = 0.12186 \text{ mol}$$

$$PV = nRT \quad \text{and} \quad V = \frac{nRT}{P}$$

$$V = \frac{(0.12186 \text{ mol})(0.08206 \text{ L atm/mol K})(273 \text{ K})}{1 \text{ atm}} = 2.73 \text{ L}$$

47. A sample of carbon monoxide was collected over water at a total pressure of 756 torr and a temperature of 18°C. What is the pressure of the carbon monoxide? (See Table 10.1 for the vapor pressure of water.)

Solution

The total pressure of the system is the sum of the partial pressures of CO and H_2O at the stated temperature.

$$P_T = P_{CO} + P_{H_2O} \text{ at } 18°C$$

$$P_{CO} = P_T - P_{H_2O} = 756 \text{ torr} - 15.5 \text{ torr} = 740 \text{ torr}$$

49. The volume of a sample of a gas collected over water at 30.0°C and 0.932 atm is 627 mL. What will the volume of the dried gas be at STP?

Solution

The pressure exerted by the sample of gas is the sum of the partial pressures of the gas and of water vapor. To obtain the volume of dried gas composing the sample, the partial pressure of water must be subtracted.

$$P_T = 0.932 \text{ atm} = P_{H_2O} + P_{gas}$$

$$P_{gas} = 0.932 \text{ atm} - \left(3.18 \text{ mm Hg} \times \frac{1 \text{ atm}}{760 \text{ mm Hg}}\right) = 0.89016$$

Chapter 10 The Gaseous State and the Kinetic-Molecular Theory   135

List all the data and use the combined gas laws to calculate V.

V_1: 627 mL P_1: 0.8916 T_1: 30.0°C = 303 K

V_2: ? P_2: 1 atm T_2: 273 K

$$V_2 = 627 \text{ mL} \times \frac{0.89016 \text{ atm}}{1 \text{ atm}} \times \frac{273 \text{ K}}{303 \text{ K}} = 503 \text{ mL}$$

52. (a) What is the concentration of the atmosphere in molecules per milliliter at 25°C and 1.00 atm?

Solution

(a) We assume that the atmosphere behaves ideally and that the total pressure is simply the sum of the partial pressures of the gases. Therefore, $PV = nRT$ applies to the atmosphere. Calculate the volume of one mole of air at the stated conditions and divide to determine molecules per milliliter.

$$V = \frac{nRT}{P} = \frac{(1 \text{ mol})(0.08206 \text{ L atm/mol K})(298.15 \text{ K})}{1.00 \text{ atm}} = 24.466 \text{ L}$$

$$\text{Concentration} = \frac{6.022 \times 10^{23} \text{ molecules}}{24.466 \text{ L} \times 1000 \text{ mL/L}} = 2.46 \times 10^{19} \text{ molecules/mL}$$

COMBINING VOLUMES, MOLECULAR WEIGHTS, STOICHIOMETRY

53. What volume of O_2 at STP is required to oxidize 20.0 L of CO at STP to CO_2? What volume of CO_2 is produced at STP?

Solution

Write the balanced equation for the reaction.

$$\begin{array}{ccc} 2\text{CO} & + \quad \text{O}_2 \longrightarrow & 2\text{CO}_2 \\ 2 \text{ mol} & 1 \text{ mol} & 2 \text{ mol} \\ 2(22.4 \text{ L}) & 22.4 \text{ L} & 2(22.4 \text{ L}) \end{array}$$

The balanced equation shows that 2 moles of CO requires one mole of O_2 and that 2 moles of CO_2 are produced. Gay-Lussac's law states that gases react in simple proportions by volume.

20.0 L CO : 10.0 L O_2 : 20 L CO_2

55. Calculate the volume of oxygen required to burn 5.00 L of propane gas, C_3H_8, to produce carbon dioxide and water, if the volumes of both the propane and oxygen are measured under the same conditions.

Solution

Write the balanced equation and apply Gay-Lussac's law.

$$\begin{array}{cc} 5.00 \text{ L} & V \\ C_3H_8 \quad + \quad 5O_2 & \longrightarrow 3CO_2 + 4H_2O \\ 1 \text{ volume} \quad 5 \text{ volumes} & \end{array}$$

$$V(O_2) = 5.00 \text{ L } C_3H_8 \times \frac{5 \text{ L } O_2}{1 \text{ L } C_3H_8} = 25.0 \text{ L}$$

57. An acetylene tank for an oxyacetylene welding torch will provide 9340 L of acetylene gas, C_2H_2, at STP. How many tanks of oxygen, each providing 7.00×10^3 L of O_2 at STP, will be required to burn the acetylene, forming CO_2 and H_2O?

Solution

Write the balanced equation and apply Gay-Lussac's law.

$$\begin{array}{cc} 9340 \text{ L} & V \\ 2C_2H_2 + 5O_2 \longrightarrow 4CO_2 + 2H_2O \\ 2 \text{ mol} + 5 \text{ mol} + 4 \text{ mol} \quad 2 \text{ mol} \end{array}$$

$$V(O_2) = 9340 \text{ L } C_2H_2 \times \frac{5 \text{ L } O_2}{2 \text{ L } C_2H_2} = 23{,}350 \text{ L}$$

$$\text{No. tanks } O_2 = 23{,}350 \text{ L} \times \frac{1 \text{ tank}}{7.00 \times 10^3 \text{ L}} = 3.34 \text{ tanks}$$

59. What is the molecular weight of a gas if 125 mL of the gas at a pressure of 99.5 kPa at 22°C has a mass of 0.157 g?

Solution

Use the ideal gas equation to calculate the number of moles of gas required to produce a pressure of 99.5 kPa in a 125 mL vessel. Then calculate the molecular weight from the stated mass.

$$PV = nRT$$

$$n = \frac{PV}{RT} = \frac{(99.5 \text{ kPA})(0.125 \text{ L})}{(8.314 \text{ L kPA/mol K})(295.15)} = 0.005068 \text{ mol}$$

Since 0.005068 mol = 0.157 g,

$$1 \text{ mol gas} = \frac{0.157 \text{ g}}{0.005068} = 31.0 \text{ g}$$

61. A sample of an oxide of nitrogen isolated from the exhaust of an automobile was found to weigh 0.571 g and to occupy 1.00 L at 356 torr and 27°C. Calculate the molecular weight of this oxide, and determine if it was N_2O, NO, NO_2, N_2O_4, or N_2O_5.

Solution

Calculate the molecular weight of the gas by using the ideal gas equation. The formula can be determined by comparing molecular weights of the various N-O compounds to the experimental values.

$$PV = nRT, \quad \text{where} \quad n = \frac{\text{mass}}{\text{g-mol. wt}}$$

Substitution and rearranging to solve for g-mol. wt yields

$$PV = nRT \quad \text{and} \quad PV = \left(\frac{m}{\text{g-mol. wt}}\right)RT$$

$$\text{g-mol. wt} = \frac{mRT}{PV} = \frac{(0.571 \text{ g})(0.08206 \text{ L atm/mol K})(300.15 \text{ K})}{\left(356 \text{ torr} \times \dfrac{1 \text{ atm}}{760 \text{ torr}}\right)(1.00 \text{ L})}$$

Chapter 10 The Gaseous State and the Kinetic-Molecular Theory

g-mol. wt = 30.0

The molecular weight of NO is 30.0.

63. Cyclopropane is a gas containing only carbon and hydrogen and is often used as an anaesthetic for major surgery. If 250 mL of cyclopropane at 120°C and 0.72 atm reacts with O_2 to give 750 mL of CO_2 and 750 mL of $H_2O(g)$ at the same temperature and pressure, what is the molecular formula of cyclopropane?

Solution

This reaction is an excellent example of Gay-Lussac's law in which the reacting gas produces product in a ratio of one to three.

$$\begin{array}{cccc} 250 \text{ mL} & & 750 \text{ mL} & 750 \text{ mL} \\ \text{cyclopropane} + O_2 & \longrightarrow & CO_2 & + \quad H_2O \\ 1 \text{ volume} & & 3 \text{ volumes} & + 3 \text{ volumes} \end{array}$$

Since the ratio of C to H in CO_2 and H_2O is 1C : 2H, the ratio of C to H in cyclopropane must be 1C : 2H and the molecule of cyclopropane must contain 3C : 6H, hence, C_3H_6.

66. Hydrogen gas will reduce Fe_3O_4 in the reverse of the reaction shown in exercise 65.

(a) What volume of H_2 at 200 atm and 19°C is required to reduce 1.00 metric ton (1000 kg = 1 metric ton) of Fe_3O_4 to Fe?

Solution

(a) Write the balanced equation for the reaction, calculate the number of moles of H_2 required at STP and convert moles of H_2 at STP to volume at reaction conditions.

$$\begin{array}{cc} 1000 \text{ kg} & V \\ Fe_3O_4(s) + 4H_2(g) \longrightarrow 4H_2O(g) + 3Fe(s) \\ 1 \text{ mol} \;+\; 4 \text{ mol} \qquad 4 \text{ mol} \;+\; 3 \text{ mol} \end{array}$$

g-form. wt (Fe_3O_4) = 231.517 g

$$\text{No. mol } H_2 = \left(1000 \text{ kg} \times \frac{1000 \text{ g}}{\text{kg}} \times \frac{1 \text{ mol}}{231.517 \text{ g}}\right)\left(\frac{4 \text{ mol } H_2}{1 \text{ mol } Fe_3O_4}\right)$$

$$= 1.7277 \times 10^4 \text{ mol}$$

$$PV = nRT \quad \text{and} \quad V = \frac{nRT}{P}$$

$$V = \frac{(1.7277 \times 10^4 \text{ mol})(0.08206 \text{ L atm/mol K})(292.15 \text{ K})}{200 \text{ atm}}$$

$$= 2.07 \times 10^3 \text{ L}$$

70. What volume of oxygen at 423.0 K and a pressure of 105.4 kPa will be produced by the decomposition of 192.7 g of BaO_2 to BaO and O_2?

Solution

Write the balanced equation for the decomposition, calculate the number of moles of oxygen produced at STP and convert this amount to volume under the stated conditions.

138   Chapter 10   The Gaseous State and the Kinetic-Molecular Theory

$$2BaO_2(s) \longrightarrow 2BaO(s) + O_2(g)$$

192.7 g, 2 mol → 2 mol; No. mol, 1 mol

g-form. wt BaO_2 = 169.3288 g

$$\text{No. mol } O_2 = 192.7 \text{ g} \times \frac{1 \text{ mol}}{169.3288 \text{ g}} \times \frac{1 \text{ mol } O_2}{2 \text{ mol } BaO_2} = 0.56901 \text{ mol}$$

From $PV = nRT$, $V = nRT/P$,

$$V = \frac{(0.56901 \text{ mol})(8.314 \text{ L kPa/mol K})(423.0 \text{ K})}{105.4 \text{ kPa}} = 18.98 \text{ L}$$

73. As 1.00 g of the radioactive element radium decays over 1 yr, it produces 1.16×10^{18} alpha particles (helium nuclei). Each alpha particle becomes an atom of helium gas. What is the pressure, in Pa, of the helium gas produced if it occupies a volume of 125 mL at a temperature of 25°C?

Solution

Calculate the number of moles of helium atoms produced, then apply the ideal gas equation to calculate the pressure at the stated conditions.

$$n(\text{He}) = 1.16 \times 10^{18} \text{ atoms} \times \frac{1 \text{ mol}}{6.022 \times 10^{23} \text{ atoms}} = 1.9263 \times 10^{-6} \text{ mol}$$

$$P = \frac{nRT}{V} = \frac{(1.9263 \times 10^{-6} \text{ mol})(8.314 \text{ L kPa/mol K})(298.15 \text{ K})}{(0.125 \text{ L})}$$

$$= 3.82 \times 10^{-2} \text{ kPa} = 38.2 \text{ Pa}$$

KINETIC-MOLECULAR THEORY, GRAHAM'S LAW

76. Describe what happens to the average kinetic energy of ideal gas molecules when the conditions are changed as follows:

  (a) The volume of gas is increased by decreasing the pressure at constant temperature.
  (b) The volume of gas is increased by increasing the temperature at constant pressure.
  (c) The average velocity of the molecules is decreased by a factor of 0.5.

Solution

(a) No change in kinetic energy.
(b) Kinetic energy increases as a consequence of increasing the temperature.
(c) The relationship of kinetic energy to velocity is $KE = \frac{1}{2}mu^2$.
  Since mass is constant, kinetic energy changes proportionally to the square of velocity, that is,

$$\frac{u_1^2}{u_2^2} = \frac{2KE_1/m}{2KE_2/m} \quad \text{and} \quad \frac{u_1^2}{u_2^2} = \frac{KE_1}{KE_2}$$

When $u_2 = \frac{1}{2}u_1$, the equation becomes

$$\left(\frac{u_1}{0.5u_1}\right)^2 = \frac{KE_1}{KE_2} \quad \text{and} \quad (2)^2 = \frac{KE_1}{KE_2}$$

$$KE_2 = \frac{KE_1}{4}$$

Chapter 10 The Gaseous State and the Kinetic-Molecular Theory 139

80. The root-mean-square speed of H₂ molecules at 25°C is about 1.6 km/s. What is the root-mean-square speed of a N₂ molecule at 25°C?

Solution

Given that speed is inversely proportional to the square root of the molecular weight of H₂ and N₂, the rms equation can be used to calculate the rms of N₂ based on the speed and molecular weight of H₂.

$$\frac{u_{rms}(H_2)}{u_{rms}(N_2)} = \frac{\sqrt{\text{mol. wt }(N_2)}}{\sqrt{\text{mol. wt }(H_2)}}$$

$$\frac{1.6 \text{ km/s}}{u_{rms}(N_2)} = \frac{\sqrt{(14.0067)(2)}}{\sqrt{(1.0079)(2)}}$$

$$\frac{1.6 \text{ km/s}}{u_{rms}(N_2)} = \sqrt{13.8969} = 3.728$$

$$u_{rms}(N_2) = \frac{1.6 \text{ km/s}}{3.728} = 0.43 \text{ km/s}$$

82. Heavy water, D₂O (mol. wt = 20.03), can be separated from ordinary water, H₂O (mol. wt = 18.01), as a result of the difference in the relative rate of diffusion of the molecules in the gas phase. Calculate the relative rates of diffusion of H₂O and D₂O.

Solution

Graham's law applies to this situation. D₂O being slightly heavier than H₂O will diffuse more slowly. Diffusion rates are inversely proportional to the square root of their respective molecular weights.

$$\frac{\text{Rate }(H_2O)}{\text{Rate }(D_2O)} = \frac{\sqrt{\text{mol. wt }(D_2O)}}{\sqrt{\text{mol. wt }(H_2O)}}$$

$$\frac{R(H_2O)}{R(D_2O)} = \frac{\sqrt{20.03}}{\sqrt{18.01}} = \sqrt{1.1216} = 1.055$$

$$R(H_2O) = 1.055 \text{ times } R(D_2O)$$

84. Calculate the relative rate of diffusion of H₂ compared to D₂ (at. wt D = 2.0) and of O₂ compared to O₃.

Solution

Use Graham's law:

$$\frac{R(H_2)}{R(D_2)} = \frac{\sqrt{\text{mol. wt }(D_2)}}{\sqrt{\text{mol. wt }(H_2)}} \quad \text{and} \quad \frac{R(H_2)}{R(D_2)} = \sqrt{\frac{4}{2}} = \sqrt{2} = 1.4$$

$$R(H_2) = 1.4 \, R(D_2)$$

$$\frac{R(O_3)}{R(O_2)} = \sqrt{\frac{31.9988}{47.9982}} = 0.667 \quad \text{and} \quad R(O_3) = 0.667 \, R(O_2)$$

86. When two cotton plugs, one moistened with ammonia and the other with hydrobromic acid, are simultaneously inserted into opposite ends of a glass tube 87.0 cm long, a white ring of NH₄Br forms where gaseous NH₃ and gaseous HBr first come into contact.

140 Chapter 10 The Gaseous State and the Kinetic-Molecular Theory

$$NH_3(g) + HBr(g) \longrightarrow NH_4Br(s)$$

At what distance from the ammonia-moistened plug does this occur?

Solution

The distance from each end of the tube is proportional to the diffusion rates of the gases. Ammonia, the lighter gas, will travel farther down the tube than will HBr. Their distances can be calculated from the Graham's law equation.

$$\frac{R(NH_3)}{R(HBr)} = \frac{\sqrt{\text{mol. wt (HBr)}}}{\sqrt{\text{mol. wt (NH}_3)}} = \sqrt{\frac{80.9111}{17.0304}} = \sqrt{4.75103} = 2.1797$$

$$R(NH_3) = 2.1797\ R(HBr)$$

Distance: HBr = 27.4 cm; NH$_3$ = 59.6 cm

NONIDEAL BEHAVIOR OF GASES

89. For which of the following sets of conditions will a real gas behave most like an ideal gas, and for which conditions would a real gas be expected to deviate from ideal behavior? Explain.

(a) low volume, high pressure
(b) low temperature, high pressure
(c) low pressure, high temperature

Solution

(a) Bonding will develop in this situation. Certain gases will liquify under these conditions.
(b) This is a more likely situation than (a) for strong bonds to develop. Even hydrogen and helium can be liquified under these extremes.
(c) This situation promotes ideal behavior. Molecules have high speeds and move through greater distances between collisions; they also have shorter contact times. Little bonding develops under these conditions.

ADDITIONAL EXERCISES

91. A commercial mercury vapor analyzer can detect concentrations of gaseous Hg atoms (which are poisonous) in air as low as 2×10^{-6} mg/L of air. What is the partial pressure of gaseous mercury if the atmospheric pressure is 733 torr at 26°C?

Solution

The total pressure is the sum of the normal atmospheric components plus that due to mercury. Use the ideal gas equation to calculate the total number of moles of gas required to produce the stated conditions. The pressure of mercury is the mole fraction of mercury times the total pressure.

$$\text{No. mol (Hg)} = 2 \times 10^{-6}\ \text{mg} \times \frac{1\ \text{g}}{1000\ \text{mg}} \times \frac{1\ \text{mol}}{200.59\ \text{g}} = 9.9706 \times 10^{-12}\ \text{mol}$$

$$n_T = \frac{PV}{RT} = \frac{(733/760\ \text{atm})(1.0\ \text{L})}{(0.08206\ \text{L atm/mol K})(299.15\ \text{K})} = 0.03929\ \text{mol}$$

$$P(\text{Hg}) = \frac{9.9706 \times 10^{-12}\ \text{mol}}{0.03929\ \text{mol}} \times 733\ \text{torr} = 2 \times 10^{-7}\ \text{torr}$$

Chapter 10 The Gaseous State and the Kinetic-Molecular Theory

93. Butene, an important petrochemical used in the production of synthetic rubber, is composed of carbon and hydrogen. A 0.0124-g sample of this compound produces 0.0390 g of CO_2 and 0.0159 g of H_2O when burned in an oxygen atmosphere. If the volume of a 0.125-g gaseous sample of the compound at 735 torr and 45°C is 60.2 mL, what is the molecular formula of the compound?

Solution

Determine the empirical formula of the hydrocarbon from the masses of CO_2 and H_2O produced. Then based on the data from the gas measurements, use $PV = nRT$ to calculate the molecular weight. The molecular formula is a multiple of the empirical formula.

$$\underset{0.0124 \text{ g}}{C_xH_y} + O_2 \longrightarrow \underset{0.0390 \text{ g}}{CO_2} + \underset{0.0159 \text{ g}}{H_2O}$$

$$\text{Mass (C)} = \frac{12.011}{43.999} \times 0.0390 \text{ g} = 0.010646 \text{ g}$$

$$\text{Mass (H)} = 0.0124 \text{ g} - 0.010646 = 0.001754 \text{ g}$$

Relative number of moles of atoms:

C: $0.010646/12.011 = 0.0008864$ $0.0008864/0.0008864 = 1$

H: $0.001754/1.0079 = 0.0017399$ $0.0017399/0.0008864 = 1.963$

Empirical formula = C_1H_2
From $PV = nRT$, $n = PV/RT$

$$n = \frac{(735/760 \text{ atm})(0.0602 \text{ L})}{(0.08206 \text{ L atm/mol K})(318.15 \text{ K})} = 0.00223 \text{ mol}$$

0.00223 mol $= 0.125$ g   or 1 mol $= 56.05$ g

$(C_1H_2)_x = 56.05$ g   or $(12.011 \text{ g} + 2(1.0079 \text{ g}))_x = 56.05$

$$x = \frac{56.05 \text{ g}}{14.03 \text{ g}} = 4$$

Molecular formula = C_4H_8

95. Thin films of amorphous silicon for electronic applications are prepared by decomposing silane gas, SiH_4, on a hot surface at low pressures.

$$SiH_4(g) \xrightarrow{\Delta} Si(s) + 2H_2(g)$$

What volume of SiH_4 at 150 Pa and 825 K is required to produce a 20.0-cm-by-20.0-cm film that is 200 Å thick (1 Å = 10^{-8} cm)? The density of amorphous silicon is 1.9 g/cm³.

Solution

From the balanced equation, one mole of SiH_4 produces 1 mole of Si. First calculate the mass of the layer of silicon. Then use $PV = nRT$ to calculate the volume of SiH_4 required.

$$\text{Mass (Si)} = 20.0 \text{ cm} \times 20.0 \text{ cm} \times 200 \times 10^{-8} \text{ cm} \times 1.9 \text{ g/cm}^3 = 1.52 \times 10^{-3} \text{ g}$$

$$\text{Mole (Si)} = 1.52 \times 10^{-3} \text{ g} \times \frac{1 \text{ mol}}{28.0855 \text{ g}} = 5.412 \times 10^{-5} \text{ mol}$$

$$V = \frac{nRT}{P} = \frac{(5.412 \times 10^{-5} \text{ mol})(8.314 \text{ L kPa/mol K})(825 \text{ K})}{0.150 \text{ kPa}} = 2.47 \text{ L}$$

97. Ethanol, C_2H_5OH, is often produced by the fermentation of sugars. For example, the preparation of ethanol from the sugar glucose is represented by the unbalanced equation

$$C_6H_{12}O_6(aq) \xrightarrow{yeast} C_2H_5OH(aq) + CO_2(g)$$

What volume of CO_2 at STP is produced by the fermentation of 1.00 metric ton (1 metric ton = 1000 kg) of glucose if the reaction has a yield of 95.2%?

Solution

Balance the equation and apply Gay-Lussac's law, and assume 100% reaction efficiency at STP.

$$\begin{array}{ccc} C_6H_{12}O_6 & \rightarrow 2C_2H_5OH & + & 2CO_2 \\ 1 \text{ mol} & & & 2 \text{ mol} \\ 180.1572 \text{ g} & & & 2(22.414 \text{ (L/mol)}) \end{array}$$

$$V(CO_2) = 1 \times 10^3 \text{ kg} \times \frac{1 \times 10^3 \text{ g}}{\text{kg}} \times \frac{1 \text{ mol } C_6H_{12}O_6}{180.1572 \text{ g}} \times \frac{2 \text{ mol } CO_2}{1 \text{ mol } C_6H_{12}O_6}$$

$$\times \frac{22.414 \text{ L}}{\text{mol}} = 2.49 \times 10^5 \text{ L}$$

At 95.2% efficiency: $V = 2.49 \times 10^5 \text{ L} \times 0.952 = 2.37 \times 10^5 \text{ L}$

99. One method of analysis of amino acids is the van Slyke method. The characteristic amino groups ($-NH_2$) in protein material are allowed to react with nitrous acid, HNO_2, to form N_2 gas. From the volume of the gas, the amount of amino acid can be determined. A 0.0604-g sample of a biological material containing glycine, $CH_2(NH_2)COOH$, was analyzed by the van Slyke method, giving 3.70 mL of N_2 collected over water at a pressure of 735 torr and 29°C. What was the percentage of glycine in the sample?

$$CH_2(NH_2)COOH + HNO_2 \rightarrow CH_2(OH)COOH + H_2O + N_2$$

Solution

First calculate the volume of dry N_2 at STP produced from the reaction. Use this value to determine the mass of glycine required for the production.
Vapor pressure (H_2O) = 30.0 mm Hg at 29°C

$$P(N_2) = 735 \text{ torr} - 30.0 \text{ torr} = 705 \text{ torr}$$

$$V(N_2) = 3.70 \text{ mL} \times \frac{705 \text{ torr}}{760 \text{ torr}} \times \frac{273.15}{302.15} = 3.103 \text{ mL}$$

$$\begin{array}{ccc} \text{mass?} & & 3.103 \text{ mL} \\ CH_2(NH_2)COOH & + HNO_2 \rightarrow CH_2(OH)COOH + H_2O + & N_2 \\ 1 \text{ mol} & & 1 \text{ mol} \\ 75.07 & & 22,414 \text{ mL} \end{array}$$

$$\text{Mass (Gly)} = 3.103 \text{ mL } N_2 \times \frac{1 \text{ mol}}{22,414 \text{ mL}} \times \frac{1 \text{ mol Gly}}{1 \text{ mol } N_2} \times \frac{75.07}{\text{mol}} = 0.01039$$

$$\text{Percent Gly} = \frac{0.01039 \text{ g}}{0.0604 \text{ g}} \times 100\% = 17.2\%$$

Chapter 10 The Gaseous State and the Kinetic-Molecular Theory

101. One step in the production of sulfuric acid (Section 9.9, Part 1) involves oxidation of $SO_2(g)$ to $SO_3(g)$ using air at 400°C with vanadium(V) oxide as a catalyst. Assuming that air is 21% oxygen by volume, what volume of air at 31°C and 0.912 atm is required to oxidize enough SO_2 to give 1.00 metric ton (1000 kg = 1 metric ton) of sulfuric acid, H_2SO_4?

Solution

Using the balanced equation for the oxidation, calculate the volume of O_2 required at STP. Then convert to air percentage at the stated conditions.

$$\begin{array}{ccc} & V & 1.00 \times 10^3 \text{ g} \\ 2SO_2 + & O_2 \longrightarrow & 2SO_3 \\ 2 \text{ mol} + & 1 \text{ mol} & 2 \text{ mol} \end{array}$$

$$V(O_2) \text{ at STP} = 1.00 \times 10^3 \text{ kg} \times \frac{10^3 \text{ g}}{\text{kg}} \times \frac{1 \text{ mol } SO_3}{80.0582 \text{ g}} \times \frac{1 \text{ mol } O_2}{2 \text{ mol } SO_3}$$

$$\times 22.414 \frac{L}{\text{mol}} = 1.3999 \times 10^5 \text{ L}$$

At 21% of air at STP: $V = 1.3999 \times 10^5 \text{ L}/0.21 = 6.666 \times 10^5 \text{ L air}$

At stated conditions,

$$V \text{ (air)} = 6.666 \times 10^5 \text{ L} \times \frac{1.0 \text{ atm}}{0.912 \text{ atm}} \times \frac{304.15}{273.15} = 8.14 \times 10^5 \text{ L}$$

103. A 0.500-L bottle contains 250.0 mL of a 3.0% (by mass) solution of hydrogen peroxide. How much will the pressure (in atm) in the bottle increase if the H_2O_2 decomposes to $H_2O(l)$ and $O_2(g)$ at 24°C? Assume that the density of the solution is 1.00 g/cm^3, that the volume of the liquid does not change during the decomposition, and that the solubility of O_2 in water can be neglected. Does this calculation explain why hydrogen peroxide solutions are sold in bottles with pressure relief valves? Explain.

Solution

The confined space in the container contains air and water vapor at 24°C and ambient pressure. Any increase in pressure will be due to the release of O_2 by the reaction in which the O_2 is confined to the 250.0 mL space. To calculate the change in pressure, determine the number of moles of O_2 produced by the reaction, then use the ideal gas equation to calculate the pressure due to O_2 at 24°C.

$$\begin{array}{ccc} (0.03)(250.0 \text{ g}) = 7.50 \text{ g} & & n \\ 2H_2O_2 & \longrightarrow 2H_2O + & O_2 \\ 2 \text{ mol} & & 1 \text{ mol} \end{array}$$

$$n(O_2) \text{ STP} = 7.5 \text{ g } H_2O_2 \times \frac{1 \text{ mol}}{34.0146 \text{ g}} \times \frac{1 \text{ mol } O_2}{2 \text{ mol } H_2O_2} = 0.11025 \text{ mol}$$

$$P(O_2) \text{ at } 24°C = \frac{nRT}{V} = \frac{(0.11025 \text{ mol})(0.08206 \text{ L atm/mol K})(297.15 \text{ K})}{0.250 \text{ L}}$$

$$= 11 \text{ atm}$$

11
Condensed Matter: Liquids and Solids

INTRODUCTION

There are three ordinary states of matter: solid, liquid, and gas. We encounter solids in many forms, such as lumber, metal, ice, rock, and so forth. Similarly, liquids of various types occur as greases, syrups, water, alcohol, and even as a metal, mercury. Most gases, such as nitrogen in the air we breathe, are invisible; generally, it is the effect of gases that we experience, although some gases have color, such as chlorine, and some are visible when present in sufficient concentration. (Smokestack and other exhaust plumes are often visible, but this is because the plumes consist of dispersions of particulate matter in the air.)

Different states of aggregation are possible for a substance; these states are called *phases* (from a Greek word meaning "appearance"). The term *phase* is used in chemistry to indicate a homogeneous region, all parts of which are the same. Water is the most familiar example of a substance that may exist in several different forms or phases. Under certain conditions water may exist as solid ice, liquid water, or gaseous water vapor.

Depending on the controlled conditions that are imposed, another kind of variation in form or phase characterizes certain solids. For example, carbon may appear as two different solid phases: diamond and graphite.

Typically, a sufficient change in temperature will cause a change in phase or state. When solid ice is heated, for example, the temperature rises and the ice experiences a phase change as it is converted to liquid water.

Liquids have properties that are intermediate between those of the solids and gases; they are the molten or fluid form of substances. As a substance is transformed from a solid to a liquid, the rigid lattice structure of the solid is lost. Some structure does exist in liquids, but the structure becomes fluid in the sense that structures are continually forming and dispersing. On the other hand, in the gaseous state all structure is lost. Liquids have an ability to flow because of the continual shifting of their internal structures, although some liquids, such as glass, molten polymers, long-chain hydrocarbons, and proteins, tend to flow very slowly or not at all.

This chapter examines some of the regular structural forms exhibited in solids, some of the reasons why these forms exist, and the energies associated with transformations among liquids, solids, and gases.

READY REFERENCE

Bragg equation (11.22)   X-ray techniques are used to determine crystal patterns caused by the location of atoms in crystalline substances. The equation $\lambda = 2d \sin \theta$, developed by Bragg, relates the wavelength λ of reflected X rays to the product of the distance between atomic planes *d* in a crystal times the sine of the angle θ of the reflected X rays.

Energy and change of state   The transformation of any substance from one phase or state to another involves a change in the total energy of the substance. Depending on the direction of the phase change, energy is either absorbed or released by the substance during the change. This diagram represents the possible types of transformations a substance may undergo.

$$\text{Solid} \underset{\text{energy}-}{\overset{+\text{energy}}{\rightleftarrows}} \text{Liquid} \underset{\text{energy}-}{\overset{+\text{energy}}{\rightleftarrows}} \text{Gas}$$

The processes indicated in the diagram are reversible; that is, the energy absorbed in each forward process equals the energy released in the reverse process. The temperature of the substance remains constant during a phase change. The energies associated with these changes at the melting point and boiling point are the heat of fusion and heat of vaporization.

Heat of fusion (11.7)   The quantity of heat required to transform a substance, at its melting point, from a solid to a liquid. The heat of fusion of water is 333.6 J/g, or about 80 cal/g.

Heat of vaporization (11.6)   The quantity of heat required to transform a substance, at its boiling point, from a liquid to a gas. The heat of vaporization of water, at 100°C, is 2258 J/g, or about 540 cal/g.

Pythagorean theorem   This theorem relates the hypotenuse of a right triangle to the other two sides; the square of the hypotenuse equals the sum of the squares of the other two sides.

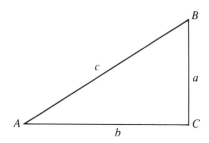

Radius ratio (r^+/r^-) (11.17)   The radius ratio is computed for a compound by dividing the radius of the cation (positive ion) by the radius of the anion (negative ion). Based on empirical data, the ratio is useful for predicting the structural geometry of a crystalline substance.

SOLUTIONS TO CHAPTER 11 EXERCISES

KINETIC-MOLECULAR THEORY AND THE CONDENSED STATE

1. In what ways are liquids similar to solids? In what ways are liquids similar to gases?

Solution

Liquids and solids both have relatively strong intermolecular bonds. Both have fixed volumes at constant temperature. Gases and liquids are similar in that both have shapes which are controlled by their respective containers. Gases, though, occupy the total volume of the container and must be contained in all directions.

3. Describe how the motion of the molecules changes as a substance changes from a solid to a liquid and from a liquid to a gas.

Solution

Molecules in a solid occupy fixed positions and have very little mobility. As the liquid forms, molecules assume greater mobility and greater kinetic energy. Bonds between molecules are weaker and less directional. In the gas phase, molecules are moving at high speeds and essentially no intermolecular bonding exists at elevated temperatures.

5. Explain why a liquid will assume the shape of any container into which it is poured, while a solid is rigid and retains its shape.

Solution

Molecules in solids occupy highly organized and fixed structures, which come about through stronger and more directional bonds as the solid forms from the liquid phase. Molecules in the liquid phase simply do not exhibit sufficient bonding to retain a fixed structure.

7. Explain why a sample of amorphous boric oxide, (B_2O_3, Fig. 11.2) is considered as a form of a liquid even though it looks like a solid, is rigid, and will shatter if struck with a hammer.

Solution

Boric oxide, like ordinary glass, exhibits properties of solids, but liquid structures are more like plastics in that they will flow over extended periods of time. Molecules of boric oxide have strong intermolecular bonds, but the bonds lack directionality to give them lattice structures which would keep them in fixed positions.

MELTING, EVAPORATION, AND BOILING

9. What feature characterizes the equilibrium between a liquid and its vapor in a closed container as dynamic?

Solution

Molecules are continually exchanged between the liquid and the vapor. This exchange of position and energy maintains a constant vapor pressure and a constant temperature.

11. Explain why the vapor pressure of a liquid decreases as its temperature decreases.

Chapter 11 Condensed Matter: Liquids and Solids

Solution

Vapor pressure is a function of the temperature of the liquid and specifically a function of the kinetic energy of the molecules composing the liquids. Therefore, lower temperatures produce lower kinetic energies among molecules and as a consequence, lower vapor pressures.

13. How does the boiling point of a liquid differ from its normal boiling point?

Solution

The boiling point of a liquid is a function of the atmospheric pressure above the liquid. The boiling point of a liquid varies as the pressure varies. The normal boiling point is the temperature at which a liquid will boil when the atmospheric pressure is exactly one atmosphere.

15. The liquid C_4H_{10} in a butane lighter has a boiling point of $-1°C$; the octane (C_8H_{18}) in gasoline has a boiling point of $125°C$. Explain why the boiling point of $C_4H_{10}(l)$ is lower than that of $C_8H_{18}(l)$ even though the types of intermolecular forces are the same in both liquids.

Solution

Short range and short term London force bonds are the primary intermolecular bonds in these compounds. The longer carbon chain of C_8H_{18} allows formation of more London force bonds per molecule, thus more of these must be broken to vaporize the molecule.

17. Why does iodine sublime more rapidly at 110°C than at 25°C?

Solution

The melting point of iodine is 113.5°C. At 110°C the high temperature near the melting point causes the I_2 molecules to have a much higher kinetic energy resulting in weaker intermolecular attractions. This allows molecules near the surface to pass directly into the vapor state more rapidly than at the lower temperature.

19. What is the relationship between the intermolecular forces in a solid and its melting temperature?

Solution

Substances having strong intermolecular bonds have high melting points.

HEAT OF FUSION AND HEAT OF VAPORIZATION

21. A syringe like that shown in Fig. 11.13(a) at a temperature of 20°C is filled with liquid ether so that there is no space for any vapor. In one experiment the temperature is kept constant and the plunger is withdrawn somewhat, forming a volume that can be occupied by vapor. (a) What is the approximate pressure of the vapor produced? (b) If the system were insulated in a second experiment so that no heat could enter or leave, would the temperature of the liquid increase, decrease, or remain constant as the plunger is withdrawn? Explain?

Solution

(a) The pressure will be the normal vapor pressure of ether at 20°C, 450 torr.

(b) As the plunger is withdrawn, the temperature will decrease as the liquid evaporates to provide more ether vapor to fill the void formed by withdrawing the plunger.

23. Explain why steam produces much more severe burns than does the same mass of boiling water.

Solution

As steam condenses to boiling water, it produces heat equal to its heat of vaporization, 40.64 kJ/mol.

25. How much heat in kilojoules is required to change 100 g of water at 100°C to steam at 100°C?

Solution

This is an example of a phase transition that occurs at constant temperature.

Heat = mass × heat of vaporization

Heat = 100 g × 2258 J/g = 225,800 J = 226 kJ

27. In hot, dry climates water is cooled by allowing some of it to evaporate slowly. How much water must evaporate to cool 1.0 L (1.0 kg) from 39°C to 21°C? Assume that the heat of vaporization of water is constant between 21° and 39° and is equal to the value at 25°C.

Solution

Each gram of water evaporated removes 2443 J from the system. Calculate the amount of heat change required, and divide by the amount of heat per gram of water evaporated.

$$\text{Heat required} = 1.0 \text{ kg} \times \frac{4.18 \text{ kJ}}{\text{kg °C}} \times 18°C = 75.24 \text{ kJ}$$

$$\text{Mass of water} = 75.24 \text{ kJ} \times \frac{1 \text{ g H}_2\text{O}}{2.443 \text{ kJ}} = 31 \text{ g}$$

29. How much water at 0°C could be converted to ice at 0°C by the cooling action of vaporizing 2.0 mol of liquid ammonia at its boiling point?

Solution

Each gram of water converted to ice at 0°C requires a heat loss of 333.6 J. Given that the heat of vaporization of ammonia is 23.31 kJ/mol, calculate the heat required, then divide by 333.6 J/g to obtain the mass of water.

Heat (NH$_3$) = 2(23.31 kJ/mol) = 46.62 kJ

$$\text{Mass (ice)} = 46.62 \text{ kJ} \times \frac{1 \text{ g ice}}{0.336 \text{ kJ}} = 140 \text{ g}$$

Chapter 11 Condensed Matter: Liquids and Solids

PHASE DIAGRAMS, CRITICAL TEMPERATURE AND PRESSURE

31. Is it possible to liquify oxygen at room temperature (about 20°C)? Is it possible to liquify ammonia at room temperature? Explain your answers.

Solution

The critical temperature of oxygen is 154.3 K (−119°C). Therefore, oxygen cannot be liquified at 20°C. The critical temperature of ammonia is above 20°C (132°C), therefore, it can be liquified.

33. From the phase diagram for water (Fig. 11.14), determine the physical state of water at:

(a) 800 torr and −10°C
(c) 400 torr and 50°C
(d) 400 torr and 90°C
(f) 2 torr and −10°C

Solution

(a) Solid
(c) Liquid
(d) Gas
(f) Gas

35. What phase changes can water undergo as the pressure changes if the temperature is held at 0.005°C? If the temperature is held at −20°C? At 20°C?

Solution

At 0.005°C, solid, liquid. At −20°C, solid, gas. At 20°C, solid, liquid, gas.

37. If one continues beyond point C on the line BC in Fig. 11.14, at what temperature and pressure does the line end?

Solution

The line beyond C ends at the critical point of pressure and temperature, where it becomes impossible to tell the difference between the gas and liquid phase, namely 374°C and a pressure of 165,500 torr (217.7 atmospheres).

39. Will a sample of carbon dioxide in a cylinder at a pressure of 20 atm and a temperature of −30°C exist as a solid, liquid, or gas? (A portion of the phase diagram of carbon dioxide is shown.)

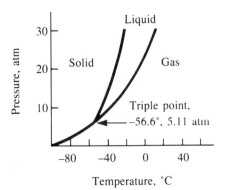

Solution

The intersection of a horizontal line from 20 atm with a vertical line from −30° falls in the liquid region.

41. What phase changes can carbon dioxide undergo as the temperature changes if the pressure is held at 20 atm? If the pressure is held a 1 atm? (See exercise 39 for a portion of the phase diagram for CO_2.)

Solution

At 20 atm, as the temperature increases from −80°C through −40°C to 0°C, the CO_2 goes through solid, liquid, and gas phases.

At 1 atm, as the temperature increases, the CO_2 goes through the solid phase directly into the gas phase (it sublimes).

ATTRACTIVE FORCES IN LIQUIDS AND SOLIDS

43. Why do the boiling points of the noble gases increase in the order He < Ne < Ar < Kr < Xe?

Solution

With greater numbers of electrons and protons, the larger atoms have greater potential for developing London force-type bonds. Greater bonding potential leads to higher boiling points.

45. Predict the member of each of the following pairs that has the stronger attractive forces. Verify your answers using the heats of vaporization calculated in exercise 30.

   (a) $Al(s)$ or $I_2(s)$
   (b) $CHCl_3(l)$ or $CCl_4(l)$
   (c) $CH_3OH(l)$ or $C_2H_5OH(l)$
   (d) $C_2H_5Cl(l)$ or $C_2H_5OH(l)$

Solution

(a) The metallic bonds in aluminum are much stronger than the covalent bonds in iodine. Heats of vaporization are Al, 326 kJ/mol and I_2, 62.4 kJ/mol.
(b) $CHCl_3$ is a polar molecule and the dipole-dipole attractions are expected to be stronger than the London forces in the nonpolar CCl_4. However, CCl_4 is larger than $CHCl_3$ so CCl_4 has larger London forces than $CHCl_3$, so the expected difference in attractive forces is less. The heats of vaporization are $CHCl_3$, 31.4 kJ/mol and CCl_4, 32.5 kJ/mol.
(c) Ethanol, C_2H_5OH, has the capacity to form London force bonds in addition to hydrogen bonding occurring at the C—OH site in both CH_3OH and C_2H_5OH. Heats of vaporization are CH_3OH, 38.0 kJ/mol and C_2H_5OH, 42.6 kJ/mol.
(d) C_2H_5OH can form hydrogen bonds in addition to the London forces in both C_2H_5OH and C_2H_5Cl. Heats of vaporization are C_2H_5Cl, 24.3 kJ/mol and C_2H_5OH, 42.6 kJ/mol.

47. The heat of vaporization of $CO_2(l)$ is 226 J/g. Would you expect the heat of vaporization of $CS_2(l)$ to be 364 J/g, 226 J/g, or 119 J/g? Explain your answer.

Solution

Due to the greater mass of CS_2 and the smaller electronegativity of S, we would expect a greater intermolecular bond strength between CS_2 molecules. In fact, CS_2 is a liquid at room temperature. It does, however, evaporate readily from an open container: 364 J/g.

49. The melting point of $H_2O(s)$ is 0°C. Would you expect the melting point of $H_2S(s)$ to be −85°C, 0°C, or 185°C? Explain your answer.

Solution

Water is a very polar molecule and develops extraordinarily strong hydrogen bonds between neighboring molecules. These strong bonds give rise to the high melting and boiling points of water. In contrast, H_2S is a gas at room temperature in part due to the larger size of the sulfur atom which limits its capacity for hydrogen bonding. We would predict −85°C.

51. Explain why a hydrogen bond between two water molecules, $HOH\cdots OH_2$, is stronger than a hydrogen bond between two ammonia molecules, $H_2NH\cdots NH_3$.

Solution

The greater electronegativity of oxygen compared to nitrogen leads to a more polar O—H bond and stronger dipole-dipole attraction (hydrogen bond) in water than in ammonia.

53. Arrange each of the following sets of compounds in order of increasing boiling temperature:

(a) HF, H_2O, CH_4
(c) CH_4, C_2H_6, C_3H_8

Solution

(a) $CH_4 < HF < H_2O$
(c) $CH_4 < C_2H_6 < C_3H_8$

PROPERTIES OF CRYSTALLINE SOLIDS

55. Magnesium metal crystallizes in a hexagonal closest packed structure. What is the coordination number of a magnesium atom?

Solution

Coordination number refers to the number of nearest neighbors. A Mg atom is surrounded by and in contact with six other Mg atoms in a layer. It is also in contact with three Mg atoms in the layer above and with three in the layer below. Consequently, it is in contact with twelve other atoms, so its coordination number is 12.

57. Describe the crystal structure of copper, which crystallizes with four equivalent metal atoms in a cubic unit cell.

Solution

A cubic cell having four equivalent atoms is face-centered cubic: $\frac{1}{2}$ atom to each face (3) plus $\frac{1}{8}$ to each corner (1), a total of four.

59. Barium crystallizes in a body-centered cubic unit cell with an edge length of 5.025 Å.

   (a) What is the atomic radius of barium in this structure?
   (b) Calculate the density of barium.

Solution

(a) In a body-centered cubic unit cell, the metal atoms are in contact along the diagonal of the cube. The diagonal of the cube forms a right triangle with the unit cell edge and the diagonal of a face. Use the Pythagorean theorem to determine the length of the diagonal, d, on the face of the cube in terms of e.

$$d^2 = e^2 + e^2 = 2e^2$$
$$d = \sqrt{2}e$$

The diagonal of the cube is the length of four atomic radii and can be calculated by again using the Pythagorean theorem.

$$(\text{diagonal})^2 = (4r)^2 = (2e)^2 + e^2$$
$$= 16r^2 = 3e^2$$
$$\text{diagonal} = 4r = \sqrt{3}e$$
$$r = \frac{\sqrt{3}}{4}e = \frac{\sqrt{3}}{4}(5.025 \text{ Å}) = 2.176 \text{ Å}$$

(b) Given a body-centered cubic structure, each unit cell contains two atoms. Use the unit cell edge length to calculate the unit cell volume and the volume occupied by each atom. Multiply to obtain the molar volume and divide the gram atomic weight by this value to obtain density (e = edge length).

$$V(\text{cell}) = e^3 = (5.025 \times 10^{-8} \text{ cm})^3 = 1.26884 \times 10^{-22} \text{ cm}^3$$
$$V(\text{atom}) = 1.26884 \times 10^{-22} \text{ cm}^3 \text{ atom}^{-1}/2 \text{ atoms} = 6.3442 \times 10^{-23} \text{ cm}^3$$
$$V(\text{mole}) = 6.3442 \times 10^{-23} \text{ cm}^3 \times 6.022 \times 10^{23} \text{ atoms/mol} = 38.204 \text{ cm}^3$$
$$D(\text{Ba}) = 137.33 \text{ g}/38.204 \text{ cm}^3 = 3.595 \text{ g/cm}^3$$

61. Aluminum crystallizes in a face-centered cubic unit cell with an aluminum atom on each lattice point with the edge length of the unit cell equal to 4.050 Å.

   (a) Calculate the atomic radius of aluminum.
   (b) Calculate the density of aluminum.

Solution

(a) Aluminum atoms are in contact along the diagonal of the cube face: diagonal = $4r(\text{Al})$.

Chapter 11 Condensed Matter: Liquids and Solids 153

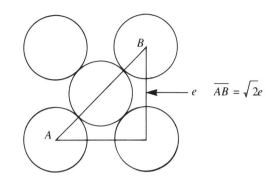

$$r\,(\text{Al}) = \frac{\sqrt{2}(4.050\ \text{Å})}{4} = 1.432\ \text{Å}$$

(b) Follow the same procedure for density as used in exercise 59(b), but noting that a face-centered cubic cell contains four atoms per cell instead of two.

$$V\,(\text{cell}) = e^3 = (4.050 \times 10^{-8}\ \text{cm})^3 = 6.643 \times 10^{-23}\ \text{cm}^3$$

$$V\,(\text{mol}) = 6.643 \times 10^{-23}\ \frac{\text{cm}^3}{\text{cell}} \times \frac{1\ \text{cell}}{4\ \text{atoms}} \times 6.022 \times 10^{23}\ \frac{\text{atoms}}{\text{mol}}$$

$$= 10.001\ \text{cm}^3/\text{mol}$$

$$D\,(\text{Al}) = 26.9815 \frac{\text{g}}{\text{mol}} \times \frac{1\ \text{mol}}{10.001\ \text{cm}^3} = 2.698\ \text{g/cm}^3$$

63. Thallous iodide crystallizes in a simple cubic array of iodide ions with thallium ions in all of the cubic holes. What is the formula of thallous iodide? Explain your answer.

Solution

In a cubic array there is one cubic hole that can be occupied for each anion. The ratio of thallium to iodide must be 1 to 1: TlI

65. Although it is a covalent compound, silicon carbide can be described as a cubic closest packed array of silicon atoms, with carbon atoms in $\frac{1}{2}$ of the tetrahedral holes. What is the coordination number of carbon in silicon carbide? What is the hybridization of the silicon and carbon atoms? Explain your answer.

Solution

A tetrahedral hole is bound by four silicon atoms in silicon carbide: Coordination number is 4; hybridization is sp^3.

67. A compound containing zinc, aluminum, and sulfur crystallizes with a closest packed array of sulfide ions. Zinc ions are found in $\frac{1}{8}$ of the tetrahedral holes, and aluminum ions in $\frac{1}{2}$ of the octahedral holes. What is the empirical formula of the compound? Explain your answer.

Solution

The ratio of Zn:Al:S is determined from the ratio of holes to the anion, S. In this case, 2 tetrahedral:1 octahedral: 1 S. Given the filling order of $\frac{1}{8}:\frac{1}{2}:1$, the empirical formula is:

Zn $\frac{1}{8}$ × 2 : Al $\frac{1}{2}$ × 1 : S   or   Zn $\frac{1}{4}$ Al $\frac{1}{2}$ S$_1$

or ZnAl$_2$S$_4$

69. Explain why the chemically similar alkali metal chlorides NaCl and CsCl have different structures, whereas chemically different NaCl and MnS have the same structure.

Solution

Structures are determined from charge and size and are really not related to chemical similarity. In this case, the r^+/r^- value for CsCl is sufficiently greater than that of NaCl that a different structure is required.

71. A cubic unit cell contains iodide ions at the corners and a thallium ion in the center. What is the formula of the compound? Explain your answer.

Solution

This is essentially a CsCl structure with a ratio of one thallium ion to one iodide ion.

73. A cubic unit cell contains a cobalt ion in the center of the cell, potassium ions at the corners of the cell, and fluoride ions in the center of each face. What is the empirical formula of this compound? What is the oxidation number of the cobalt ion? Explain your answer.

Solution

Each cell will contain a cobalt ion at the center, $8(\frac{1}{8})$ potassium at the corners and $(6)\frac{1}{2}$ fluoride ions from the faces: 1 K : 1 Co : 3 F implies the empirical formula KCoF$_3$. The oxidation state is found from the electroneutrality balance: K = +1, 3F = −3. Their sum is −2 which must be balanced by the Co charge, +2.

75. NaH crystallizes with the same crystal structure as NaCl. The edge length of the cubic unit cell of NaH is 4.880 Å.

   (a) Calculate the ionic radius of H$^-$. (The ionic radius of Na$^+$ may be found on the inside of the back cover of the main text.)
   (b) Calculate the density of NaH.

Solution

(a) Assume a face-centered arrangement with the hydride ions in contact along the diagonal of the face. Sodium ions fill the space along the edge giving an edge length of $2r_{Na^+} + 2r_{H^-}$. The radius of the hydride is computed from the edge length and the reported radius of Na$^+$.

$$e = 4.880 \text{ Å} = 2(0.95 \text{ Å}) + 2r_{H^-}$$

$$r_{H^-} = \frac{4.880 \text{ Å} - 1.90 \text{ Å}}{2} = 1.49 \text{ Å}$$

(b) Assume 4 NaH units per cell.

$$V \text{(cell)} = (4.880 \times 10^{-8} \text{ cm})^3 = 1.162 \times 10^{-22} \text{ cm}^3$$

$$V \text{(NaH unit)} = 1.162 \times 10^{-22} \text{ cm}^3/4 = 2.905 \times 10^{-23} \text{ cm}^3$$

$V/\text{mol} = 2.905 \times 10^{-23} \text{ cm}^3 \times 6.022 \times 10^{23} \text{ units/mol} = 17.496 \text{ cm}^3$

$\text{Density} = \dfrac{23.99767 \text{ g}}{17.496 \text{ cm}^3} = 1.372 \text{ g/cm}^3$

77. The lattice energy of LiF is 1023 kJ/mol, and the Li-F distance is 2.008 Å. MgO crystallizes in the same structure as LiF but with a Mg—O distance of 2.05 Å. Which of the following values most closely approximates the lattice energy of MgO: 255, 890, 1023, 2046, or 4008 kJ/mol? Explain your choice.

Solution

The lattice energy is given by $U = C(z^+z^-/R_0)$ where z^+z^- is the product of the ion charges and R_0 is the interatomic distance. For LiF, $z^+z^- = (+1)(-1) = -1$, while in MgO, $z^+z^- = (+2)(-2) = -4$. Given a small difference in R_0, we could expect nearly a factor of four difference in U. More precisely, assume that C_{LiF} and C_{MgO} are equal, then

$$U(\text{MgO}) = 1023 \, \dfrac{\text{kJ}}{\text{mol}} \left(\dfrac{2.008}{2.05} \right)(4) = 4008 \text{ kJ/mol}$$

79. What X-ray wavelength would give a second-order reflection ($n = 2$) with a θ angle of 20.40° from planes with a spacing of 4.00 Å?

Solution

The X-ray wavelength is calculated from the Bragg equation: $n\lambda = 2d \sin \theta$ where d is the distance between atomic planes, the angle of reflection, θ, and n is the order of reflection.

$$\lambda = \dfrac{2d \sin \theta}{n} = \dfrac{2(4.00 \text{ Å}) \sin 20.40°}{2}$$

$$= \dfrac{2(4.00 \text{ Å})(0.348572)}{2} = 1.39 \text{ Å}$$

81. Gold crystallizes in a face-centered cubic unit cell. The second-order reflection ($n = 2$) of X rays for the planes that make up the tops and bottoms of the unit cells is at $\theta = 22.20°$. The wavelength of the X rays is 1.54 Å. What is the density of metallic gold?

Solution

The unit cell edge is d (the Bragg equation), the spacing between planes. Calculate d and use it to compute the unit cell volumes. Given that the face-centered cell contains 4 atoms, calculate the molar volume and then the density.

$$e = d = \dfrac{n\lambda}{2 \sin \theta} = \dfrac{2(1.54 \times 10^{-8} \text{ cm})}{2 \sin 22.20°} = 4.0758 \times 10^{-8} \text{ cm}$$

$V(\text{cell}) = (4.0758 \times 10^{-8} \text{ cm})^3 = 6.7707 \times 10^{-23} \text{ cm}^3$

$V(\text{atom}) = V(\text{cell})/4 = 1.6927 \times 10^{-23} \text{ cm}^3$

$V/\text{mol} = 1.6927 \times 10^{-23} \text{ cm}^3/\text{atom} \times 6.022 \times 10^{23} \text{ atom/mol} = 10.1933 \text{ cm}^3$

$D(\text{Au}) = 196.9665 \text{ g}/10.1933 \text{ cm}^3 = 19.3 \text{ g/cm}^3$

ADDITIONAL EXERCISES

83. Explain why some molecules in a solid may have enough energy to sublime away from the solid even though the solid does not contain enough energy to melt.

    Solution

    Only surface particles escape; they escape primarily because they are not bonded to the solid in the plane away from the solid mass: Less energy is required.

85. How much energy is required to convert 135.0 g of ice at $-8.0°C$ to steam at $225°C$? (See exercise 84 for additional information.)

    Solution

    The required transitions are

    $$H_2O(s) \text{ at } -8.0°C \xrightarrow{1} H_2O(s) \text{ at } 0°C \xrightarrow{2} H_2O(l) \text{ at } 0°C$$
    $$\xrightarrow{3} H_2O(l) \text{ at } 100°C \xrightarrow{4} H_2O(g) \text{ at } 100°C \xrightarrow{5} H_2O(g) \text{ at } 225°C$$

    Heat $= q_1 + q_2 + q_3 + q_4 + q_5$

    $q_1 = 135 \text{ g} \times 2.04 \text{ J/g K} \times 8.0 \text{ K} = 2.2032 \times 10^3 \text{ J}$

    $q_2 = 135 \text{ g} \times 333.6 \text{ J/g} = 4.5036 \times 10^4 \text{ J}$

    $q_3 = 135 \text{ g} \times 4.18 \text{ J/g K} \times 100 \text{ K} = 5.643 \times 10^4 \text{ J}$

    $q_4 = 135 \text{ g} \times 2258 \text{ J/g} = 3.0483 \times 10^5 \text{ J}$

    $q_5 = 135 \text{ g} \times 2.00 \text{ J/g K} \times 125 \text{ K} = 3.375 \times 10^4 \text{ J}$

    Heat $= 4.42 \times 10^5 \text{ J} = 442 \text{ kJ}$

87. (a) A river is 20 ft wide and has an average depth of 5 ft and a current of 3 mi/h. A power plant dissipates 2.1×10^5 kJ of waste heat into the river every second. What is the temperature difference between the water upstream and downstream from the plant?

    Solution

    (a) The amount of heat transferred to the river is known, 2.1×10^5 kJ per second, but the amount of water that is to absorb this heat must be calculated from the data. To calculate the amount of water flowing past the point of discharge, assume that the volume equals that of a rectangular container having dimensions

    $$W = 20 \text{ ft} \times 12\frac{\text{in}}{\text{ft}} \times 2.54\frac{\text{cm}}{\text{in}} = 609.6 \text{ cm}$$

    $$D = 5 \text{ ft} \times 12\frac{\text{in}}{\text{ft}} \times 2.54\frac{\text{cm}}{\text{in}} = 152.4 \text{ cm}$$

    L = distance traveled at 3 mi/h in 1 s

    $$= 3\frac{\text{mi}}{\text{h}} \times 5280\frac{\text{ft}}{\text{mi}} \times 12\frac{\text{in}}{\text{ft}} \times 2.54\frac{\text{cm}}{\text{in}} \times \frac{1 \text{ h}}{3600 \text{ s}} = 134.1 \text{ cm}$$

    $V(L \times W \times D) = 134.1 \text{ cm} \times 609.6 \text{ cm} \times 152.4 \text{ cm} = 1.24 \times 10^7 \text{ cm}^3$

Assume that river water has a density of 1 g/cm³.

Mass (H₂O) = 1.24 × 10⁷ g = 1.24 × 10⁴ kg

The temperature change is calculated from:

Heat = mass × sp. heat × ΔT

$$\Delta T = \frac{\text{heat}}{mc} = \frac{2.1 \times 10^5 \text{ kJ}}{1.24 \times 10^4 \text{ kg} \times 4.18 \text{ kJ/kg°C}} = 4°C$$

89. Is work done on a liquid when it evaporates to give a gas or is work done by the liquid?

Solution

When a liquid evaporates, it expands and does work by pushing back the atmosphere.

91. How much heat must be removed to condense 2.00 L of HCN(g) at 25°C and 1.0 atm to a liquid under the same conditions?

Solution

Calculate the number of moles of HCN under these conditions. Using standard enthalpy values, compute the heat required to condense the HCN.

$PV = nRT$ and $n = PV/RT$

n = (1 atm)(2.00 L)/(0.08206 L atm/mol K)(298.15 K) = 0.08176 mol

HCN(g) ⟶ ½H₂(g) + C(s) + ½N₂(g)	$\Delta H°$	= −135 kJ
C(s) + ½N₂(g) + ½H₂(g) ⟶ HCN(l)	ΔH	= 108.9 kJ
HCN(g) ⟶ HCN(l)	$\Delta H°$	= −135 kJ + 108.9 kJ = −26.1 kJ/mol

Heat to be removed = 0.08176 mol × 26.1 $\frac{\text{kJ}}{\text{mol}}$ = 2.1 kJ

93. (a) To break each hydrogen bond in ice requires 3.5 × 10⁻²⁰ J. The measured heat of fusion of ice is 6.01 kJ/mol. Essentially all of the energy involved in the heat of fusion goes to break hydrogen bonds. What percentage of the hydrogen bonds are broken when ice is converted to liquid water?
(b) How much additional heat would be required, per mole, to break the remaining hydrogen bonds?

Solution

(a) Assume that each water molecule is bonded to two others in ice; there are two hydrogen bonds per molecule. Breaking all the bonds in one mole of ice would require breaking two moles of hydrogen bonds.

Total H-bond energy = 2 × 6.022 × 10²³ $\frac{\text{bonds}}{\text{mol}}$ × 3.5 × 10⁻²⁰ $\frac{\text{J}}{\text{bond}}$

= 4.22 × 10¹ kJ

No. H-bonds required to equal heat of fusion:

No. bonds needed = 6.01 $\frac{\text{kJ}}{\text{mol}}$ × $\frac{1 \text{ bond}}{3.5 \times 10^{-23} \text{ kJ}}$ = 1.72 × 10²³

$$\text{Total H-bonds} = 2 \times 6.022 \times 10^{23} \frac{\text{bonds}}{\text{mol}} = 1.20 \times 10^{24}$$

$$\text{Percent H-bonds broken} = \frac{1.72 \times 10^{23}}{1.20 \times 10^{24}} \times 100\% = 14\%$$

(b) Additional energy: 42.2 kJ − 6.01 kJ = 36 kJ

95. The carbon atoms in the unit cell of diamond occupy the same positions as both the zinc and sulfur atoms in cubic zinc sulfide. How many carbon atoms are found in the unit cell of diamond? The bonds between the carbon atoms are covalent. What is the hybridization of a carbon atom in diamond?

Solution

The ZnS structure has 2 tetrahedral holes per anion, one half of which are occupied by zinc ions. Each unit cell contains four atoms each of Zn and of S. In diamond there are 8 atoms per unit cell; the hybridization is sp^3 with each C surrounded by 4 other carbons.

97. In terms of its internal structure, explain the high melting point and hardness of diamond. (The structure of diamond is described in exercise 95 and shown in Fig. 11.22.)

Solution

The carbon-carbon bonds are very strong (high melting point), and the tetrahedral lattice provides a very rigid (hard) structure.

99. What is the percent by mass of titanium in rutile, a mineral that contains titanium and oxygen, if the structure of rutile can be described as a closest packed array of oxygen atoms with titanium in $\frac{1}{2}$ of the octahedral holes? What is the oxidation number of titanium?

Solution

The ratio of octahedral holes to oxygen anions is 1 to 1 in a closest packed array. Only $\frac{1}{2}$ of the octahedral holes are occupied. This means that the titanium to oxygen ratio is 1 to 2 or that the formula is TiO_2. The percentage by mass of Ti in the structure is:

$$\text{Percent Ti} = \frac{47.90}{47.90 + 2(15.9994)} \times 100\% = 59.95\%$$

and the oxidation number of Ti is + 4.

12

Solutions; Colloids

INTRODUCTION

The combination of two or more nonreacting substances forms a mixture. Mixtures are very common in our surroundings and play very important roles in life processes. In this chapter a special kind of mixture, called a *solution,* is considered. A solution is a mixture that has two or more components; one component, the solvent, serves to dissolve the other component, the solute. Usually the solvent is considered to be the substance in greater amount. Many solutions are quite familiar. A few of these are listed as specific cases and then generalized:

1. The juice of sorghum: solid (sugar) dissolved in liquid (water)
2. Stainless steel: solid (chromium) dissolved in solid (iron)
3. Air: gas (oxygen, carbon dioxide, and other gases) dissolved in gas (nitrogen)
4. Antifreeze solution: liquid (ethylene glycol) dissolved in liquid (water)

Because matter exists in the solid, liquid, and gaseous states, it is possible for substances to form nine different types of solutions, with each state serving as a solute or as a solvent.

Solutions are homogeneous mixtures; that is, the solute is distributed evenly throughout the solvent. Some substances mix with solvents and form suspensions but do not form solutions; examples include vinegar and oil as in salad dressings, butterfat in water as in homogenized milk, and dust particles in air. Whether a solute will form a solution with a solvent depends primarily on three factors. These are (1) the size of the solute particles; (2) the capacity of the solvent to form strong bonds with the solute, thereby lowering the energy of the system; and (3) the tendency of the solute particles to form a more random distribution within the solvent, further lowering the energy of the system.

Solvents composed of dipolar molecules dissolve ionic substances and other dipolar substances. Thus water, a dipolar substance, dissolves ionic compounds such as NaCl and KI as well as dipolar substances such as HCl and HNO_3. Conversely, solvents composed of nonpolar molecules dissolve nonpolar substances in which the intermolecular bonding in both solvent and solute is primarily of the van der Waals type. Examples of nonpolar solvent-nonpolar solute solutions include oil in gasoline and grease in dry-cleaning fluid. In general, solvents tend to dissolve solutes that share bonding types. This fact gives rise to an often quoted phrase, "like dissolves like."

READY REFERENCE

Boiling-point elevation; freezing-point depression (12.18, 12.20)   The addition of a nonvolatile solute to a solvent to form a solution affects the boiling point and also the freezing point of the solvent. These values are changed by amounts proportional to the number of solute particles dissolved in solution. This proportionality is usually expressed in terms of the solution molality and the boiling-point elevation constant, K_b, or the freezing-point depression constant, K_f. Both K_b and K_f have specific values for each solvent but are independent of the particular solute. The boiling-point elevation of a solvent is given by

$$\Delta bp = K_b m$$

and the freezing-point depression of a solvent is given by

$$\Delta fp = K_f m$$

For water, K_b and K_f are 0.512°C per molal and 1.86°C per molal, respectively.

Electrolytes (12.6) Substances known as electrolytes dissolve in solvents to produce solutions that conduct electric current. Acids, bases, and salts that form ions in solutions are electrolytes. Other substances are considered electrolytes when, in either a molten or supercooled state, they conduct electricity. The conduction of electric current in aqueous solutions depends on the existence of ions; in molten or supercooled systems it depends on "free" electrons. Strong electrolytes are substances that form solutions that are good electrical conductors. Strong electrolytes are extensively ionized in solution. Some, like sodium chloride dissolved in water, are nearly 100% ionized:

$$NaCl(s) + H_2O(l) \xrightarrow{\sim 100\%} Na^+(aq) + Cl^-(aq)$$

Weak electrolytes are substances that form solutions that are poor electrical conductors. Only a small fraction of a weak electrolyte's particles are ionized in solution. Acetic acid in water is such a case:

$$CH_3COOH(l) + H_2O(l) \xrightarrow{1 \text{ to } 5\%} CH_3COO^-(aq) + H^+(aq)$$

Molality, m (12.14) A unit that expresses the concentration of the solute as a ratio of moles of solute to the mass of solvent expressed in kilograms.

$$\text{molality} = \frac{\text{amount of solute (mol)}}{\text{mass of solvent (kg)}}$$

At first glance molarity and molality appear to be quite similar. A major difference in the two units is that molarity is defined in terms of the volume of solution while molality is defined in terms of the mass of solvent. The volume of a fixed quantity of a solution is temperature dependent, but the mass of solution is not. Molarity, therefore, varies with temperature, but molality does not.

Molarity, M (12.13) A unit that expresses concentration as a ratio of moles of solute to the volume of solution expressed in liters.

$$\text{molarity} = \frac{\text{amount of solute (mol)}}{\text{volume of solution (L)}}$$

Mole fraction (12.15) A unit that expresses the concentration of a component in a solution as a ratio of the number of its moles to the total number of moles of all components in that solution. The mole fraction of component A, X_A, in a solution containing components A through D, is

$$X_A = \frac{\text{amount of A (mol)}}{\text{sum of amount (mol) of each component}} = \frac{n_A}{n_A + n_B + n_C + n_D}$$

Chapter 12 Solutions; Colloids

Most solutions used in laboratories are composed of two or more components, with the mole fractions of their components calculated from the masses of the components used. Mole fractions can also be calculated from molalities and mass percentage.

Percent composition (12.12) Concentration is often expressed in terms of a percentage, because percentage depends only on the easily measured quantities mass and/or volume of the components of the solution. Percent composition is commonly expressed in two ways.

The biological, medical, and allied health professions mostly use percentage as a ratio of mass or volume of solute to volume of solution. In the first instance, if the mass of the solute is used, we have

$$\text{Mass-volume percent} = \frac{\text{mass of solute (g)}}{100 \text{ mL solution}} \times 100\%$$

Example: A 5% solution of sodium chloride contains 5 g of NaCl in 100 mL of solution, or 1000 mL contains 50 g of NaCl.

If the volume of the solute is to be compared to the volume of the solution, we have

$$\text{Volume-volume percent} = \frac{\text{volume solute (mL)}}{100 \text{ mL solution}} \times 100\%$$

Example: A 5% ethanol solution contains 5 mL of ethanol in 100 mL of solution; 1000 mL of this solution contains 50 mL of ethanol.

These percentage units are not readily adaptable to other units normally used in chemistry. They are, therefore, not commonly used in chemical calculations and are not used in the textbook.

In some areas of chemistry, percentage is expressed as a ratio of the mass of solute to the mass of the solvent. Sometimes referred to as weight-weight percentage, the percent by mass unit can be related to other concentration units including mole fraction and molality.

$$\text{Percent by mass unit} = \frac{\text{mass solute}}{\text{mass solution}} \times 100\%$$

Example: A 5% solution of sodium chloride in water contains 5 g of NaCl in 95 g H_2O. The ratio of NaCl to solution is 5 g : 100 g.

Solute and solvent (12.1) A solution consists of two parts, a solute and the solvent it is dissolved in. The solute generally is the substance in smallest concentration in a solution. The solvent is the substance in the greatest amount in the solution. The solute may be solid, liquid, or gas; the solvent may be solid, liquid, or gas. Although a solid dissolved in a liquid usually comes to mind when one thinks of a solution, there are other possibilities, as discussed in the text.

Solution (12.1) A solution is a special kind of mixture in which all of the particles are of ionic or molecular dimensions not exceeding about 1.0 pm (10 Å). In addition, the solute particles are uniformly distributed throughout the solvent. This mixture is referred to as a *homogeneous* mixture and is a true solution. Solutions are distinguished from other types of mixtures primarily by properties arising from the small size of solute particles in a solution. The particles may be molecular, ionic, or atomic; they may form small clusters in solution, which usually will not exceed diameters of more than a few angstroms.

SOLUTIONS TO CHAPTER 12 EXERCISES

THE NATURE OF SOLUTIONS

1. How do solutions differ from compounds? From ordinary mixtures?

   Solution

   Solutions are formed from two or more compounds: e.g., water (compound) and sucrose (compound, also known as table sugar). A solution is a special kind of mixture in which the component compounds are subdivided to the molecular or ionic state. In addition all components exist in the same phase— there is no separation into layers in a solution.

3. Why are the majority of chemical reactions most readily carried out in solution?

   Solution

   Chemical reactions occur at surfaces between molecules or ions. A solution offers several practical benefits. Among these are: (1) Reacting particles already exist in the molecular or ionic state thereby maximizing reaction surface area; (2) solutions, especially liquids, are easy to stir or blend and are convenient to transfer from one vessel to another; (3) reaction quantities and reaction rates are easier to control; and (4) components move about freely in a solution and can come into contact and react.

5. Describe the effect on the solubility of a gas of

   (a) the pressure of the gas,
   (b) the temperature of the solution, and
   (c) the reaction of the gas with the solvent.

   Solution

   (a) The solubility of a gas increases with an increase in the gas pressure above the solvent surface.
   (b) The solubility of a gas decreases with an increase in the temperature of the solution.
   (c) Certain gases react with solvents to produce compounds which are stable in solution. The formation of a compound has the effect of increasing the solubility of the gas.

7. In order to prepare supersaturated solutions of most solids, saturated solutions are cooled. Supersaturated solutions of gases are prepared by warming saturated solutions. Explain the reasons for the difference in the two procedures.

   Solution

   The solubility of most common substances increases with an increase in temperature. Cooling a saturated solution, therefore, creates a mixture containing a greater solute-solvent ratio than in a saturated state. Conversely, gas solubility in a solvent increases as the solvent is cooled. Heating a gas-solvent solution will cause the mixture to exceed the saturated solute-solvent ratio.

Chapter 12 Solutions; Colloids 163

9. Explain why ionic compounds that are fused (melted) are good conductors but solid ionic compounds are not.

   Solution

   Conductivity requires that ions have mobility. In ionic solids, ions exist in fixed lattice positions and are essentially immobile. The molten state allows mobility.

11. Explain why HCl is a nonelectrolyte when dissolved in carbon tetrachloride and an electrolyte when dissolved in water.

    Solution

    Strong hydrogen bonds develop between water and HCl. Hydrogen chloride reacts with water, forming H_3O^+ and Cl^- ions, thus ionization of HCl molecules occurs freely in the mixture. No such reaction occurs between nonpolar CCl_4 molecules and the polar HCl molecules.

THE PROCESS OF DISSOLUTION

13. Explain why the ions Na^+ and Cl^- are strongly solvated in water but not in benzene.

    Solution

    Strong ion-dipole bonds exist between NaCl and water with the total Na^+—H_2O and Cl^-—H_2O bond energy being greater than the NaCl lattice energy. As a consequence NaCl dissolves in water. Benzene (C_6H_6) is a nonpolar liquid, and the potential for strong ion-benzene molecule bonds is negligible.

15. Which intermolecular attractions (Sections 11.11, 11.12, 12.9, 12.10) should be most important in each of the following solutions:

    (a) NH_3 in water
    (b) methanol, CH_3OH, in ethanol, C_2H_5OH
    (c) the strong acid, sulfuric acid, H_2SO_4, in water
    (d) HCl in benzene, C_6H_6
    (e) carbon tetrachloride, CCl_4, in freon, CF_2Cl_2
    (f) $NaNO_3$ in water
    (g) The weak acid, acetic acid, CH_3CO_2H, in water

    Solution

    (a) Both are polar; dipole-dipole (hydrogen bonds)
    (b) Both methanol and ethanol hydrogen-bond to each other and to themselves.
    (c) Covalent bonds form in H_3O^+ then strong hydrogen bonds develop between H_3O^+ and HSO_4^- and water.
    (d) The strong polar nature of HCl tends to induce polarity in the nonpolar benzene which creates weak London forces between molecules and allows dissolution to occur.
    (e) London forces
    (f) ion-dipole
    (g) hydrogen bonds

17. Compare the processes that occur when ethanol, hydrogen chloride, ammonia, and sodium hydroxide dissolve in water. Write equations and prepare sketches showing the form in which each of these compounds is present in its respective solution.

Solution

Water and ethanol mix through hydrogen-bonding. Water and HCl react according to the equation

$$HCl(g) + H_2O(l) \xrightarrow{100\%} H_3O^+(aq) + Cl^-(aq)$$

then the ions are solvated by water molecules through very strong hydrogen bonds. Ammonia is very soluble in water because hydrogen bonding occurs between H_2O and NH_3. It undergoes ionization at about one percent according to the equation

$$NH_3(g) + H_2O(l) \xrightarrow{1\%} NH_4^+(aq) + OH^-(aq)$$

Sodium hydroxide is an ionic solid that bonds strongly to water by ion-dipole attraction between Na^+ and H_2O, and ion-dipole and H-bonds between OH^- and H_2O. It dissolves producing a strongly electrolytic solution.

$$NaOH(s) + H_2O(l) \xrightarrow{100\%} Na^+(aq) + OH^-(aq)$$

19. Explain, in terms of the intermolecular attractions between solute and solvent, why methanol, CH_3OH, is miscible with water in all proportions, but only 0.11 g of butanol, $CH_3CH_2CH_2CH_2OH$, will dissolve in 100 g of water.

Solution

Methanol and water are miscible in all proportions because their individual hydrogen bonds are about as strong as their hydrogen bonds to one another. Butanol molecules are much less polar, more fat-like and substantially larger which greatly reduces the potential for hydrogen bonding. Quantitatively, methanol has 3 hydrogen bonds per 6 atoms, whereas butanol has 3 hydrogen bonds per 15 atoms. A larger fraction of the butanol molecule will not hydrogen bond, hence the reduced solubility.

UNITS OF CONCENTRATION

21. Distinguish between 1 M and 1 m solutions.

Solution

Molarity (*M*) is defined as moles of solute per volume of solution whereas molality (*m*) is defined as moles of solute per kilogram of solvent. A 1 M solution contains one mole of solute dissolved in one liter of solution. A 1 m solution contains one mole of solute dissolved in one kilogram of solvent. The two solution volumes may be dramatically different.

23. There are about 10 g of calcium as Ca^{2+}, in 1.0 L of milk. What is the molarity of Ca^{2+} in milk?

Chapter 12 Solutions; Colloids

Solution

$$\text{Molarity} = \frac{\text{mol Ca}^{2+}}{\text{L of milk}} = \frac{10 \text{ g} \times \dfrac{1 \text{ mol Ca}^{2+}}{40.08 \text{ g}}}{1.0 \text{ L}} = 0.25 \text{ M}$$

25. Calculate the molarity of each of the following solutions:

    (a) 1.82 kg of H_2SO_4 per liter of concentrated sulfuric acid

    (b) 1.9×10^{-4} g of NaCN per 100 mL, the minimum lethal concentration of sodium cyanide in blood serum

    (c) 27 g of glucose, $C_6H_{12}O_6$, in 500 mL of solution used for intravenous injection

    (d) 2.20 kg of formaldehyde, H_2CO, in 5.50 L of a solution used to "fix" tissue samples

    (e) 0.029 g of I_2 in 0.100 L solution, the solubility of I_2 in water at 20°C

Solution

(a) $\text{Molarity} = \dfrac{\text{mol } H_2SO_4}{L} = \dfrac{1.82 \text{ kg} \times 1000 \dfrac{g}{kg} \times \dfrac{1 \text{ mol } H_2SO_4}{98.0734 \text{ g}}}{1 \text{ L}} = 18.6 \text{ M}$

(b) $M = \dfrac{\text{mol NaCN}}{L} = \dfrac{1.9 \times 10^{-4} \text{ g} \times \dfrac{1 \text{ mol NaCN}}{49.00647 \text{ g}}}{0.100 \text{ L}} = 3.9 \times 10^{-5} \text{ M}$

(c) $M = \dfrac{\text{mol } C_6H_{12}O_6}{L} = \dfrac{27 \text{ g} \times \dfrac{1 \text{ mol glucose}}{180.1572 \text{ g}}}{0.500 \text{ L}} = 0.30 \text{ M}$

(d) $M = \dfrac{\text{mol } H_2CO}{L} = \dfrac{2.20 \text{ kg} \times 1000 \dfrac{g}{kg} \times \dfrac{1 \text{ mol } H_2CO}{30.0262 \text{ g}}}{5.50 \text{ L}} = 13.3 \text{ M}$

(e) $M = \dfrac{\text{mol } I_2}{L} = \dfrac{0.029 \text{ g} \times \dfrac{1 \text{ mol } I_2}{253.809 \text{ g}}}{0.100 \text{ L}} = 1.1 \times 10^{-3} \text{ M}$

27. Calculate the mole fractions of solute and solvent in each of the solutions in exercise 26.

Solution

The mole fraction of a specific component in a solution is its ratio in moles to the total number of moles of all components in the solution. The solutions in this exercise all have two components—a solute and a solvent. Calculate the number of moles of each component, then divide each component factor by the total moles to obtain the fraction.

(a) $\text{mol (NaHCO}_3) = 69 \text{ g} \times \dfrac{1 \text{ mol}}{84.00687 \text{ g}} = 0.82136 \text{ mol}$

$\text{mol (H}_2\text{O}) = 1.00 \times 10^3 \text{ g} \times \dfrac{1 \text{ mol}}{18.0152 \text{ g}} = 55.50868 \text{ mol}$

$n_{\text{Total}} = 56.3300 \text{ mol}$

$$X\,(\text{H}_2\text{O}) = \frac{55.50868 \text{ mol}}{56.3300 \text{ mol}} = .985$$

$$X\,(\text{NaHCO}_3) = \frac{0.82136 \text{ mol}}{56.3300 \text{ mol}} = .015$$

(b) $\text{mol (H}_2\text{SO}_4) = 583 \text{ g} \times \dfrac{1 \text{ mol}}{98.0734 \text{ g}} = 5.9445 \text{ mol}$

$\text{mol (H}_2\text{O}) = 1.50 \times 10^3 \text{ g} \times \dfrac{1 \text{ mol}}{18.0152 \text{ g}} = 83.26302 \text{ mol}$

$n_{\text{Total}} = 89.2075 \text{ mol}$

$$X\,(\text{H}_2\text{O}) = \frac{83.26302 \text{ mol}}{89.2075 \text{ mol}} = 0.933$$

$$X\,(\text{H}_2\text{SO}_4) = \frac{5.9445 \text{ mol}}{89.2075 \text{ mol}} = 0.0666$$

(c) $\text{mol (NH}_4\text{NO}_3) = 120 \text{ g} \times \dfrac{1 \text{ mol}}{80.0432 \text{ g}} = 1.49919 \text{ mol}$

$\text{mol (H}_2\text{O}) = 250 \text{ g} \times \dfrac{1 \text{ mol}}{18.0152 \text{ g}} = 13.8772 \text{ mol}$

$n_{\text{Total}} = 15.37639 \text{ mol}$

$$X\,(\text{NH}_4\text{O}_3) = \frac{1.49919 \text{ mol}}{15.37639 \text{ mol}} = 0.0975$$

$$X\,(\text{H}_2\text{O}) = \frac{13.8772 \text{ mol}}{15.37639 \text{ mol}} = 0.902$$

(d) $\text{mol (NaCl)} = 0.86 \text{ g} \times \dfrac{1 \text{ mol}}{58.443 \text{ g}} = 0.0147 \text{ mol}$

$\text{mol (H}_2\text{O}) = 100 \text{ g} \times \dfrac{1 \text{ mol}}{18.0152 \text{ g}} = 5.551 \text{ mol}$

$n_{\text{Total}} = 5.566 \text{ mol}$

$X\,(\text{NaCl}) = 2.64 \times 10^{-3}$

$X\,(\text{H}_2\text{O}) = 0.997$

(e) $\text{mol (codeine)} = 46.85 \text{ g} \times \dfrac{1 \text{ mol}}{299.3688 \text{ g}} = 0.156496 \text{ mol}$

$\text{mol (ethanol)} = 125.5 \text{ g} \times \dfrac{1 \text{ mol}}{46.0688 \text{ g}} = 2.724186 \text{ mol}$

$n_{\text{Total}} = 2.88068 \text{ mol}$

$$X\,(\text{codeine}) = \frac{0.156496 \text{ mol}}{2.88068 \text{ mol}} = 0.05433$$

$$X\,(\text{ethanol}) = \frac{2.724186 \text{ mol}}{2.88068 \text{ mol}} = 0.9457$$

Chapter 12 Solutions; Colloids

(f) $\text{mol } (I_2) = 25 \text{ g} \times \dfrac{1 \text{ mol}}{253.809 \text{ g}} = 0.098499 \text{ mol}$

$\text{mol (ethanol)} = 125 \text{ g} \times \dfrac{1 \text{ mol}}{46.0688 \text{ g}} = 2.71333 \text{ mol}$

$n_{\text{Total}} = 2.81183 \text{ mol}$

$X (I_2) = \dfrac{0.098499 \text{ mol}}{2.81183 \text{ mol}} = 0.035$

$X (\text{ethanol}) = \dfrac{2.71333 \text{ mol}}{2.81183 \text{ mol}} = 0.965$

29. What mass of sodium hydroxide (50.2% by mass) is needed to prepare 200.0 g of a 20.0% solution of NaOH by mass?

Solution

The solution must contain 20.0% by mass NaOH. Therefore, 200.0 g of solution contains:

$\text{Mass (NaOH)} = 0.200 \times 200.0 \text{ g} = 40.0 \text{ g}$

The 40.0 g mass is to be obtained from a 50.2% pure substance. The amount of substance required is:

$0.502 \text{ (mass of sample)} = 40.0 \text{ g}$

$\text{Mass (sample)} = \dfrac{40.0 \text{ g}}{0.502} = 79.7 \text{ g}$

31. What mass of HCl is contained in 45.0 mL of an HCl solution with a density of 1.19 g/mL and containing 37.21% HCl by mass?

Solution

The solution contains 37.21% HCl by mass and the remainder is water. Calculate the mass of the 45.0 mL sample, then multiply by the percentage to obtain the mass of HCl.

$\text{Mass (sample)} = 45.0 \text{ mL} \times 1.19 \text{ g/mL} = 53.55 \text{ g}$

$\text{Mass (HCl)} = 53.55 \text{ g} \times 0.3721 = 19.9 \text{ g}$

33. The hardness of water (hardness count) is usually expressed as parts per million (by mass of $CaCO_3$), which is equivalent to milligrams of $CaCO_3$ per liter of water. What is the molar concentration of Ca^{2+} ions in a water sample with a hardness count of 175?

Solution

Since $CaCO_3$ contains 1 mol Ca^{2+} per mol of $CaCO_3$, the molar concentration of Ca^{2+} equals the molarity of $CaCO_3$.

$M (Ca^{2+}) = \dfrac{\text{mol } CaCO_3}{L} = \dfrac{175 \text{ mg} \times \dfrac{1 \text{ mol}}{100.0792 \text{ g}} \times \dfrac{1 \text{ g}}{1000 \text{ mg}}}{1 \text{ L}} = 1.75 \times 10^{-3} \text{ M}$

35. The concentration of glucose, $C_6H_{12}O_6$, in normal spinal fluid is 75 mg/100 g. What is the molal concentration?

Solution

Assume that spinal fluid is simplified to water and glucose. Calculate moles of glucose from the specified mass, subtract the mass of glucose from the 100 g sample and apply the definition of molality.

$$\text{mol (glucose)} = 75 \text{ mg} \times \frac{1 \text{ g}}{1000 \text{ mg}} \times \frac{1 \text{ mol}}{180.1572 \text{ g}} = 0.0004163 \text{ mol}$$

$$\text{Molality} = \frac{0.0004163 \text{ mol}}{(100 \text{ g} - 0.075 \text{ g})/1000 \text{ g kg}^{-1}} = 4.2 \times 10^{-3} \text{ m}$$

37. A 1.577 M solution of $AgNO_3$ has a density of 1.220 g/cm^3. What is the molality of the solution?

Solution

Assume a volume of one liter for the calculations. The amount of $AgNO_3$ is 1.577 mol and the remaining mass is that of water.

$$\text{Mass (1.0 L of solution)} = 1.220 \text{ g/cm}^3 \times 1000 \text{ cm}^3/\text{L} = 1220 \text{ g}$$

$$\text{Mass (AgNO}_3) = 1.577 \text{ mol} \times \frac{169.8729 \text{ g}}{\text{mol}} = 267.8896 \text{ g}$$

$$\text{Mass (H}_2\text{O}) = 1220 \text{ g} - 267.8896 \text{ g} = 952 \text{ g} = .952 \text{ kg}$$

$$\text{Molality} = \frac{1.577 \text{ mol AgNO}_3}{0.952 \text{ kg H}_2\text{O}} = 1.656 \text{ m}$$

39. A 15.0% solution of K_2CrO_4 by mass has a density of 1.129 g/cm^3. Calculate the molarity of the solution.

Solution

Assume a volume of one liter. Calculate the number of moles of K_2CrO_4 from the percentage in solution. The molarity follows from the definition.

$$\text{Mass K}_2\text{CrO}_4 \text{ per liter} = 0.15 \times 1.129 \frac{\text{g}}{\text{cm}^3} \times 1000 \frac{\text{cm}^3}{\text{L}} = 169.35 \text{ g}$$

$$\text{mol (K}_2\text{CrO}_4) = 169.35 \text{ g} \times \frac{1 \text{ mol}}{194.1902 \text{ g}} = 0.872 \text{ mol/L}$$

41. What volume of 0.333 M HNO_3 is required to react completely with 1.25 g of sodium hydrogen carbonate?

$$NaHCO_3 + HNO_3 \longrightarrow NaNO_3 + CO_2 + H_2O$$

Solution

The reactants react on a 1 mol : 1 mol basis. Calculate the number of moles of $NaHCO_3$ available, then using the definition of molarity, calculate the volume of HNO_3 required.

mol $NaHCO_3$ available = mol HNO_3 required

Chapter 12 Solutions; Colloids

$$\text{mol (NaHCO}_3\text{)} = 1.25 \text{ g} \times \frac{1 \text{ mol}}{84.00687 \text{ g}} = 0.0148797 \text{ mol}$$

$$M \text{ (HNO}_3\text{)} = \frac{\text{mol HNO}_3}{V(L)} \quad \text{or} \quad \text{mol (HNO}_3\text{)} = M \times V(L)$$

$$\text{mol (HNO}_3\text{)} = 0.0148797 \text{ mol} = 0.333 \text{ M} \times V(L)$$

$$V(L) = \frac{0.0148797 \text{ mol}}{0.333 \text{ M}} = 0.0447 \text{ L} = 44.7 \text{ mL}$$

43. What volume of a 0.33 M solution of hydrochloric acid, HCl, would be required to neutralize completely 1.00 L of 0.215 M barium hydroxide, $Ba(OH)_2$?

Solution

Write the balanced equation. The equation shows

$$2HCl + Ba(OH)_2 \longrightarrow BaCl_2 + 2H_2O$$

that 2 mol of HCl are required for each mol of $Ba(OH)_2$ used. Both reactants are in solution, and the amount of each component must be calculated by using the definition of molarity.

$$\text{mol Ba(OH)}_2 \text{ available} = 1.00 \text{ L} \times 0.215 \frac{\text{mol}}{\text{L}} = 0.215 \text{ mol}$$

$$\text{mol (HCl) required} = 2(0.215 \text{ mol}) = 0.430 \text{ mol}$$

$$V \text{ (HCl) solution} = 0.430 \text{ mol} \times \frac{1.0 \text{ L}}{0.33 \text{ mol HCl}} = 1.3 \text{ L}$$

45. A gaseous solution was found to contain 25% H_2, 20% CO, and 55% CO_2 by mass. What is the mole fraction of each component?

Solution

Assume that the solution weighs 100 g, then calculate the mass of each component. Next calculate moles of each component, then obtain each mole fraction.

Masses: H_2 = 25 g; CO = 20 g; CO_2 = 55 g

n_{H_2} = 25 g × 1 mol/2.0158 g = 12.402024 mol

n_{CO} = 20 g × 1 mol/28.0104 g = 0.7140205 mol

n_{CO_2} = 55 g × 1 mol/44.0098 g = 1.2497217 mol

n_T = 14.365766

X_{H_2} = 12.402024 mol/14.365766 mol = 0.863

X_{CO} = 0.7140205 mol/14.365766 mol = 0.050

X_{CO_2} = 1.2497217 mol/14.365766 mol = 0.087

47. Calculate the mole fractions of methanol, CH_3OH, ethanol, C_2H_5OH, and water in a solution that is 50% methanol, 30% ethanol, and 20% water by mass.

Solution

Assume that the total solution mass is 100 g. Calculate the mass and the number of moles of each component, then calculate the respective mole fractions.

$$\text{mol (CH}_3\text{OH)} = 0.50 \times 100 \text{ g} \times 1 \text{ mol}/32.042 \text{ g} = 1.56045 \text{ mol}$$

$$\text{mol (C}_2\text{H}_5\text{OH)} = 0.30 \times 100 \text{ g} \times 1 \text{ mol}/46.0688 \text{ g} = 0.651199 \text{ mol}$$

$$\text{mol (H}_2\text{O)} = 0.20 \times 100 \text{ g} \times 1 \text{ mol}/18.0152 \text{ g} = \underline{1.1101736 \text{ mol}}$$

$$n_T = 3.32182$$

$X \text{(CH}_3\text{OH)} = 1.56045 \text{ mol}/3.32182 \text{ mol} = 0.47$
$X \text{(C}_2\text{H}_5\text{OH)} = 0.651199 \text{ mol}/3.32182 \text{ mol} = 0.20$
$X \text{(H}_2\text{O)} = 1.1101736 \text{ mol}/3.32182 \text{ mol} = 0.33$

49. Calculate

    (a) the percent composition and
    (b) the molality of an aqueous solution of sucrose, $C_{12}H_{22}O_{11}$, if the mole fraction of sucrose is 0.0677.

Solution

For simplicity, assume that the total number of moles from both components is exactly one. The number of moles of sucrose and of water is:

$$\text{mol (sucrose)} = 0.0677 \times 1 \text{ mol} = .0677 \text{ mol}$$

$$\text{mol (H}_2\text{O)} = 1 - .0677 = 0.9323 \text{ mol}$$

(a) Mass (sucrose) = $0.0677 \text{ mol} \times 342.2992 \text{ g/mol} = 23.17366 \text{ g}$

  Mass (H$_2$O) = $0.9323 \text{ mol} \times 18.0152 \text{ g/mol} = 16.7956 \text{ g}$

  $$\text{Percent sucrose} = \frac{23.17366 \text{ g}}{39.9692 \text{ g}} = 58.0\%$$

(b) $$\text{Molality} = \frac{\text{mol sucrose}}{\text{kg H}_2\text{O}} = \frac{0.0677 \text{ mol}}{0.0167956 \text{ kg}} = 4.03 \text{ m}$$

51. What volume of a sulfuric acid solution that is 96.0% by mass with a specific gravity of 1.84 at 25°C is required to prepare 8.00 L of a 1.50 M solution of sulfuric acid at 25°C? The density of water at 25°C is 0.99707 g/mL.

Solution

The dilute solution (1.50 M) is to be prepared from the stock concentrate (96.0% pure). First calculate the mass of H$_2$SO$_4$ required to prepare the dilute solution. Next calculate the mass of H$_2$SO$_4$ in 1 mL of the concentrate. Divide to obtain the volume required.

Mass H$_2$SO$_4$ needed = $8.00 \text{ L} \times 1.50 \text{ mol/L} \times 98.0734 \text{ g/mol} = 1176.88 \text{ g}$

Mass H$_2$SO$_4$/mL concentrate = $0.96 \times 1.84 \text{ g/mL} = 1.7664 \text{ g/mL}$

$$V \text{(conc)} = 1176.88 \text{ g} \times \frac{1 \text{ mL}}{1.7664 \text{ g}} = 666 \text{ mL}$$

Chapter 12 Solutions; Colloids

53. (a) What volume of 0.75 M HBr is required to prepare 1.0 L of 0.33 M HBr?
    (b) What volume of 3.50% KOH by mass (density 1.012 g/mL) can be prepared from 0.150 L of 30.0% KOH by mass (density 1.288 g/mL)?

    Solution

    (a) Apply the dilution formula to calculate the volume of concentrated solution needed.

    $M_C V_C = M_D V_D$

    $(0.75 \text{ M})(V_C) = (0.33 \text{ M})(1.0 \text{ L})$

    $V_C = 0.44 \text{ L}$

    (b) The dilution formula applies here except that the density of each solution must be included in order to compare each solution.

    $(30.0\% \times 1.288 \text{ g/mL})(0.150 \text{ L}) = (3.50\% \times 1.012 \text{ g/mL})(V_D)$

    $V_D = \dfrac{0.300 \times 1.288 \text{ g/mL} \times 0.150 \text{ L}}{.0350 \times 1.012 \text{ g/mL}} = 1.64 \text{ L}$

COLLIGATIVE PROPERTIES

55. Which will evaporate faster under the same conditions, 50 mL of distilled water or 50 mL of sea water?

    Solution

    Because dissolved minerals in sea water lower its vapor pressure and increase its boiling point, water will evaporate faster from the distilled water.

57. Why will a mole of hydrogen chloride depress the freezing point of 1 kg of water by almost twice as much as a mole of ethanol?

    Solution

    Freezing point depression is determined by the number of moles of solute particles present in solution. Hydrogen chloride ionizes in water to create approximately two moles of ions (H_3O^+ + Cl^-) from each mole of HCl dissolved. Ethanol, by comparison, ionizes only slightly. Therefore, an equimolar amount of HCl will depress the freezing point by about twice as much as ethanol.

59. Explain what is meant by osmotic pressure. How is the osmotic pressure of a solution related to its concentration?

    Solution

    Osmotic pressure is the pressure required to prevent diffusion of solvent across a semipermeable membrane that separates a solution from pure solvent. The pressure is directly proportional to the concentration of solute.

61. A 1 m solution of HCl in benzene has a freezing point of 0.4°C. Is HCl an electrolyte in benzene? Explain.

Solution

If no dissociation of HCl occurs in benzene, the freezing point should be depressed by 5.12°C given a 1 m solution. Since the observed freezing point of the solution is 0.4°C, the freezing point has been depressed by:

Normal f.p. (Table 12.2) = 5.48°C

Δf.p. = 5.48°C − 0.4°C = 5.08°C

Given only one significant figure in the observed freezing point and since a depression of 5.08°C approximately equals the depression constant, 5.12°C, we an assume that little, if any, dissociation occurs: HCl is a nonelectrolyte.

63. Some mammals, including humans, excrete hypertonic urine (Section 12.22) in order to conserve water. In humans, some parts of the ducts of the kidney (which contain the urine) pass through a fluid that contains a much more concentrated salt solution than that normally found in the body. Explain how this could help conserve water in the body.

Solution

The hypertonic solution causes water to be drawn from the urine, thus decreasing the volume of water eliminated.

65. A solution contains 15.00 g of urea, $CO(NH_2)_2$, per 0.200 kg of water. If the vapor pressure of water at 25°C is 23.7 torr, what is the vapor pressure of the solution at 25°C?

Solution

The change in vapor pressure is an illustration of Raoult's law. The new vapor pressure is calculated from the mole fraction of water in the solution by using:

P solvent in solution = P pure solvent $\times$ X solvent

mol (urea) = 15.00 g $\times$ 1 mol/60.0554 g = 0.24977 mol

mol (H_2O) = 200 g $\times$ 1 mol/18.0152 g = 11.101736 mol

n_{Total} = 11.351506 mol

$$X_{H_2O} = \frac{11.101736 \text{ mol}}{11.351506 \text{ mol}} = 0.97799$$

$P_{solution}$ = 23.7 torr $\times$ 0.97799 = 23.2 torr

67. A sample of an organic compound (a nonelectrolyte) weighing 1.350 g lowered the freezing point of 10.0 g of benzene by 3.66°C. Calculate the molecular weight of the organic compound.

Solution

The molecular weight of the sample is calculated by using the relationship of molality to the change in freezing point.

$$\Delta fp = K_f m = K_f \frac{\text{mol solute}}{\text{kg solvent}}$$

$K_f = 5.12°C/m$

Rearranging, solving for mol solute and substitution gives

$$\text{mol solute} = \frac{(\Delta fp)(\text{kg solvent})}{K_f} = \frac{(3.66°C)(0.0100 \text{ kg})}{5.12°C/m}$$

mol solute = 0.0071484 mol = 1.350 g solute

1 mol = 189 g

69. What is the approximate freezing point of a 0.37 m aqueous solution of sodium bromide? Assume complete dissociation of this electrolyte.

Solution

Sodium bromide is a strong electrolyte and will be extensively dissociated in solution; assume 100% dissociation in this example.

$$\text{NaBr}(s) + \text{H}_2\text{O} \xrightarrow{100\%} \text{Na}^+(aq) + \text{Br}^-(aq)$$
0.37 mol   0.37 mol   0.37 mol

$\Delta fp = K_f m = 1.86°C/m \times 2 \times 0.37 \text{ m} = 1.38°C$

fp = 0°C − 1.4°C = −1.4°C

71. If 26.4 g of the nonelectrolyte dibromobenzene, $C_6H_4Br_2$, is dissolved in 0.250 kg of benzene, what is

(a) the freezing point of the solution and
(b) the boiling point of the solution at 1 atm?

Solution

Calculate the molality of the solution, the change in freezing point and finally, the new freezing point.

g-mol. wt $(C_6H_4Br_2) = 235.9056$ g

(a) $\Delta fp = K_f m = 5.12°C/m \dfrac{\left(26.4 \text{ g} \times \dfrac{1 \text{ mol}}{235.9056 \text{ g}}\right)}{0.250 \text{ kg}} = 2.2919°C$

fp = 5.48°C − 2.29°C = 3.19°C

(b) $\Delta bp = K_b m = 2.53°C/m \ (0.44764 \text{ m}) = 1.13°C$

bp = 80.1°C + 1.13°C = 81.2°C

73. What is the boiling point, at 1 atm, of a solution containing 115.0 g of sucrose, $C_{12}H_{22}O_{11}$, in 350.0 g of water?

Solution

$\Delta bp = K_b m = 0.512°C/m \dfrac{\left(115.0 \text{ g} \times \dfrac{1 \text{ mol}}{342.2992 \text{ g}}\right)}{0.3500 \text{ kg}} = 0.491°C$

bp = 100.0°C + 0.491°C = 100.5°C

75. The osmotic pressure of solution containing 7.0 g of insulin per liter is 23 torr at 25°C. What is the molecular weight of insulin?

Solution

Rearrange the algebraic definition of osmotic pressure to give the molecular weight in terms of π and mass of insulin.

$$\pi = MRT = \frac{\left(\frac{\text{mass(insulin)}}{\text{g-mol. wt}}\right)}{L} RT$$

$$\text{g-mol. wt} = \frac{(\text{mass(insulin)})(R)(T)}{\pi \times L}$$

$$\text{g-mol. wt} = \frac{7.0 \text{ g} \times 0.082 \text{ L atm/mol K} \times 298.15 \text{ K}}{23/760 \text{ atm} \times 1 \text{ L}} = 5700 \text{ g}$$

COLLOIDS

77. What is the difference between dispersion methods and condensation methods for preparing colloidal systems?

Solution

Dispersion is normally accomplished by taking finely divided, powdered substances and blending them with a high speed stirrer into a suspending medium. Condensation techniques often involve molecules or ions that can become hydrated on contact with water and in turn bond to adjacent hydrated units forming gels. Water purification processes often involve the addition of alum, which in turn hydrates to form a gelatinous precipitate.

79. Identify the dispersed phase and the dispersion medium in each of the following colloidal systems: starch dispersion, smoke, fog, pearl, whipped cream, floating soap, jelly, milk, and opal.

Solution

Starch-water; unburned hydrocarbons-air; water-air; water-$CaCO_3$; air-cream; air-soap; fruit solids and sugar-water; butterfat-water; water-SiO_2

81. Discuss the similarities and differences of soaps and detergents.

Solution

Common soaps are the alkali metal salts of fatty acids derived from animal or vegetable fats. Detergents are salts derived through the reaction of fatty alcohols with sulfuric acid followed by reaction with common bases. Both soaps and detergents have similar actions in dissolving grease and dirt. Detergents, however, function well in hard water while soaps tend to precipitate. Because of this definite advantage, modern personal care cleansing products are made from detergents such as sodium lauryl sulfate.

Chapter 12 Solutions; Colloids

83. What is the structure of a gel?

Solution

Gels are formed through the absorption of a solvent into a fibrous network, which takes on three-dimensional structure that apparently involves strong bonds between the solvent and the dispersed phase. Gelatin, for example, is a protein which when dispersed in hot water forms a highly organized system involving multiple bond types.

ADDITIONAL EXERCISES

85. For which of the following gases is solubility in water not directly proportional to the pressure of the gas: HCl, H_2, SO_2, NH_3, CH_4, N_2? Explain.

Solution

HCl, SO_2, and NH_3. These gases dissolve in water and undergo reaction to form products as shown:

$$HCl(g) + H_2O(l) \xrightarrow{100\%} H_3O^+(aq) + Cl^-(aq)$$

$$SO_2(g) + H_2O(l) \xrightarrow{100\%} H_2SO_3(aq)$$

$$NH_3(g) + H_2O(l) \xrightarrow{1\%} NH_4^+(aq) + OH^-(aq)$$

87. What is a constant boiling solution? Describe two ways to prepare a constant boiling solution of hydrochloric acid.

Solution

A solution boils at a constant temperature when the solution and the vapor have the same composition.

Method 1: Start with a dilute solution containing HCl and water. Begin heating the solution and start it boiling. The boiling point will increase until the HCl-H$_2$O vapor and the solution have the same concentration. The boiling temperature will remain at 110°C (760 torr).

Method 2: Saturate a solution with HCl gas. Follow the procedure as in Method 1. The initial vapor will contain an excess of HCl. As the boiling continues, the ratio will contain more water. Eventually the ratio will stabilize and the boiling temperature will remain constant.

89. The approximate radius of a water molecule, assuming a spherical shape, is 1.40 Å (1 Å = 10^{-8} cm). Assume that water molecules cluster around each metal ion in a solution so that the water molecules essentially touch both the metal ion and each other. On this basis, and assuming that 4, 6, 8, and 12 are the only possible coordination numbers, what is the maximum number of water molecules that can hydrate each of the following ions?

(a) Mg^{2+} (radius 0.65 Å)
(b) Al^{3+} (0.50 Å)
(c) Rb^+ (1.48 Å)
(d) Sr^{2+} (1.13 Å)

176 Chapter 12 Solutions; Colloids

Solution

Use the radius-ratio rule discussed in Chapter 11.

(a) $r(Mg^{2+})/r(H_2O) = 0.65/1.40 = 0.464$ 6
(b) $r^+/r(H_2O) = 0.50/1.40 = 0.357$ 4
(c) $r^+/r(H_2O) = 1.48/1.40 = 1.06$ 12
(d) $r^+/r(H_2O) = 1.13/1.40 = 0.807$ 8

91. How many liters of HCl(g) at 25°C and 1.26 atm are required to prepare 2.50 L of a 1.50 M solution of HCl?

Solution

Calculate the number of moles of HCl required by applying the definition of molarity. Next use the ideal gas equation to calculate the volume required under the stated conditions.

No. mol (HCl) = 1.50 M × 2.50 L = 3.75 mol

$$V = \frac{nRT}{P} = \frac{(3.75 \text{ mol})(0.08206 \text{ L atm/mol K})(298.15 \text{ K})}{1.26 \text{ atm}} = 72.8 \text{ L}$$

93. A 0.300-L volume of gaseous ammonia measured at 28.0°C and 754 torr was absorbed in 0.100 L of water. How many milliliters of 0.200 M hydrochloric acid are required in the neutralization of this aqueous ammonia? What is the molarity of the aqueous ammonia solution? (Assume no change in volume when the gaseous ammonia is added to the water.)

Solution

Calculate moles of ammonia in the sample.

$$n = PV/RT = \frac{(754/760 \text{ atm})(0.300 \text{ L})}{(0.08206 \text{ L atm/mol K})(301.15 \text{ K})} = 0.01204 \text{ mol}$$

The reaction with HCl occurs on a 1 mol : 1 mol basis.

$$NH_3(aq) + HCl(aq) \longrightarrow NH_4^+(aq) + Cl^-(aq)$$

No. mol (HCl) = 0.01204 mol = M × L

$$V(L) = \frac{0.01204 \text{ mol}}{0.200 \text{ mol/L}} = 0.0602 \text{ L} = 60.2 \text{ mL}$$

$$M(NH_3) = \frac{0.01204 \text{ mol}}{0.100 \text{ L}} = 0.120 \text{ M}$$

95. Calculate the volume of the concentrated acid and the volume of water required to produce 0.525 L of 0.105 M nitric acid by diluting 10.00 M nitric acid. Assume that the volumes of nitric acid and water are additive.

Solution

Use the dilution formula to determine the volume concentrate required.

$$M_C V_C = M_D V_D$$

Chapter 12 Solutions; Colloids   177

$$V_C = \frac{M_D V_D}{M_C} = \frac{(0.105 \text{ M})(0.525 \text{ L})}{(10.00 \text{ M})} = 0.00551 \text{ L} = 5.51 \text{ mL}$$

Difference (H_2O) = 525 mL − 5.51 mL = 519 mL

97. A solution of sodium carbonate having a volume of 0.250 L was prepared from 2.032 g of $Na_2CO_3 \cdot 10H_2O$. Calculate the molarity of this solution.

Solution

$$M = \frac{\text{moles }(Na_2CO_3 \cdot 10H_2O)}{0.250 \text{ L}} = \frac{2.032 \text{ g} \times 1 \text{ mol}/285.99 \text{ g}}{0.250 \text{ L}} = 0.0284 \text{ M}$$

99. A sample of $HgCl_2$ weighing 9.41 g is dissolved in 32.75 g of ethanol, C_2H_5OH ($K_b = 1.20°C/m$). The boiling-point elevation of the solution is 1.27°C. Is $HgCl_2$ an electrolyte in ethanol? Show your calculations.

Solution

$$\Delta bp = K_b m = (1.20°C/m)\left(\frac{9.41 \text{ g} \times 1 \text{ mol } HgCl_2/271.496 \text{ g}}{0.03275 \text{ kg}}\right) = 1.27°C$$

The observed change equals the theoretical change; therefore, no dissociation occurs.

101. A solution of 0.045 g of an unknown organic compound in 0.550 g of camphor melts at 158.4°C. The melting point of pure camphor is 178.75°C. K_f for camphor is 37.7°C/m. The solute contains 93.46% C and 6.54% H by mass. What is the molecular formula of the solute? Show your calculations.

Solution

Calculate the molecular weight of the substance. Then, using methods for calculating the empirical formula of the compound, compare the empirical formula weight to the calculated molecular weight to determine the formula of the compound.

$$\Delta fp = K_f m = K_f \left(\frac{\text{mass solute/g-mol. wt}}{\text{mass camphor (kg)}}\right)$$

$$\text{g-mol. wt} = \frac{(K_f)(\text{mass solute})}{(\Delta fp)(\text{No. kg camphor})} = \frac{(37.7°C/m)(0.045 \text{ g})}{(20.0°C)(0.000550 \text{ kg})} = 154.2 \text{ g/mol}$$

Empirical formula:

C: 93.46%/12.011 = 7.781   7.781/6.4887 = 1.199 or 1.2

H: 6.54%/1.0079 = 6.4887   6.4887/6.4887 = 1

Ratio: $C_{1.2}H_1$ or $C_{12}H_{10}$

The formula weight of $C_{12}H_{10}$ agrees with the calculated formula weight.

103. The sugar fructose contains 40.0% C, 6.7% H, and 53.3% O by mass. A solution of 11.7 g of fructose in 325 g of ethanol has a boiling point of 78.59°C. The boiling point of ethanol is 78.35°C, and K_b for ethanol is 1.20°C/m. What is the molecular formula of fructose?

Solution

Apply the same techniques as in exercise 101.

$$\Delta bp = K_b m \text{ and g-mol. wt} = (K_b)\frac{\text{(mass fructose)}}{\text{(mass ethanol)}}$$

$$\text{g-mol. wt} = \frac{1.20°C/m \times 11.7 \text{ g}}{(0.24°C)(0.325 \text{ kg})} = 180 \text{ g/mol}$$

Empirical formula:

    C: 40.0%/12.011 = 3.330   3.330/3.33 = 1

    H: 6.7%/1.0079 = 6.647   6.647/3.33 = 2 $C_1H_2O_1$

    O: 53.3%/15.9994 = 3.331   3.331/3.33 = 1

Given the g-mol. wt (180), the molecular formula is $C_6H_{12}O_6$.

105. Calculate the molarity, the molality, and the percent by mass of the following solutions of commercially available acids at 25°C. The density of water at 25°C is 0.99707 g/mL.

(a) a solution of 452 g of HCl in 1.0 L (specific gravity 1.19)
(b) a solution of 994 g of HNO_3 in 1.0 L (specific gravity 1.42)
(c) a solution of 1.75 kg of H_2SO_4 in 1.0 L of solution (specific gravity 1.84)

Solution

(a) $M = \dfrac{\text{No. mol HCl}}{L} = \dfrac{452 \text{ g} \times 1 \text{ mol}/36.4609 \text{ g}}{1.0 \text{ L}} = 12 \text{ M}$

$m = \dfrac{\text{No. mol HCl}}{\text{No. kg } H_2O} = \dfrac{452 \text{ g} \times 1 \text{ mol}/36.4609 \text{ g}}{\left(1.19 \text{ g/mL} \times 1000 \dfrac{mL}{L} \times \dfrac{1 \text{ kg}}{1000 \text{ g}}\right) - 0.452 \dfrac{kg}{L}} = 17 \text{ m}$

Percent by mass = $\dfrac{452 \text{ g HCl}}{1190 \text{ g solution}} \times 100\% = 38\%$

(b) $M = \dfrac{\text{No. mol } HNO_3}{L} = \dfrac{994 \text{ g} \times 1 \text{ mol}/63.0128 \text{ g}}{1.0 \text{ L}} = 16 \text{ M}$

$m = \dfrac{\text{No. mol } HNO_3}{\text{No. kg } H_2O} = \dfrac{994 \text{ g} \times 1 \text{ mol}/63.0128 \text{ g}}{\left(1.42 \text{ g/mL} \times \dfrac{1000 \text{ mL}}{L} \times \dfrac{kg}{1000 \text{ g}}\right) - 0.994 \dfrac{kg}{L}} = 37 \text{ m}$

Percent by mass = $\dfrac{994 \text{ g } HNO_3}{1420 \text{ g solution}} \times 100\% = 70\%$

(c) $M = \dfrac{\text{No. mol } H_2SO_4}{L} = \dfrac{1750 \text{ g} \times 1 \text{ mol}/98.07 \text{ g}}{1.0 \text{ L}} = 18 \text{ M}$

$m = \dfrac{\text{No. mol } H_2SO_4}{\text{No. kg } H_2O} = \dfrac{1750 \text{ g} \times 1 \text{ mol}/98.07 \text{ g}}{\left(1.84 \text{ g/mL} \times \dfrac{1000 \text{ mL}}{L} \times \dfrac{1 \text{ kg}}{1000 \text{ g}}\right) - 1.75 \dfrac{kg}{L}} = 2.0 \times 10^2 \text{ m}$

Percent by mass = $\dfrac{1750 \text{ g } H_2SO_4}{1840 \text{ g solution}} \times 100\% = 95\%$

13

The Active Metals

INTRODUCTION

The active metals include the alkali metals, alkaline earth metals except beryllium, and aluminum. They are all reactive to acids (forming hydrogen), and to water (forming hydroxides). The adherent oxide coating on aluminum and magnesium masks this reactivity, sometimes with devastating consequences when conditions permit oxygen or acids and bases to come into contact with the pure material at elevated temperatures.

The stoichiometry of the different metals reflects their positions in the periodic table. Physical properties reflect the differences in their ionic charge and the relative sizes of their ions.

READY REFERENCE

Fractional crystallization (13.5)   This is a process in which two or more salts are separated from each other by taking advantage of the difference in solubility of the salts. The first salt to crystallize will contain only a small amount of the other(s). The process may be repeated with the recovered salt to prepare a still purer material.

Metathetical reactions (13.9)   A metathetical reaction is one in which the counter ion(s) change partners between the initial reactant state and the final product state. The nature of the positive and negative ions does not change in the reaction.

SOLUTIONS TO CHAPTER 13 EXERCISES

THE ALKALI METALS

1. Why does the reactivity of the alkali metals increase from Li to Cs?

   Solution

   The distance of the single, outermost (valence) electron from the nucleus increases in the family from Li to Cs. This increase, in addition to shielding effects of inner electrons in the heavier elements, makes the electron easier to remove from the heavier elements and, therefore, the reactivity increases from Li to Cs.

3. Why should alkali metals never be handled with the fingers?

Solution

The metals react vigorously, even explosively, with water. Handling them could result in severe burns.

5. By analogy with its reaction with water, suggest a chemical formula for the product of the reaction between sodium metal and $H_2S(g)$.

Solution

$2Na + 2H_2S \longrightarrow 2NaSH$ (sodium hydrosulfide) $+ H_2$

7. Outline the chemistry of the Solvay process for the production of sodium carbonate.

Solution

The Solvay process produces both $NaHCO_3$ and Na_2CO_3. The overall reaction is summarized by:

$$Na^+ + Cl^- + NH_4^+ + HCO_3^- \longrightarrow NaHCO_3 + NH_4^+ + Cl^- \qquad (1)$$

($NaHCO_3$ forms because it is the least soluble compound.)

$$2NaHCO_3 \xrightarrow{\Delta} Na_2CO_3 + H_2O + CO_2 \qquad (2)$$

The NH_4CO_3 for reaction 1 is prepared in the reaction

$$NH_3(aq) + CO_2(aq) + H_2O(l) \longrightarrow NH_4HCO_3(aq)$$

The NH_3 is recovered from the product of reaction 1 by the reaction

$$Ca(OH)_2 + 2NH_4^+ \longrightarrow Ca^{2+} + 2NH_3 + 2H_2O$$

The $Ca(OH)_2$ is prepared from limestone, which is also the source of CO_2.

$$CaCO_3 \xrightarrow{\Delta} CaO + CO_2(g)$$

$$CaO + H_2O \longrightarrow Ca(OH)_2$$

9. What is the principal reaction involved when baking powder acts as a leavening agent?

Solution

$$KHC_4H_4O_6(aq) + NaHCO_3(aq) \longrightarrow NaKC_4H_4O_6(aq) + CO_2(g) + H_2O(l)$$

11. Cite evidence that hydride ions are strongly basic, and that peroxide ions are strongly basic.

Solution

They both accept H^+ from H_2O, a very weak acid.

$H^- + H_2O \longrightarrow H_2 + OH^-$ (basic)

$O_2^{2-} + H_2O \longrightarrow O_2H^- + OH^-$ (basic)

13. What is the difference that makes LiCl soluble in water whereas LiF is insoluble?

Chapter 13 The Active Metals

Solution

The smaller ionic radius of F^- leads to a larger lattice energy for LiF over LiCl and to a greatly reduced solubility.

15. What does the existence of large surface deposits of Chile saltpeter indicate about the climate of Chile in the regions where these deposits are found?

Solution

Nitrates are soluble in water, especially the nitrates of alkali metals. Therefore, for deposits of nitrates to exist in nature they must be kept free of water—low rainfall and shielding in rock formations.

17. How much anhydrous sodium carbonate contains the same number of moles of sodium as in 100 g of $Na_2CO_3 \cdot 10H_2O$?

Solution

1 mol of $Na_2CO_3 \cdot 10H_2O$ contains 1 mol of Na_2CO_3

$$\text{No. mol } Na_2CO_3 \cdot 10H_2O = 100 \text{ g} \times \frac{1 \text{ mol}}{286.14084 \text{ g}} = 0.3495 \text{ mol}$$

$$\text{Mass } Na_2CO_3 = 0.3495 \text{ mol} \times \frac{105.9887 \text{ g}}{\text{mol}} = 37.0 \text{ g}$$

19. Give balanced equations for the reactions occurring at the electrodes and the overall reaction for the electrolysis of lithium chloride.

Solution

Cathode (reduction):

$$2Li^+ + 2e^- \longrightarrow 2Li(l)$$

Anode (oxidation):

$$2Cl^- \longrightarrow Cl_2(g) + 2e^-$$

Overall reaction:

$$2Li^+ + 2Cl^- \longrightarrow 2Li(l) + Cl_2(g)$$

THE ALKALINE EARTH METALS

21. What property or properties of beryllium or its compounds keeps it from being classified as an active metal?

Solution

Beryllium compounds have predominantly covalent character as opposed to the ionic character of other alkaline earth elements.

23. Write balanced equations for the steps in the extraction of magnesium from sea water.

Solution

$$Mg^{2+} + 2OH^- \longrightarrow Mg(OH_2)(s)$$

$$Mg(OH)_2 + 2HCl \xrightarrow{H_2O} MgCl_2 \cdot 6H_2O$$

$$MgCl_2 \cdot 6H_2O \xrightarrow{\Delta} MgCl_2 \cdot 1\tfrac{1}{2}H_2O + 4\tfrac{1}{2}H_2O$$

$$MgCl_2 \cdot 1\tfrac{1}{2}H_2O \xrightarrow[\Delta]{H_2O} MgCl_2 + 1\tfrac{1}{2}H_2O$$

$$MgCl_2 \xrightarrow[700°C]{\text{electrolysis}} Mg(l) + Cl_2(g)$$

Cathode (reduction):

$$Mg^{2+} + 2e^- \longrightarrow Mg$$

Anode (oxidation):

$$2Cl^- - 2e^- \longrightarrow Cl_2$$

25. Why cannot a magnesium fire be extinguished by either water or carbon dioxide? Suggest a method of extinguishing such a fire.

Solution

The extreme temperature of a magnesium fire combined with the tremendous affinity for oxygen decomposes water, releasing combustible hydrogen; and magnesium combines with CO_2 to release elemental carbon, which also burns. To extinguish such a fire, smother it with sand or clay.

27. Identify each of the following: quicklime, slaked lime, milk of lime.

Solution

Quicklime—CaO
Slaked lime—Ca(OH)$_2$
Milk of lime—a suspension of Ca(OH)$_2$

29. Write a balanced equation describing the dehydrating action of calcium metal.

Solution

$$Ca(s) + 2H_2O(l) \longrightarrow Ca(OH)_2(aq) + H_2(g)$$

31. What is the essential reaction involved in the setting of mortar?

Solution

$$Ca(OH)_2 + CO_2 \longrightarrow CaCO_3 + H_2O$$

33. Based solely on the lattice energy of the product, should you expect Mg or Ba to be more reactive with oxygen?

Solution

The smaller magnesium ion leads to a substantially higher lattice energy and to a greater reactivity toward oxygen.

Chapter 13 The Active Metals

35. Crushed limestone is used in the treatment of acidic soils. Write balanced equations for the reactions involved.

Solution

$$CaCO_3(s) + 2H^+(aq) \longrightarrow Ca^{2+}(aq) + CO_2 + H_2O$$

37. Write balanced equations for the reaction of barium with each of the following: oxygen, hydrogen chloride, hydrogen, sulfur, and water.

Solution

$$2Ba + O_2 \longrightarrow 2BaO \text{ or } Ba + O_2 \longrightarrow BaO_2$$

$$Ba + 2HCl \longrightarrow BaCl_2 + H_2$$

$$Ba + H_2 \longrightarrow BaH_2$$

$$Ba + S \longrightarrow BaS$$

$$Ba + 2H_2O \longrightarrow Ba(OH)_2 + H_2$$

39. When $MgNH_4PO_4$ is heated to 1000°C, it is converted to $Mg_2P_2O_7$. A 1.203-g sample containing magnesium yielded 0.5275 g of $Mg_2P_2O_7$ after precipitation of $MgNH_4PO_4$ and heating. What percent by mass of magnesium was present in the original sample?

Solution

Two moles of $MgNH_4PO_4$ yield one mole of $Mg_2P_2O_7$ when decomposed and one mole of $MgNH_4PO_4$ contains one mole of magnesium. Calculate the mass of magnesium in the sample from the number of moles of $MgNH_4PO_4$ required to produce 0.5275 g of $Mg_2P_2O_7$.

$$\text{No. mol } MgNH_4PO_4 \text{ present} = \frac{0.5275 \text{ g } Mg_2P_2O_7}{222.5533 \text{ g/mol}} \times \frac{2 \text{ mol } MgNH_4PO_4}{1 \text{ mol } Mg_2P_2O_7}$$

$$= 0.0047404 \text{ mol}$$

$$\text{Mass (Mg)} = 0.0047404 \text{ mol} \times 24.305 \frac{g}{mol} = 0.1152 \text{ g}$$

$$\text{Percent Mg} = \frac{0.1152 \text{ g}}{1.203 \text{ g}} \times 100\% = 9.577\%$$

GROUP IIIA

41. Describe the production of metallic aluminum by electrolytic reduction.

Solution

The net reactions upon the purification of aluminum ore to Al_2O_3 are as follows:

Extract from ore with NaOH:

$$AlO(OH) \xrightarrow{NaOH} Na[Al(OH)_4]$$

Recover:

$$2Na[Al(OH)_4] + H_2SO_4 \longrightarrow 2Al(OH)_3 + Na_2SO_4 + 2H_2O$$

Sinter:

$$Al(OH)_3 \xrightarrow{\Delta} Al_2O_3 + H_2O$$

Dissolve in Na_3AlF_6 and electrolyze.

Cathode:

$$Al^{3+} + 3e^- \longrightarrow Al(s)$$

Anode:

$$2O^{2-} - 2e^- \longrightarrow O_2$$
$$2C + O_2 \longrightarrow 2CO$$
$$C + O_2 \longrightarrow CO_2$$

43. Illustrate the amphoteric nature of aluminum hydroxide by suitable equations.

    Solution

    $$Al(OH)_3 + 3H^+ \longrightarrow Al^{3+} + 3H_2O$$
    $$Al(OH)_3 + OH^- + 2H_2O \longrightarrow [Al(H_2O)_2(OH)_4]^-$$

45. Why is it impossible to prepare anhydrous aluminum chloride by heating the hexahydrate to drive off the water?

    Solution

    The hydrate decomposes with the liberation of water, which, in a heated environment, reacts to produce HCl and the oxide because of the greater lattice energy of the oxide.

47. Write balanced chemical equations for the following reactions:

    (a) Gaseous hydrogen fluoride is bubbled through a suspension of bauxite in molten sodium fluoride.
    (b) Metallic aluminum is burned in air.
    (c) Aluminum is heated in an atmosphere of chlorine.
    (d) Aluminum sulfide is added to water.
    (e) Aluminum hydroxide is added to a solution of nitric acid.

    Solution

    (a) $Al_2O_3 \cdot 2H_2O + 6HF + 6NaF \longrightarrow 2Na_3AlF_6 + 5H_2O(g)$
    (b) $4Al(s) + 3O_2 \longrightarrow 2Al_2O_3$
    (c) $2Al(s) + 3Cl_2 \longrightarrow 2AlCl_3$
    (d) $Al_2S_3 + 6H_2O \longrightarrow 2Al(OH)_3 + 3H_2S$
    (e) $Al(OH)_3 + 3HNO_3 \longrightarrow Al(NO_3)_3(aq) + 3H_2O$

49. What properties of thallium or its compounds keep it from being classified as an active metal?

Chapter 13 The Active Metals

Solution

Thallium exhibits two oxidation states and has three valence electrons after a just-completed set of d orbitals, which makes it distinctly different from other active metals.

ADDITIONAL EXERCISES

51. (a) A current of 1000 A flowing for 96.5 s contains 1 mol of electrons. How long will it take to produce 100 kg of sodium metal when a current of 50,000 A is passed through a Down's cell like that shown in Figure 13.6 if the yield of sodium is 100% of the theoretical yield?
(b) What volume of chlorine at 25°C and 1.00 atm is produced?

Solution

(a) In the reduction of Na^+, one mole of Na metal is produced for each mole of electrons transferred. Calculate the number of moles of sodium metal in 100 kg and from that the time required for a 50,000-A current to produce this amount.

$$\text{No. mol Na} = 100 \text{ kg} \times \frac{10^3 \text{g}}{\text{kg}} \times \frac{1 \text{ mol}}{22.98977 \text{ g}} = 4349.76 \text{ mol}$$

$$\text{mol } e^- \text{ required} = 4349.76 \text{ mol}$$

A current of 50,000 A will transfer 1 mol of electrons 50 times faster than a current of 1000 A.

$$1000 \text{ A} \times 96.5 \text{ s} = 50,000 \text{ A} \times (\text{time})$$

$$t = 1.93 \text{ s at 50,000 A for 1 mol } e^-$$

$$\text{Time} = 4349.76 \text{ mol} \times 1.93 \frac{\text{s}}{\text{mol}} \times \frac{1 \text{ h}}{3600 \text{ s}} = 2.33 \text{ h}$$

(b) $2Cl^- - 2e^- \longrightarrow Cl_2$

No. mol Cl_2 produced = 4349.76 mol/2 = 2174.88 mol

$$V(25°C) = 2174.88 \times 22.4 \text{ L/mol} \times \frac{298.15 \text{ K}}{273.15 \text{ K}} = 5.32 \times 10^4 \text{ L}$$

53. Peroxides, like oxides, are basic. What volume of 0.250 M HCl solution is required to neutralize a solution containing 5.00 g of BaO_2?

Solution

$$BaO_2 + 2HCl \longrightarrow BaCl_2 + H_2O_2$$

No. mol HCl required = 2 × No. mol BaO_2 present

$$= 2(5.00 \text{ g } BaO_2 \times 1 \text{ mol}/169.3288 \text{ g})$$

$$= 0.05906 \text{ mol}$$

$V(\text{HCl}) = 0.05906 \text{ mol}/0.250 \text{ M} = 236 \text{ mL}$

55. Determine the oxidation number of oxygen in aluminum oxide, barium peroxide, and cesium superoxide.

Solution

Al$_2$O$_3$ (-2); BaO$_2$ (-1); CsO$_2$ ($-\frac{1}{2}$)

57. What volume of oxygen at 25°C and 1.00 atm is required to prepare 1.00 kg of Na$_2$O? To prepare 1.00 kg of Na$_2$O$_2$?

Solution

(a) $\quad\quad\quad\quad\quad V \quad\quad\quad$ 1.00 kg
$\quad\quad$ 4Na $+$ O$_2$ $\longrightarrow$ 2Na$_2$O
$\quad\quad$ 4 mol : 1 mol : 2 mol

$$V(\text{O}_2 \text{ at STP}) = 1.00 \text{ kg Na}_2\text{O} \times \frac{1000 \text{ kg}}{\text{kg}} \times \frac{1 \text{ mol}}{61.9789 \text{ g}}$$

$$\times \frac{1 \text{ mol O}_2}{2 \text{ mol Na}_2\text{O}} \times \frac{22.41 \text{ L}}{\text{mol}} = 180.8 \text{ L}$$

$$V(25°\text{C}) = 180.8 \text{ L} \times \frac{298.15°\text{C}}{273.15°\text{C}} = 197 \text{ L}$$

(b) $\quad\quad\quad\quad\quad V \quad\quad$ 1.00 kg
$\quad\quad$ 2Na $+$ O$_2$ $\longrightarrow$ Na$_2$O$_2$
$\quad\quad$ 2 mol : 1 mol $\quad$ 1 mol

$$V(\text{O}_2 \text{ at STP}) = 1000 \text{ g Na}_2\text{O}_2 \times \frac{1 \text{ mol}}{77.9783 \text{ g}} \times \frac{1 \text{ mol O}_2}{1 \text{ mol Na}_2\text{O}_2}$$

$$\times \frac{22.41 \text{ L}}{\text{mol}} = 287.39 \text{ L}$$

$$V(25°\text{C}) = 287.39 \text{ L} \times \frac{298.15°\text{C}}{273.15°\text{C}} = 314 \text{ L}$$

59. Calculate the density of vapor above a sample of aluminum trichloride at 171°C. The vapor pressure of AlCl$_3$ at 171°C is 400 torr.

Solution

Use the ideal gas equation to calculate the mass of 1.0 L of AlCl$_3$ at the stated conditions. Remember that AlCl$_3$ exists as a dimer, Al$_2$Cl$_6$, in the gas phase.

$$n = \frac{PV}{RT} = \frac{(400/760 \text{ atm})(1.0 \text{ L})}{(0.08206 \text{ L atm/mol K})(444.15 \text{ K})} = 0.01444 \text{ mol}$$

$$\frac{\text{Mass}}{\text{L}} = 0.01444 \text{ mol} \times \frac{266.68 \text{ g}}{\text{mol}} = 3.85 \text{ g/L}$$

61. Alpha-alumina crystallizes with an approximately closest packed array of oxide ions, with aluminum ions in the octahedral holes. What fraction of the octahedral holes are occupied?

Solution

Formula Al$_2$O$_3$: The ratio of octahedral holes to anions is 1:1. Therefore, Al^{3+} ions must fill $\frac{2}{3}$ of the holes: Al$\frac{2}{3}$: O.

14

Chemical Kinetics

INTRODUCTION

The concept of speed, velocity, or rate is so familiar that it may come as a surprise that the idea is relatively modern and can be traced to Sir Isaac Newton (1642–1727). So, too, the study of the rate of a chemical reaction did not originate until recently. In the 1850s the Norwegian scientists Guldberg and Waage were the first to write rate equations. This early work was quickly followed by studies of the factors on which rates depend, such as temperature and the concentrations of reactants.

Today, measurement of the rates of chemical reactions plays a key role in understanding reaction mechanisms (the detailed way in which the reactants come together to form the products). Indeed, the object of most of the work in kinetics today is to elucidate these mechanisms.

The importance of mechanisms is easy to understand. Most reactions proceed in several steps and are said to be *complex*; *elementary* reactions, on the other hand, occur in only one step. Complex reactions often occur by free-radical mechanisms. (A free radical is a reactive, neutral species that contains an odd number of electrons. The hydroxyl radical

$$H{:}\ddot{O}{\cdot}$$

is an example.) Industry is acutely interested in the study of complex reactions, because the industrial procedures that produce the wide variety of chemical intermediates and final products involve, for the most part, such complex processes. Examples are numerous and include such diverse processes as the cracking of heavy oil fractions to usable chemicals, the process of explosion, and even the effects of enzymes. Indeed, enzyme kinetics, an area of study in kinetics, has been instrumental in the development of the understanding of biochemical processes.

READY REFERENCE

Activated complex-activation energy (14.8–14.9)   Whether a product forms as a result of an interaction of reacting particles in a chemical reaction depends on several factors. The collision theory of reactions states that reacting species must collide and, if conditions are suitable, form an unstable intermediate species called an *activated complex*. The complex proceeds to form the desired product. The energy required to give the reactants sufficient energy to form activated complexes is called the *activation energy*.

Arrhenius equation (14.9)   One of the most important relationships in chemical kinetics is the Arrhenius equation, which relates the rate constant, k, of a reaction to the temperature. Its form, as developed by Svante Arrhenius, is

$$k = A \times 10^{-(E_a/2.303\ RT)}$$

where A is the frequency factor, E_a is the activation energy, R is the gas constant, 8.314 J mol^{-1} K^{-1}, and T is the Kelvin temperature. Many natural processes, even though they are complex, obey this law because they are controlled by chemical reactions. For example, the law is obeyed by the chirring of crickets and the flashing of fireflies.

Catalyst (14.15)   A catalyst is a substance that changes the rate of a chemical reaction but is itself unchanged and, in theory, can be recovered at the end of the process. If a substance increases the rate of a catalyzed reaction, it is called an *accelerator;* and if it slows the rate of the reaction, it is called an *inhibitor*.

Chain reaction (14.14)   This term refers to a reaction that, once started, occurs in a series of steps in which one or more reactions are repeated, thereby providing additional reagent to keep the overall reaction continuing. Such a reaction is characterized by an *initiation step,* a *propagation step,* and a *termination step.* The initiation step generates the species (reactive intermediate) necessary for the propagation step to continue. Thus the reaction continues until the termination step finally consumes all of the reactive intermediate.

Elementary reaction (14.10)   Reactions that occur in one step are termed *elementary*. By contrast, the *molecularity of the reaction* is referred to as the number of molecules that participate in the single step. Thus, *unimolecular* refers to a reaction in which only one molecule is needed. Similarly, *bimolecular* means two molecules participate, and *termolecular* refers to a simultaneous reaction of three molecules. This last type of reaction is comparatively rare.

Half-life of a reaction (14.7)   The half-life, $t_{1/2}$ (time required for one-half of the original amount of limiting reactant to be converted to product), can be expressed in simple terms for first-order reactions only as

$$t_{1/2} = \frac{0.693}{k}$$

where k is a rate constant. Elements undergoing radioactive decay are one class of substances that decompose via first-order kinetics.

Order of a reaction (14.6)   In the general reaction $aA + bB \rightleftharpoons cC + dD$, if the rate of the reaction is $R = k[C]^c[D]^d$, then the order of the reaction is the sum of the exponents, $c + d$. Several common cases occur: If the value of the sum is 1, the reaction is first order and the rate depends on the concentration of only one species. If the value is 2, the reaction is second order and depends on the concentration of two species. If the sum is 0 (that is, if the rate is independent of the concentration), the reaction is zeroth order.

Rate of chemical reaction (14.1–14.5)   Rate of reaction describes the amount of a substance (moles) that disappears or is formed by a reaction in unit time. In a reaction in which the stoichiometric coefficients are all 1 for the products and reactants, the rate R is directly proportional to the concentrations of the reacting substances. For the reaction $A + B \rightleftharpoons C + D$, the rate equation is written as $R_1 = k_1[A][B]$, where the brackets indicate molar concentrations and k_1 is the rate constant for the formation of product. The opposing reaction is the formation of reactants; its rate is $R_2 = k_2[C][D]$. At equilibrium, the opposing rates are equal. Setting the rates equal gives

Chapter 14 Chemical Kinetics 189

$$k_1[A][B] = k_2[C][D] \quad \text{or} \quad \frac{k_1}{k_2} = \frac{[C][D]}{[A][B]}$$

Because the ratio of the two constants is a constant, $K_e = k_1/k_2$, where K_e is the equilibrium constant.

SOLUTIONS TO CHAPTER 14 EXERCISES

REACTION RATES AND RATE EQUATIONS

1. State the factors that determine the rate of reaction and explain how each is responsible for changing the rate.

   Solution

   Four factors affect the rate: temperature, concentration, state of subdivision, and use of catalysts. An increase in temperature causes an increase in rate (see the Arrhenius equation) because it increases the kinetic energy of the reactants and the frequency of collisions. An increase in concentration causes an increase in rate, because there are more molecules to react per unit volume and more collisions per unit time. The more finely a reactant is subdivided, the greater the surface area available for reaction to occur. Catalysts help the reaction to attain a state of equilibrium faster by providing a path with a lower activation energy.

4. Nitrosyl chloride, NOCl, decomposes to NO and Cl_2.

   $$2NOCl(g) \longrightarrow 2NO(g) + Cl_2(g)$$

   Determine the rate equation and the rate constant for this reaction from the following data:

[NOCl], M	0.10	0.20	0.30
Rate, mol L^{-1} s^{-1}	8.0×10^{-10}	3.20×10^{-9}	7.2×10^{-9}

   Solution

   The object of this problem is to use the general rate expression
   Rate = $k[NOCl]^m$, first to determine the value of m and then, by substituting data from one experiment into the equation, to find the value of k. The data listed as substituted into the rate equation give

   Exp. 1: 8.0×10^{-10} mol L^{-1} s^{-1} = $k[0.10 \text{ mol } L^{-1}]^m$

   Exp. 2: 3.20×10^{-9} mol L^{-1} s^{-1} = $k[0.20 \text{ mol } L^{-1}]^m$

   Exp. 3: 7.2×10^{-9} mol L^{-1} s^{-1} = $k[0.30 \text{ mol } L^{-1}]^m$

   The value of m can be found by inspection. Examining Exp. 1 and 2 it is found that the rate increases by a factor of four as the concentration increases by a factor of 2; from Exp. 1 and 3 the rate increases by a factor of 9 while the concentration increases by a factor of 3. This can only happen if m is 2. The value of k as calculated from the first set of data is

   $$k = \frac{8.0 \times 10^{-10} \text{ mol } L^{-1} \text{ s}^{-1}}{[0.10 \text{ mol } L^{-1}]^2} = 8.0 \times 10^{-8} \text{ L mol}^{-1} \text{ s}^{-1}$$

6. The rate constant for the decomposition at 45°C of nitrogen(V) oxide, N_2O_5, dissolved in chloroform, $CHCl_3$, is 6.2×10^{-4} min^{-1}.

$$2N_2O_5 \longrightarrow 4NO_2 + O_2$$

The decomposition is first order in N_2O_5.

(a) What is the rate of the reaction when $[N_2O_5] = 0.25$ M?
(b) What is the concentration of N_2O_5 remaining at the end of 3.5 h if the initial concentration of N_2O_5 was 0.25 M?

Solution

(a) The rate of reaction for a first-order reaction in N_2O_5 may be written as

Rate = $k[N_2O_5]$

where k, the rate constant at 45°C, is 6.2×10^{-4} min^{-1}. When $[N_2O_5] = 0.25$ M,

Rate = 6.2×10^{-4} min^{-1} (0.25 mol/L)

= $1.55 \times 10^{-4} = 1.6 \times 10^{-4}$ mol L^{-1} min^{-1}

(b) In 210 min (3.5 h), the amount of N_2O_5 that decomposes is

1.6×10^{-4} mol L^{-1} min^{-1} (210 min) = 3.4×10^{-2} mol L^{-1}

The amount remaining is

0.25 M − 0.034 M = 0.22 M

7. Hydrogen reacts with nitrogen(II) oxide to form nitrogen(I) oxide, laughing gas, according to the equation

$$H_2(g) + NO(g) \longrightarrow N_2O(g) + H_2O(g)$$

Determine the rate equation and the rate constant for the reaction from the following data:

[NO], M	0.40	0.80	0.80
[H$_2$], M	0.35	0.35	0.70
Rate, mol L^{-1} s^{-1}	5.040×10^{-3}	2.016×10^{-2}	4.032×10^{-2}

Solution

It is intended that the data for each separate experiment given in this problem be substituted into the general rate equation (rate = $k[NO]^m [H_2]^n$) to determine how n and m, the order of the reaction with respect to each component, varies. The three reactions, after substituting the data, are

Exp. 1: 5.040×10^{-3} mol L^{-1} s^{-1} = $k[0.40$ mol L$^{-1}]^m [0.35$ mol L$^{-1}]^n$

Exp. 2: 2.016×10^{-2} mol L^{-1} s^{-1} = $k[0.80$ mol L$^{-1}]^m [0.35$ mol L$^{-1}]^n$

Exp. 3: 4.032×10^{-2} mol L^{-1} s^{-1} = $k[0.80$ mol L$^{-1}]^m [0.70$ mol L$^{-1}]^n$

Because k, m, and n refer to the same reaction, they must retain the same values throughout the three experiments. Going from Experiment 1 to Experiment 2, holding [H$_2$] constant, shows that doubling the concentration of NO causes a fourfold increase in the rate. This can only occur if the value of m is

2 since $[2]^2 = 4$. In a similar manner, holding the concentration of NO constant in Experiment 2 and Experiment 3 with subsequent doubling of the concentrations of H_2 doubles the rate. The rate equation holds for this latter group of two experiments only if n equals 1. Therefore, the overall rate is second order in NO and first order in H_2.

To determine the value of the rate constants, the data of experiment 1 is used in the rate law.

$$\text{Rate} = k[NO]^2 [H_2]^1$$

$$5.040 \times 10^{-3} \text{ mol L}^{-1} \text{ s}^{-1} = k[0.40 \text{ mol L}^{-1}]^2 [0.35 \text{ mol L}^{-1}]$$

$$k = \frac{5.040 \times 10^{-3} \text{ mol L}^{-1} \text{ s}^{-1}}{(0.16 \text{ mol}^2 \text{ L}^{-2})(0.35 \text{ mol L}^{-1})} = 9.0 \times 10^{-2} \text{ L}^2 \text{ mol}^{-2} \text{ s}^{-1}$$

9. Most of the 13.1 billion pounds of HNO_3 produced in the United States during 1986 was prepared by the following sequence of reactions, each run in a separate reaction vessel.

$$4NH_3(g) + 5O_2(g) \longrightarrow 4NO(g) + 6H_2O(g) \tag{1}$$
$$2NO(g) + O_2(g) \longrightarrow 2NO_2(g) \tag{2}$$
$$3NO_2(g) + H_2O(l) \longrightarrow 2HNO_3(aq) + NO(g) \tag{3}$$

The first reaction is run by burning ammonia in air over a platinum catalyst. This reaction is fast. The reaction in Equation (3) is also fast. The second reaction limits the rate at which nitric acid can be prepared from ammonia. If Equation (2) is second order in NO and first order in O_2, what is the rate of formation of NO_2 when the oxygen concentration is 0.45 M and the nitric oxide concentration is 0.70 M? The rate constant for the reaction is $5.8 \times 10^{-6} \text{ L}^2 \text{ mol}^{-2} \text{ s}^{-1}$.

Solution

The rate law governing the formation of HNO_3 is

$$\text{Rate} = k[NO]^2 [O_2]$$

From the data given,

$$\text{Rate} = 5.8 \times 10^{-6} \text{ L}^2 \text{ mol}^{-2} \text{ s}^{-1} [0.70 \text{ mol/L}]^2 [0.45 \text{ mol/L}]$$

$$= 1.3 \times 10^{-6} \text{ mol L}^{-1} \text{ s}^{-1}$$

KEYSTROKES: 5.8 [EE] [+/−] 6 × .70 [x^2] × .45 = 1.2789 − 06

12. A liter of a 1 M solution of H_2O_2 slowly decomposes into H_2O and O_2. If 0.50 mol of H_2O_2 decomposes during the first 6 h of the reaction, explain why only 0.25 mol decomposes during the next 6-h period.

Solution

$$H_2O_2(l) \longrightarrow H_2O(l) + O_2(g)$$

This is a first-order reaction in which the time required to consume a given fraction of starting reagent is always constant. It takes 6 h to decompose half of the original material and an additional 6 h to decompose the same fraction, that is half of 0.50 mol where 0.50 mol can be thought of as the starting amount at the beginning of the second 6-h period.

13. The decomposition of SO_2Cl_2 to SO_2 and Cl_2 is a first-order reaction with $k = 2.2 \times 10^{-5}$ s^{-1} at 320°C.

    (a) Determine the half-life of the reaction.
    (b) At 320°C, how much $SO_2Cl_2(g)$ would remain in a 1.00-L flask 2.5 h after the introduction of 0.0215 mol of SO_2Cl_2? Assume that the rate of the reverse reaction is so slow that it can be ignored.

Solution

(a) The half-life of a first-order reaction is determined from the expression

$$t_{1/2} = \frac{0.693}{k} = \frac{0.693}{2.2 \times 10^{-5} \text{ s}^{-1}} = 3.2 \times 10^4 \text{ s}$$

(b) For a first-order reaction,

$$\log \frac{[A_0]}{[A]} = \frac{kt}{2.303}$$

The concentration of A_0 is 0.0215 M, and substitution gives

$$\log \frac{[0.0215]}{[A]} = \frac{2.2 \times 10^{-5} \text{ s}^{-1} [2.5 \text{ h} \times 60 \text{ min/h} \times 60 \text{ s/min}]}{2.303}$$

$$\log 0.0215 - \log A = 0.08597$$

$$-\log A = 0.08597 + 1.66756$$

$$\log A = -1.7535$$

$A = 0.0176$ M. In the 1.00-L flask there are 1.8×10^{-2} mol.

KEYSTROKES: 1.7535 [+/−] [INV] [log] = 0.0176

14. Radioactive materials decay by a first-order process. The very dangerous isotope strontium-90 decays with a half-life of 28 years.

    $^{90}_{38}\text{Sr} \longrightarrow\ ^{90}_{39}\text{Y} + e^-$

    What is the rate constant for this decay?

Solution

The half-life is $t_{1/2} = \dfrac{0.693}{k}$, where k is the rate constant.

$$k = \frac{0.693}{t_{1/2}} = \frac{0.693}{28 \text{ yr}} = 2.5 \times 10^{-2} \text{ yr}^{-1}$$

17. The reaction of compound A to give compounds C and D was found to be second order in A. The rate constant for the reaction was determined to be 2.42 L mol^{-1} s^{-1}. If the initial concentration is 0.0500 mol/L, what is the value of $t_{1/2}$?

Solution

For a second-order reaction the half-life is concentration-dependent.

$$t_{1/2} = \frac{1}{k[A_0]} = \frac{1}{2.42 \text{ L mol}^{-1} \text{ s}^{-1} \times 0.0500 \text{ mol L}^{-1}} = 8.26 \text{ s}$$

Chapter 14 Chemical Kinetics

18. The half-life of a reaction of compound A to give compounds D and E is 8.50 min when the initial concentration of A is 0.150 mol/L. How long will it take for the concentration to drop to 0.0300 mol/L if the reaction is (a) first order with respect to A? (b) second order with respect to A? In your calculations, give careful attention to the units for each quantity, showing how the units cancel to provide the proper units in each answer.

Solution

(a) In a first-order reaction, the half-life is given by $t_{1/2} = 0.693/k$. Knowing $t_{1/2}$, the value of k is found:

$$k = \frac{0.693}{8.50 \text{ min}} = 0.0815 \text{ min}^{-1}$$

then, from

$$\log \frac{[A_0]}{[A]} = \frac{kt}{2.303}$$

$$t = \log\left[\frac{0.150 \text{ mol/L}}{0.0300 \text{ mol/L}}\right] \times \frac{2.303}{0.0815 \text{ min}^{-1}} = 0.69897 \times 28.258 \text{ min} = 19.8 \text{ min}$$

(b) In a second-order reaction, the rate is concentration dependent;

$$t_{1/2} = \frac{1}{k[A_0]}$$

$$k = \frac{1}{t_{1/2}[A_0]} = \frac{1}{8.50 \text{ min } [0.150 \text{ mol/L}]} = 0.784 \text{ L mol}^{-1} \text{ min}^{-1}$$

Then substitution into

$$\frac{1}{[A]} - \frac{1}{[A_0]} = kt$$

gives

$$t = \frac{\dfrac{1}{0.0300 \text{ mol L}^{-1}} - \dfrac{1}{0.150 \text{ mol L}^{-1}}}{0.784 \text{ L mol}^{-1} \text{ min}^{-1}} = \frac{33.333 \text{ L mol}^{-1} - 6.667 \text{ L mol}^{-1}}{0.784 \text{ L mol}^{-1} \text{ min}^{-1}}$$

$$= 34.0 \text{ min}$$

COLLISION THEORY OF REACTION RATES

19. Chemical reactions occur when reactants collide. For what reasons may a collision fail to produce a chemical reaction?

Solution

If orientation is not favorable a collision may not produce a product. If the energy of collision is not sufficient, the activation energy needed to put a reactant into a transition state will be lacking, and thus no reaction will occur.

21. What is the activation energy of a reaction, and how is this energy related to the activated complex of the reaction?

Solution

The activation energy of a reaction is the energy required to raise the energy of the reactants to the energy of the activated state. Unless the reactant forms an activated complex, no reaction is possible.

23. If the rate of a reaction doubles for every 10°C rise in temperature, how much faster would the reaction proceed at 45°C than at 25°C? at 95°C than at 25°C?

Solution

The rate doubles for each 10°C rise in temperature; 45° is a 20° increase over 25°. Thus, the rate doubles two times, or 2^2 (rate at 25°) = 4 times faster. Ninety-five degrees is a 70° increase over 25°. Thus, the rate doubles 7 times, or 2^7 (rate at 25°) = 128 times faster.

25. In an experiment, a sample of $NaClO_3$ was 90% decomposed in 48 min. Approximately how long would this decomposition have taken if the sample had been heated 20° higher?

Solution

The rate doubles for each 10°C rise in temperature. A 20° increase would increase the rate 4 times, thereby decreasing the time required to one fourth of its original value: 48 min/4 = 12 min.

27. The rate constant at 325°C for the reaction $C_4H_8 \longrightarrow 2C_2H_4$ (Section 14.11) is 6.1×10^{-8} s^{-1}, and the activation energy is 261 kJ per mole of C_4H_8. Determine the frequency factor for the reaction.

Solution

The rate constant k is related to the activation energy E_a by a relationship known as Arrhenius' law. Its form is

$$k = A \times 10^{-(E_a/2.303\ RT)}$$

where A is the frequency factor. Using the data above, and converting kJ to joules,

$$6.1 \times 10^{-8}\ s^{-1} = A \times 10^{-[+261,000\ J/2.303(8.314\ J\ K^{-1})\ (325\ +\ 273)K]}$$

$$= A \times 10^{-22.8}$$

Convert $10^{-22.8}$ to its equivalent expressed as a whole power of 10:

KEYSTROKES: 10 $\boxed{y^x}$ 22.8 $\boxed{+/-}$ = 1.5848932 − 23

$$A = \frac{6.1 \times 10^{-8}\ s^{-1}}{1.58 \times 10^{-23}} = 3.8 \times 10^{15}\ s^{-1}$$

ELEMENTARY REACTIONS, REACTION MECHANISMS, CATALYSTS

29. Which of the following equations, as written, could describe elementary reactions?

   (a) $Cl_2 + CO \longrightarrow Cl_2CO$;
   rate = $k[Cl_2]^{3/2}[CO]$

Chapter 14 Chemical Kinetics

(b) $PCl_3 + Cl_2 \longrightarrow PCl_5$;
   rate $= k[PCl_3][Cl_2]$
(c) $2NO + 2H_2 \longrightarrow N_2 + 2H_2O$;
   rate $= k[NO][H_2]$
(d) $2NO + O_2 \longrightarrow 2NO_2$;
   rate $= k[NO]^2[O_2]$
(e) $NO + O_3 \longrightarrow NO_2 + O_2$;
   rate $= k[NO][O_3]$

Solution

(a) For an elementary reaction it is unlikely to have a collision involving more than two reactants. Thus it would be improbable to find the concentration in rate equation raised to a power other than 1 or 2. See also (c).
(b) The rate expression indicates that both reactants are involved in the reaction. A binary collision is likely, leading to the possibility of an elementary reaction.
(c) The rate equation does not correspond to the stoichiometry of the overall equation and, therefore, the reaction cannot be elementary.
(d) This could correspond to a ternary collision process, one not highly likely, but possible as an elementary process.
(e) This corresponds to a simple binary collision and could be an elementary reaction.

31. Why are elementary reactions involving three or more reactants very uncommon?

Solution

Such reactions are very uncommon because of the very small probability of having three or more different reagents simultaneously colliding with the proper orientation.

34. Account for the increase in reaction rate brought about by a catalyst.

Solution

The general mode of action for a catalyst is to provide a mechanism by which the reactants can unite more readily by a path with a lower reaction energy. The rates of both the forward and reverse reactions are increased, thus leading to a faster attainment of equilibrium.

ADDITIONAL EXERCISES

37. An elevated level of the enzyme alkaline phosphatase (ALP) in the serum is an indication of possible liver or bone disorder. The level of serum ALP is so low that it is very difficult to measure directly. However, ALP catalyzes a number of reactions, and its relative concentration can be determined by measuring the rate of one of these reactions under controlled conditions. One such reaction is the conversion of p-nitrophenyl phosphate (PNPP) to p-nitrophenoxide ion (PNP) and phosphate ion. Control of temperature during the test is very important; the rate of the reaction increases 1.47 times if the temperature changes from 30°C to 37°C. What is the activation energy for the ALP-catalyzed conversion of PNPP to PNP and phosphate?

Solution

Changes in rate brought about by temperature changes are governed by the Arrhenius equation: $k = A \times 10^{-E_a/2.303RT}$. In this particular reaction, k increases by 1.47 as T changes from 30°C (303 K) to 37°C (310 K). The Arrhenius equation may be solved for A under both sets of conditions and then the A eliminated between the two equations.

$$A = \frac{k}{10^{-E_a/2.303R(303\ K)}} = \frac{1.47\,k}{10^{-E_a/2.303R(310\ K)}}$$

Eliminating k from both sides, taking logs, and rearranging,

$$\frac{-E_a}{2.303 \times 8.314\ \text{J mol}^{-1}\ \text{K}^{-1}\ (310\ K)} = \log 1.47 - \frac{E_a}{2.303 \times 8.314\ \text{J mol}^{-1}\ \text{K}^{-1}\ (303\ K)}$$

$$\frac{-E_a}{5935.6\ \text{J mol}^{-1}} = 0.1673 - \frac{E_a}{5801.6\ \text{J mol}^{-1}}$$

$$\frac{E_a}{5801.6} - \frac{E_a}{5935.6} = 0.1673\ \text{J mol}^{-1}$$

$$E_a(1.72366 \times 10^{-4} - 1.68474 \times 10^{-4}) = 0.1673\ \text{J mol}^{-1}$$

$$3.892 \times 10^{-6}\ E_a = 0.1673\ \text{J mol}^{-1}$$

$$E_a = 42986\ \text{J mol}^{-1} = 43\ \text{kJ mol}^{-1}$$

15

An Introduction to Chemical Equilibrium

INTRODUCTION

Guldberg and Waage extended their rate equations to develop the equilibrium expression for a chemical reaction. Although the method they used proved to be faulty, the validity of the final equilibrium expression remains unchallenged. Knowledge of the equilibrium constant expression has proven extremely important to chemists. The amount of any substance in a chemical reaction at equilibrium may be determined from the expression under a given set of conditions.

Depending upon how the equilibrium expression is viewed, it may be applied to solubility problems and to acid-base problems, as well as those treated here. A knowledge of equilibrium is essential as a theoretical base for studying most areas of chemistry.

READY REFERENCE

Equilibrium (15.1)   When the two opposing (forward and reverse) reactions in a system occur simultaneously at the same rate, the system is said to be at equilibrium. Chemical systems at equilibrium are dynamic, but no net change occurs in the amounts of reactants or products.

Law of mass action (15.2)   For a general equation of a chemical reaction written as

$$a\text{A} + b\text{B} + \cdots \rightleftharpoons c\text{C} + d\text{D} + \cdots$$

the law of mass action is, at equilibrium,

$$\frac{[\text{C}]^c[\text{D}]^d \cdots}{[\text{A}]^a[\text{B}]^b \cdots} = K_e$$

This law of chemical equilibrium or mass action applies at a particular temperature, and the brackets indicate the molar concentrations of the different chemical species raised to a power equal to their stoichiometric coefficients. Pressure is proportional to concentration, so an equilibrium constant expressed in terms of pressure, K_p, is defined as

$$K_p = \frac{(p_\text{C})^c(p_\text{D})^d}{(p_\text{A})^a(p_\text{B})^b}$$

Le Châtelier's principle (15.4)   If some applied action causes a stress (such as a change in concentration, pressure, or temperature) to occur in a chemical system at equilibrium, then the equilibrium shifts in a way that tends to undo the effect of the stress.

SOLUTIONS TO CHAPTER 15 EXERCISES

EQUILIBRIUM CONSTANTS

1. Using the law of mass action, derive the mathematical expression of the equilibrium constant for the following reversible reactions:

    (a) $N_2(g) + O_2(g) \rightleftharpoons 2NO(g)$
    (b) $NH_4Cl(s) \rightleftharpoons NH_3(g) + HCl(g)$
    (d) $S_8(g) \rightleftharpoons 8S(g)$
    (f) $4NH_3(g) + 5O_2(g) \rightleftharpoons 4NO(g) + 6H_2O(g)$
    (h) $BaSO_3(s) \rightleftharpoons BaO(s) + SO_2(g)$
    (i) $CO_2(g) + H_2(g) \rightleftharpoons CO(g) + H_2O(g)$

    Solution

    (a) $K = \dfrac{[NO]^2}{[N_2][O_2]}$

    (b) $K = [NH_3][HCl]$

    (d) $K = \dfrac{[S]^8}{[S_8]}$

    (f) $K = \dfrac{[NO]^4[H_2O]^6}{[NH_3]^4[O_2]^5}$

    (h) $K = [SO_2]$

    (i) $K = \dfrac{[CO][H_2O]}{[CO_2][H_2]}$

3. A sample of PCl_5 is put into a 1.00-L vessel and heated. At equilibrium, the vessel contains 0.40 mol of $PCl_3(g)$ and 0.40 mol of $Cl_2(g)$. The value of the equilibrium constant for the decomposition of PCl_5 to PCl_3 and Cl_2 at this temperature is 0.50 mol L^{-1}. Calculate the concentration of PCl_5 at equilibrium.

    Solution

    The reaction is $PCl_5(g) \rightleftharpoons PCl_3(g) + Cl_2(g)$. The equilibrium constant expression is

    $$K = \frac{[PCl_3][Cl_2]}{[PCl_5]} = 0.50 \text{ mol L}^{-1} = \frac{0.40 \text{ mol L}^{-1} \times 0.40 \text{ mol L}^{-1}}{[PCl_5]}$$

    $$[PCl_5] = \frac{0.16 \text{ mol}^2 \text{ L}^{-2}}{0.50 \text{ mol L}^{-1}} = 0.32 \text{ mol L}^{-1} = 0.32 \text{ M}$$

5. A sample of $NH_3(g)$ was formed from $H_2(g)$ and $N_2(g)$ at 500°C. If the equilibrium mixture was found to contain 1.35 mol H_2 per liter, 1.15 mol N_2 per liter, and 4.12×10^{-1} mol NH_3 per liter, what is the value of the equilibrium constant for the formation of NH_3?

    Solution

    The reaction may be written as

    $$N_2(g) + 3H_2(g) \rightleftharpoons 2NH_3(g)$$

Chapter 15 An Introduction to Chemical Equilibrium

The equilibrium constant for the reaction is

$$K = \frac{[NH_3]^2}{[N_2][H_2]^3} = \frac{[4.12 \times 10^{-1} \text{ mol L}^{-1}]^2}{[1.15 \text{ mol L}^{-1}][1.35 \text{ mol L}^{-1}]^3}$$

$$= \frac{0.170 \text{ mol}^2 \text{ L}^{-2}}{(1.15 \text{ mol L}^{-1})(2.46 \text{ mol}^3 \text{ L}^{-3})}$$

$$= 0.0600 \text{ mol}^{-2} \text{ L}^2 = 6.00 \times 10^{-2} \text{ mol}^{-2} \text{ L}^2$$

7. The vapor pressure of water is 0.0728 atm at 40°C. What is the equilibrium constant for the transformation $H_2O(l) \rightleftharpoons H_2O(g)$?

Solution

The equilibrium constant may be expressed in terms of concentrations or in terms of pressures. For the transformation $H_2O(l) \rightleftharpoons H_2O(g)$ the equilibrium may be expressed most easily in terms of pressures. The equilibrium constant depends only on the substance in the gaseous phase at equilibrium. Thus,

$$K_p = P_{H_2O(g)} = 0.0728 \text{ atm}$$

8. In general, the equilibrium constant for a reaction in the gas phase has different units and a different numerical value when pressures rather than concentrations are used to evaluate it. Show that such is not the case for the decomposition of HI into H_2 and I_2; show that the numerical value of the equilibrium constant for this particular reaction is independent of the units in which concentration is expressed.

Solution

For the reaction $2HI(g) \rightleftharpoons H_2(g) + I_2(g)$,

$$K = \frac{[H_2][I_2]}{[HI]^2}$$

The units of H_2 and I_2 in the numerator will be squared. The units of HI must be the same as those in the numerator. Consequently, when the HI units are squared, they will equal the units squared from the amounts of H_2 and I_2.

9. (a) Calculate the equilibrium concentration of NO_2 in a solution prepared by dissolving 0.20 mol of N_2O_4 in 400 mL of chloroform. For the reaction $N_2O_4(g) \rightleftharpoons 2NO_2(g)$, in chloroform, $K = 1.07 \times 10^{-5}$.
   (b) What will be the percent decomposition of the original N_2O_4?

Solution

(a) For the reaction, the equilibrium constant is

$$K = \frac{[NO_2]^2}{[N_2O_4]} = 1.07 \times 10^{-5}$$

Let x be the amount of N_2O_4 that decomposes from the initial 0.20 mol in 400 mL of chloroform.

Then

$$\left(\frac{(2x)^2}{\frac{0.20-x}{0.400}}\right) = 1.07 \times 10^{-5}$$

Clearing and putting into standard form

$$1.6x^2 + 1.07 \times 10^{-5}x - 2.14 \times 10^{-6} = 0$$

From the quadratic equation

$$x = \frac{-b \pm \sqrt{b^2 - 4ac}}{2a}$$

$$x = \frac{-1.07 \times 10^{-5} \pm \sqrt{(1.07 \times 10^{-5})^2 - 4(1.6)(-2.14 \times 10^{-6})}}{8}$$

$$x = +1.1529 \times 10^{-3} \quad \text{or} \quad -1.1598 \times 10^{-3}$$

The positive root is the concentration of N_2O_4 and $2x$ or 2.3×10^{-3} M is the equilibrium concentration of NO_2.

(b) $\quad$ Percent decomposition $= \dfrac{1.155 \times 10^{-3}}{0.5 - 1.155 \times 10^{-3}} \times 100\% = 0.23\%$

11. The rate of the reaction $H_2(g) + I_2(g) \longrightarrow 2HI(g)$ at 25°C is given by

$$\text{Rate} = 1.7 \times 10^{-18} [H_2][I_2]$$

The rate of decomposition of gaseous HI to $H_2(g)$ and $I_2(g)$ at 25°C is given by

$$\text{Rate} = 2.4 \times 10^{-21} [HI]^2$$

What is the equilibrium constant for the formation of gaseous HI from the gaseous elements at 25°C?

Solution

The equilibrium constant may be written as

$$K = \frac{k_{\text{forward}}}{k_{\text{reverse}}}$$

since at equilibrium, $\text{Rate}_1 = \text{Rate}_2$, or

$$k_{\text{forward}}[H_2][I_2] = k_{\text{reverse}}[HI]^2$$

This may be expressed as

$$\frac{k_{\text{forward}}}{k_{\text{reverse}}} = \frac{[HI]^2}{[H_2][I_2]}$$

which is recognized as an equilibrium constant. Substitution gives

$$K = \frac{k_{\text{forward}}}{k_{\text{reverse}}} = \frac{1.7 \times 10^{-18}}{2.4 \times 10^{-21}} = 7.1 \times 10^2$$

Chapter 15 An Introduction to Chemical Equilibrium

13. Acetic acid (CH_3CO_2H) and ethanol (C_2H_5OH) react in dioxane solvent to produce ethyl acetate ($CH_3CO_2C_2H_5$) and water.

$$CH_3CO_2H + C_2H_5OH \rightleftharpoons CH_3CO_2C_2H_5 + H_2O$$

The equilibrium constant is 4.00. If 1.00 mol of acetic acid and 2.00 mol of ethanol are mixed in enough dioxane to make 1.00 L of solution, how many moles of each reactant and product will be present in the solution when equilibrium is established?

Solution

The molecular weight of acetic acid is 60.052; that of ethanol is 46.069. Let x be the amount of acetic acid that forms product.

$$K = 4.00 = \frac{[CH_3CO_2C_2H_5][H_2O]}{[CH_3CO_2H][C_2H_5OH]} = \frac{(x)(x)}{(1.00 - x)(2.00 - x)}$$

In standard form $3.00x^2 - 12.00x + 8.00 = 0$

From the quadratic equation

$$x = \frac{12 \pm \sqrt{144 - 96}}{6} = 3.15 \text{ or } 0.845$$

Only 0.845 mol L^{-1} makes chemical sense, so the concentrations are:

$[CH_3CO_2C_2H_5] = [H_2O] = 0.845$ mol L^{-1};

$[CH_3CO_2H] = 0.155$ mol L^{-1}; $[C_2H_5OH] = 1.155$ mol L^{-1}

14. (a) Write the mathematical expression of the law of chemical equilibrium for the reversible reaction $2NO_2 \rightleftharpoons 2NO + O_2$.

(b) At 1 atm and 25°C, NO_2 with an initial concentration of 1.00 M is 3.3×10^{-3}% decomposed into NO and O_2. Calculate the equilibrium constant.

Solution

(a) $K = \dfrac{[NO]^2[O_2]}{[NO_2]^2}$

(b) $K = \dfrac{[3.3 \times 10^{-5} \text{ mol L}^{-1}]^2[1.65 \times 10^{-5} \text{ mol L}^{-1}]}{[0.999967 \text{ mol L}^{-1}]^2} = 1.8 \times 10^{-14}$ mol L^{-1}

EFFECT OF CHANGES IN CONCENTRATION, PRESSURE, OR TEMPERATURE ON EQUILIBRIUM

17. For each of the following reactions between gases at equilibrium, determine the effect of the equilibrium concentrations of the products when the temperature is decreased and when the external pressure on the system is decreased.

(a) $2H_2O(g) \rightleftharpoons 2H_2(g) + O_2(g)$ $\Delta H = 484$ kJ
(b) $N_2(g) + O_2(g) \rightleftharpoons 2NO(g)$ $\Delta H = 181$ kJ
(c) $N_2(g) + 3H_2(g) \rightleftharpoons 2NH_3(g)$ $\Delta H = -92.2$ kJ
(d) $2O_3(g) \rightleftharpoons 3O_2(g)$ $\Delta H = 285$ kJ
(e) $H_2(g) + F_2(g) \rightleftharpoons 2HF(g)$ $\Delta H = 541$ kJ

Solution

(a) $2H_2O(g) \rightleftharpoons 2H_2(g) + O_2(g)$ $\Delta H = 484$ kJ
This is an endothermic process; as T decreases, less heat is available to force the reaction to the right so products will decrease. A decrease in external pressure will favor formation of more product since the larger number of moles of product will have more space available.

(b) $N_2(g) + O_2(g) \rightleftharpoons 2NO(g)$ $\Delta H = 181$ kJ
A decrease in T for the endothermic reaction will favor formation of more reactants at the expense of the concentration of the products. A decrease in pressure will not change the equilibrium, since both sides have the same number of moles of gaseous reactants.

(c) $N_2(g) + 3H_2(g) \rightleftharpoons 2HN_3(g)$ $\Delta H = -92.2$ kJ
A decrease in T for the exothermic reaction will cause more product to be formed. A decrease in pressure favors formation of reactants.

(d) $2O_3(g) \rightleftharpoons 3O_2(g)$ $\Delta H = -285$ kJ
A decrease in temperature shifts the equilibrium to produce more products. A decrease in pressure shifts the equilibrium to produce more products.

(e) $H_2(g) + F_2(g) \rightleftharpoons 2HF(g)$ $\Delta H = 541$ kJ
A decrease in temperature decreases the amount of products. A decrease in external pressure has no effect on the equilibrium.

19. For the reaction in exercise 17(b), the equilibrium constant at 2000°C is 6.2×10^{-4}. Consider each of the following situations, and decide whether or not a reaction will occur and, if so, in which direction it will predominantly proceed.

(a) A 5.0-L flask contains 0.26 mol of N_2, 0.0062 mol of O_2, and 0.0010 mol of NO at 2000°C.

Solution

(a) $[N_2] = 0.26$ mol/5.0 L $= 0.052$

$[O_2] = 0.0062$ mol/5.0 L $= 1.24 \times 10^{-3}$

$[NO] = 0.0010$ mol/5.0 L $= 2.0 \times 10^{-4}$

$$\frac{[NO]^2}{[N_2][O_2]} = \frac{(2.0 \times 10^{-4})^2}{(0.052)(1.24 \times 10^{-3})} = 6.2 \times 10^{-4}$$

The system is at equilibrium; no further reaction will occur.

21. How will an increase in temperature affect each of the following equilibria? An increase in pressure?

(a) $N_2(g) + 3H_2(g) \rightleftharpoons 2NH_3(g)$ $\Delta H = -92.2$ kJ
(c) $N_2(g) + O_2(g) \rightleftharpoons 2NO(g)$ $\Delta H = 181$ kJ

Solution

	Direction of Equilibrium Shift	
Reaction	Temp. Increase	Pressure Increase
(a)	to the left	to the right
(c)	to the right	no change

Chapter 15 An Introduction to Chemical Equilibrium   203

22. **(a)** For the system

$$H_2(g) + CO_2(g) \rightleftharpoons H_2O(g) + CO(g)$$

the equilibrium constant is 1.6 at 990°C. Calculate the number of moles of each component in the equilibrium mixture that results from placing 1.00 mol of H_2 and 1.00 mol of CO_2 in a sealed 5.00-L reactor at 990°C.

(b) Calculate the number of moles of each component at a second equilibrium if 1.00 moles of additional H_2 is added to the equilibrium mixture obtained in part (a).

(c) Compare the answers in parts (a) and (b). Are they consistent with equilibrium theory and Le Châtelier's principle? Explain your conclusions.

Solution

(a) Let x = number of moles of CO_2 and H_2 that decompose

$$K = 1.6 = \frac{[H_2O][CO]}{[H_2][CO_2]} = \frac{(x)(x)}{(1.00 - x)(1.00 - x)}$$

Since the volume appears an equal number of times in numerator and denominator, it cancels:

$$x^2 = 1.6 - 3.2x + 1.6x^2 \qquad 0.6x^2 - 3.2x + 1.6 = 0$$

Using the quadratic formula to solve for x,

$$x = \frac{-b \pm \sqrt{b^2 - 4ac}}{2a} = \frac{3.2 \pm \sqrt{10.24 - 3.84}}{2(.6)} = \frac{3.20 \pm 2.53}{1.2}$$

$$x = 0.56, 4.8$$

Only 0.56 is chemically feasible. Substitution gives
$[H_2O] = [CO] = 0.56$ mol: $[H_2] = [CO_2] = 0.44$ mol

(b) 1.00 moles of H_2 is added to the equilibrium mixture. Let x now equal the amount of H_2 that forms products.

$$K = 1.6 = \frac{[0.56 + x][0.56 + x]}{[1.44 - x][0.44 - x]} = \frac{0.313 + 1.12x + x^2}{0.634 - 1.88x + x^2}$$

$$1.014 - 3.008x + 1.6x^2 = 0.313 + 1.12x + x^2$$

$$0.6x^2 - 4.128x + 0.701 = 0$$

$$x = \frac{+4.128 \pm \sqrt{17.040 - 1.682}}{2(0.6)} = \frac{4.128 \pm 3.919}{1.2} = 0.174$$

$[H_2O] = [CO] = 0.56 + 0.174 = 0.73$ mol;

$[H_2] = 1.44 - 0.174 = 1.3$ mol;

$[CO_2] = 0.44 - 0.174 = 0.27$ mol

(c) Yes. The added H_2 forces more CO_2 to decompose forming more CO and H_2O, thereby decreasing the amount of H_2 added. The system has adjusted to remove the stress (added H_2) according to Le Châtelier's principle.

24. An equilibrium mixture of N_2, H_2, and NH_3 in a 1.00-L vessel is found to contain 0.300 mole of N_2, 0.400 mole of H_2, and 0.100 mole of NH_3. How many moles of H_2 must be introduced into the vessel in order to double the equilibrium concentration of NH_3 if the temperature remains unchanged?

Solution

The reaction is $N_2(g) + 3H_2(g) \rightleftharpoons 2NH_3(g)$. The equilibrium constant is calculated from the equilibrium data.

$$K = \frac{[NH_3]^2}{[N_2][H_2]^3} = \frac{[0.100]^2}{[0.300][0.400]^3} = 0.5208$$

Now the H_2 is changed to cause the concentration of NH_3 to double. To form an additional 0.100 mol of NH_3, 0.050 mol of N_2 must decompose, leaving 0.300 mol − 0.050 mol = 0.250 mol N_2 at the new equilibrium. Let $x = [H_2]$ at equilibrium.

$$K = \frac{[NH_3]^2}{[N_2][H_2]^3} = \frac{(0.200)^2}{(0.250)(x)^3} = 0.5208$$

$$0.5208 x^3 = \frac{0.400}{0.250} = 0.160$$

$$x^3 = 0.307$$

$$x = 0.675 \text{ mol L}^{-1}$$

This is the equilibrium amount of H_2. To form 0.100 mol NH_3 requires an additional 0.150 mol H_2. Therefore, the amount of H_2 added has to be 0.675 mol − 0.400 mol + 0.150 mol = 0.425 mol.

26. A 1.00-L vessel at 400°C contains the following equilibrium concentrations: N_2, 1.00 M; H_2, 0.50 M; and NH_3, 0.50 M. How many moles of hydrogen must be removed from the vessel in order to increase the concentration of nitrogen to 1.2 M?

Solution

The reaction is $N_2(g) + 3H_2(g) \rightleftharpoons 2NH_3(g)$. The equilibrium constant for this equilibrium is calculated from the original data.

$$K = \frac{[NH_3]^2}{[N_2][H_2]^3} = \frac{(0.50)^2}{(1.00)(0.50)^3} = 2.0 \text{ mol}^{-2} \text{ L}^2$$

The concentration of N_2 increases by 1.2 M − 1.0 M = 0.2 M. From the balanced equation, 1 mol $N_2 \rightarrow$ 2 mol NH_3. Thus, 0.2 M N_2 must have come from the decomposition of 2(0.2 M NH_3) = 0.4 M NH_3. The amount of NH_3 remaining is 0.5 M − 0.4 M = 0.1 M NH_3. In addition to the nitrogen formed, hydrogen was also formed. Because $NH_3 \rightarrow \frac{3}{2} H_2$, the decomposition of 0.4 M NH_3 results in the formation of 0.60 M H_2. Thus, a total of 0.50 M + 0.60 M = 1.10 M H_2 is the total amount of H_2.

In order to maintain equilibrium, hydrogen must be removed from the reaction mixture to allow the concentration of N_2 to increase to 1.2 M. Let x = concentration of H_2 in the final mixture. Then

$$K = \frac{[0.1 \text{ mol L}^{-1}]^2}{[1.20 \text{ mol L}^{-1}][x]^3} = 2.0 \text{ mol}^{-2} \text{ L}^2$$

$$2.4 \, x^3 \text{ mol}^{-1} \text{ L}^1 = 0.01 \text{ mol}^2 \text{ L}^{-2}$$

$$x^3 = 4.17 \times 10^{-3} \text{ mol}^3 \text{ L}^{-3}$$

$$x = 0.16 \text{ mol L}^{-1}$$

Only 0.16 M remains, so

$$1.10 \text{ M} - 0.16 \text{ M} = 0.94 \text{ M H}_2$$

were removed. This is not unreasonable even though more hydrogen was removed than originally present, because the decomposition of NH_3 formed additional H_2.

28. The equilibrium constant for the reaction

$$CO + H_2O \rightleftharpoons CO_2 + H_2$$

is 5.0 at a given temperature.

(a) Upon analysis, an equilibrium mixture of the substances present was found to contain 0.20 mol of CO, 0.30 mol of water vapor, and 0.90 mol of H_2 in a liter. How many moles of CO_2 were there in the equilibrium mixture?

(b) Maintaining the same temperature, additional H_2 was added to the system, and some water vapor was removed by drying. A new equilibrium mixture was thereby established containing 0.40 mol of CO, 0.30 mol of water vapor, and 1.2 mol of H_2 in a liter. How many moles of CO_2 were in the new equilibrium mixture? Compare this with the quantity in part (a) and discuss whether the second value is reasonable. Explain how it is possible for the water vapor concentration to be the same in the two equilibrium solutions even though some vapor was removed before the second equilibrium was established.

Solution

(a) For the above reaction,

$$K = \frac{[CO_2][H_2]}{[CO][H_2O]} = 5.0$$

The concentrations at equilibrium are 0.20 M CO, 0.30 M H_2O, 0.90 M H_2. Substitution gives

$$K = 5.0 = \frac{[CO_2][0.90]}{[0.20][0.30]}$$

$$[CO_2] = \frac{5.0(0.20)(0.30)}{0.90} = 0.33 \text{ M}$$

Amount of CO_2 = 0.33 mol L^{-1} × 1 L = 0.33 mol

(b) At the particular temperature of reaction, K remains constant at 5.0. The new concentrations are 0.40 M CO, 0.30 M H_2O, 1.2 M H_2.

$$\frac{[CO_2][1.2]}{[0.40][0.30]} = 5.0$$

$$[CO_2] = [0.50 \text{ M}]$$

Amount of CO_2 = 0.50 mol L^{-1} × 1 L = 0.50 mol

Added H_2 forms some water to compensate for the removal of water vapor.

30. In a 3.0-L vessel, the following equilibrium partial pressures are measured: N_2, 190 torr; H_2, 317 torr; NH_3, 1000 torr. Hydrogen is removed from the

206   Chapter 15   An Introduction to Chemical Equilibrium

vessel until the partial pressure of nitrogen, at equilibrium, is equal to 250 torr. Calculate the partial pressures of the other substances under the new conditions.

Solution

The reaction is $N_2(g) + 3H_2(g) \rightleftharpoons 2NH_3(g)$

$$K_p = \frac{(P_{NH_3})^2}{(P_{N_2})(P_{H_2})^3} = \frac{(1000 \text{ torr})^2}{(190 \text{ torr})(317 \text{ torr})^3} = \frac{10^6}{(190)(3.185 \times 10^7)} = 1.65 \times 10^{-4}$$

From the balanced equation, 1 mol N_2 produces 2 mol NH_3. The pressure, which is equivalent to a concentration of N_2, increases: 250 torr − 190 torr = 60 torr. Therefore, the pressure of NH_3 must decrease by 2(60 torr); its pressure is 1000 − 120 = 880 torr. In order to maintain equilibrium, hydrogen must be removed from the reaction mixture. Let x = pressure of H_2 in the final mixture. Then,

$$K = \frac{(880)^2}{(250)(x)^3} = 1.65 \times 10^{-4}$$

$0.041x^3 \times 774,400$; $x^3 = 1.889 \times 10^7$; $x = P_{H_2} = 266$ torr

HETEROGENEOUS EQUILIBRIA

32. A sample of ammonium chloride was heated in a closed container.

$$NH_4Cl(s) \rightleftharpoons NH_3(g) + HCl(g)$$

At equilibrium the pressure of $NH_3(g)$ was found to be 1.75 atm. What is the equilibrium constant for the decomposition at this temperature?

Solution

Since the decomposition must generate the same pressure of HCl as NH_3, there must be 1.75 atm of HCl present.

$$K_p = P_{NH_3} P_{HCl} = (1.75 \text{ atm})(1.75 \text{ atm}) = 3.06 \text{ atm}^2$$

33. Sodium sulfate 10-hydrate, $Na_2SO_4 \cdot 10H_2O$, dehydrates according to the equation

$$Na_2SO_4 \cdot 10H_2O(s) \rightleftharpoons Na_2SO_4(s) + 10H_2O(g)$$

with $K = 4.08 \times 10^{-25}$ at 25°C. What is the pressure of water vapor in equilibrium with a sample of $Na_2SO_4 \cdot 10H_2O$?

Solution

Because two of the substances involved in the equilibrium are solids, their concentrations are constant and do not appear in the equilibrium expression. Thus,

$$K = 4.08 \times 10^{-25} = [P_{H_2O}]^{10}$$

$$P_{H_2O} = \sqrt[10]{4.08 \times 10^{-25}} = 3.64 \times 10^{-3} \text{ atm}$$

KEYSTROKES: 4.08 $\boxed{EE}$ 25 $\boxed{+/-}$ $\boxed{INV}$ $\boxed{x^{1/y}}$ 10 = 3.639703 − 03

Chapter 15 An Introduction to Chemical Equilibrium

ADDITIONAL EXERCISES

39. The hydrolysis of the sugar sucrose to the sugars glucose and fructose

$$C_{12}H_{22}O_{11} + H_2O \longrightarrow C_6H_{12}O_6 + C_6H_{12}O_6$$

follows a first-order rate equation for the disappearance of sucrose.

$$\text{Rate} = k[C_{12}H_{22}O_{11}]$$

(The products of the reaction, glucose and fructose, have the same molecular formulas but differ in the arrangement of the atoms in their molecules.)

(a) In neutral solution, $K = 2.1 \times 10^{-11}$ s^{-1} at 27°C and 8.5×10^{-11} s^{-1} at 37°C. Determine the activation energy, the frequency factor, and the rate constant for this equation at 47°C.

Solution

(a) The text demonstrates that the value of E_a may be determined from a plot of log K against $1/T$ that gives a straight line whose slope is $-E_a/2.303R$. This is based on the equation

$$\log K = \log A - \frac{E_a}{2.303\,RT}$$

Only two data points are given and these must determine a straight line when log K is plotted against $1/T$. The values needed are

$K_1 = 2.1 \times 10^{-11}$ $\log K_1 = -10.6778$

$K_2 = 8.5 \times 10^{-11}$ $\log K_2 = -10.0706$

$T_1 = 27°C = 300$ K $1/T_1 = 3.3333 \times 10^{-3}$

$T_2 = 37°C = 310$ K $1/T_2 = 3.2258 \times 10^{-3}$

The slope of the line determined by these points is given by

$$\text{Slope} = \frac{\Delta(\log K)}{\Delta(1/T)} = \frac{(-10.0706) - (-10.6778)}{(3.2258 \times 10^{-3}) - (3.3333 \times 10^{-3})}$$

$$= \frac{0.6072}{-0.1075 \times 10^{-3}} = -5648$$

$$E_a = -2.303(8.314 \text{ J mol}^{-1})(-5648)$$

$$= 108{,}100 \text{ J} = 108 \text{ kJ}$$

Whenever differences of very small numbers are taken, such as the reciprocals of T above, an inherent problem occurs. In order to have accurate differences, a larger number of significant figures than justified by the data must be used. Thus five figures were used to obtain the value $E_a = 108$ kJ. This difficulty may be alleviated by the approach shown below.

For only two data points Arrhenius' law

$$K = A \times 10^{-E_a/2.303\,RT}$$

may be used in an equally accurate, analytical solution for E_a. This is possible because the value of A will be the same throughout the course of the reaction. Once the value of E_a is determined, the value of A may be determined from either equations (1) or (2) shown below. Then K at 47°C may be determined using the value of E_a and A so determined. The procedure is as follows:

$$K = A \times 10^{-E_a/2.303\,RT}$$

$$2.1 \times 10^{-11}\ \text{s}^{-1} = A \times 10^{-E_a/2.303(8.314\ \text{J K}^{-1})(300\ \text{K})} \tag{1}$$

$$8.5 \times 10^{-11}\ \text{s}^{-1} = A \times 10^{-E_a/2.303(8.314\ \text{J K}^{-1})(310\ \text{K})} \tag{2}$$

Equating the values of A as solved from (1) and (2),

$$2.1 \times 10^{-11}\ \text{s}^{-1} \times 10^{+E_a/2.303(8.314\ \text{J K}^{-1})(300\ \text{K})}$$
$$= 8.5 \times 10^{-11}\ \text{s}^{-1} \times 10^{+E_a/2.303(8.314\ \text{J K}^{-1})(310\ \text{K})}$$

or

$$2.1 \times 10^{-11} \times 10^{+E_a/5744} = 8.5 \times 10^{-11} \times 10^{+E_a/5936}$$

Taking common logs of both sides gives

$$(\log 2.1 \times 10^{-11}) + \frac{E_a}{5744} = (\log 8.5 \times 10^{-11}) + \frac{E_a}{5936}$$

$$-10.68 + \frac{E_a}{5744} = -10.07 + \frac{E_a}{5936}$$

$$E_a\left(\frac{1}{5744} - \frac{1}{5936}\right) = -10.07 + 10.68$$

$$E_a\,(1.741 \times 10^{-4} - 1.685 \times 10^{-4}) = 0.61$$

$$E_a = \frac{0.61}{0.056 \times 10^{-4}} = 109\ \text{kJ}$$

The value of A may be found from either equation (1) or (2). Using (1),

$$2.1 \times 10^{-11}\ \text{s}^{-1} = A \times 10^{-109,000/2.303(8.314)(300)} = A \times 10^{-18.98}$$

$$A = 2.1 \times 10^{-11}\ \text{s}^{-1} \times 10^{+18.98} = 2.1 \times 10^{-11}(9.55 \times 10^{18}\ \text{s}^{-1}) = 2.0 \times 10^{8}\ \text{s}^{-1}$$

KEYSTROKES: 2.1 [EE] [+/−] 11 × 10 [y^x] 18.98 = 2.0054844   08

The value of K at 47°C may be determined from the Arrhenius equation now that the value of E_a and A have been calculated.

$$K = A \times 10^{-E_a/2.303\,RT}$$

$$= 2.0 \times 10^{8}\ \text{s}^{-1} \times 10^{-109,000\ \text{J}/2.303(8.314\ \text{J K}^{-1})(320\ \text{K})}$$

$$= 2.0 \times 10^{8}\ \text{s}^{-1} \times 10^{-17.79} = 2.0 \times 10^{8}\ \text{s}^{-1}(1.62 \times 10^{-18})$$

$$= 3.2 \times 10^{-10}\ \text{s}^{-1}$$

41. The reaction between hydrogen, $H_2(g)$, and sulfur, $S_8(g)$, to produce H_2S is exothermic at 25°C. How should the pressure and temperature be adjusted in order to improve the equilibrium yield of H_2S, assuming that all reactants and products are in the gaseous state? How will these conditions affect the rate of attainment of equilibrium?

Solution

$$8H_2(g) + S_8(g) \rightleftharpoons 8H_2S(g)$$

Chapter 15 An Introduction to Chemical Equilibrium

An increase in pressure will force the reaction to the right. However, too much pressure may cause $S_8(g)$ to condense. Therefore, the reaction should be run at as high a pressure as possible and as low a temperature as feasible to prevent condensation and yet keep the rate of reaction fast enough to produce at a reasonable rate.

42. One possible mechanism for the reaction of H_2O_2 with I^- in acid solution (Section 14.14) is

$$H_3O^+ + I^- \longrightarrow HI + H_2O \quad \text{(fast)}$$

$$H_2O_2 + HI \longrightarrow H_2O + HOI \quad \text{(slow)}$$

$$HOI + H_3O^+ + I^- \longrightarrow 2H_2O + I_2 \quad \text{(fast)}$$

$$I_2 + I^- \longrightarrow I_3^- \quad \text{(fast)}$$

(a) Write the rate equation for the slow elementary reaction.
(b) Write the equilibrium constant expression for the first elementary reaction.
(c) Solve the equilibrium constant expression from part (b) for [HI], and substitute this concentration into the rate equation from part (a) to obtain the overall rate equation for this mechanism.

Solution

(a) The slow step is $H_2O_2 + HI \longrightarrow H_2O + HOI$

$$\text{Rate} = k\,[H_2O_2][HI]$$

(b) Since water can be incorporated into the equilibrium constant for the step,

$$K = \frac{[HI]}{[H_3O^+][I^-]}$$

(c) $[HI] = K\,[H_3O^+][I^-]$

Substitution into the rate equation gives

$$\text{Rate} = kK\,[H_2O_2][H_3O^+][I^-]$$

$$= K'\,[H_2O_2][H_3O^+][I^-]$$

16

Acids and Bases

INTRODUCTION

Scientists have speculated on the nature of acids and bases for some two hundred years. Early attempts to define acids were based on easily discernible properties. Substances that tasted sour, caused vegetable dyes to change color, and reacted with metals to produce hydrogen were called *acids*. *Bases* were defined as those substances that had properties that were the opposite of acids: they tasted bitter, felt slippery, and caused vegetable dyes to change to different colors from those produced by acids.

The first classification of acids and bases according to their chemical properties was done by Svante Arrhenius. He said that acids caused the increase of hydrogen ion in aqueous solution and bases caused the increase of hydroxide ion in solution. This was a rather limited view, because substances other than hydrogen ion and hydroxide ion that reacted in the same way as these did were not allowed in the classification scheme.

A useful but also somewhat limited definition was proposed independently by Johannes Brønsted and Thomas Lowry. According to this theory, acids are defined as proton donors and bases as proton acceptors. Acids are thought to contain ionizable protons that can be transferred to bases. This definition is broader than Arrhenius' definition, because a base can be anything that accepts a hydrogen ion, not just a hydroxide ion.

G. N. Lewis proposed a much more comprehensive definition of acids and bases in 1923. According to his definition, an acid is *any species that can accept a pair of electrons* from a donor, the base. The Lewis classification scheme includes all ionizable protonic substances and all the donors in the Brønsted-Lowry scheme. The Lewis theory has the advantage of being able to explain acidic or basic characteristics of compounds not containing protons, including metal and nonmetal oxides. In contrast to the previous schemes, many reactions in organic chemistry also can be explained in terms of the Lewis concept.

A more recent definition of acids and bases involves classifying substances by their capacity to form cations or anions characteristic of a solvent. A cation of the solvent is said to be acidic; an anion of the solvent is said to be basic. This definition is particularly appropriate for work in nonaqueous solvents.

READY REFERENCE

Dilution formulas A volume of concentrated solution can be diluted to form a solution of lesser concentration. A larger volume of solution is formed thereby in which the number of moles or equivalents of solute has remained constant; only the volume of solution has changed. The following two relations involving molarity and normality are extremely useful for preparing solutions.

Chapter 16 Acids and Bases

Molarity:

No. mol solute in $soln_1$ = no. mol solute in $soln_2$

$$[M_{soln_1}][V_{soln_1}] = [M_{soln_2}][V_{soln_2}]$$

$$\left[\frac{mol}{L}\right]_1 (no.\ L_1) = \left[\frac{mol}{L}\right]_2 (no.\ L_2)$$

Example involving molarity:

Calculate the volume of 0.15 M H_2SO_4 solution that can be prepared by diluting 20.0 mL of 6.00 M H_2SO_4.

$$(M_1)(V_1) = (M_2)(V_2)$$

$$V_2 = \frac{(M_1)(V_1)}{M_2} = \frac{(6.00\ mol/L)(0.0200\ L)}{0.15\ mol/L} = 0.80\ L$$

Normality:

No. equiv solute in $soln_1$ = no. equiv solute in $soln_2$

$$[N_{soln_1}][V_{soln_1}] = [N_{soln_2}][V_{soln_2}]$$

$$\left[\frac{equiv}{L}\right]_1 (no.\ L_1) = \left[\frac{equiv}{L}\right]_2 (no.\ L_2)$$

Example involving normality:

Calculate the volume of 6.0 N H_2SO_4 required to prepare 4.00 L of 0.10 N H_2SO_4 solution.

$$N_1 V_1 = N_2 V_2$$

$$V_1 = \frac{N_2 V_2}{N_1} = \frac{(0.10\ equiv/L)(4.00\ L)}{6.0\ equiv/L} = 0.067\ L = 67\ mL$$

End point (16.13) This term, normally used with reactions taking place in aqueous solution, defines the point in a reaction at which an equal number of equivalents of the reactants have been consumed. Some type of indicator, either chemical or electronic, is used to monitor the reaction and to determine the end point.

Equivalents (16.14) An equivalent is the mass of a substance that donates or accepts 1 mol (6.022×10^{23}) of electrons (e^-), protons (H^+), or hydroxide ions (OH^-), or that neutralizes 1 mole of negative or positive charges. One equivalent (equiv) in a protonic acid-base reaction is the molar mass (molecular weight) of the substance divided by the number of protons donated or consumed per molecule of acid or base used, respectively:

$$\text{Equivalent} = \frac{\text{molar mass of acid or base}}{\text{number of replaceable } H^+ \text{ or } OH^- \text{ or } e^- \text{ transferred}}$$

The mass of an equivalent can be deduced only by examination of the chemical equation representing a specific reaction — not merely from the formula of the acid or base. The need to know the balanced chemical equation for the reaction is illustrated in the variability of the mass of the equivalent for the acids and bases in the following examples:

Example 1: $H_2SO_4(aq) + 2NaOH(aq) \rightarrow Na_2SO_4(aq) + 2H_2O(l)$

In this reaction, 1 mol of H_2SO_4 donates 2 mol of protons, which are accepted by 2 mol of NaOH. The mass of the equivalents for the acid and base are

$$\text{Mass of equivalent } (H_2SO_4) = \frac{1 \text{ mol}}{2 \text{ equiv}} \times \text{molar mass} = \frac{1 \text{ mol}}{2 \text{ equiv}} \frac{98.07 \text{ g}}{\text{mol}} = 49.04 \text{ g}$$

$$\text{Mass of equivalent } (NaOH) = \frac{1 \text{ mol}}{1 \text{ equiv}} \times \text{molar mass} = \frac{1 \text{ mol}}{1 \text{ equiv}} \frac{39.997 \text{ g}}{\text{mol}} = 39.997 \text{ g}$$

Example 2: $2H_2SO_4(aq) + Ca(OH)_2(aq) \longrightarrow Ca(HSO_4)_2(aq) + H_2O(l)$

In this reaction, only one of the protons from H_2SO_4 is reacted, while both hydroxide ions from $Ca(OH)_2$ are reacted. The mass of an equivalent is

$$\text{Mass of equivalent } (H_2SO_4) = \frac{1 \text{ mol}}{1 \text{ equiv}} \times \frac{98.07 \text{ g}}{\text{mol}} = 98.07 \text{ g}$$

$$\text{Mass of equivalent } (Ca(OH)_2) = \frac{1 \text{ mol}}{2 \text{ equiv}} \times \frac{74.09 \text{ g}}{\text{mol}} = 74.09 \text{ g}$$

Neutralization reaction (16.12) Any reaction of an acid and a base in which equal equivalents of acid and base are reacted is known as a *neutralization reaction*. The resulting solution may be neutral, acidic, or basic depending on the H^+ and OH^- concentrations (or their counterparts in nonaqueous systems) that remain. These concentrations, in turn, depend on the relative strengths of the reacting substances.

Normality (N) (16.14) The normality of a solution is the number of equivalents of solute dissolved per liter of solution:

$$\text{Normality} = \frac{\text{equivalents of solute}}{V \text{ of solution in liters}}$$

Related equations (units only):

$$\text{Equivalents} = \text{normality} \times V(L) = \frac{\text{equiv}}{L} \times L = \text{equivalents}$$

$$V \text{ (soln)} = \frac{\text{equiv}}{\text{equiv}/L} = L$$

Because the mass of an equivalent of a substance is specific to a reaction, the normality of a solution must be determined from a known reaction.

A useful relationship involving normality and equivalents follows from the definition of mass of an equivalent. The number of equivalents of all reactants in a reaction is equal to the number of equivalents of product:

no. g-equiv. wt reactant 1 = no. g-equiv. wt reactant 2 = $\cdots$ = no. g-equiv. wt reactant n

= no. g-equiv. wt product 1 = $\cdots$ = no. g-equiv. wt product n

If the reactants and/or products are in solution, their quantities are related through their normalities as

$$N_{R_1}V_{R_1} = N_{R_2}V_{R_2} = \cdots = N_{R_n}V_{R_n}$$
$$= N_{P_1}V_{P_1} = N_{P_2}V_{P_2} = \cdots = N_{P_n}V_{P_n}$$

Chapter 16 Acids and Bases 213

SOLUTIONS TO CHAPTER 16 EXERCISES

BRØNSTED ACIDS AND BASES

2. Show by suitable equations that each of the following species can act as a Brønsted acid:

   (a) H_3O^+
   (b) HCl
   (f) HSO_4^-

   Solution

   In order to act as a Brønsted acid, the species must act as a proton donor.

   (a) $H_3O^+ + OH^- \longrightarrow 2H_2O$
   (b) $HCl + 2H_2O \longrightarrow H_3O^+ + Cl^-$
   (f) $HSO_4^- + S^{2-} \longrightarrow SO_4^{2-} + HS^-$

3. Show by suitable equations that each of the following species can act as a Brønsted base:

   (a) H_2O
   (d) CN^-
   (f) $H_2PO_4^-$

   Solution

   In order to act as a Brønsted base, the species must be able to accept a proton.

   (a) $H_2O + H_2SO_4 \longrightarrow H_3O^+ + HSO_4^-$
   (d) $CN^- + H^+ \longrightarrow HCN$
   (f) $H_2PO_4^- + HClO_4 \longrightarrow H_3PO_4 + ClO_4^-$

4. Write equations for the reaction of each of the following with water:

   (a) HCl
   (c) NH_3
   (d) NH_2^-
   (g) NH_4^+

   What is the role played by water in each of these acid-base reactions?

   Solution

   (a) $HCl + H_2O \longrightarrow H_3O^+ + Cl^-$. Water acts as a base.
   (c) $NH_3 + H_2O \longrightarrow NH_4^+ + OH^-$. Water acts as an acid.
   (d) $NH_2^- + H_2O \longrightarrow NH_3 + OH^-$. Water acts as an acid.
   (g) $NH_4^+ + H_2O \longrightarrow NH_3 + H_3O^+$. Water acts as a base.

6. What is the conjugate acid formed when each of the following reacts as a base?

   (a) OH^-
   (c) H_2O
   (e) HCO_3^-
   (h) H^-

   Is each a strong or weak acid?

Solution

(a) OH$^-$, H$_2$O, weak
(c) H$_2$O, H$_3$O$^+$, strong
(e) HCO$_3^-$, H$_2$CO$_3$, weak
(h) H$^-$, H$_2$, weak

8. Identify and label the Brønsted acid, its conjugate base, the Brønsted base, and its conjugate acid in each of the following equations:

   (a) HNO$_3$ + H$_2$O $\longrightarrow$ H$_3$O$^+$ + NO$_3^-$
   (b) CN$^-$ + H$_2$O $\rightleftharpoons$ HCN + OH$^-$
   (e) O^{2-} + H$_2$O $\longrightarrow$ 2OH$^-$
   (f) [Cu(H$_2$O)$_3$(OH)]$^+$ + [Al(H$_2$O)$_6$]$^{3+}$ $\rightleftharpoons$ [Cu(H$_2$O)$_4$]$^{2+}$ + [Al(H$_2$O)$_5$(OH)]$^{2+}$

Solution

In the following solutions,

acid$_1$ = Brønsted acid

base$_1$ = its conjugate base

base$_2$ = Brønsted base

acid$_2$ = its conjugate acid

(a) HNO$_3$ + H$_2$O $\longrightarrow$ H$_3$O$^+$ + NO$_3^-$
    acid$_1$ base$_2$ acid$_2$ base$_1$

(b) CN$^-$ + H$_2$O $\rightleftharpoons$ HCN + OH$^-$
    base$_2$ acid$_1$ acid$_2$ base$_1$

(e) O^{2-} + H$_2$O $\longrightarrow$ OH$^-$ + OH$^-$
    base$_2$ acid$_1$ acid$_2$ base$_1$

(f) [Cu(H$_2$O)$_3$(OH)]$^+$ + [Al(H$_2$O)$_6$]$^{3+}$ $\longrightarrow$ [Cu(H$_2$O)$_4$]$^{2+}$ + [Al(H$_2$O)$_5$(OH)]$^{2+}$
    base$_2$ acid$_1$ acid$_2$ base$_1$

11. Gastric juice, the digestive fluid produced in the stomach, contains hydrochloric acid, HCl. Milk of magnesia, a suspension of solid Mg(OH)$_2$ in an aqueous medium, is sometimes used to neutralize excess stomach acid. Write a complete balanced equation for the neutralization reaction, and identify the conjugate acid-base pairs.

    Solution

    Mg(OH)$_2$ + HCl $\longrightarrow$ Mg^{2+}(*aq*) + 2Cl$^-$(*aq*) + 2H$_2$O
    base$_2$ acid$_1$ base$_1$ acid$_2$

12. Nitric acid reacts with insoluble copper(II) oxide to form soluble copper(II) nitrate, Cu(NO$_3$)$_2$. Write the balanced chemical equation for the reaction of an aqueous solution of HNO$_3$ with CuO, and indicate the conjugate acid-base pairs.

    Solution

    CuO + 2HNO$_3$ $\longrightarrow$ Cu^{2+}(*aq*) + 2NO$_3^-$(*aq*) + H$_2$O
    base$_2$ acid$_1$ base$_1$ acid$_2$

Chapter 16 Acids and Bases

14. State which of the following species are amphiprotic and write chemical equations illustrating the amphiprotic character of the species:

    (b) $H_2PO_4^-$
    (c) S^{2-}
    (d) CH_4

    Solution

    (b) $H_2PO_4^-$ is amphiprotic.

    $$H_2PO_4^- + H^+ \rightleftharpoons H_3PO_4$$
    $$H_2PO_4^- \rightleftharpoons H^+ + HPO_4^{2-}$$

    (c) S^{2-} is not amphiprotic; only one expression can be written.

    $$S^{2-} + H^+ \rightleftharpoons HS^-$$

    (d) CH_4 is not amphiprotic.

    $$CH_4 \rightleftharpoons CH_3^- + H^+$$

STRENGTHS OF ACIDS AND BASES

16. Predict which acid in each of the following pairs is the stronger:

    (a) H_2O or HF
    (b) NH_3 or H_2O
    (c) PH_3 or HI
    (d) NH_3 or H_2S

    Explain your reasoning for each.

    Solution

    (a) HF; F is more electronegative than O.
    (b) H_2O; O is more electronegative than N.
    (c) HI; PH_3 is weaker than HCl; HCl is weaker than HI. Thus PH_3 is weaker than HI.
    (d) H_2S; NH_3 is weaker than H_2O; H_2O is weaker than H_2S. Thus NH_3 is weaker than H_2S.

18. Rank the compounds in each of the following groups in order of increasing acidity, and explain the order you assign:

    (a) HCl, HBr, HI
    (c) $NaHSO_3$, $NaHSeO_3$, $NaHSO_4$
    (e) HF, H_2O, NH_3, CH_4

    Solution

    (a) HCl < HBr < HI. H—X bond energy decreases down a group. This is more significant than the decrease in electronegativity.
    (c) $NaHSeO_3$ < $NaHSO_3$ < $NaHSO_4$. In polyoxy acids the more electronegative central element, S in this case, forms the stronger acid. The larger the number of oxygens on the central atom (the higher its oxidation number) also causes a greater release of hydrogen atoms, hence resulting in a stronger acid.

(e) $CH_4 < NH_3 < H_2O < HF$. This series can be thought of in terms of increasing electronegativity of the atom attached to hydrogen. The greater the electronegativity, the stronger a binary acid in the same period of the Periodic Table.

19. Rank the bases in each of the following groups in order of increasing base strength. Explain.

    (b) H_2O, OH^-, H^-, Cl^-
    (d) ClO_4^-, ClO^-, ClO_2^-, ClO_3^-

    Solution

    (b) Cl^-, H_2O, OH^-, H^-. Cl^- is the anion of a strong acid and has no tendency to act as a base. H^- is very strong.

    (d) ClO_4^-, ClO_3^-, ClO_2^-, ClO^-. Again the one with the largest number of oxygens (that is, the highest oxidation number) releases the proton most easily and is the poorest base.

21. Some porcelain cleaners contain sodium hydrogen sulfate, $NaHSO_4$. Is a solution of $NaHSO_4$ acidic, neutral, or basic? These cleansers remove the deposits due to hard water (primarily calcium carbonate, $CaCO_3$) very effectively. Write a balanced chemical equation for one reaction between $NaHSO_4$ and $CaCO_3$.

    Solution

    HSO_4^- is a moderate acid, so a solution of $NaHSO_4$ is acidic.

    $$2NaHSO_4 + CaCO_3 \rightarrow Na_2SO_4 + CaSO_4 + CO_2 + H_2O$$

23. How can the relative strengths of strong acids be measured?

    Solution

    Water accepts a proton from all strong acids with equal ease. It is too strong a base to differentiate between acids. Therefore, a weaker base, such as ethanol, should be used.

25. Soaps are sodium and potassium salts of a family of acids called fatty acids, which are isolated from animal fats. These acids, which are related to acetic acid, CH_3CO_2H, all contain the carboxyl group, $—CO_2H$, and have about the same strength as acetic acid. Examples include palmitic acid, $C_{15}H_{31}CO_2H$, and stearic acid, $C_{17}H_{35}CO_2H$.

    (a) Write a balanced chemical equation indicating the formation of sodium palmitate, $C_{15}H_{31}CO_2Na$, from palmitic acid and sodium carbonate, and a corresponding equation for the formation of sodium stearate.
    (b) Is a soap solution acidic, basic, or neutral?

    Solution

    (a) $2C_{15}H_{31}CO_2H + Na_2CO_3 \rightarrow 2C_{15}H_{31}CO_2^-Na^+ + H_2CO_3$

    $2C_{17}H_{35}CO_2H + Na_2CO_3 \rightarrow 2C_{17}H_{35}CO_2^-Na^+ + H_2CO_3$

    (b) Soap is basic because the sodium salt can ionize subtracting H^+ ion through the dissociation of water, then leaving free OH^- ions in solution.

Chapter 16 Acids and Bases

SOLUTIONS OF BRØNSTED ACIDS AND BASES

26. Write the equation for the essential reaction between aqueous solutions of strong acids and of strong bases.

Solution

$$H^+(aq) + OH^-(aq) \longrightarrow H_2O(aq)$$

28. Containers of $NaHCO_3$, sodium hydrogen carbonate (sometimes called sodium bicarbonate), are often kept in chemical laboratories for use in neutralizing any acids or bases spilled. Write balanced chemical equations for the neutralization by $NaHCO_3$ of (a) a solution of HCl and (b) a solution of KOH.

Solution

(a) $HCl(aq) + NaHCO_3(aq) \longrightarrow NaCl(aq) + H_2CO_3(aq)$
(b) $KOH(aq) + NaHCO_3(aq) \longrightarrow NaKCO_3(aq) + HOH$

30. In aqueous solution arsenic acid, H_3AsO_4, ionizes in three steps. Write an equation for each step, and label the conjugate acid-base pairs. Which arsenic-containing species are amphiprotic?

Solution

$$H_3AsO_4 \rightleftharpoons H^+ + H_2AsO_4^-$$
$$H_2AsO_4^- \rightleftharpoons H^+ + HAsO_4^{2-}$$
$$HAsO_4^{2-} \rightleftharpoons H^+ + AsO_4^{3-}$$

$H_2AsO_4^-$ and $HAsO_4^{2-}$ are amphiprotic.
$H_2AsO_4^-$ is the conjugate base of the acid H_3AsO_4.
$H_2AsO_4^-$ also is the conjugate acid of its conjugate base $HAsO_4^{2-}$.
Likewise, $HAsO_4^{2-}$ is the conjugate acid of its conjugate base AsO_4^{3-}.

32. Describe the chemical reaction, using equations where appropriate, that you would expect from the combination of a solution of $HClO_4$ with each of the following:

(a) a potassium hydroxide solution
(c) aluminum
(d) solid aluminum sulfide, Al_2S_3
(g) calcium carbonate

Solution

(a) $KOH(aq) + HClO_4(aq) \longrightarrow KClO_4(aq) + H_2O$
(c) $2Al(s) + 6HClO_4(aq) \longrightarrow 3H_2(g) + 2Al(ClO_4)_3(aq)$
(d) $Al_2S_3(s) + 6HClO_4(aq) \longrightarrow 3H_2S(g) + 2Al(ClO_4)_3(aq)$
(g) $CaCO_3(s) + 2HClO_4(aq) \longrightarrow Ca(ClO_4)_2(aq) + CO_2(g) + H_2O(l)$

35. Define the term *salt*, and give examples of normal salts, hydrogen salts, hydroxysalts, and oxysalts.

Solution

A *salt* is the product, other than water, of the reaction of an acid with a base. A variety of possibilities exist. A normal salt has no H^+ or OH^- units in the structure. An example is NaCl. A hydrogen salt is a salt in which one or more protons remain, such as $NaHCO_3$. A hydroxysalt is a salt with one or more hydroxy functions unreacted, for example, basic copper(II) iodate, $Cu(OH)IO_3$. Oxysalts contain oxygen as well as one of the more commonly thought of salt formers, for example, gallium oxychloride, $6GaOCl \cdot 14H_2O$.

pH AND pOH

37. Calculate the pH and pOH of each of the following solutions. (Each is a strong acid or a strong base.)

    (a) 0.100 M HBr
    (c) 0.0071 M $Ba(OH)_2$

Solution

(a) pH is defined as $-\log[H^+]$. For 0.100 M HBr,
$$pH = -\log 0.100 = -(-1.000) = 1.000$$
$$pH + pOH = 14 \quad pOH = 14 - pH = 14.000 - 1.000 = 13.000$$

(c) For 0.0071M $Ba(OH)_2$, $[OH^-] = 2(0.0071\ M) = 0.0142\ M$
$$pOH = -\log[OH^-] = -\log 0.0142 = 1.85$$
$$pH = 14 - pOH = 14 - 1.85 = 12.15$$

TITRATION; NORMALITY

38. Calculate the mass of an equivalent of each of the reactants in the following equations to one decimal place:

    (a) $H_2SO_4 + 2LiOH \longrightarrow Li_2SO_4 + 2H_2O$
    (e) $H_2S + HgCl_2 \longrightarrow HgS + 2HCl$

Solution

(a) Each sulfuric acid molecule reacts with two hydroxy groups. So
$$\frac{\text{mol. wt}}{2} = \frac{98.0\ \text{g mol}^{-1}}{2\ \text{equiv mol}^{-1}} = 49.0\ \text{g equiv}^{-1}$$

Each LiOH reacts its single hydroxy group and therefore the mass of an equivalent is the same as its formula weight, 23.9 g.

(e) In H_2S each hydrogen reacts so that the mass of an equivalent is the molecular weight divided by 2:
$$32.06 + 2(1.008) = 34.08\ \text{g mol}^{-1}$$
$$\text{Mass of equivalent} = \frac{34.08\ \text{g mol}^{-1}}{2\ \text{equiv mol}^{-1}} = 17.0\ \text{g}$$

For $HgCl_2$ the mass of an equivalent is equal to its gram-molecular weight divided by 2.
$$\text{Mass of equivalent} = \frac{200.59\ \text{g mol}^{-1} + 2(35.453\ \text{g mol}^{-1})}{2\ \text{equiv mol}^{-1}} = \frac{271.5\ \text{g mol}^{-1}}{2\ \text{equiv mol}^{-1}} = 135.7\ \text{g}$$

Chapter 16 Acids and Bases

39. Calculate the normality of each of the following solutions:

(a) 4.0 equivalents of HCl in 2.0 liters of solution.

(c) 0.30 equivalents of H_2SO_4 in 400 mL of solution.

Solution

(a) The normality of HCl is calculated from the relation

$$N = \frac{\text{equiv}}{L} = \frac{4.0 \text{ equiv}}{2.0 \text{ L}} = 2.0 \text{ N}$$

(c) $N = \dfrac{0.30 \text{ equiv}}{0.400 \text{ L}} = 0.75 \text{ N}$

41. (a) How many grams of H_2SO_4 are contained in 1.2 L of 0.50 M solution?

(b) How many grams of H_2SO_4 are contained in 1.2 L of 0.50 N solution?

Solution

(a) Let x = mass of H_2SO_4. From the definition

$$M = \frac{\text{mol of substance}}{\text{volume}} = \frac{x/\text{molecular weight}}{\text{volume (L)}}$$

$$0.50 \text{ M} = 0.50 \frac{\text{mol}}{\text{L}} = \frac{x/98.07 \text{ g mol}^{-1}}{1.2 \text{ L}}$$

$$x = 1.2 \times 0.50 \times 98.07 \text{ g} = 59 \text{ g}$$

(b) Assume both hydrogens in H_2SO_4 are replaceable.

$$N = \frac{\text{equiv solute}}{V \text{ (L)}} = \frac{x \text{ g solute} \times \dfrac{1 \text{ mol}}{98.07 \text{ g}} \times \dfrac{2 \text{ equivalents}}{1 \text{ mol}}}{1.2 \text{ L}} = 0.5 \text{ N}$$

$$x = 1.2 \times 0.50 \times 98.07 \text{ g}/2 \text{ equivalent} = 29 \text{ g}$$

44. A 0.366 N solution of KOH is titrated with H_2SO_4, producing K_2SO_4. If a 32.00-mL sample of the KOH solution is used, what volume of 0.366 M H_2SO_4 is required to reach the end point? What volume of 0.366 N H_2SO_4? (Assume both protons react.)

Solution

The end point is defined as the point in a reaction at which an equal number of equivalents of both reactants have reacted. In this case no. equiv KOH reacted = no. equiv H_2SO_4 reacted. The number of equivalents of KOH reacted is

$$\text{KOH} = \frac{0.366 \text{ equivalent}}{L} \times 0.03200 \text{ L} = 0.0117 \text{ equivalent}$$

This number of equivalents must be equal to the number of equivalents of H_2SO_4 reacted. Because 1 M H_2SO_4 equals 2N H_2SO_4 (both protons react),

$$0.366 \text{ M } H_2SO_4 = 0.732 \text{ N } H_2SO_4$$

Then the 0.0117 equivalent H_2SO_4 needed to react with KOH must have come from the 0.732 N solution. Thus,

220   Chapter 16 Acids and Bases

$$0.0117 \text{ equivalent } H_2SO_4 = \frac{0.732 \text{ equivalent}}{L} \times V$$

$$V = \frac{0.0117 \text{ equiv}}{0.732 \text{ equiv/L}} = 0.01598 \text{ L} = 16.0 \text{ mL}$$

For 0.366 N H_2SO_4, follow the same procedure

$$V = \frac{0.0117 \text{ equiv}}{0.366 \text{ equiv/L}} = 0.03197 \text{ L} = 32.0 \text{ mL}$$

47. What volume of 0.75 N phosphoric acid, H_3PO_4, would be required to neutralize completely 0.500 L of 0.70 M potassium hydroxide?

Solution

Neutralization is the point in an acid-base reaction when equal equivalents of a protonic acid and a hydroxyl base have reacted. In most cases the resulting salt solutions are not neutral; the concentrations of hydronium ion and hydroxide ion in solution are unequal because of the relative strengths of the acid and base.

$$(N_{acid})(V_{acid}) = (N_{base})(V_{base})$$

$$(0.75 \text{ N})(V_{acid}) = (0.70 \text{ N})(0.500 \text{ L})$$

$$V_{acid} = \frac{(0.70)(0.500) \text{ L}}{(0.75)} = 0.47 \text{ L}$$

49. Titration of 0.4500 of an acid requires 282.00 mL of 6.00×10^{-3} N NaOH to reach the end point. What is the mass of an equivalent of the acid?

Solution

At the end point an equal number of equivalents of acid (in 0.4500 g) has reacted with the same number of equivalents of base.

$$\text{No. equiv acid} = \text{no. equiv base} = \left(\frac{6.00 \times 10^{-3} \text{ equiv}}{L}\right)(0.28200 \text{ L})$$

$$= 1.69 \times 10^{-3}$$

Because 0.4500 g = 1.69 equiv $\times 10^{-3}$ equiv

$$1 \text{ equiv} = \frac{0.4500 \text{ g}}{1.69 \times 10^{-3} \text{ equiv}} = 266 \text{ g}$$

LEWIS ACIDS AND BASES

52. Write the Lewis structures of the reactants and product of each of the following equations,, and identify the Lewis acid and the Lewis base in each:

(a) $CO_2 + OH^- \longrightarrow HCO_3^-$
(c) $I^- + I_2 \longrightarrow I_3^-$
(e) $O^{2-} + SO_3 \longrightarrow SO_4^{2-}$

Chapter 16 Acids and Bases

Solution

(a) $:O=C=\ddot{O}: + [:\ddot{O}-H]^- \longrightarrow \left[\begin{array}{c}:\ddot{O}: \quad :\ddot{O}: \\ \diagdown \diagup \\ C \\ | \\ :\ddot{O}-H \end{array}\right]^-$ and resonance form

acid · base

(c) $[:\ddot{I}:]^- + :\ddot{I}-\ddot{I}: \longrightarrow [:\ddot{I}-\ddot{I}-\ddot{I}:]^-$

base · acid

(e) $[:\ddot{O}:]^{2-} + \left[\begin{array}{c}:\ddot{O}: \quad :\ddot{O}: \\ \diagdown \diagup \\ S \\ | \\ :\ddot{O}: \end{array}\right]^{2-} \longrightarrow \left[\begin{array}{c} :\ddot{O}: \\ | \\ :\ddot{O}-S-\ddot{O}: \\ | \\ :\ddot{O}: \end{array}\right]^{2-}$

base · and 2 additional resonance forms acid

55. The reaction of SO_3 with H_2SO_4 to give pyrosulfuric acid, $H_2S_2O_7$ (Section 9.9), is a Lewis acid-base reaction that results in the formation of S—O—S bonds. Write the chemical equation for this reaction showing the Lewis structures of the reactants and product, and identify the Lewis acid and Lewis base.

Solution

$$\underset{\text{acid}}{\begin{array}{c}:\ddot{O}: \quad :\ddot{O}: \\ \diagdown \diagup \\ S \\ | \\ :\ddot{O}:\end{array}} + \underset{\text{base}}{\begin{array}{c}:\ddot{O}: \\ | \\ :\ddot{O}-S-\ddot{O}-H \\ | \\ :\ddot{O}: \\ | \\ H\end{array}} \longrightarrow \begin{array}{c} :\ddot{O}: \quad\quad :\ddot{O}: \\ | \quad\quad\quad | \\ :\ddot{O}-S-\ddot{O}-S-\ddot{O}-H \\ | \quad\quad\quad | \\ :\ddot{O}: \quad\quad :\ddot{O}: \\ | \\ H\end{array}$$

57. Each of the species given below may be considered to be a Lewis acid-base adduct of a proton with a Lewis base. For each of the following pairs, indicate which adduct is the stronger acid:

(a) HCl or HCN
(b) HBr or H_3O^+
(d) H_2O or NH_3

Solution

(a) HCl
(b) H_3O^+
(d) H_2O

ADDITIONAL EXERCISES

58. The concentration of hydrochloric acid secreted by the stomach after a meal is about 1.2×10^{-3} M. What is the pH of stomach acid?

Solution

$$pH = -\log[H^+] = -\log[1.2 \times 10^{-3}] = -(-2.92) = +2.92$$

60. In papermaking the use of wood pulp as a source of cellulose and of alum-rosin sizing (which keeps ink from soaking into the pages and blurring) produces a paper that deteriorates in 25-50 years. This is quite satisfactory for most purposes, but it is not suitable for the needs of libraries and archives. The deterioration is due to formation of acids. Alum-rosin slowly produces sulfuric acid under humid conditions. Also, wood pulp contains lignin, which forms carboxylic acids as it ages. These acids are similar to acetic acid in that they contain the acidic —CO_2H group. In order to stop this acidification, books are sometimes soaked in magnesium hydrogen carbonate solution; treated with cyclohexylamine, $C_6H_{11}NH_2$, a base like ammonia but with one hydrogen atom replaced by a —C_6H_{11} group; and treated with gaseous diethyl zinc, $(C_2H_5)_2Zn$, a covalent molecule containing Zn—C bonds. Diethyl zinc is a source of $C_2H_5^-$, which is very much like CH_3^- in its properties. Write balanced equations for these reactions.

Solution

$$H_2SO_4 + Mg(HCO_3)_2 \rightarrow MgSO_4 + 2CO_2 + 2H_2O$$

$$2RCO_2H + Mg(HCO_3)_2 \rightarrow Mg(RCO_2)_2 + 2CO_2 + 2H_2O$$

$$RCO_2H + C_6H_{11}NH_2 \rightarrow (C_6H_{11}NH_3^+)(RCO_2^-)$$

$$H_2SO_4 + Zn(C_2H_5)_2 \rightarrow ZnSO_4(s) + 2C_2H_6(g)$$

$$2R{-}CO_2H(s) + Zn(C_2H_5)_2(g) \rightarrow 2C_2H_6(g) + (R{-}CO_2)_2Zn(s)$$

63. Predict the products of the following reactions, and write balanced equations for each. There may be more than one reasonable equation depending on the stoichiometry assumed for the reactants.

(a) $Fe_2O_3(s) + H_2SO_4(l) \rightarrow$
(c) $HCN(g) + Na \rightarrow$
(e) $Li_3N(s) + NH_3(l) \rightarrow$
(h) $MgH_2(s) + H_2S(g) \rightarrow$
(j) $KHCO_3(s) + KHS(s) \rightarrow$

Solution

Correct (but not the only) equations are as follows:

(a) $Fe_2O_3(s) + 3H_2SO_4(l) \rightarrow Fe_2(SO_4)_3(s) + 3H_2O(l)$
(c) $2HCN(g) + 2Na \rightarrow 2NaCN(s) + H_2(g)$
(e) $Li_3N(s) + 2NH_3(l) \rightarrow 3LiNH_2(s)$
(h) $MgH_2(s) + H_2S(g) \rightarrow MgS(s) + 2H_2(g)$
(j) $KHCO_3(s) + KHS(s) \rightarrow H_2S(g) + K_2O(s) + CO_2(g)$

64. How many milliliters of a 0.1500 M solution of KOH will be required to titrate 40.00 mL of a 0.0656 M solution of H_3PO_4?

$$H_3PO_4 + 2KOH \rightarrow K_2HPO_4 + 2H_2O$$

Solution

For KOH the molarity and normality are equal. From the equation in this reaction, H_3PO_4 reacts as though it had two equivalents of protons. Therefore, the normality is twice its molarity.

$$N = 2 \times 0.0656 \text{ M} = 0.1312 \text{ N}$$

Equivalent (acid) = equivalent (base)

$$0.1312 \text{ N} \times 40.00 \text{ mL} = 0.1500 \text{ N} \times V(\text{KOH})$$

$$V(\text{KOH}) = \frac{0.1312 \text{ N} \times 40.00 \text{ mL}}{0.1500 \text{ N}} = 35.0 \text{ mL}$$

66. A 0.3420-g sample of potassium acid phthalate, $KHC_8H_4O_4$, reacts with 35.73 mL of a NaOH solution in a titration. What is the molar concentration of the NaOH?

$$KHC_8H_4O_4 + NaOH \longrightarrow KNaC_8H_4O_4 + H_2O$$

Solution

Potassium acid phthalate (KHP) is monoacidic, so its normality and molarity are the same.

$$\text{Equiv }(KHC_8H_4O_4) = \frac{0.3420 \text{ g}}{204.2234 \text{ g equiv}^{-1}} = 1.675 \times 10^{-3} \text{ equiv}$$

$$\text{Equiv (KHP)} = 1.675 \times 10^{-3} \text{ equiv} = \text{equiv (NaOH)} = N \times 0.03573 \text{ L}$$

$$N = M(\text{NaOH}) = \frac{1.675 \times 10^{-3} \text{ equiv}}{0.03573 \text{ L}} = 0.04687 \text{ M}$$

67. Trichloroacetic acid, CCl_3CO_2H, is amphiprotic.

$$CCl_3CO_2H + B \longrightarrow CCl_3CO_2^- + BH^+$$

$$HA + CCl_3CO_2H \longrightarrow CCl_3CO_2H_2^+ + A^-$$

Write equations for the reaction of pure $CCl_3CO_2H(l)$ with $H_2O(l)$, $HClO_4(l)$, $HBr(g)$, $NH_3(g)$, and $CH_3CO_2H(l)$. Trichloroacetic acid has an acid strength between those of H_3O^+ and HSO_4^-.

Solution

$$CCl_3CO_2H(l) + H_2O(l) \rightleftharpoons CCl_3CO_2^- + H_3O^+$$

$$CCl_3CO_2H(l) + HClO_4(l) \rightleftharpoons CCl_3CO_2H_2^+ + ClO_4^-$$

$$CCl_3CO_2H(l) + HBr(g) \rightleftharpoons CCl_3CO_2H_2^+ + Br^-$$

$$CCl_3CO_2H(l) + NH_3(g) \rightleftharpoons CCl_3CO_2^- + NH_4^+$$

$$CCl_3CO_2H(l) + CH_3CO_2H(l) \rightleftharpoons CCl_3CO_2^- + CH_3CO_2H_2^+$$

70. The reaction of 0.871 g of sodium with an excess of liquid ammonia containing a trace of $FeCl_3$ as a catalyst produced 0.473 L of pure H_2 measured at 25°C and 745 torr. What is the equation for the reaction of sodium with liquid ammonia? Show your calculations.

Solution

In the reaction with liquid ammonia, sodium functions as a Lewis base by donating electrons to hydrogen ions formed through the ionization of ammonia:

$$NH_3 \longrightarrow NH_2^- + H^+$$

To write the equation for the overall reaction, first determine the mole ratio of sodium used to hydrogen produced.

$$\text{No. mol Na} = 0.871 \text{ g} \times \frac{1 \text{ mol}}{22.98977} = 0.0379 \text{ mol}$$

From the ideal gas equation, the number of moles of hydrogen produced equals

$$PV = nRT, \qquad n = \frac{PV}{RT},$$

$$n_{H_2} = \frac{\left(\dfrac{745}{760}\ \text{atm}\right)(0.473\ \text{L})}{\left(\dfrac{0.08206\ \text{L atm}}{\text{mol K}}\right)(298\ \text{K})} = 0.0190\ \text{mol}$$

Thus, 0.0379 mol of Na reacts with 0.0190 mol of H_2. The ratio of Na to H_2 is

0.0379 : 0.0190   or   2Na : 1H_2

Therefore, the balanced equation must be

2Na + 2NH_3 ⟶ 2$NaNH_2$ + H_2

17

Ionic Equilibria of Weak Electrolytes

INTRODUCTION

Weak electrolytes are substances that only slightly conduct an electric current in aqueous solution. Examples include the carbonates in the oceans, in ground water, and even in blood; organic acids that make up the fats of plants and animals; and the vast number of sulfides and sulfites formed by the oxidation of sulfur. One important aspect of weak electrolytes is their major role in the control of acidity in our natural environment.

Some weak electrolytes in water function as either weak acids or weak bases. These undergo dissociation, reaching an equilibrium that involves both ions of the electrolyte and of the undissociated electrolyte molecules. Aqueous solutions of weak acids, HA, and weak bases, WB, are represented by the equilibria

$$HA(aq) + H_2O(l) \rightleftharpoons H_3O^+(aq) + A^-(aq) \qquad K_a \ll 1$$

$$WB(aq) + H_2O(l) \rightleftharpoons WBH^+(aq) + OH^-(aq) \qquad K_b \ll 1$$

In general, the fraction of electrolyte molecules so dissociated at equilibrium ranges from a high of about 20% to a low of less than 1%.

The salts formed by the reaction of weak acids with weak bases are themselves weak bases or weak acids; they are called the *conjugate base* or *conjugate acid* of the respective acid or base. In combination with the parent acid or base, these salts form solutions, called *buffer solutions,* that are resistant to changes in acidity. If the salts are dissolved by themselves in water, they react with water to form solutions that are acidic or basic depending on their origin.

READY REFERENCE

Acid and base ionization constants (K_a for weak acids and K_b for weak bases) (17.4) An ionization constant is a measure of the dissociation of an acid or base in aqueous solution into $H_3O^+(aq)$ or $OH^-(aq)$. In general, aqueous solutions of weak acids and weak bases are ionized according to the reactions

$$HA(aq) + H_2O(l) \rightleftharpoons H_3O^+(aq) + A^-(aq)$$

$$WB(aq) + H_2O(l) \rightleftharpoons WBH^+(aq) + OH^-(aq)$$

where HA stands for a weak acid and WB stands for a weak base. The corresponding equilibrium expressions are

$$K_a = K[H_2O] = \frac{[H_3O^+][A^-]}{[HA]} \qquad K_b = K'[H_2O] = \frac{[WBH^+][OH^-]}{[WB]}$$

in which the concentration of water has been multiplied by the equilibrium constant and has been redefined as K_a or K_b. The brackets indicate molar concentrations of the ions or molecular species in solution. Typical values for K_a and K_b are considerably less than 1 (see Appendixes F and G). The larger the value of K_a or K_b, the greater the extent of ionization or dissociation; hence the acid or base is correspondingly stronger.

Buffer solution (17.8) A buffer solution contains either a weak acid or a weak base that is in solution with a salt having an ion common to the weak acid or weak base. Such solutions resist changes in acidity produced by the addition of more acid or base. This is accomplished by reaction of the added acid or base with either the ions A^-, WBH^+, or with the molecular acid or base already in solution. The reactions that can occur upon the addition of acid or base to a buffer solution are

Weak acid–salt:

$$A^-(aq) + H^+(aq) \longrightarrow HA(aq)$$
$$HA(aq) + OH^-(aq) \longrightarrow A^-(aq) + H_2O$$

Weak base–salt:

$$WBH^+(aq) + OH^-(aq) \longrightarrow WB(aq) + H_2O$$
$$WB(aq) + H^+(aq) \longrightarrow WBH^+(aq)$$

Buffer solutions are described algebraically in terms of the K_a or K_b defining the weak acid or weak base. The general expressions defining buffers are

Weak acid–salt:

$$K_a = [H^+] \frac{[\text{salt}]}{[\text{acid}]}$$

Weak base–salt:

$$K_b = [OH^-] \frac{[\text{salt}]}{[\text{base}]}$$

Typical buffer solutions are composed of weak acids and their sodium salts or weak bases and their chlorides. Examples are acetic acid–sodium acetate and ammonia–ammonium chloride.

Ion-product constant for water (K_w) (17.1) The equilibrium constant that describes the relation between the hydrogen ion and hydroxyl ion concentrations in aqueous solution is

$$K_w = [H^+][OH^-] = 1.0 \times 10^{-14} \text{ at } 25°C$$

In pure water, the concentration of H^+ and of OH^- equals 1.0×10^{-7} M at 25°C.

pH (17.1) The pH of a solution is a measure of the hydrogen ion concentration in solution and is defined as

$$pH = -\log[H^+]$$

The pH values of most aqueous solutions range from a low of about 0 to a high of about 14; at a pH of 0.0, $[H^+] = 1.0$ M, and at 14.0, $[H^+] = 1.0 \times 10^{-14}$ M.

Example 1: Calculate the pH of a solution that has a hydrogen ion concentration of 0.0025 M.

Chapter 17 Ionic Equilibria of Weak Electrolytes

Solution: If you use a calculator with logarithm capability, merely enter the concentration value, take the log, and change the sign to get pH. However, if you use log tables, the [H$^+$] should be rewritten in scientific notation and the log taken as follows:

$$pH = -\log 0.0025 = -\log 2.5 \times 10^{-3}$$
$$= -[\log 2.5 + \log 10^{-3}] = -[0.3979 + (-3)]$$
$$= -(-2.60) = 2.60$$

The numbers to the right of the decimal point are the significant figures in the pH value; the 2, in this case, represents the power of 10 in the log function.

Hydrogen ion concentrations are frequently needed in calculations involving equilibrium constants. Given the pH of a solution, the [H$^+$] can be calculated by using the definition of pH.

$$pH = -\log[H^+]$$
$$\log[H^+] = -pH$$
$$[H^+] = 10^{-pH}$$

By using a calculator with log capability, you should be able to enter the exponent, $-pH$, and calculate directly. Log tables, however, contain only positive values, and negative log values must be rewritten as a product of two exponential base-10 terms for conversion to a rational term.

Example 2: Calculate the hydrogen ion concentration in a solution with a pH of 9.26.

Solution: $[H^+] = 10^{-pH} = 10^{-9.26}$. To evaluate this exponential, rewrite $10^{-9.26}$ as a product of two log base-10 exponentials — one with a positive decimal exponent, the other with a negative integer exponent. In order to maintain the same value as the original term, calculate the difference between -9.26 and the next smaller integer exponent. The result gives

$$10^{-9.26} = 10^{0.74} \times 10^{-10}$$

The [H$^+$] is the antilog of 0.74 times 10^{-10}:

$$10^{0.74} = 5.5$$

and therefore

$$[H^+] = 5.5 \times 10^{-10}$$

pOH (17.1) The pOH of a solution is a measure of the hydroxide ion (OH$^-$) concentration in solution and is defined as

$$pOH = -\log[OH^-]$$

The expression for K_w relates pOH and pH through the concentration of OH$^-$ and H$^+$ in the following way:

$$K_w = [H^+][OH^-] = 1.0 \times 10^{-14}$$

Take logs of both sides and multiply by (-1).

$$-\log[H^+] + (-\log[OH^-]) = -(-14.00)$$

Substitution from the definitions above yields

$$pH + pOH = 14.00$$

This relation is especially useful when working with basic solutions in which the [OH⁻] is known or is to be calculated.

Quadratic equations A quadratic equation is defined as any equation that can be written in the form

$$ax^2 + bx + c = 0$$

in which a, b, and c are constants, $a \neq 0$. Because these equations are second order, in x^2, they may have two solutions, one (or more) of which may be zero or even negative. In chemistry, reactions involve actual amounts of substances; equations rarely are written in such a way that negative or zero values of x are plausible.

SOLUTIONS TO CHAPTER 17 EXERCISES

IONIC EQUILIBRIA OF ACIDS AND BASES; pH AND pOH

1. Calculate the pH and pOH of each of the following acidic solutions:

    (a) 0.0092 M HOCl ([H⁺] = 1.8 × 10⁻⁵ M)
    (c) 0.0992 M HC₂O₄⁻ (percent ionization is 2.5%)

    Solution

    (a) pH = −log [H⁺] = −log 1.8 × 10⁻⁵ = 4.74
     pOH = 14.00 − pH = 14.00 − 4.74 = 9.26
    (c) The amount of H⁺ produced by the ionization gives rise to the acidity.
     0.0992 M HC₂O₄⁻ × 0.025 = 2.48 × 10⁻³ M = [H⁺]
     pH = −log [H⁺] = −log 2.48 × 10⁻³ = 2.60
     pOH = 14.00 − pH = 14.00 − 2.60 = 11.40

2. Calculate the pH and pOH of each of the following basic solutions:

    (a) 0.0784 M C₆H₅NH₂ ([OH⁻] = 6.0 × 10⁻⁶ M)
    (c) 0.222 M CN⁻ (percent of ions that produce OH⁻ is 1.1%)

    $$CN^- + H_2O \rightleftharpoons HCN + OH^-$$

    Solution

    (a) pOH = −log [OH⁻] = −log 6.0 × 10⁻⁶ = 5.22
     pH = 14.00 − pOH = 14.00 − 5.22 = 8.78
    (c) 0.222 M CN⁻ × 0.011 = 2.44 × 10⁻³ M OH⁻
     pOH = −log [OH⁻] = −log 2.44 × 10⁻³ = 2.61
     pH = 14.00 − pOH = 14.00 − 2.61 = 11.39

3. Calculate the [H⁺] and the percent ionization of each of the following solutions:

    (a) 0.19 M HNO₂ (pH = 2.04)
    (d) 0.407 M HC₂O₄⁻ (pH = 2.29)

Chapter 17 Ionic Equilibria of Weak Electrolytes

Solution

(a) Find the [H$^+$] and divide it by 0.19 M times 100 to find the percent ionization.

$$pH = 2.04 = -\log [H^+]$$

$$[H^+] = \text{antilog} (-2.04) = 9.1 \times 10^{-3} \text{ M}$$

KEYSTROKES: 2.04 $\boxed{\pm}$ $\boxed{\text{2nd}}$ $\boxed{\log}$ = 9.1201 − 03

$$\text{Percent ionization} = \frac{9.1 \times 10^{-3} \text{ M}}{0.19 \text{ M}} \times 100 = 4.8\%$$

(d) pH = 2.29 = −log [H$^+$]; [H$^+$] = antilog (−2.29) = 5.1 × 10^{-3} M

$$\text{Percent ionization} = \frac{5.1 \times 10^{-3} \text{ M}}{0.407 \text{ M}} \times 100 = 1.3\%$$

6. What ionic and molecular species are present in an aqueous solution of hydrogen fluoride, HF? A solution of sulfuric acid? A solution of SO$_2$ in water (sulfurous acid)?

Solution

HF: Hydrogen fluoride is present principally as HF molecules in aqueous solution. It is a weak acid. As such, it ionizes to a slight extent to form ions as shown by the equation:

$$HF(aq) \rightleftharpoons H^+(aq) + F^-(aq)$$

H$_2$SO$_4$: Sulfuric acid is normally considered as a strong acid. It does, however, ionize in two steps in which the first step is complete in relatively dilute solutions. The second ionization is inhibited by the existence of hydrogen ion from the first reaction, and it occurs only to about 10%, while the HSO$_4^-$ ion in water without H$^+$ as the counter ion would ionize to about 15%. The successive ionizations occur according to the following reactions.

$$H_2SO_4(aq) \xrightarrow{100\%} H^+(aq) + HSO_4^-(aq)$$

$$HSO_4^-(aq) \xrightleftharpoons{10\%} H^+(aq) + SO_4^{2-}(aq)$$

SO$_2$: Sulfur dioxide is the acid anhydride of sulfurous acid, H$_2$SO$_3$. Mixed with water, SO$_2$ reacts to form H$_2$SO$_3$ as shown by the equation

$$SO_2(g) + H_2O(l) \rightleftharpoons H_2SO_3(aq)$$

Sulfurous acid is a weak acid that ionizes to about 35% in a 0.1 M solution. It is fairly strong among the weak acids and, like H$_2$SO$_4$, ionizes in two steps.

$$H_2SO_3(aq) \xrightleftharpoons{\sim 35\%} H^+(aq) + HSO_3^-(aq) \qquad K_a = 1.2 \times 10^{-2}$$

The ionization of HSO$_3^-$ occurs only slightly.

$$HSO_3^-(aq) \xrightleftharpoons{\leq 1\%} H^+(aq) + SO_3^{2-}(aq)$$

In solutions of polyprotic acids, all possible ions are present, but those generated by the second and third ionizations in sequence are present in much smaller amounts than those from the first step.

9. Calculate the concentration of each of the ions in the following solutions of strong electrolytes:

   (a) 1.642 M HNO_3
   (c) 0.107 M $[Al(H_2O)_6]_2(SO_4)_3$

Solution

(a) Nitric acid, HNO_3, is a strong electrolyte and is fully ionized in aqueous solution. In the ionization, one molecule of HNO_3 ionizes to produce one ion each of hydrogen ion, H^+, and nitrate ion, NO_3^-. Each ion concentration, therefore, equals the molarity of HNO_3.

$$HNO_3(aq) \xrightarrow{100\%} H^+(aq) + NO_3^-(aq)$$
$$1.642 \text{ M} \longrightarrow 1.642 \text{ M} + 1.642 \text{ M}$$

(c) Some metal ions in aqueous solution bind readily to one or more water molecules to form complex salts. The solid form of the complex, when placed in water, dissociates or ionizes fully to form the hydrated metal ions and the associated anions. The complex salt, aluminum(III) sulfate, dissociates according to the following equation:

$$[Al(H_2O)_6]_2(SO_4)_3(aq) \longrightarrow 2[Al(H_2O)_6]^{3+}(aq) + 3SO_4^{2-}(aq)$$

1 mol	2 mol	3 mol
0.107 M	2(0.107 M)	3(0.107 M)
	0.214 M	0.321 M

10. From the equilibrium concentrations given, calculate K for each of the following weak acids or weak bases:

    (a) HCN: $[H^+] = 1.6 \times 10^{-5}$ M; $[CN^-] = 1.6 \times 10^{-5}$ M; $[HCN] = 0.6$ M
    (d) OCN^-: $[OH^-] = 4.00 \times 10^{-5}$ M; $[HOCN] = 2.63 \times 10^{-7}$ M; $[OCN^-] = 0.364$ M

Solution

(a) The reaction is $HCN \rightleftharpoons H^+ + CN^-$. The value of K_a is calculated from

$$K_a = \frac{[H^+][CN^-]}{[HCN]} = \frac{(1.6 \times 10^{-5})(1.6 \times 10^{-5})}{0.6} = 4 \times 10^{-10}$$

(d) The reaction is $OCN^- + H_2O \rightleftharpoons HOCN + OH^-$

$$K_b = \frac{[OH^-][HOCN]}{[OCN^-]} = \frac{(4.00 \times 10^{-5})(2.63 \times 10^{-7})}{0.364} = 2.89 \times 10^{-11}$$

11. From the following data, calculate the value of the equilibrium constant and the pOH:

    (a) 0.0088 M HClO: 0.20% ionization
    (d) 0.0992 M $HC_2O_4^-$: $[OH^-] = 3.97 \times 10^{-12}$ M

Solution

(a) $0.0088 \text{ M} \times 0.0020 = 1.76 \times 10^{-5}$ M H^+

$$K_a = \frac{[H^+][ClO^-]}{[HClO]} = \frac{(1.76 \times 10^{-5})(1.76 \times 10^{-5})}{(0.0088 - 1.76 \times 10^{-5})} = 3.5 \times 10^{-8}$$

Chapter 17 Ionic Equilibria of Weak Electrolytes   231

$$pH = -\log [H^+] = -\log 1.76 \times 10^{-5} = 4.75$$

$$pOH = 14.00 - pH = 14.00 - 4.75 = 9.25$$

(d) $pOH = -\log [OH^-] = -\log 3.97 \times 10^{-12} = 11.40$

$$[H^+] = \frac{K_w}{[OH^-]} = \frac{1.00 \times 10^{-14}}{3.97 \times 10^{-12}} = 2.52 \times 10^{-3}$$

$$K_a = \frac{[H^+][C_2O_4^{2-}]}{[HC_2O_4^-]} = \frac{(2.52 \times 10^{-3})(2.52 \times 10^{-3})}{0.0992} = 6.40 \times 10^{-5}$$

12. Calculate the hydroxide ion concentration, the percent ionization of the weak base, the pH, and the pOH in each of the following solutions:

   (a) 0.222 M CN^- ($K_b = 2.5 \times 10^{-5}$)
   (e) 0.0784 M $C_6H_5NH_2$ ($K_b = 4.6 \times 10^{-10}$)

Solution

(a) $CN^-(aq) + H_2O(l) \rightleftharpoons HCN(aq) + OH^-(aq)$

$$K_b = \frac{[HCN][OH^-]}{[CN^-]} = 2.5 \times 10^{-5}$$

Let x = moles of CN^- that react with water to form HCN and OH^-

$$x = [OH^-] = [HCN]$$

$$[CN^-] = 0.222 - x$$

Substitution into the expression for K_b gives

$$K_b = \frac{(x)(x)}{(0.222 - x)} = 2.5 \times 10^{-5}$$

Can x in $(0.222 - x)$ be neglected? Yes, because the concentration is more than 3 orders of magnitude greater than K. Solving for the positive root of x from the expression gives

$$x^2 = (2.5 \times 10^{-5})(0.222) = 5.55 \times 10^{-6}$$

$$x = 2.4 \times 10^{-3} \text{ M}$$

$$\text{Percent ionization} = \frac{2.4 \times 10^{-3}}{0.222} \times 100\% = 1.1\%$$

$$pOH = -\log [OH^-] = -\log 2.4 \times 10^{-3} = 2.62$$

$$pH = 14.00 - 2.62 = 11.38$$

(e) The hydroxide ion concentration is calculated from the equilibrium expression for K_b by using the same approach as in (a). Use of the quadratic formula may be necessary.

$$C_6H_5NH_2(aq) + H_2O(l) \rightleftharpoons C_6H_5NH_3^+(aq) + OH^-(aq)$$

$$K_b = K[H_2O] = 4.6 \times 10^{-10} = \frac{[C_6H_5NH_3^+][OH^-]}{[C_6H_5NH_2]}$$

Let x = moles of $C_6H_5NH_2$ that ionize at equilibrium.

$$x = [OH^-] = [C_6H_5NH_3^+]$$

$$[C_6H_5NH_2] = 0.0784 - x$$

Substitution into the expression for K_b gives

$$K_b = 4.6 \times 10^{-10} = \frac{(x)(x)}{0.0784 - x}$$

Can x in $(0.0784 - x)$ be neglected? Yes, 10^{-10} is more than 3 orders of magnitude smaller than 7.8×10^{-2}. Solving for the positive root of x gives

$$x^2 = (0.0784)(4.6 \times 10^{-10}) = 3.61 \times 10^{-11}$$

$$x = [OH^-] = 6.0 \times 10^{-6} \text{ M}$$

$$\text{Percent ionization} = \frac{6.0 \times 10^{-6}}{0.0784} \times 100\% = 7.6 \times 10^{-3}\%$$

$$pOH = -\log [OH^-] = -\log 6.0 \times 10^{-6} = 5.22$$

$$pH = 14.00 - 5.22 = 8.78$$

13. Calculate the percent ionization of each of the following solutions. Calculate (1) using the simplifying assumption that the amount of dissociation is negligible compared to the concentration of acid originally present and (2) without using simplifying assumption and with use of successive approximations or the quadratic formula. Compare the values.

   (a) 0.020 M HCO_2H $(K_a = 1.8 \times 10^{-4})$
   (c) 1.0 M hydrazoic acid, HN_3 $(K_a = 1 \times 10^{-4}$ for $HN_3 \rightleftharpoons H^+ + N_3^-)$

Solution

(a) With simplifying assumption:

$$K_a = \frac{[H^+][HCO_2^-]}{[HCO_2H]} = \frac{(x)(x)}{0.020} = 1.8 \times 10^{-4}$$

$$x^2 = 3.6 \times 10^{-6}; \quad x = [H^+] = 1.9 \times 10^{-3}$$

$$\text{Percent ionization} = \frac{1.9 \times 10^{-3} \times 100}{0.020} = 9.5\%$$

Using the quadratic equation:

$$K_a = \frac{(x)(x)}{0.020 - x} = 1.8 \times 10^{-4}$$

$$x^2 + 1.8 \times 10^{-4} x - 3.6 \times 10^{-6} = 0$$

$$x = \frac{-b \pm \sqrt{b^2 - 4ac}}{2a} = \frac{-1.8 \times 10^{-4} \pm \sqrt{3.24 \times 10^{-8} + 1.44 \times 10^{-5}}}{2}$$

$$= \frac{-1.8 \times 10^{-4} \pm 3.8 \times 10^{-3}}{2} = 1.8 \times 10^{-3} \, [H^+]$$

$$\text{Percent ionization} = \frac{1.8 \times 10^{-3}}{0.020} \times 100 = 9.0\%$$

(c) With the simplifying assumption:

$$K_a = \frac{[H^+][N_3^-]}{[HN_3]} = \frac{(x)(x)}{1.0} = 1 \times 10^{-4}$$

$$x^2 = 1 \times 10^{-4}; \quad x = [H^+] = 1 \times 10^{-2}$$

$$\text{Percent ionization} = \frac{1 \times 10^{-2}}{1} \times 100 = 1\%$$

Chapter 17 Ionic Equilibria of Weak Electrolytes   233

Using the quadratic equation:

$$K_a = \frac{(x)(x)}{1.0 - x} = 1 \times 10^{-4}; \quad x^2 + 1 \times 10^{-4}x - 1 \times 10^{-4} = 0$$

$$x = \frac{-b \pm \sqrt{b^2 - 4ac}}{2a} = \frac{-1 \times 10^{-4} \pm \sqrt{10^{-8} + 4 \times 10^{-4}}}{2}$$

$$= \frac{-1 \times 10^{-4} \pm 2 \times 10^{-2}}{2} = 1 \times 10^{-2}$$

This gives the same percent ionization as with the simplification.

16. Calculate the hydrogen ion concentration and percent ionization of the weak acid in each of the following solutions. Note that the ionization constants may be such that the change in electrolyte concentration cannot be neglected, and the quadratic formula or successive approximations may be required.

   (b) 0.0184 M HCNO   ($K_a = 3.46 \times 10^{-4}$)
   (e) 0.02173 M CH$_2$ClCO$_2$H   ($K_a = 1.4 \times 10^{-3}$)

Solution

(b) The procedure for calculating the percent ionization for the weak acids in this exercise is the same as in exercise 13; the same rules apply.

$$\text{HCNO}(aq) \rightleftharpoons \text{H}^+(aq) + \text{CNO}^-(aq)$$

$$K_a = 3.46 \times 10^{-4} = \frac{[\text{H}^+][\text{CNO}^-]}{[\text{HCNO}]}$$

Let x equal the concentration of HCNO that ionizes. At equilibrium the concentrations of the solution components are

$$x = [\text{H}^+] = [\text{CNO}^-]$$

$$[\text{HCNO}] = 0.0184 - x$$

Substitution gives

$$K_a = 3.46 \times 10^{-4} = \frac{(x)(x)}{0.0184 - x}$$

Can the x in $(0.0184 - x)$ be neglected?

$$[\text{HCNO}] = 1.84 \times 10^{-2} \quad \text{and} \quad K_a = 3.46 \times 10^{-4}$$

Exponential difference equals 2, so x cannot be neglected. Therefore, solve the equation for x by using the quadratic formula. The process is as follows:

$$x^2 = 3.46 \times 10^{-4} (0.0184 - x)$$

$$= 6.37 \times 10^{-6} - 3.46 \times 10^{-4} x$$

Rearrangement to the standard quadratic form gives

$$x^2 + 3.46 \times 10^{-4} x - 6.37 \times 10^{-6} = 0$$

The equation is now in the form $ax^2 + bx + c = 0$, and the quadratic formula applies:

$$x = \frac{-b \pm \sqrt{b^2 - 4ac}}{2a}$$

Substitution into the formula gives

$$x = \frac{-3.46 \times 10^{-4} \pm \sqrt{(3.46 \times 10^{-4})^2 - 4(1)(-6.37 \times 10^{-6})}}{2(1)}$$

and

$$x = \frac{-3.46 \times 10^{-4} \pm \sqrt{1.197 \times 10^{-7} + 2.548 \times 10^{-5}}}{2}$$

$$x = \frac{-3.46 \times 10^{-4} \pm \sqrt{2.56 \times 10^{-5}}}{2} = \frac{-3.46 \times 10^{-4} \pm 5.06 \times 10^{-3}}{2}$$

The positive root of x is

$$x = \frac{4.71 \times 10^{-3}}{2} = 2.36 \times 10^{-3} \text{ M} = [H^+] = [CNO^-]$$

If the simplification had been made, the value of $[H^+]$ would have been 2.52×10^{-3}.

$$\text{Percent ionization} = \frac{2.36 \times 10^{-3}}{0.0184} \times 100\% = 12.8\%$$

(e) $CH_2ClCO_2H(aq) \rightleftharpoons H^+(aq) + CH_2ClCO_2^-(aq)$

$$K_a = 1.4 \times 10^{-3} = \frac{[H^+][CH_2ClCO_2^-]}{[CH_2ClCO_2H]}$$

Let x = moles of CH_2ClCO_2H that ionize at equilibrium; in 1 L, the number of moles is the numerical value of the concentration.

$x = [H^+] = [CH_2ClCO_2^-]$

$[CH_2ClCO_2H] = 0.02173 - x$

and

$$K_a = 1.4 \times 10^{-3} = \frac{(x)(x)}{0.02173 - x}$$

Can x in $(0.02173 - x)$ be neglected? No! Solving the equation of x via the quadratic formula gives

$$x^2 = 1.4 \times 10^{-3}(0.02173 - x) = 3.042 \times 10^{-5} - 1.4 \times 10^{-3}x$$

$$x^2 + 1.4 \times 10^{-3}x - 3.042 \times 10^{-5} = 0$$

$$x = \frac{-1.4 \times 10^{-3} \pm \sqrt{(1.4 \times 10^{-3})^2 - 4(1)(-3.042 \times 10^{-5})}}{2(1)}$$

$$x = \frac{-1.4 \times 10^{-3} \pm \sqrt{1.236 \times 10^{-4}}}{2} = \frac{-1.4 \times 10^{-3} \pm 1.1 \times 10^{-2}}{2}$$

The positive root of x is

$x = 4.8 \times 10^{-3}$ M $= [H^+]$

$$\text{Percent ionization} = \frac{4.8 \times 10^{-3}}{0.02173} \times 100\% = 22\%$$

Chapter 17 Ionic Equilibria of Weak Electrolytes   235

18. Calculate [OH⁻] and pH for each of the following:

   (b) 0.00253 M $C_6H_5O^-$ ($K_b = 7.81 \times 10^{-5}$)

   Solution

   (b) Since K and the concentration do not differ by more than 10^3, use the quadratic equation, letting $x = $ [OH⁻].

   $$K_b = \frac{[C_6H_5OH][OH^-]}{[C_6H_5O^-]} = \frac{(x)(x)}{0.00253 - x} = 7.81 \times 10^{-5}$$

   $$x = \frac{-b \pm \sqrt{b^2 - 4ac}}{2a} = \frac{-7.81 \times 10^{-5} \pm \sqrt{6.1 \times 10^{-9} + 7.90 \times 10^{-7}}}{2}$$

   $$= \frac{-7.81 \times 10^{-5} \pm 8.92 \times 10^{-4}}{2} = 4.1 \times 10^{-4} = [OH^-]$$

   $$pOH = -\log[OH^-] = -\log 4.1 \times 10^{-4} = 3.39$$

   $$pH = 14.00 - pOH = 14.00 - 3.39 = 10.61$$

19. How would the extent of ionization of acetic acid in a solution be changed by the addition of potassium acetate? By the addition of hydrogen chloride?

   Solution

   Addition of the common ion, acetate, would depress the ionization of acetic acid. Addition of the hydrogen chloride puts more hydrogen ions into solution, thus depressing ionization.

21. What is the fluoride ion concentration in 0.675 L of a solution that initially contains 0.1400 g of HF? (See note in exercise 20.)

   Solution

   First calculate the molarity, then from K_a the concentration of F⁻. Let $x = $ [F⁻] = [H⁺]

   $$M\ HF = \frac{0.1400\ g}{20.0063\ g\ mol^{-1} \times 0.675\ L} = 0.01037\ M$$

   $$K_a = \frac{[H^+][F^-]}{[HF]} = \frac{(x)(x)}{[HF]} = \frac{x^2}{0.01037 - x} = 7.2 \times 10^{-4}$$

   $$x^2 + 7.2 \times 10^{-4} x - 7.46 \times 10^{-6} = 0$$

   $$x = \frac{-b \pm \sqrt{b^2 - 4ac}}{2a} = \frac{-7.2 \times 10^{-4} \pm \sqrt{5.184 \times 10^{-7} + 2.986 \times 10^{-5}}}{2}$$

   $$= \frac{-7.2 \times 10^{-4} \pm \sqrt{3.038 \times 10^{-5}}}{2} = \frac{-7.2 \times 10^{-4} \pm 5.51 \times 10^{-3}}{2}$$

   $$= 2.4 \times 10^{-3} = [F^-]$$

23. A solution that contains 20.0 g of lactic acid ($CH_3CHOHCO_2H$) in 1.00 L of solution has a hydrogen ion concentration of 5.49×10^{-3} M. What is the value of the ionization constant?

Solution

Calculate the molarity of the lactic acid, then substitute into the equilibrium expression to find K.

Molar mass
$CH_3CHOHCO_2H = 3(12.011) + 6(1.0079) + 3(15.9994) = 90.079$ g

M lactic acid $= 20.0$ g $\times \dfrac{1 \text{ mol}}{90.079 \text{ g}} \times \dfrac{1}{1.00 \text{ L}} = 0.222$ M

The ionization of lactic acid is

$$CH_3CHOHCO_2H(aq) \rightleftharpoons CH_3CHOHCO_2^-(aq) + H^+(aq)$$

$$LA \rightleftharpoons LA^- + H^+$$

$$K_a = \dfrac{[H^+][LA^-]}{[LA]} = \dfrac{(5.49 \times 10^{-3})(5.49 \times 10^{-3})}{(0.222 - 5.49 \times 10^{-3})}$$

$$= \dfrac{3.01 \times 10^{-5}}{0.2165} = 1.39 \times 10^{-4}$$

25. The artificial sweetener sodium saccharide (often referred to as sodium saccharin on soft drink cans) contains the basic ion $C_7H_4NSO_3^-$ ($K_b = 4.8 \times 10^{-3}$). What is the hydroxide ion concentration in a liter of water sweetened with 0.100 g of $Na(C_7H_4NSO_3)$?

Solution

Find the molar mass of the salt and calculate the molarity of 0.100 g of Na saccharide. Then calculate the concentration of the ion form from the K_b.

Molar mass:

$C_7H_4NSO_3Na = 7(12.011) + 5(1.0079) + 3(15.9994) + 14.0067 + 32.06 + 22.98977$

$= 206.17$ g/mol

M salt $= 0.100$ g$/206.17$ g mol$^{-1} = 4.85 \times 10^{-4}$ M in one liter

The amount of OH$^-$ is the same as the salt form in solution. Let

$$x = [C_7H_4NSO_3^-] = [OH^-]$$

Then

$$K_b = 4.8 \times 10^{-3} = \dfrac{[H^+][\text{Sacc}^-]}{[\text{Sacc}]} = \dfrac{x^2}{4.85 \times 10^{-4} - x}$$

The quadratic formula is required.

$$x^2 + 4.8 \times 10^{-3} x - 2.328 \times 10^{-6} = 0$$

$$x = \dfrac{-b \pm \sqrt{b^2 - 4ac}}{2a} = \dfrac{-4.8 \times 10^{-3} \pm \sqrt{2.304 \times 10^{-5} + 9.31 \times 10^{-6}}}{2}$$

$$= \dfrac{-4.8 \times 10^{-3} \pm 5.688 \times 10^{-3}}{2} = 4.4 \times 10^{-4} \text{ M} = [OH^-]$$

where the positive root is chosen. If the simplifying assumption were made to ignore the amount dissociated, the answer would be 1.5×10^{-3} M; clearly this would be too large an error.

Chapter 17 Ionic Equilibria of Weak Electrolytes   **237**

27. A 0.010 M solution of HF is 23% ionized. Calculate [H$^+$], [F$^-$], [HF], and the ionization constant for HF.

Solution

Calculate the molar concentrations of [H$^+$] = [F$^-$] from the percent ionized.

[H$^+$] = [F$^-$] = 0.23(0.010 M) = 0.0023 M

Then the [HF] is equal to the original amount less that which ionized.

$$K_a = \frac{[H^+][F^-]}{[HF]} = \frac{(0.0023)(0.0023)}{(0.010 - 0.0023)} = \frac{5.29 \times 10^{-6}}{0.0077} = 6.9 \times 10^{-4}$$

28. Calculate the ionization constants for each of the following solutes from the percent ionization and the concentration of the solute:

(d) 0.050 M HClO, 8.4 × 10^{-2}% ionized

(f) 0.010 M HNO$_2$, 19% ionized

Solution

The ionization constant for a weak acid or base is calculated from the equilibrium concentration of the solution components. The initial concentration of the ionizing substance is given in each part along with the percent ionized. Multiplication of the initial amount by the percentage gives the equilibrium amount of ionized species. Subtraction of the equilibrium amount from the initial amount of material gives its equilibrium amount. Then the equilibrium constant can be calculated.

(d) HClO(*aq*) ⇌ H$^+$(*aq*) + ClO$^-$(*aq*)

At equilibrium:

[H$^+$] = 8.4 × 10^{-2}% of 0.050 M = [ClO$^-$]

= (0.00084)(0.050) = 4.2 × 10^{-5} M

Since 1 mol of HClO ionizes to produce 1 mol each of H$^+$ and ClO$^-$, the equilibrium concentration of HClO is

[HClO] = [0.050 − 4.2 × 10^{-5}] M

$$K_a \text{ (HClO)} = \frac{[H^+][ClO^-]}{[HClO]}$$

Substitution of the concentrations into the expression gives

$$K_a = \frac{(4.2 \times 10^{-5})(4.2 \times 10^{-5})}{(0.050 - 4.2 \times 10^{-5})} = 3.5 \times 10^{-8}$$

(f) HNO$_2$(*aq*) ⇌ H$^+$(*aq*) + NO$_2^-$(*aq*)

$$K_a = \frac{[H^+][NO_2^-]}{[HNO_2]}$$

At equilibrium:

[H$^+$] = [NO$_2^-$] = 19% of 0.010 M

= 0.19(0.010)M = 0.0019 M

[HNO$_2$] = (0.010 − 0.0019)M

$$K_a = \frac{(0.0019)(0.0019)}{(0.010 - 0.0019)} = 4.4 \times 10^{-4}$$

30. Calculate the pH and pOH of pure water at 100°C ($K_w = 1 \times 10^{-12}$ at 100°C).

Solution

Water self-ionizes.

$$H_2O \rightleftharpoons H^+ + OH^-$$

The equilibrium constant, K, for this reaction is normally combined with the concentration of water and the product is defined as K_w. ($K_w = 1.0 \times 10^{-12}$ at 100°C).

$$K_w = [H^+][OH^-] = 1.0 \times 10^{-12}$$

According to the equilibrium, $[H^+] = [OH^-]$ and the concentrations are obtained from the expression for K_w.

$$[H^+] = [OH^-] = \sqrt{1.0 \times 10^{-12}} = 1.0 \times 10^{-6}$$

The pH of any solution or even pure water is the negative logarithm of the hydrogen ion concentration.

$$pH = -\log[H^+]$$

For pure water:

$$pH = -\log[1.0 \times 10^{-6}] \text{ at } 100°C$$

On a calculator equipped with a logarithm function, merely enter the $[H^+]$ as 1.0×10^{-6}, take the log by depressing the log key, and change the sign; this value is the pH of water at 100°C.

KEYSTROKES: 1.0 [EE] [+/−] 6 [2nd] [log] [+/−] = 6.00

However, if you are finding logs from a log table, the $[H^+]$ should be written in exponential notation and logs taken in the following way:

$$pH = -\log[H^+] = -\log 1.0 \times 10^{-6} = -[\log 1.0 + \log 10^{-6.0}]$$
$$= -[0.0000 + (-6.00)] = -(-6.00) = 6.00$$

Here the log of the rational term (1.0) has been found in the log table and added to the log of the exponential term, which is simply the value of the exponent.

The pOH of a solution or of pure water is defined as the negative logarithm of the hydroxide ion concentration.

$$pOH = -\log[OH^-] = -\log[1.0 \times 10^{-6}]$$

Compute the pOH in the same manner as was done for the pH.

$$pOH = -\log 1.0 \times 10^{-6} = -(-6.00) = 6.0$$

Alternatively, the pH plus the pOH for an aqueous solution at 100°C equals 12.0. By knowing either pH or the pOH, the other factor can be calculated simply by subtracting the known value from 12.0 to obtain the unknown. Given that the pH is 6.0, for example, the pOH is

$$pH + pOH = 12.0$$
$$pOH = 12.0 - pH = 12.0 - 6.0 = 6.0$$

32. Calculate the pH of a solution resulting from mixing 0.10 L of 0.10 M NaOH with 0.40 L of 0.025 M HF.

Solution

This solution is a combination of the weak acid, HF, and the strong base, NaOH. At equilibrium there are 0.10 L × 0.10 M = 0.01 mol OH⁻ from NaOH and 0.40 L × 0.025 M = 0.01 mol HF. The OH⁻ from NaOH will exactly neutralize the H⁺ of HF. However, since HF is a weak acid, the 0.01 mol F⁻ or 0.02 M F⁻ will abstract some H⁺ from water, leaving OH⁻ in solution, thus making the solution basic. The reaction is

$$F^- + H_2O \rightleftharpoons HF + OH^-$$

This is just the ionization of the base F⁻

$$K_b = \frac{[HF][OH^-]}{[F^-]}$$

Substitute $K_w/[H^+]$ for $[OH^-]$ in the expression.

$$K_b = \frac{[HF]\,K_w}{[F^-][H^+]} = \frac{K_w}{K_a} = \frac{1.0 \times 10^{-11}}{7.2 \times 10^{-4}} = 1.389 \times 10^{-11}$$

Let x = the concentration of HF formed. Then $x = [OH^-]$. In tabular form:

	[F⁻]	[HF]	[OH⁻]
initial conc., M	0.020	0	~0
equil. conc., M	0.020 − x	x	x

$$K_b = \frac{x^2}{0.020 - x} = 1.389 \times 10^{-11}$$

Since x is very small relative to 0.020, we may drop it from the denominator. Then

$$x^2 = 0.020 \times 1.389 \times 10^{-11}$$

$$x = 5.27 \times 10^{-7}$$

$$pOH = -\log 5.27 \times 10^{-7} = 6.28$$

$$pH = 14.00 - 6.28 = 7.72$$

33. Calculate the pH of a solution prepared by mixing 10.0 mL of 0.10 M LiOH and 20.0 mL of a 0.500 M benzoic acid solution. K_a for benzoic acid is 6.7×10^{-5}.

Solution

When 0.0100 L of 0.10 M LiOH (0.001 mol) reacts with 0.0200 L of 0.500 M benzoic acid (0.01 mol), 0.001 mol of benzoate ion will be formed; in addition there remains 0.009 mol of benzoic acid. This constitutes a buffer solution, a total of 0.0300 L.

	[benzoic acid]	[H⁺]	[benzoate ion]
initial conc., M	0.333	0	0
equil. conc., M	0.300 − x	x	0.033 + x

$$K_a = 6.7 \times 10^{-5} = \frac{[H^+][\text{benzoate ion}]}{[\text{benzoic acid}]} = \frac{x(0.033 + x)}{0.0300 - x}$$

Since x is small in comparison to 0.300 or 0.033, delete x in numerator and denominator.

$$6.7 \times 10^{-5} (0.300) = 0.033 \, x$$

$$x = [H^+] = 6.09 \times 10^{-4}$$

$$pH = -\log [H^+] = 3.22$$

35. Calculate the pH of a solution made by mixing equal volumes of 0.30 M NH_3 and 0.030 M HNO_3.

Solution

The reaction is

$$NH_3(aq) + HNO_3(aq) \rightleftharpoons NH_4^+(aq) + NO_3^-(aq)$$

Upon mixing there will remain 0.30 M − 0.030 M = 0.270 M NH_3. The ammonium ion, the conjugate acid of the weak base ammonia, reacts with water.

$$NH_4^+ + H_2O \rightleftharpoons NH_3 + H_3O^+$$

$$K_a = \frac{[NH_3][H_3O^+]}{[NH_4^+]} = \frac{K_w}{K_b} = \frac{1.0 \times 10^{-14}}{1.8 \times 10^{-5}} = 5.6 \times 10^{-10}$$

	$[NH_3]$	$[NH_4^+]$	$[H_3O^+]$
initial conc., M	0.270	0	~0
equil. conc., M	0.270 + x	0.03 − x	x

Then

$$K_a = 5.6 \times 10^{-10} = \frac{(0.270 + x)(x)}{(0.03 - x)}$$

Since x is small the additive and subtractive x's are dropped.

$$5.6 \times 10^{-10} (0.03) = 0.270x$$

$$x = 6.22 \times 10^{-11} = [H_3O^+]$$

$$pH = -\log[H^+] = 10.21$$

BUFFERS

37. What constitutes a buffer solution?

Solution

Any solution of a weak acid or a weak base and its salt will function as a buffer solution.

40. Calculate the pH of each of the following buffer solutions, containing equal volumes of two solutes in the concentrations indicated:

(b) 0.30 M CH_3CO_2H, 0.30 M $NaCH_3CO_2$

(e) 0.25 M H_3PO_4; 0.15 M NaH_2PO_4

Solution

Values for K are obtained from Appendix F or G.

(b) This solution is a buffer solution containing the weak acid, CH_3CO_2H, and a salt. The salt contains the anion of the acid. The acetic acid equilibrium describes the solution; the added salt merely causes a shift in the equilibrium.

$$CH_3CO_2H(aq) \rightleftharpoons H^+(aq) + CH_3CO_2^-(aq)$$

Let x = moles of CH_3CO_2H that ionize. At equilibrium:

$$[H^+] = x$$

The total amount of $CH_3CO_2^-$ in solution equals the amount of $CH_3CO_2^-$ added by way of $NaCH_3CO_2$ plus the amount x produced by the ionization of CH_3CO_2H:

$$[CH_3CO_2^-] = (x + 0.30)M$$

$$[CH_3CO_2H] = (0.30 - x)M$$

$$K_a = \frac{[H^+][CH_3CO_2^-]}{[CH_3CO_2H]} = 1.8 \times 10^{-5}$$

Substitution of the concentration into the equation gives

$$K_a = 1.8 \times 10^{-5} = \frac{(x)(x + 0.30)}{(0.30 - x)}$$

The rules for neglecting x in the denominator also apply to buffer or hydrolysis problems; that is, problems in which the salt reacts with water. x can be neglected; because x is small relative to 0.30 in $(0.30 - x)$, it also is small relative to 0.30 in $(x + 0.30)$. The expression is simplified to

$$1.8 \times 10^{-5} = \frac{(x)(0.30)}{(0.30)}$$

$$x = 1.8 \times 10^{-5} = [H_3O^+]$$

Also,

$$pH = -\log[H^+] = -\log(1.8 \times 10^{-5}) = -[\log 1.8 + \log 10^{-5}]$$
$$= -[0.2553 + (-5)] = 4.74$$

(e) Assume that NaH_2PO_4 ionizes completely to give

$$NaH_2PO_4(aq) \rightarrow Na^+(aq) + H_2PO_4^-(aq)$$

The $H_2PO_4^-$ is the ion in common with the first ionization of H_3PO_4; it serves as the first ionization of H_3PO_4 and as the salt-acid pair required for the solution to have buffering capacity. The amount of hydrogen ion in solution is controlled by the equilibrium of H_3PO_4 and is calculated from the expression for K_a of H_3PO_4.

$$H_3PO_4(aq) \rightleftharpoons H^+(aq) + H_2PO_4^-(aq) \qquad K_a = 7.5 \times 10^{-3}$$

Let x = moles of H_3PO_4 that ionize.
At equilibrium

$$x = [H^+]$$

The total amount of $H_2PO_4^-$ in solution equals the amount of $H_2PO_4^-$ added plus that produced by the ionization of H_3PO_4.

$$[H_2PO_4^-] = (x + 0.15)M$$

$$[H_3PO_4] = (0.25 - x)M$$

Substitution into the expression for K_a gives

$$K_a = \frac{[H^+][H_2PO_4^-]}{[H_3PO_4]} = \frac{(x)(x + 0.15)}{(0.25 - x)} = 7.5 \times 10^{-3}$$

We can assume that x is very small relative to 0.25 and to 0.15 and can be ignored. The expression simplifies to

$$K_a = 7.5 \times 10^{-3} = \frac{(x)(0.15)}{(0.25)}$$

Solving for x gives

$$x = [H^+] = \frac{(7.5 \times 10^{-3})(0.25)}{0.15} = 1.25 \times 10^{-2}$$

$$pH = -\log[H^+] = -\log 1.25 \times 10^{-2} = -(-1.90) = 1.90$$

43. In venous blood the following equilibrium is set up by dissolving carbon dioxide:

$$H_2CO_3 \rightleftharpoons H^+ + HCO_3^-$$

If the pH of the blood is 7.4, what is the ratio of $[HCO_3^-]$ to $[H_2CO_3]$?

Solution

$pH = 7.4; [H^+] = 3.98 \times 10^{-8}$

$$K_a = \frac{[H^+][HCO_3^-]}{[H_2CO_3]} = 4.3 \times 10^{-7}$$

$$\frac{4.3 \times 10^{-7}}{3.98 \times 10^{-8}} = \frac{[HCO_3^-]}{[H_2CO_3]} = 11$$

The ratio is 11 to 1.

44. How many moles of NH_4Cl must be added to 1.0 L of 1.0 M NH_3 solution to prepare a buffer solution with a pH of 9.00? of 9.50?

Solution

pH = 9.00: The equilibrium expression for ammonia defines this buffer system; NH_4^+ from NH_4Cl is the common ion.

$$NH_3(aq) + H_2O(l) \rightleftharpoons NH_4^+(aq) + OH^-(aq) \quad K_a = 1.8 \times 10^{-5}$$

$$K_a = \frac{[NH_4^+][OH^-]}{[NH_3]} = 1.8 \times 10^{-5}$$

Given a pH of 9.00, the pOH equals 5.00 and the $[OH^-]$ equals 1.0×10^{-5}. Substituting this OH^- concentration into the expression for K_a gives

$$\frac{[NH_4^+](1.0 \times 10^{-5})}{(1.0)} = 1.8 \times 10^{-5}$$

Chapter 17 Ionic Equilibria of Weak Electrolytes 243

Solving for [NH$_4^+$] gives

[NH$_4^+$] = 1.8 M

For 1.0 L, 1.8 moles of NH$_4$Cl must be added.

pH = 9.50: The pOH equals 4.5 and thus the [OH$^-$] = 3.16 × 10^{-5}. Substitution gives

$$\frac{[NH_4^+](3.16 \times 10^{-5})}{1.0} = 1.8 \times 10^{-5}$$

[NH$_4^+$] = 0.570 M

For 1.0 L, 0.57 mol must be added.

45. How many moles of sodium acetate must be added to 1.0 L of a 1.0 M acetic acid solution to prepare a buffer solution with a pH of 5.08? Of 4.20?

Solution

The equilibrium expression for acetic acid defines this buffer system.

$$CH_3CO_2H(aq) + H_2O(l) \rightleftharpoons H^+(aq) + CH_3CO_2^-(aq)$$

$$K_a = 1.8 \times 10^{-5} = \frac{[H^+][OAc^-]}{[HOAc]}$$

For a pH = 5.08, [H$^+$] = 8.32 × 10^{-6}. The ratio

$$\frac{[OAc^-]}{[HOAc]} = \frac{1.80 \times 10^{-5}}{8.32 \times 10^{-6}} = 2.16$$

Starting with 1.0 mol HOAc, 2.2 mol OAc$^-$ are required. For a pH = 4.20, [H$^+$] = 6.31 × 10^{-5}. The ratio

$$\frac{[OAc^-]}{[HOAc]} = \frac{1.80 \times 10^{-5}}{6.31 \times 10^{-5}} = 0.29$$

Starting with 1.0 mol HOAc, 0.29 mol OAc$^-$ are required.

IONIC EQUILIBRIA OF POLYPROTIC ACIDS AND BASES

48. Explain how successive ionization constants for polyprotic acids differ.

Solution

The second and successive ionization constants are progressively smaller because, as the first hydrogen ion is removed, there is left a negative ion from which it is harder to remove a second or third hydrogen ion. Thus, the ionic attraction between the negative ion and positive proton trying to leave tends to keep the second and higher ionization constants progressively smaller.

50. Calculate the concentration of each species present in a 0.050 M solution of H$_2$S.

Solution

Hydrogen sulfide is a weak acid that ionizes in two steps:

1. $H_2S(aq) \rightleftharpoons H^+(aq) + HS^-$ $K_1 = 1.0 \times 10^{-7}$
2. $HS^-(aq) \rightleftharpoons H^+(aq) + S^{2-}$ $K_2 = 1.0 \times 10^{-19}$

The total hydrogen ion concentration approximately equals the H^+ generated from the first ionization. K_2 is smaller than K_1 by 10^{12} times, so very little of the HS^- formed in reaction 1 ionizes to give additional hydrogen ions and S^{2-}. The second step is further depressed by the existence of H^+ from reaction 1. It can be assumed that the concentrations of H^+ and of HS^- are practically equal.

From Reaction 1:

$$K_1 = 1.0 \times 10^{-7} = \frac{[H^+][HS^-]}{[H_2S]}$$

Let x = the concentration or amount (mol) of H_2S that ionizes.

$$x = [H^+] = [HS^-]; \quad 0.05 - x = [H_2S]$$

$$\frac{(x)(x)}{(0.05 - x)} = 1.0 \times 10^{-7}$$

Simplify to get

$$x^2 = (1.0 \times 10^{-7})(0.05) = 5.0 \times 10^{-9}$$

$$x = [H^+] = [HS^-] = 7.1 \times 10^{-5} \text{ M}$$

Next, substitute this data into the expression for K_2.

$$K_2 = 1.0 \times 10^{-19} = \frac{[H^+][S^{2-}]}{[HS^-]}$$

The concentration of S^{2-} equals K_2 because $[H^+]$ equals $[HS^-]$, and $[H_2S] = 0.050$ M because the amount that dissociates is so small. The $[OH^-]$ is calculated from K_w as

$$[OH^-] = \frac{K_w}{[H^+]} = \frac{1.0 \times 10^{-14}}{7.1 \times 10^{-5}} = 1.4 \times 10^{-10} \text{ M}$$

51. Calculate the concentration of each species present in a 0.0250 M solution of Na_3PO_4.

$$PO_4^{3-} + H_2O \rightleftharpoons HPO_4^{2-} + OH^- \quad (K_{b1} = 2.8 \times 10^{-2})$$

$$HPO_4^{2-} + H_2O \rightleftharpoons H_2PO_4^- + OH^- \quad (K_{b2} = 1.6 \times 10^{-7})$$

$$H_2PO_4^- + H_2O \rightleftharpoons H_3PO_4 + OH^- \quad (K_{b3} = 1.3 \times 10^{-12})$$

Solution

The total hydrogen ion concentration approximately is equal to the H^+ generated on the first ionization. In an aqueous solution of trisodium phosphate there are seven ionic and molecular species present:

PO_4^{3-}, HPO_4^{2-}, $H_2PO_4^-$, H_3PO_4, OH^-, H^+, and Na^+.

Besides the equilibria listed another equilibrium is present, namely:

$$H_2O(l) + H_2O(l) \rightleftharpoons H_3O^+(aq) + OH^-(aq) \quad K_w = 1.00 \times 10^{-14} \text{ M}^2$$

There is a mass balance of phosphorus,

$$0.0250 \text{ M} = \left[PO_4^{3-}\right] + \left[HPO_4^{2-}\right] + \left[H_2PO_4^-\right] + \left[H_3PO_4\right]$$

and an electroneutrality condition which must be met.

$$\left[H_3O^+\right] + \left[Na^+\right] = \left[OH^-\right] + 3\left[PO_4^{3-}\right] + 2\left[HPO_4^{2-}\right] + \left[H_2PO_4^-\right]$$

We also have the relation $0.025 \text{ M} = 3[\text{Na}^+]$. This gives a total of seven equations and seven unknowns which can be solved. Fortunately some approximations may be made to simplify matters. First, the value of K_{b1} is much greater than K_{b2} which in turn is much greater than K_{b3}. K_{b3} still is larger than K_w. Therefore, we may neglect the second and third associations as well as K_w. A table gives the concentrations in the first association to form HPO_4^{2-}.

	$\text{PO}_4^{3-}(aq)$	$+ \text{H}_2\text{O} \rightleftharpoons$	$\text{OH}^-(aq) +$	$\text{HPO}_4^{2-}(aq)$
initial conc.	0.0250 M		0	0
equil. conc.	$0.0250 \text{ M} - [\text{HPO}_4^{2-}]$		$[\text{OH}^-]$	$[\text{HPO}_4^{2-}]$
or	$0.0250 \text{ M} - [\text{OH}^-]$		$[\text{OH}^-]$	$[\text{OH}^-]$

since

$[\text{OH}^-] \approx [\text{HPO}_4^{2-}]$.

$$K_{b1} = \frac{[\text{HPO}_4^{2-}][\text{OH}^-]}{[\text{PO}_4^{3-}]} = \frac{[\text{OH}^-]^2}{0.0250 - [\text{OH}^-]} = 2.8 \times 10^{-2}$$

Let $x = [\text{OH}^-]$. Then, $x^2 + 2.8 \times 10^{-2} x - 0.0250(2.8 \times 10^{-2}) = 0$. Use the quadratic equation to solve for $[\text{OH}^-]$.

$$x = \frac{-2.8 \times 10^{-2} \pm \sqrt{7.84 \times 10^{-4} + 2.8 \times 10^{-3}}}{2}$$

$$= \frac{-2.8 \times 10^{-2} \pm \sqrt{3.58 \times 10^{-3}}}{2} = \frac{-2.8 \times 10^{-2} + 0.05986}{2}$$

$$= 0.0159 = 0.016 \text{ M} = [\text{OH}^-]$$

The concentration of $\text{HPO}_4^{2-} = 0.016$ M and $[\text{PO}_4^{3-}] = 0.0250 - 0.0159 \approx 0.0091$ M. Now calculate the concentration of the remaining species by using the just calculated results. Using K_{b2} calculate $[\text{H}_2\text{PO}_4^-]$.

$$K_{b2} = \frac{[\text{H}_2\text{PO}_4^-][\text{OH}^-]}{[\text{HPO}_4^{2-}]}$$

$$[\text{H}_2\text{PO}_4^-] = \frac{K_{b2}[\text{HPO}_4^{2-}]}{[\text{OH}^-]}$$

$$[\text{H}_2\text{PO}_4^-] = \frac{1.6 \times 10^{-7}(0.016)}{0.016} = 1.6 \times 10^{-7} \text{ M}$$

Using the K_{b3} expression, calculate $[\text{H}_3\text{PO}_4]$.

$$\text{H}_3\text{PO}_4 = \frac{1.3 \times 10^{-12}[\text{H}_2\text{PO}_4^-]}{[\text{OH}^-]}$$

$$= \frac{1.3 \times 10^{-12}(1.6 \times 10^{-7})}{0.016} = 1.3 \times 10^{-17} \text{ M}$$

Then

$$[\text{H}^+] = \frac{K_w}{[\text{OH}^-]} = \frac{1.00 \times 10^{-14}}{0.016} = 6.3 \times 10^{-13} \text{ M}$$

And finally from the mass balance, since $[\text{H}^+]$ makes almost no contribution,

$$[\text{Na}^+] = 3[\text{PO}_4^{3-}] + 2[\text{HPO}_4^{2-}] + [\text{H}_2\text{PO}_4^-] + [\text{OH}^-]$$

$$= 3(0.0091) + 2(0.016) + 1.6 \times 10^{-7} + 0.016$$

$$= 0.027 + 0.032 + 0.016 = 0.075 \text{ M}$$

TITRATION CURVES AND INDICATORS

54. Calculate the pH after the addition of 20.0, 24.9, 25.0, 25.1, and 30.0 mL of NaOH in the titration of HCl described in Table 17.7 in Section 17.17.

Solution

Use the relations

$$V \times M = \text{mol (NaOH)}$$

$$|(\text{mol HCl} - \text{mol NaOH})|/V = M\,[H^+] \text{ or } [OH^-]$$

$$[H^+] = \frac{1 \times 10^{-14}}{[OH^-]} \qquad pH = -\log[H^+]$$

Volume (L)	M NaOH	mol (added)	mol $(HCl)_i$ 0.025 L × 0.100 M	moles remaining
0.0200	0.100	0.00200	0.00250	0.00050 (H^+)
0.0249	0.100	0.00249	0.00250	0.00001 (H^+)
0.0250	0.100	0.00250	0.00250	0.00000 (H^+)*
0.0251	0.100	0.00251	0.00250	0.00001 (OH^-)
0.0300	0.100	0.00300	0.00250	0.00050 (OH^-)

*in aqueous solution $[H^+] = [OH^-] = 1.0 \times 10^{-7}$ M

V of soln (L)	$M(OH^-)$	$M(H^+)$	pH
0.0450		0.0111	1.95
0.0499		2.00×10^{-4}	3.70
0.0500		1.00×10^{-7}	7.00
0.0501	1.996×10^{-4}	5.01×10^{-11}	10.30
0.0550	9.09×10^{-3}	1.10×10^{-12}	11.96

57. Determine a theoretical titration curve for the titration of 25.0 mL of 0.150 M NH_3 with 0.150 M HCl.

Solution

As in Exercise 54, calculate the pH for varying additions of acid to the ammonia. Then plot the pH against the volume of acid added. The amount of NH_3 is 0.025 L × 0.150 M = 3.75×10^{-3} mol NH_3. As acid is added the amount of NH_3 is reduced, but the salt NH_4^+ is formed, which reacts with water forming a buffer solution. Consequently, the equation for a buffer solution must be used to calculate the concentration of OH^- remaining.

The original pH is determined from the molarity of NH_3 and its K_b.

$$K_b = 1.8 \times 10^{-5} = \frac{[NH_4^+][OH^-]}{[NH_3]}$$

Let $x = [OH^-] = [NH_4^+]$. Then

$$\frac{x^2}{0.150 - x} = 1.8 \times 10^{-5}$$

$$x^2 = 2.7 \times 10^{-6}$$

$$x = 1.64 \times 10^{-3}$$

$$pOH = 2.78 \qquad pH = 11.2$$

Upon addition of 5.0 mL of 0.150 M HCl, 7.50×10^{-4} mol H⁺ is available to react with 7.5×10^{-4} mol of NH_3. 3.75×10^{-3} mol $- 7.50 \times 10^{-4}$ mol $= 3.00 \times 10^{-3}$ mol NH_3 remains and 0.75×10^{-3} mol NH_4^+ are in solution. The molarity is calculated as mol OH⁻/combined volume. Then pOH is calculated from rearrangement of K_b.

$$[OH^-] = 1.8 \times 10^{-5} = \frac{[NH_3]}{[NH_4^+]}$$

where $[NH_4^+]$ is calculated from the moles of HCl used and the new volume.
The final step is the calculation of $[OH^-]$ to $[H^+]$ to pH:

$$[H^+] = \frac{K_w}{[OH^-]}; \quad pH = -\log[H^+]$$

Representative volumes are listed.

V(L) of 0.150 M HCl added	Moles of HCl added	Moles remaining		Volume (Liters)
0.0000	0.0	0.00375	(NH₃)	0.0250
0.0050	0.00075	0.00300		0.0300
0.0100	0.00150	0.00225		0.0350
0.0150	0.00225	0.00150		0.0400
0.0200	0.00300	0.00075		0.0450
0.0220	0.00330	0.00045		0.0470
0.0240	0.00360	0.00015		0.0490
0.0245	0.00368	0.00007		0.0495
0.0249	0.00374	0.00001		0.0499
0.0250	0.00375	0.00	equivalence*	0.0500
0.0251	0.00376	0.00001	(H⁺)**	0.0501
0.0255	0.00382	0.00007	excess	0.0505
0.0260	0.00390	0.00015		0.0510
0.0280	0.00420	0.00045		0.0530
0.0300	0.00450	0.00075		0.0550
0.0350	0.00525	0.00150		0.0600
0.0400	0.00600	0.00225		0.0650
0.0450	0.00675	0.00300		0.0700
0.0500	0.00750	0.00375		0.0750

*At the equivalence point, only NH_4^+ is present, and the reaction is

$$NH_4^+ + H_2O \rightleftharpoons NH_3 + H_3O^+$$

**After the equivalence point the pH is dependent on the excess hydrogen ion.

M(NH₃) or M(H⁺)		M(NH₄⁺)	M(OH⁻) or M(H⁺)		pH
0.150	[NH₃]	1.64×10^{-3}	1.64×10^{-3}	[OH⁻]	11.2
0.100		0.025	7.20×10^{-5}		9.86
0.064		0.043	2.68×10^{-5}		9.43
0.038		0.056	1.22×10^{-5}		9.09
0.0167		0.067	4.48×10^{-6}		8.65
0.0096		0.070	2.47×10^{-6}		8.39
0.0031		0.073	7.64×10^{-7}		7.88
0.0014		0.0743	3.39×10^{-7}		7.53
0.0002	↓	0.0749	4.81×10^{-8}	↓	6.68
6.45×10^{-6}	[H⁺]	0.075	6.45×10^{-6}	[H⁺]	5.19
1.996×10^{-4}					3.70
1.386×10^{-3}					2.86
2.941×10^{-3}					2.53
8.491×10^{-3}					2.07
0.0136					1.87
0.025					1.60
0.0346					1.46
0.0429					1.37
0.0500	↓			↓	1.30

58. Why does an acid-base indicator change color over a range of pH values rather than at a specific pH?

Solution

The color change in an indicator depends upon the ratio of nonionized and ionized forms. The predominant color in acid requires at least a 1 to 10 ratio of forms whereas the predominant color in base requires a 10 to 1 ratio of forms. The transition from one condition to the other comes about by addition of acid or base resulting, in general, in a change of 2 pH units between colors.

60. Using the data in Table 17.6, arrange the following indicators (which are weak acids) in increasing order of acid strength: methyl orange, litmus, thymol blue.

Solution

The indicator that changes at the lowest pH is the strongest acid (that is, it releases hydrogen ions most easily). The order is litmus (pH range 4.7-8.2) < methyl orange (pH range 3.1-4.4) < thymol blue (pH range 1.2-2.8).

61. A 0.5000-g sample of an impure monobasic amine is titrated to the equivalence point with 75.00 mL of 0.100 M HCl. The pH after 40.00 mL of acid is added is 10.65. The molecular weight of the pure base is known to be 59.1. (a) Calculate the ionization constant of this base and (b) its percent purity. (c) Which indicator from the following list would be most suitable for this titration? State your reasoning.

   Orange IV (red-yellow; pH 1.2-2.6)
   Methyl orange (red-yellow; pH 3.1-4.4)
   Methyl red (red-yellow); pH 4.8-6.0)
   Bromthymol blue (yellow-blue; pH 6.0-7.6)
   Phenolphthalein (colorless-pink; pH 8.3-10.0)
   Thymolphthalein (colorless-blue; pH 9.3-10.5)

Chapter 17 Ionic Equilibria of Weak Electrolytes   249

Solution

(a) The amine is a weak base, and titration to the equivalence point means that 75.00 mL of 0.100 M HCl = 7.5×10^{-3} moles of amine were initially present. When 40.00 mL of acid (0.100 M) is added, 4.00×10^{-3} mol H$^+$ has been added; 7.5×10^{-3} mol amine $- 4.00 \times 10^{-3}$ mol (HCl) = 3.5×10^{-3} mol (amine) unreacted. In 40.00 mL of solution this is 0.0875 M (amine) and 0.100 M protonated amine.

$$\text{Amine} + \text{H}^+ \longrightarrow \text{protonated amine}$$

However, amine + H$_2$O $\rightleftharpoons$ protonated amine + OH$^-$. Therefore,

$$K_b = \frac{[\text{protonated amine}][\text{OH}^-]}{[\text{amine}]}$$

From the pH = 10.65, [H$^+$] = 2.24×10^{-11}, [OH$^-$] = 4.46×10^{-4}

$$K_b = \frac{0.100}{0.0875}(4.46 \times 10^{-4}) = 5.1 \times 10^{-4}$$

(b) The molecular weight is 59.1 g/mol.

$$(0.0875 \text{ M} + 0.10 \text{ M})\left(\frac{40 \text{ mL}}{1000 \text{ mL}}\right)(59.1 \text{ g/mol}) = 0.44325 \text{ g}$$

of pure amine is in the sample.

$$\frac{0.44325 \text{ g}}{0.5000 \text{ g}} = .8865 = 88.7\%$$

(c) The pH at which the color changes is calculated from the equivalence information and the K_a of the protonated species.

$$K_a = \frac{K_w}{K_b} = \frac{1.0 \times 10^{-14}}{5.1 \times 10^{-4}} = 1.96 \times 10^{-11} = \frac{[\text{amine}][\text{H}^+]}{[\text{protonated amine}]}$$

Let $x = [\text{H}^+]$. At the equivalence point 7.5×10^{-3} mol of protonated amine are initially present. In 75.00 mL, this corresponds to 7.5×10^{-3} mol/0.075 L = 0.1 M of protonated amine which can react to form [H$^+$].

$$\frac{x^2}{0.1 - x} = 1.96 \times 10^{-11}$$

Neglect x in the denominator.

$$x^2 = 1.96 \times 10^{-12}$$
$$x = [\text{H}^+] = 1.4 \times 10^{-6}$$
$$\text{pH} = -\log[\text{H}^+] = 5.85$$

This falls in the range of the methyl red indicator.

63. The indicator dinitrophenol is an acid with a K_a of 1.1×10^{-4}. In a 1.0×10^{-4} M solution, it is colorless in acid and yellow in base. Calculate the pH range over which it goes from 10% ionized (colorless) to 90% ionized (yellow).

Solution

$$\text{HIn (colorless)} \rightleftharpoons \text{H}^+ + \text{In}^- \text{ (yellow)}$$

$$[H^+]_{\text{acid form}} = \frac{K_a[\text{HIn}]}{[\text{In}^-]} = 1.1 \times 10^{-4} \frac{(9.0 \times 10^{-5})}{(1.0 \times 10^{-5})} = 9.9 \times 10^{-4}; \quad pH = 3.00$$

$$[H^+]_{\text{base form}} = 1.1 \times 10^{-4} \frac{(1.0 \times 10^{-5})}{(9.0 \times 10^{-5})} = 1.22 \times 10^{-5}; \quad pH = 4.91$$

REACTION OF SALTS WITH WATER

64. Why does the salt of a strong acid and a strong base give an approximately neutral solution?

Solution

Both counter ions being generated from either a strong acid or strong base have little tendency to abstract a proton or hydroxide ion from water. Consequently, the solution should be neutral.

67. Does the salt of a weak acid and a weak base give an acidic solution, a basic solution, or a neutral solution? Explain your answer.

Solution

The answer depends upon the relative values of K_a and K_b of the weak acid and base involved. If their values are approximately the same the solution will be neutral. If the K_a is larger than the K_b, the solution will behave as a solution of a salt of a weak base, and the solution will be acidic. On the other hand, if K_b is larger than K_a, the solution will behave as a solution of a salt of a weak acid, and the solution will be basic.

70. Arrange the conjugate bases of the acids in Table 17.3 in order of increasing strength.

Solution

The conjugate base strength increases as the acid strength decreases. The order is:

$$CN^- > BrO^- > ClO^- > CH_3CO_2^- > HCO_2^- > NCO^- > NO_2^- > F^-$$

72. (a) Write the equation for the reaction that makes potassium cyanide, KCN, behave as a base in water.
(b) What is K_b for this reaction?
(c) What is the pH of a 0.255 M solution of KCN?

Solution

(a) $CN^- + H_2O \rightleftharpoons HCN + OH^-$

(b) $K_b = \dfrac{[\text{HCN}][\text{OH}^-]}{[\text{CN}^-]} = \dfrac{K_w}{K_a} = \dfrac{1.0 \times 10^{-14}}{4 \times 10^{-10}} = 2.5 \times 10^{-5}$

(c) $K_b = 2.5 \times 10^{-5} = \dfrac{x^2}{0.255}$

$x^2 = 6.38 \times 10^{-6}$

$x = 2.52 \times 10^{-3} = [\text{OH}^-]$

$[H^+] = \dfrac{K_w}{[\text{OH}^-]} = \dfrac{1.0 \times 10^{-14}}{2.52 \times 10^{-3}} = 3.96 \times 10^{-12}; \quad pH = 11.4$

73. Calculate the pH of each of the following solutions:

(b) 0.4735 M NaCN

(c) 0.333 M [(CH$_3$)$_2$NH$_2$]$_2$SO$_4$ [Note that (CH$_3$)$_2$NH$_2^+$ is the conjugate acid of the weak base (CH$_3$)$_2$NH, just as NH$_4^+$ is the conjugate acid of the weak base NH$_3$.]

Solution

The solute in each of these solutions is salt composed of ions in which one of the ions originated from either a weak acid or a weak base. These ions undergo reaction with water to produce acidic or basic solutions depending on their origin.

(b) 0.4735 M NaCN. This salt dissociates in water, yielding

$$\text{NaCN}(aq) \rightleftharpoons \text{Na}^+(aq) + \text{CN}^-(aq)$$

Reaction of the CN$^-$ with water occurs because it is the anion of a *weak* acid. Similar reaction of Na$^+$ does not occur because Na$^+$ is a cation of a *strong* base and is fully dissociated in aqueous solution. Thus

$$\text{CN}^-(aq) + \text{H}_2\text{O}(l) \rightleftharpoons \text{HCN}(aq) + \text{OH}^-(aq)$$

The hydroxide ion is a strong base and HCN is a weak acid; the reaction increases the concentration of hydroxide ion, thereby causing the solution to be basic. The equilibrium constant for this process is K_b. Its value is

$$K_b = \frac{[\text{HCN}][\text{OH}^-]}{[\text{CN}^-]}$$

Since [OH$^-$] = K_w/[H$^+$], substitution gives

$$K_b = \frac{[\text{HCN}] K_w}{[\text{CN}^-][\text{H}^+]} = \frac{K_w}{K_a (\text{HCN})}$$

where

$$K_a(\text{HCN}) = \frac{[\text{H}^+][\text{CN}^-]}{[\text{HCN}]} = 4 \times 10^{-10}$$

Therefore,

$$K_b = \frac{1 \times 10^{-14}}{4 \times 10^{-10}} = 2.5 \times 10^{-5}$$

If in our reaction x equals the moles of CN$^-$ that react with water, then, at equilibrium,

$$x = [\text{HCN}] = [\text{OH}^-]$$

and

$$[\text{CN}^-] = (0.4735 - x) \text{ M}$$

$$K_b = 2.5 \times 10^{-5} = \frac{(x)(x)}{0.4735 - x}$$

Neglect x in (0.4735 − x).

$$x^2 = 1.18 \times 10^{-5} \text{ or } x = 3.44 \times 10^{-3} = [\text{OH}^-]$$

$$\text{pOH} = -\log 3.44 \times 10^{-3} = 2.46$$

$$\text{pH} = 14.00 - \text{pOH} = 14.00 - 2.46 = 11.5$$

(c) 0.333 M [(CH$_3$)$_2$NH$_2$]$_2$SO$_4$ (dimethylamine sulfate). The dimethyl-ammonium ion will react with water as

$$(CH_3)_2NH_2^+(aq) + H_2O(l) \rightleftharpoons (CH_3)_2NH(aq) + H_3O^+(aq)$$

$$K_a = \frac{K_w}{K_b\,[(CH_3)_2NH]} = \frac{1.0 \times 10^{-14}}{7.4 \times 10^{-4}} = \frac{[(CH_3)_2NH][H_3O^+]}{[(CH_3)_2NH_2^+]} = 1.35 \times 10^{-11}$$

Let x = moles of (CH$_3$)$_2$NH$_2^+$ that react with water.
At equilibrium:

$$x = [H_3O^+] = [(CH_3)_2NH]$$

Because the dissociation of the sulfate produces 2 ions of (CH$_3$)$_2$NH$_2^+$,

$$[(CH_3)_2NH_2^+] = 2(0.333) - x$$

$$K_a = 1.35 \times 10^{-11} = \frac{(x)(x)}{(0.666 - x)}$$

x is sufficiently small compared to 0.666 that it can be neglected.

$$1.35 \times 10^{-11} = \frac{x^2}{0.666}$$

$$x^2 = 8.99 \times 10^{-12}$$

$$x = 3.00 \times 10^{-6} = [H_3O^+]$$

$$pH = -\log(3.00 \times 10^{-6}) = 5.52$$

74. Calculate the ionization constant (K_a or K_b) for each of the following acids or bases.

   (d) F$^-$
   (g) NH$_4^+$

Solution

(d) The fluoride ion reacts with water as a base in the following way:

$$F^-(aq) + H_2O(l) \rightleftharpoons HF(aq) + OH^-(aq)$$

An expression for the equilibrium is written in the usual way,

$$K_b = \frac{[HF][OH^-]}{[F^-]}$$

This expression is related to the ionization of HF:

$$HF(aq) \rightleftharpoons H^+(aq) + F^-(aq)$$

$$K_a = \frac{[H^+][F^-]}{[HF]} = 7.2 \times 10^{-4}$$

Inspection of the expressions for K_b and for K_a indicates an inverse relationship between K_a and K_b with the substitution of $[H^+] = K_w/[OH^-]$ in the expression for K_a.

$$K_a = \frac{[H^+][F^-]}{[HF]} = \frac{\frac{K_w}{[OH^-]} \times [F^-]}{[HF]} = \frac{K_w[F^-]}{[HF][OH^-]}$$

$$K_b = K_w \times \frac{1}{K_a} = \frac{K_w}{K_a}$$

Chapter 17 Ionic Equilibria of Weak Electrolytes   253

This final relationship turns out to be the general case for the reaction of an ion from either a weak acid or a weak base.

For the F⁻ ion:

$$K_b = \frac{1.0 \times 10^{-14}}{7.2 \times 10^{-4}} = 1.4 \times 10^{-11}$$

(g) This ion is the conjugate acid of the weak base ammonia. Reaction of NH_4^+ gives

$$NH_4^+(aq) + H_2O(l) \rightleftharpoons NH_3(aq) + H_3O^+(aq)$$

$$K_a = \frac{[NH_3][H_3O^+]}{[NH_4^+]} = \frac{K_w}{K_b(NH_3)} = \frac{1.0 \times 10^{-14}}{1.8 \times 10^{-5}} = 5.6 \times 10^{-10}$$

76. In a 0.0010 M solution of KCN, the KCN reacts with water to the extent of 14.0%. Calculate the value of the ionization constant of HCN.

Solution

The cyanide ion, CN⁻, is the conjugate base of the weak acid, HCN, and reacts with water as

$$CN^-(aq) + H_2O(l) \rightleftharpoons HCN(aq) + OH^-(aq)$$

$$K_b = \frac{[HCN][OH^-]}{[CN^-]} = \frac{K_w}{K_a(HCN)}$$

Each concentration factor in the expression for K_b can be determined from the degree of hydrolysis and the initial concentration of KCN.

	[CN⁻]	[HCN]	[OH⁻]
Initial:	0.0010	0	0
	0.0010 − (0.14)(0.0010)	(0.14)(0.0010)	(0.14)(0.0010)
Final:	0.00086	0.00014	0.00014

$$K_b = \frac{(0.00014)(0.00014)}{0.00086} = 2.28 \times 10^{-5}$$

$$K_b = \frac{K_w}{K_a(HCN)} \text{ or } K_a(HCN) = \frac{K_w}{K_b} = \frac{1.0 \times 10^{-14}}{2.28 \times 10^{-5}} = 4.4 \times 10^{-10}$$

79. Sodium nitrite, $NaNO_2$, is used in meat processing. What are [H⁺] and the pH of a 0.225 M $NaNO_2$ solution? Is the solution acidic or basic?

Solution

The nitrite ion comes from a weak acid HNO_2, $K_a = 4.5 \times 10^{-4}$. The reaction of the nitrite ion with water is

$$NO_2^- + H_2O \rightleftharpoons HNO_3 + OH^-$$

$$K_b = \frac{[HNO_3][OH^-]}{[NO_2^-]} = \frac{K_w}{K_a} = \frac{1.0 \times 10^{-14}}{4.5 \times 10^{-4}} = 2.22 \times 10^{-11}$$

Let $x = [OH^-]$. Then

$$K_b = \frac{x^2}{(0.225 - x)}.$$

Neglecting x in the denominator,

$$x^2 = 0.225 \times 2.22 \times 10^{-11} = 5.00 \times 10^{-12}$$

$$x = [OH^-] = 2.24 \times 10^{-6}$$

$$[H^+] = \frac{1.00 \times 10^{-14}}{2.24 \times 10^{-6}} = 4.46 \times 10^{-9}$$

$$pH = -\log[H^+] = 8.35$$

The solution is basic.

80. The ion HTe⁻ is an amphiprotic species; it can act as either an acid or a base. What is K_a for the acid reaction of HTe⁻ with H₂O? What is K_b for the reaction in which HTe⁻ functions as a base in water?

Solution

As an acid

$$HTe^- + H_2O \rightleftharpoons Te^{2-} + H_3O^+$$

$$K_a = \frac{[Te^{2-}][H^+]}{[HTe^-]}$$

As a base

$$HTe^- + H_2O \rightleftharpoons H_2Te + OH^-$$

$$K_b = \frac{[H_2Te][OH^-]}{[HTe^-]}$$

From Appendix F, $K_a = 1 \times 10^{-5}$. To find K_b where HTe⁻ acts as a base, use K_a for H₂Te. $K_a = 2.3 \times 10^{-3}$.

$$K_b = \frac{K_w}{K_a} = \frac{1.0 \times 10^{-14}}{2.3 \times 10^{-3}} = 4.3 \times 10^{-12}$$

ADDITIONAL EXERCISES

82. By calculation, check the values given in Section 16.3 for the percent ionization of 0.10 M solutions of the weak acids HSO₄⁻, HNO₂, and CH₃CO₂H.

Solution

In each of the following let $x = [H^+]$.

(a) For percent ionization of 0.10 M HSO₄⁻

$$K_a = 1.2 \times 10^{-2} = \frac{[SO_4^{2-}][H^+]}{[HSO_4^-]} = \frac{x^2}{0.10 - x}$$

Use the quadratic equation on $x^2 + 1.2 \times 10^{-2}x - 1.2 \times 10^{-3} = 0$:

$$x = \frac{-1.2 \times 10^{-2} \pm \sqrt{(1.2 \times 10^{-2})^2 + 4.8 \times 10^{-3}}}{2} = \frac{-0.012 \pm 0.0703}{2} = 0.0291 \text{ M}$$

(b) For percent ionization of HNO₂

$$K_a = 4.5 \times 10^{-4} = \frac{[NO_2^-][H^+]}{[HNO_2]} = \frac{x^2}{0.10 - x}$$

Chapter 17 Ionic Equilibria of Weak Electrolytes 255

Use the quadratic equation on $x^2 + 4.5 \times 10^{-4}x - 4.5 \times 10^{-5} = 0$

$$x = \frac{-4.5 \times 10^{-4} \pm \sqrt{2.025 \times 10^{-7} + 1.8 \times 10^{-4}}}{2} = \frac{-4.5 \times 10^{-4} \pm 0.0134}{2}$$

$$= 6.5 \times 10^{-3} \text{ M}$$

Percent ionization $= \dfrac{6.5 \times 10^{-3} \text{ M}}{0.100 \text{ M}} \times 100 = 6.5\%$

(c) For percent ionization of CH_3CO_2H,

$$K_a = 1.8 \times 10^{-5} = \frac{[CH_3CO_2^-][H^+]}{[CH_3CO_2H]} = \frac{x^2}{0.10 - x}$$

Use the quadratic equation on $x^2 + 1.8 \times 10^{-5}x - 1.8 \times 10^{-6} = 0$

$$x = \frac{-1.8 \times 10^{-5} \pm \sqrt{(1.8 \times 10^{-5})^2 + 4(1.8 \times 10^{-6})}}{2} = \frac{-1.8 \times 10^{-5} \pm 2.68 \times 10^{-3}}{2}$$

$$= 1.3 \times 10^{-3}$$

Percent ionization $= \dfrac{1.3 \times 10^{-3} \text{ M}}{0.100 \text{ M}} \times 100 = 1.3\%$

85. How many moles of hydrogen chloride must be added to 1.50 L of a 0.450 M solution of CH_3NH_2 to give a pH of 10.95?

Solution

Methylamine is a weak base: $K_b = 4.4 \times 10^{-4}$.

$$CH_3NH_2 + H_2O \rightleftharpoons CH_3NH_3^+ + OH^-$$

In 1.50 L of a 0.450 M solution there are 0.675 mol (M × L). The pH is found for the initial solution:

$$4.4 \times 10^{-4} = \frac{[CH_3NH_3^+][OH^-]}{[CH_3NH_2]} = \frac{x^2}{0.450}$$

$$x^2 = 1.98 \times 10^{-4}$$

$$x = [OH^-] = 0.0141 \text{ M}$$

$$pOH = 1.85; \ pH = 12.15$$

This indicates that HCl must be added to bring the pH to 10.95. Addition of HCl would allow the $CH_3NH_3^+$ formed to function as an acid, the K_a for which is

$$\frac{K_w}{K_b} = \frac{1.0 \times 10^{-14}}{4.4 \times 10^{-4}} = 2.27 \times 10^{-11}$$

Calculate the ratio of forms needed at pH 10.95.

$$K_a = 2.27 \times 10^{-11} = \frac{[CH_3NH_2][H^+]}{[CH_3NH_3^+]}$$

$$\frac{[CH_3NH_2]}{[CH_3NH_3^+]} = \frac{2.27 \times 10^{-11}}{[H^+]} = \frac{2.27 \times 10^{-11}}{1.12 \times 10^{-11}} = 2.027$$

All of the $CH_3NH_3^+$ had to come from CH_3NH_2.
Let $x = [CH_3NH_3^+]$:

$$\frac{0.450 - x}{x} = 2.027$$

$$x = 0.149$$

In 1.5 L, 0.149 M × 1.5 L = 0.22 mol are required.

87. Sulfuric acid ionizes in two steps:

$$H_2SO_4 \rightleftharpoons H^+ + HSO_4^-$$

$$HSO_4^- \rightleftharpoons H^+ + SO_4^{2-}$$

Since sulfuric acid is a strong acid, the first step is effectively 100% complete ($K_{H_2SO_4}$ is large). The ionization constant for the second step, $K_{HSO_4^-}$, is 1.2×10^{-2}. Check the assumption that the first step goes completely to the right, based on the approximation that for a polyprotic acid, successive ionization constants often differ by a factor of 1×10^6 and calculate (a) $K_{H_2SO_4}$, (b) the concentration of H_2SO_4, (c) the percent ionization of H_2SO_4 into H^+ and HSO_4^- in a 0.100 M solution of H_2SO_4.

Solution

(a) If $K_{H_2SO_4}$ differs by 1×10^6 from the value of $K_{HSO_4^-}$,

$$K_{H_2SO_4} = 1.2 \times 10^{-2} \times 1 \times 10^6 = 1 \times 10^4$$

(b) The concentration of HSO_4^- and H^+ in a 0.100 M solution would both be equal to 0.100 M. Substitution into $K_{H_2SO_4}$ gives

$$K_{H_2SO_4} = \frac{[H^+][HSO_4^-]}{[H_2SO_4]} = \frac{(0.100)(0.100)}{[H_2SO_4]} = 1 \times 10^4$$

$$[H_2SO_4] = 1 \times 10^{-2}/1 \times 10^4 = 1 \times 10^{-6} \text{ M}$$

(c) The percent ionization is the amount dissociated divided by 0.1 M, then multiplied by 100 to convert to percentage.

$$\frac{(0.1 - 1 \times 10^{-6}) \text{ M}}{(0.1) \text{ M}} \times 100 = 99.999\%$$

89. Lime juice is among the most acidic of fruit juices, with a pH of 1.92. If the acidity is due to citric acid, which we can abbreviate as H_3Cit, what is the ratio of each of the following to $[Cit^{3-}]$: $[H_3Cit]$; $[H_2Cit^-]$; $[HCit^{2-}]$?

$$H_3Cit \rightleftharpoons H^+ + H_2Cit^- \qquad K_a = 8.4 \times 10^{-4} \qquad (1)$$

$$H_2Cit^- \rightleftharpoons H^+ + HCit^{2-} \qquad K_a = 1.8 \times 10^{-5} \qquad (2)$$

$$HCit^{2-} \rightleftharpoons H^+ + Cit^{3-} \qquad K_a = 4.0 \times 10^{-6} \qquad (3)$$

Solution

The pH of the solution and the K_a values for the three ionization steps are known, so a series of ratios written in terms of $[Cit^{3-}]$ can be written. Starting with Reaction (3),

$$\text{pH} = 1.92; \quad [H_3O^+] = 10^{-1.92} = 1.20 \times 10^{-2} \text{ M}$$

$$\frac{[H^+][Cit^{3-}]}{[HCit^{2-}]} = 4.0 \times 10^{-6}$$

and

$$[HCit^{2-}] = \frac{(1.20 \times 10^{-2})[Cit^{3-}]}{4.0 \times 10^{-6}} = 3.0 \times 10^3 [Cit^{3-}]$$

For Reaction (2),

$$\frac{[H^+][HCit^{2-}]}{[H_2Cit^-]} = 1.8 \times 10^{-5}$$

$$[H_2Cit^-] = \frac{(1.20 \times 10^{-2})[HCit^{2-}]}{1.8 \times 10^{-5}}$$

$$= 6.67 \times 10^2 [HCit^{2-}] = (6.67 \times 10^2)(3.0 \times 10^3)[Cit^{3-}]$$

$$= 2.0 \times 10^6 [Cit^{3-}]$$

For Reaction (1),

$$\frac{[H^+][H_2Cit^-]}{[H_3Cit]} = 8.4 \times 10^{-4}$$

$$[H_3Cit] = \frac{1.20 \times 10^{-2}[H_2Cit^-]}{8.4 \times 10^{-4}}$$

$$= 14.3[H_2Cit^-] = (14.3)(2.0 \times 10^6)[Cit^{3-}]$$

$$= 2.9 \times 10^7 [Cit^{3-}]$$

$[H_3Cit]$:	$[H_2Cit^-]$:	$[HCit^{2-}]$:	$[Cit^{3-}]$
2.9×10^7	2.0×10^6	3.0×10^3	1

90. In many detergents phosphates have been replaced with silicates as water conditioners. If 125 g of a detergent that contains 8.0% Na₂SiO₃ by weight is used in 4.0 L of water, what are the pH and the hydroxide ion concentration in the wash water?

$$SiO_3^{2-} + H_2O \rightleftharpoons SiO_3H^- + OH^- \qquad K_b = 1.6 \times 10^{-3} \qquad (1)$$

$$SiO_3H^- + H_2O \rightleftharpoons SiO_3H_2 + OH^- \qquad K_b = 3.1 \times 10^{-5} \qquad (2)$$

Solution

The hydrolysis of SiO_3^{2-} occurs in two steps. However, owing to the small amount of SiO_3H^- produced in Reaction (1) and the low K_b in Reaction (2), the amount of OH^- in the final solution is due almost entirely to the first reaction. Calculate the molarity of NaSiO₃ and let x equal the moles of SiO_3^{2-} that hydrolyze.

$$Na_2SiO_3 = 0.080(125 \text{ g}) \times \frac{1 \text{ mol}/122.1 \text{ g}}{4.0 \text{ L}} = 0.0205 \text{ M}$$

At equilibrium

$$x = [OH^-] \approx [SiO_3H^-]$$

$$[SiO_3^{2-}] \approx (0.0205 - x)$$

$$K_b = 1.6 \times 10^{-3} = \frac{(x)(x)}{(0.0205 - x)}$$

or

$$x^2 = 1.6 \times 10^{-3}(0.0205 - x)$$
$$x^2 = 1.6 \times 10^{-3}x - 3.28 \times 10^{-5} = 0$$
$$x = \frac{-1.6 \times 10^{-3} \pm \sqrt{2.56 \times 10^{-6} + 4(3.28 \times 10^{-5})}}{2}$$
$$= \frac{-1.6 \times 10^{-3} \pm 1.16 \times 10^{-2}}{2} = \frac{10.0 \times 10^{-3}}{2}$$
$$= [OH^-] = 5.0 \times 10^{-3} \text{ M}$$
$$\text{pOH} = -\log 5.0 \times 10^{-3} = 2.30$$
$$\text{pH} = 11.70$$

95. A typical urine sample contains 2.3% by mass of the base urea, $CO(NH_2)_2$

$$CO(NH_2)_2 + H_2O \rightleftharpoons CO(NH_2)(NH_3)^+ + OH^- \qquad K_b = 1.5 \times 10^{-14}$$

If the density of the urine sample is 1.06 g/cm³ and the pH of the sample is 6.35, calculate the concentration of $CO(NH_2)_2$ and of $CO(NH_2)(NH_3)^+$ in the sample.

Solution

Calculate the mass of urea and convert it to molarity. Using K_b and $[H^+]$ given, calculate the desired information.

$$2.3\% \times 1.06 \text{ g/cm}^3 \times 1000 \text{ cm}^3/\text{L} = 24.4 \text{ g/L}$$

The molecular weight of urea is 60.0556 g/mol. Then,

$$[CO(NH_2)_2] = \frac{24.4 \text{ g L}^{-1}}{60.0556 \text{ g mol}^{-1}} = 0.41 \text{ M}$$

Since pH = 6.35,

$$[H^+] = 4.47 \times 10^{-7} \qquad [OH^-] = \frac{1.0 \times 10^{-14}}{4.47 \times 10^{-7}} = 2.24 \times 10^{-8}$$

$$K_b = \frac{[CO(NH_2)(NH_3)^+][OH^-]}{[CO(NH_2)_2]} = \frac{[x][2.24 \times 10^{-8}]}{[0.41 \times x]} = 1.5 \times 10^{-14}$$

Dropping the x in the denominator,

$$x = [H^+] = 2.7 \times 10^{-7} \text{ M}$$

96. If abominable snowmen were to exist in the world in the same proportion to humans as do hydrogen ions to water molecules in pure water, how many abominable snowmen would there be in the world? Assume a world population of 4 billion people.

Solution

The ratio of H^+ to water molecules in pure water is found from K. $K_w = (55.5)K = 1 \times 10^{-14} = [H^+][OH^-]$. In 1 L of water there are 10^{-7} mol/55.5 mol $[H^+]$. Therefore, the ratio is

$$\frac{x}{10^{-7}/55.5} = \frac{4 \times 10^9}{1}$$

$$x = 7$$

18

The Solubility Product Principle

INTRODUCTION

Many ionic substances function as strong electrolytes in water even though they are only slightly soluble in it. Because of their slight solubility in water, they often form valuable mineral deposits. These substances include carbonates, such as limestone ($CaCO_3$) and sulfides, such as pyrite (FeS_2), and cinnabar (HgS). Mercury(II) sulfide, HgS, is one of the least soluble ionic substances: only about 9×10^{-28} g will dissolve in 1 L of water.

The occurrence of mineral deposits can be understood by considering the mode of formation of slightly soluble salts. In reactions involving the formation of a slightly soluble substance, the solubility is usually exceeded, and a precipitate forms. For example, the salt $BaSO_4$ exists in water as $Ba^{2+}(aq)$ and $SO_4^{2-}(aq)$, but the maximum concentration at 25°C is only 1×10^{-5} M. When a solution containing barium ions, such as $BaCl_2$, is added to one containing sulfate ions, such as H_2SO_4, the solubility limit is exceeded, and a white precipitate of $BaSO_4$ forms and settles to the bottom. The solid $BaSO_4$ is then in equilibrium with the ions in solution. This equilibrium is defined in terms of the product of ion concentrations, that is, the solubility product, K_{sp}.

$$BaSO_4(s) \rightleftharpoons Ba^{2+}(aq) + SO_4^{2-}(aq)$$

$$K_{sp} = [Ba^{2+}][SO_4^{2-}] = 1.08 \times 10^{-10}$$

Thus a precipitate will form when the product of the ion concentrations in the solubility product expression exceed the K_{sp}. Precipitation will continue until the product of the concentrations equals the K_{sp}.

Slightly soluble substances that contain ions derived from weak acids or weak bases dissolve and undergo hydrolysis or association depending on the pH of the solution. Two or more simultaneous equilibria that exist in solution may compete for the ions. In such cases all equilibria must be considered in the determination of the solubility.

A number of concepts involving solubility are treated in this chapter of the textbook. Several examples that include solubility product constants should be studied carefully before an attempt is made to solve problems at the end of the chapter. The K_{sp} concept is extended to the dissolution of precipitates through the formation of complex ions, which ultimately involve simultaneous equilibria. Many examples that involve K_{sp} also involve the concepts of weak acids, covered in Chapter 17.

READY REFERENCE

Saturated solution (18.1) A saturated solution contains a slightly soluble solute in a two-phase system in which dissolved solute is in contact with undissolved solute. The system is at equilibrium, meaning that undissolved solute continues to dissolve while ions in solution associate and crystallize simultaneously at the same rate.

Solubility product constant (K_{sp}) **(18.1)** An equilibrium constant is based on the molar concentrations of ions in a saturated solution containing a slightly soluble electrolyte. Each ion concentration is raised to the power corresponding to its stoichiometric coefficient as shown in the equilibrium expression. In general, the expression for K_{sp} is written as

$$A_xB_y(s) \rightleftharpoons xA(aq)^{(positive)} + yB(aq)^{(negative)}$$

$$K_{sp} = [A^{positive}]^x [B^{negative}]^y$$

Solubility product constants are published without units for the sake of brevity. But always keep in mind that the constants are in terms of concentrations raised to some power, and sometimes the units must be used in conjunction with equilibrium calculations.

SOLUTIONS TO CHAPTER 18 EXERCISES

SOLUBILITY PRODUCTS

1. Write the expression for the solubility product of each of the following slightly soluble electrolytes:

    (b) Ag_2S
    (d) $MgNH_4AsO_4$

    Solution

    (b) Upon dissolution, three ions will be formed, Ag^+, Ag^+, S^{2-}.

    $$K_{sp} = [Ag^+]^2 [S^{2-}]$$

    (d) $K_{sp} = [Mg^{2+}][NH_4^+][AsO_4^{3-}]$

3. How do the concentrations of Ag^+ and CrO_4^{2-} in a liter of water above 1.0 g of solid Ag_2CrO_4 change when 100 g of solid Ag_2CrO_4 is added to the system? Explain.

    Solution

    This is an example of a saturated solution in which solid solute is in contact with its ions. The solution contains the maximum possible amount of solute. Addition of Ag_2CrO_4 to this system will not change the amount that dissolves and dissociates. Any additional Ag_2CrO_4 merely precipitates and becomes part of the dynamic equilibrium.

6. Calculate the solubility product of each of the following from the solubility given:

    (b) AgBr, 5.7×10^{-7} mol L^{-1}
    (d) Ag_2CrO_4, 4.3×10^{-2} g L^{-1}

Solution

(b) Silver bromide, AgBr, is a very slightly soluble salt that dissolves in water in the following way:

$$AgBr(s) \rightleftharpoons Ag^+(aq) + Br^-(aq)$$

The K_{sp} expression includes only the ions in the solution phase and is written as

$$K_{sp} = [Ag^+][Br^-]$$

In a solution of a 1:1 electrolyte such as AgBr, the concentration of Ag^+ and of Br^- equals the solubility of AgBr. The concentration of AgBr that dissolves is 5.7×10^{-7} mol/L and, therefore, the concentration of Ag^+ and Br^- is 5.7×10^{-7} mol/L. Substitution into the K_{sp} expression gives

$$K_{sp}(AgBr) = (5.7 \times 10^{-7})(5.7 \times 10^{-7}) = 3.2 \times 10^{-13}$$

(d) Since equilibrium values are calculated in terms of molar quantities, the solubility must be in terms of molarity.

$$\text{mol. wt } Ag_2CrO_4 = 2(107.868) + 51.996 + 4(15.9994)$$
$$= 331.730 \text{ g mol}^{-1}$$

$$\text{Molar solubility } = 4.3 \times 10^{-2} \text{ g/L} \times \frac{1 \text{ mol}}{331.7 \text{ g}}$$
$$= 1.296 \times 10^{-4} \text{ M}$$

Dissolving one formula unit of Ag_2CrO_4 produces 2 Ag^+ and 1 CrO_4^{2-}. Therefore,

$$K_{sp} = [Ag^+]^2[CrO_4^{2-}]$$
$$[Ag^+] = 2(1.296 \times 10^{-4}) = 2.59 \times 10^{-4}$$
$$[CrO_4^{2-}] = 1.296 \times 10^{-4}$$
$$K_{sp} = (2.59 \times 10^{-4})^2(1.296 \times 10^{-4}) = 8.7 \times 10^{-12}$$

9. Which of the following compounds will precipitate from a solution initially containing the indicated concentrations of ions (see Appendix D for K_{sp} values)?

(a) $CaCO_3$: $[Ca^{2+}] = 0.003$ M,
              $[CO_3^{2-}] = 0.003$ M
(d) $Co(OH)_2$: $[Co^{2+}] = 0.01$ M,
               $[OH^-] = 1 \times 10^{-7}$ M

Solution

Precipitation will occur if the ion concentrations exceed the amounts allowed by the solubility product constant. To determine whether this occurs, write the expression for K_{sp}, substitute the individual concentrations into the expression, perform the indicated mathematics, and then compare the product to K_{sp}.

(a) $CaCO_3$:

$$CaCO_3(s) \rightleftharpoons Ca^{2+}(aq) + CO_3^{2-}(aq)$$
$$K_{sp} = [Ca^{2+}][CO_3^{2-}] = 4.8 \times 10^{-9}$$

262   Chapter 18   The Solubility Product Principle

Test K_{sp} against $[Ca^{2+}][CO_3^{2-}]$:

$$[Ca^{2+}][CO_3^{2-}] = (0.003)(0.003) = 9 \times 10^{-6}$$

$$K_{sp} = 4.8 \times 10^{-9} < 9 \times 10^{-6}$$

Ion product does exceed K_{sp}, so $CaCO_3$ does precipitate.

(d) $Co(OH)_2$:

$$Co(OH)_2(s) \rightleftharpoons Co^{2+}(aq) + 2OH^-(aq)$$

$$K_{sp} = [Co^{2+}][OH^-]^2 = 2 \times 10^{-16}$$

Test K_{sp} against $[Co^{2+}][OH^-]^2$

$$[Co^{2+}][OH^-]^2 = (0.01)(1 \times 10^{-7})^2 = 1 \times 10^{-16}$$

$$K_{sp} = 2 \times 10^{-16} > 1 \times 10^{-16}$$

The ion product does not exceed K_{sp}; the compound does not precipitate.

10. The *Handbook of Chemistry and Physics* gives solubilities for the compounds listed below in grams per 100 mL of water. Since these compounds are only slightly soluble, assume that the volume does not change on dissolution, and calculate the solubility product for each.

    (a) TlCl, 0.29 g/100 mL
    (b) $Ce(IO_3)_4$, 1.5×10^{-2} g/100 mL

Solution

(a) To find the solubility product, the concentration must be expressed in molarity units. The molar mass of TlCl is $(204.37) + (35.453) = 239.823$ g mol^{-1}. The concentration is

$$\frac{1000 \text{ mL}}{L} \times \frac{0.29 \text{ g}}{100 \text{ mL}} = \frac{2.9 \text{ g}}{L}$$

so

$$\frac{2.9 \text{ g/L}}{239.823 \text{ g mol}^{-1}} = 0.0121 \text{ mol/L} = 0.0121 \text{ M}$$

$$K_{sp} = [Tl^+][Cl^-] = (0.0121)(0.0121) = 1.5 \times 10^{-4}$$

(b) Find the molarity, then put the concentration into the K_{sp}. The molar mass of $Ce(IO_3)_4$ is

$$(140.12) + 4(126.9045) + 12(15.9994) = 839.73 \text{ g mol}^{-1}$$

The concentration is 1.5×10^{-1} g L^{-1} or

$$\frac{1.5 \times 10^{-1} \text{ g L}^{-1}}{839.73 \text{ g mol}^{-1}} = 1.786 \times 10^{-4} \text{ mol/L} = 1.786 \times 10^{-4} \text{ M}$$

Remember that when $Ce(IO_3)_4$ dissociates, there will be 4 times as much IO_3^- ion as Ce^{4+} ion.

$$K_{sp} = [Ce^{4+}][IO_3^-]^4 = (1.786 \times 10^{-4})(4 \times 1.786 \times 10^{-4})^4$$

$$= (1.786 \times 10^{-4})(2.605 \times 10^{-13}) = 4.7 \times 10^{-17}$$

Chapter 18 The Solubility Product Principle   263

12. Calculate the concentrations of ions in a saturated solution of each of the following (see Appendix D for solubility products):

(a) AgI
(b) Ag_2SO_4
(d) $Sr(OH)_2 \cdot 8H_2O$

Solution

(a) The dissolution of AgI produces one each of the ions Ag^+ and I^-.

$$AgI(s) \rightleftharpoons Ag^+(aq) + I^-(aq)$$

$$K_{sp} = [Ag^+][I^-] = 1.5 \times 10^{-16}$$

Let x equal the molar solubility of AgI.

$$x = [Ag^+] = [I^-]$$

$$K_{sp} = (x)(x) = 1.5 \times 10^{-16}$$

$$x^2 = 1.5 \times 10^{-16}$$

$$[Ag^+] = [I^-] = x = 1.2 \times 10^{-8} \text{ M}$$

(b) Dissolution of Ag_2SO_4 occurs as

$$Ag_2SO_4(s) \rightleftharpoons 2Ag^+(aq) + SO_4^{2-}(aq)$$

Let x equal the molar solubility of Ag_2SO_4. Dissolving Ag_2SO_4 produces 2 Ag^+ and 1 SO_4^{2-}, so the concentrations in terms of x are

$$[Ag^+] = 2x$$

$$[SO_4^{2-}] = x$$

$$K_{sp} = [Ag^+]^2[SO_4^{2-}] = 1.18 \times 10^{-5}$$

$$(2x)^2(x) = 4x^3 = 1.18 \times 10^{-5}$$

$$x^3 = 2.95 \times 10^{-6}$$

$$x = 1.43 \times 10^{-2}$$

$$[Ag^+] = 2.86 \times 10^{-2} \text{ M}; \quad [SO_4^{2-}] = 1.43 \times 10^{-2} \text{ M}$$

(d) Dissolution of $Sr(OH)_2 \cdot 8H_2O$ occurs as

$$Sr(OH)_2 \cdot 8H_2O \rightleftharpoons Sr^{2+}(aq) + 2OH^-(aq) + 8H_2O(l)$$

Let x equal the molar solubility of $Sr(OH)_2 \cdot 8H_2O$. Dissolving 1 mol of $Sr(OH)_2 \cdot 8H_2O$ produces 1 mol of Sr^{2+} and 2 mol of OH^-.

$$K_{sp} = [Sr^{2+}][OH^-]^2 = 3.2 \times 10^{-4}$$

The ion concentrations in terms of x are

$$x = [Sr^{2+}]; \quad [OH^-] = 2x$$

$$(x)(2x)^2 = 3.2 \times 10^{-4}$$

$$4x^3 = 3.2 \times 10^{-4}$$

$$x^3 = 8.0 \times 10^{-5}$$

$$[Sr^{2+}] = x = 0.043 = 4.3 \times 10^{-2} \text{ M}$$

$$[OH^-] = 8.6 \times 10^{-2} \text{ M}$$

14. **(a)** Calculate [Ag$^+$] in a saturated aqueous solution of AgBr (K_{sp} = 3.3 × 10^{-13}).

(b) What will [Ag$^+$] be when enough KBr has been added to make [Br$^-$] = 0.050 M?

(c) What will [Br$^-$] be when enough AgNO$_3$ has been added to make [Ag$^+$] = 0.020 M?

Solution

(a) The dissolution of AgBr produces one Ag$^+$ ion for each Br$^-$ ion. Let x equal the molar solubility of AgBr.

$$K_{sp} = 3.3 \times 10^{-13} = [Ag^+][Br^-] = x^2$$

$$x = [Ag^+] = [Br^-] = 5.7 \times 10^{-7} \text{ M}$$

(b) $K_{sp} = 3.3 \times 10^{-13} = [Ag^+][Br^-] = [Ag^+] \, 0.050$

$$[Ag^+] = \frac{3.3 \times 10^{-13}}{0.050} = 6.6 \times 10^{-12} \text{ M}$$

(c) $K_{sp} = 3.3 \times 10^{-13} = [Ag^+][Br^-] = 0.020 \, [Br^-]$

$$[Br^-] = \frac{3.3 \times 10^{-13}}{0.020} = 1.6 \times 10^{-11} \text{ M}$$

17. The first step in the preparation of magnesium metal is the precipitation of Mg(OH)$_2$ from sea water by the addition of Ca(OH)$_2$. The concentration of Mg^{2+}(aq) in sea water is 5.37 × 10^{-2} M. Using the solubility product for Mg(OH)$_2$, calculate the pH at which [Mg^{2+}] is reduced to 5.37 × 10^{-5} M by the addition of Ca(OH)$_2$.

Solution

The reaction is

$$Mg^{2+} + 2OH^- \rightleftharpoons Mg(OH)_2$$

$$K_{sp} = [Mg^{2+}][OH^-]^2 = 5.37 \times 10^{-5}[OH^-]^2 = 1.5 \times 10^{-11}$$

$$[OH^-] = \sqrt{1.5 \times 10^{-11}/5.37 \times 10^{-5}} = \sqrt{2.79 \times 10^{-7}} = 5.285 \times 10^{-4}$$

$$\text{pOH} = -\log[OH^-] = 3.28; \quad \text{pH} = 14.00 - \text{pOH} = 10.72$$

18. Most barium compounds are very poisonous; however, barium sulfate is often administered internally as an aid in the x-ray examination of the lower intestinal tract. This use of BaSO$_4$ is possible because of its insolubility. Calculate the molar solubility of BaSO$_4$ (K_{sp} = 1.08 × 10^{-10}) and the mass of barium present in 1.00 L of water saturated with BaSO$_4$.

Solution

The reaction is

$$BaSO_4 \rightleftharpoons Ba^{2+}(aq) + SO_4^{2-}(aq)$$

There is a one-to-one ratio of ions generated, so that [Ba^{2+}] = [SO$_4^{2-}$] in solution. The solubility product is

$$K_{sp} = [Ba^{2+}][SO_4^{2-}] = x^2 = 1.08 \times 10^{-10}$$

$$x = [Ba^{2+}] = [SO_4^{2-}] = 1.04 \times 10^{-5} \text{ M}$$

Chapter 18 The Solubility Product Principle

The atomic weight of Ba is 137.33 g/mol. The mass of Ba in 1.00 L is

$$1.04 \times 10^{-5} \text{ M} \times 137.33 \text{ g mol}^{-1} = 1.43 \times 10^{-3} \text{ g}$$

19. The solubility product of $CaSO_4 \cdot 2H_2O$ is 2.4×10^{-5}. What mass of this salt will dissolve in 1.0 L of 0.010 M K_2SO_4?

Solution

The amount of $CaSO_4 \cdot 2H_2O$ that dissolves is limited by the presence of a substantial amount of SO_4^{2-} already in solution from the K_2SO_4. This is a common-ion problem. Let x = concentration of Ca^{2+} and of SO_4^{2-} that dissociates.

$$CaSO_4(s) \rightleftharpoons Ca^{2+}(aq) + SO_4^{2-}(aq)$$

$$K_{sp} = [Ca^{2+}][SO_4^{2-}] = 2.4 \times 10^{-5}$$

Substitution of 0.010 M SO_4^{2-} generated from the complete dissociation of 0.010 M K_2SO_4, gives

$$[x][x + 0.010] = 2.4 \times 10^{-5}$$

Here, x cannot be neglected in comparison with 0.010 M; the quadratic equation must be used. In standard form,

$$x^2 + 0.010x - 2.4 \times 10^{-5} = 0$$

$$x = \frac{-0.01 \pm \sqrt{1 \times 10^{-4} + 9.6 \times 10^{-5}}}{2} = \frac{-0.01 \pm 1.4 \times 10^{-2}}{2}$$

Only the positive value will give a meaningful answer.

$$x = 2.0 \times 10^{-3} = [Ca^{2+}]$$

This is also the concentration of $CaSO_4 \cdot 2H_2O$ that has dissolved. The mass of the salt in 1 L is

$$\text{Mass } CaSO_4 \cdot 2H_2O = 2.0 \times 10^{-3} \text{ mol/L} \times 172.16 \text{ g/mol} = 0.34 \text{ g}$$

Note that the presence of the common ion, SO_4^{2-}, has caused a decrease in the concentration of Ca^{2+} that otherwise would be in solution

$$\sqrt{2.4 \times 10^{-5}} = 4.9 \times 10^{-3} \text{ M}$$

20. In one experiment a precipitate of $BaSO_4$ ($K_{sp} = 1.08 \times 10^{-10}$) was washed with 0.100 L of distilled water; in another experiment a precipitate of $BaSO_4$ was washed with 0.100 L of 0.010 M H_2SO_4. Calculate the quantity of $BaSO_4$ that dissolved in each experiment, assuming that the wash liquid became saturated with $BaSO_4$.

Solution

In each case, determine the concentration that dissolves from the solubility product. Since that concentration is per liter, multiply by 0.100 L. In pure water,

$$[Ba^{2+}][SO_4^{2-}] = x^2 = 1.08 \times 10^{-10}$$

$$x = [Ba^{2+}] = [SO_4^{2-}] = 1.04 \times 10^{-5} \text{ M}$$

In 0.100 L the amount dissolved is 1.04×10^{-5} M $\times$ 0.100 L = 1.04×10^{-6} mol. In 0.100 L of 0.010 M H_2SO_4, $K_{sp} = 1.08 \times 10^{-10} = [Ba^{2+}][0.010]$.

$$[Ba^{2+}] = 1.08 \times 10^{-10}/0.010 = 1.08 \times 10^{-8} \text{ M}$$

Thus,

$$[Ba^{2+}] = 1.08 \times 10^{-8} \text{ M} \times 0.100 \text{ L} = 1.08 \times 10^{-9} \text{ mol}$$

22. Calculate the concentration of Sr^{2+} when SrF_2 ($K_{sp} = 3.7 \times 10^{-12}$) starts to precipitate from a solution that is 0.0025 M in F^-.

Solution

Precipitation of SrF_2 will begin when the ion product of the ions Sr^{2-} and F^- exceed the K_{sp} of SrF_2.

$$SrF_2 \rightleftharpoons Sr^{2-}(aq) + 2F^-(aq)$$

$$K_{sp} = [Sr^{2+}][F^-]^2 = 3.7 \times 10^{-12}$$

The maximum concentration of Sr^{2+} that will remain in solution is

$$[Sr^{2+}] = \frac{3.7 \times 10^{-12}}{(0.0025)^2} = 5.9 \times 10^{-7} \text{ M}$$

SOLUBILITY PRODUCTS AND IONIZATION CONSTANTS

25. A solution of 0.060 M $MnBr_2$ is saturated with H_2S ([H_2S] = 0.10 M). What is the minimum pH at which MnS ($K_{sp} = 4.3 \times 10^{-22}$) will precipitate?

Solution

Two equilibria are in competition for the ions and must be considered simultaneously. Precipitation of MnS will occur when the concentration of S^{2-} in conjunction with 0.060 M Mn^{2+} exceeds the K_{sp} of MnS. But the [S^{2-}] must come from the ionization of H_2S as defined by the equilibrium

$$H_2S(aq) \rightleftharpoons 2H^+(aq) + S^{2-}(aq)$$

$$\frac{[H^+]^2[S^{2-}]}{[H_2S]} = K_1K_2(H_2S) = 1.0 \times 10^{-26}$$

Since a saturated solution of H_2S is 0.10 M, this expression becomes

$$[H^+]^2[S^{2-}] = 1.0 \times 10^{-27}$$

From the equilibrium of MnS, the minimum concentration of S^{2-} required to cause precipitation is calculated as

$$MnS(s) \rightleftharpoons Mn^{2+}(aq) + S^{2-}(aq)$$

$$K_{sp} = [Mn^{2+}][S^{2-}] = 4.3 \times 10^{-22}$$

$$[S^{2-}] = \frac{4.3 \times 10^{-22}}{0.060} = 7.17 \times 10^{-21}$$

This amount of S^{2-} will exist in solution at a pH defined by the H_2S equilibrium.

$$[H^+]^2 (7.17 \times 10^{-21}) = 1.0 \times 10^{-27}$$

$$[H^+]^2 = 1.39 \times 10^{-7}$$

$$[H^+] = 3.74 \times 10^{-4}$$

$$pH = -\log[H^+] = 3.4$$

Chapter 18 The Solubility Product Principle   267

27. Using the solubility product, calculate the molar solubility of AgCN (a) in pure water and (b) in a buffer solution with a pH of 3.00.

Solution

(a) In pure water, $K_{sp} = [\text{Ag}^+][\text{CN}^-] = 1.2 \times 10^{-16}$

$[\text{Ag}^+] = [\text{CN}^-] = \sqrt{1.2 \times 10^{-16}} = 1.1 \times 10^{-8}$ M

(b) In a buffer solution the concentration of CN⁻ is controlled by the K_a of HCN. Since the pH = 3.00, $[\text{H}^+] = 1.00 \times 10^{-3}$ M. However, the starting concentration of HCN is not known so the problem must be solved using the simultaneous equilibria involved.

$\text{AgCN}(s) \longrightarrow \text{Ag}^+ + \text{CN}^-$ (K_{sp})

$\text{CN}^- + \text{H}^+ \longrightarrow \text{HCN}$ $(K_2 = 1/K_a)$

Thus,

$\text{AgCN}(s) + \text{H}^+ \longrightarrow \text{Ag}^+ + \text{HCN}$

The equilibrium constant for this reaction is the product of $K_{sp} \times K_2$. Thus

$$K = K_{sp}K_2 = \frac{K_{sp}}{K_a} = \frac{[\text{Ag}^+][\text{HCN}]}{[\text{H}^+]} = \frac{1.2 \times 10^{-16}}{4 \times 10^{-10}} = 3 \times 10^{-7} = \frac{[\text{Ag}^+][\text{HCN}]}{[1.00 \times 10^{-3}]}$$

Because the K_a for HCN is small, essentially all the CN⁻ is in the form of HCN, so let $x = [\text{Ag}^+] = [\text{HCN}]$ if the concentrations of Ag⁺ and CN⁻ initially were approximately equal.

$[\text{Ag}^+] = x = \sqrt{3 \times 10^{-10}} = 2 \times 10^{-5}$

31. (a) What are the concentrations of Ca^{2+} and CO_3^{2-} in a saturated solution of CaCO_3 ($K_{sp} = 4.8 \times 10^{-9}$)?
    (b) What are the concentrations of Ca^{2+} and CO_3^{2-} in a buffer solution with a pH of 4.55 in contact with an excess of CaCO_3?

Solution

(a) The dissolution of CaCO_3 is highly dependent on the pH of the aqueous medium. Assume that association of CO_3^{2-} with H_3O^+ from water is insignificant; this means that the $[\text{Ca}^{2+}]$ at equilibrium equals $[\text{CO}_3^{2-}]$.

$\text{CaCO}_3(s) \rightleftharpoons \text{Ca}^{2+}(aq) + \text{CO}_3^{2-}(aq)$

$K_{sp} = 4.8 \times 10^{-9} = [\text{Ca}^{2+}][\text{CO}_3^{2-}]$

Let x equal the number of moles of CaCO_3 dissolving.

$x = [\text{Ca}^{2+}] = [\text{CO}_3^{2-}]$

$K_{sp} = (x)(x) = 4.8 \times 10^{-9}$

$x = 6.9 \times 10^{-5}$ M $= [\text{Ca}^{2+}] = [\text{CO}_3^{2-}]$

(b) The dissolution of CaCO_3 produces one ion each of Ca^{2+} and CO_3^{2-}. The carbonic acid is a weak acid and CO_3^{2-} will associate with H_3O^+ according to the following reactions:

$\text{CO}_3^{2-} + \text{H}_3\text{O}^+ \rightleftharpoons \text{HCO}_3^- + \text{H}_2\text{O}$ (1)

$\text{HCO}_3^- + \text{H}_3\text{O}^+ \rightleftharpoons \text{H}_2\text{CO}_3 + \text{H}_2\text{O}$ (2)

The concentration of Ca^{2+} at equilibrium equals the total concentration of the species containing carbonate.

$$\text{Solubility} = [Ca^{2+}] = [CO_3^{2-}] + [HCO_3^-] + [H_2CO_3] \quad (3)$$

$$K_{sp}(CaCO_3) = [Ca^{2+}][CO_3^{2-}] = 4.8 \times 10^{-9} \quad (4)$$

The solubility expression involves four unknowns, so four simultaneous equations are needed for solving the expression for $[Ca^{2+}]$. Equations (1) and (2) are merely the reverse of the successive ionizations of H_2CO_3, and the K values for H_2CO_3 can be used.

$$H_2CO_3(aq) \rightleftharpoons H^+(aq) + HCO_3^-(aq) \quad K_1 = 4.3 \times 10^{-7} \quad (5)$$

$$HCO_3^-(aq) \rightleftharpoons H^+(aq) + CO_3^{2-}(aq) \quad K_2 = 7.0 \times 10^{-11} \quad (6)$$

Equations (5) and (6) can be rearranged and expressed in terms of $[CO_3^{2-}]$. At a pH of 4.55 the $[H^+]$ is 2.8×10^{-5}.

$$\frac{[H^+][CO_3^{2-}]}{[HCO_3^-]} = 7.0 \times 10^{-11}$$

$$[HCO_3^-] = \frac{2.8 \times 10^{-5}}{7.0 \times 10^{-11}}[CO_3^{2-}] = 4.0 \times 10^5 [CO_3^{2-}]$$

and $\dfrac{[H^+][HCO_3^-]}{[H_2CO_3]} = 4.3 \times 10^{-7}$

$$[H_2CO_3] = \frac{2.8 \times 10^{-5}}{4.3 \times 10^{-7}}[HCO_3^-]$$

$$= 6.5 \times 10^1 [HCO_3^-] = 6.5 \times 10^1 (4.0 \times 10^5)[CO_3^{2-}]$$

$$= 2.6 \times 10^7 [CO_3^{2-}]$$

These values for $[HCO_3^-]$, $[H_2CO_3]$, and $[CO_3^{2-}]$ can be substituted into Equation (3), which yields

$$[Ca^{2+}] = [CO_3^{2-}] + 4.0 \times 10^5 [CO_3^{2-}] + 2.6 \times 10^7 [CO_3^{2-}]$$

$$[Ca^{2+}] = 2.6 \times 10^7 [CO_3^{2-}]$$

or $[CO_3^{2-}] = \dfrac{[Ca^{2+}]}{2.6 \times 10^7}.$

Substitution into the expression for K_{sp} yields

$$K_{sp} = [Ca^{2+}][CO_3^{2-}] = [Ca^{2+}]\frac{[Ca^{2+}]}{2.6 \times 10^7} = 4.8 \times 10^{-9}$$

$$[Ca^{2+}]^2 = 1.25 \times 10^{-1}$$

$$[Ca^{2+}] = 0.353 \text{ mol/L} = 0.35 \text{ M}$$

$$[CO_3^{2-}] = \frac{[Ca^{2+}]}{2.6 \times 10^7} = \frac{0.353}{2.6 \times 10^7} = 1.4 \times 10^{-8} \text{ M}$$

33. Calculate the concentration of Cd^{2+} resulting from the dissolution of $CdCO_3$ in a solution that is 0.250 M in CH_3CO_2H, 0.375 M in $NaCH_3CO_2$, and 0.010 M in H_2CO_3.

Solution

For $K_{sp}(CdCO_3) = [Cd^{2+}][CO_3^{2-}] = 2.5 \times 10^{-14}$, the amount of CO_3^{2-} is governed by the K_a of H_2CO_3, 4.3×10^{-7}, and the K_a of HCO_3^-, 7×10^{-11}, and

[H⁺] by the K_a of acetic acid. First, calculate the [H⁺] from K_a of acetic acid (HOAc).

$$[H^+] = \frac{K_a[HOAc]}{[OAc^-]} = \frac{1.8 \times 10^{-5}(0.250)}{(0.375)} = 1.2 \times 10^{-5} \text{ M}$$

From this and the K_a for H_2CO_3, calculate $[CO_3^{2-}]$ present. Since [H⁺] is fixed by K_a of acetic acid,

$$K_a(H_2CO_3) = \frac{[H^+][HCO_3^-]}{[H_2CO_3]}$$

$$[HCO_3^-] = \frac{K_a[H_2CO_3]}{[H^+]} = \frac{4.3 \times 10^{-7}[0.010]}{[1.2 \times 10^{-5}]} = 3.58 \times 10^{-4} \text{ M}$$

From $K_a(HCO_3^-)$ we obtain $[CO_3^{2-}]$:

$$K_a = 7 \times 10^{-11} = \frac{[H^+][CO_3^{2-}]}{[HCO_3^-]} = \frac{1.2 \times 10^{-5}[CO_3^{2-}]}{[3.58 \times 10^{-4}]}$$

$$[CO_3^{2-}] = \frac{7 \times 10^{-11} \times 3.58 \times 10^{-4}}{1.2 \times 10^{-5}} = 2.09 \times 10^{-9}$$

From the solubility product,

$$K_{sp} = [Cd^{2+}][CO_3^{2-}] = 2.5 \times 10^{-14} = Cd^{2+}(2.09 \times 10^{-9})$$

$$[Cd^{2+}] = 2.5 \times 10^{-14}/2.09 \times 10^{-9} = 1 \times 10^{-5} \text{ M}$$

34. A volume of 50 mL of 1.8 M NH_3 is mixed with an equal volume of a solution containing 0.95 g of $MgCl_2$. What mass of NH_4Cl must be added to the resulting solution to prevent the precipitation of $Mg(OH)_2$?

Solution

The hydroxide ion concentration in solution depends on two simultaneous equilibria. The maximum allowable [OH⁻] can be calculated from the K_{sp} of $Mg(OH)_2$ based on the $[Mg^{2+}]$.

$$Mg(OH)_2 \rightleftharpoons Mg^{2+}(aq) + 2OH^-(aq) \qquad K_{sp} = 1.5 \times 10^{-11}$$

$$[Mg^{2+}] = [MgCl_2]$$

$$[Mg^{2+}] = \frac{0.95 \text{ g MgCl}_2 \times \dfrac{1 \text{ mol}}{95.211 \text{ g}}}{0.10 \text{ L}} = 0.0998 \text{ M}$$

$$[Mg^{2+}][OH^-]^2 = 1.5 \times 10^{-11}$$

$$[OH^-]^2 = \frac{1.5 \times 10^{-11}}{0.0998} = 1.50 \times 10^{-10}$$

$$[OH^-] = 1.22 \times 10^{-5} \text{ M}$$

The [OH⁻] produced from NH_3 must be suppressed to 1.2×10^{-5} M by buffering the solution through the addition of NH_4Cl. The required $[NH_4^+]$ can be calculated from the equilibrium constant expression for ammonia.

$$NH_3(aq) + H_2O(l) \rightleftharpoons NH_4^+(aq) + OH^-(aq)$$

$$K_b = 1.8 \times 10^{-5} = \frac{[NH_4^+][OH^-]}{[NH_3]}$$

At equilibrium, the [NH$_3$] approximately equals 1.8 M/2 = 0.90 M, since 1.22×10^{-5} is small with respect to 0.90. Therefore, the [NH$_4^+$] is

$$1.8 \times 10^{-5} = \frac{[NH_4^+](1.22 \times 10^{-5})}{0.90}$$

$$[NH_4^+] = \frac{(0.90)(1.8 \times 10^{-5})}{1.22 \times 10^{-5}} = 1.33 \text{ M}$$

Mass NH$_4$Cl required = (1.33 mol/L)(53.5 g/mol)(0.10 L) = 7.1 g

SEPARATION OF IONS

35. A solution is 0.15 M in both Pb^{2+} and Ag$^+$. If Cl$^-$ is added to this solution, what is [Ag$^+$] when PbCl$_2$ begins to precipitate?

Solution

Lead and silver both form slightly soluble chlorides. The ion with less molar solubility will precipitate first. This precipitation will continue until the chloride ion increases sufficiently to exceed the K_{sp} of the other chloride. Since both Ag$^+$ and Pb^{2+} must exist simultaneously in solution, both K_{sp} equilibria must be satisfied.

$$PbCl_2(s) \rightleftharpoons Pb^{2+}(aq) + 2Cl^-(aq) \qquad K_{sp} = 1.7 \times 10^{-5}$$

$$AgCl(s) \rightleftharpoons Ag^+(aq) + Cl^-(aq) \qquad K_{sp} = 1.8 \times 10^{-10}$$

First calculate the maximum chloride ion concentration that can be present without causing the precipitation of Pb^{2+}. Given that [Pb^{2+}] equals 0.15 M, the [Cl$^-$] maximum is

$$[Pb^{2+}][Cl^-]^2 = 1.7 \times 10^{-5}$$

$$[Cl^-]^2 = \frac{1.7 \times 10^{-5}}{0.15} = 1.333 \times 10^{-4}$$

$$[Cl^-] = 1.06 \times 10^{-2}$$

Precipitation of AgCl will continue until the chloride ion concentration equals 1.06×10^{-2} M. At this point the [Ag$^+$] is controlled by the K_{sp} of AgCl. Substituting this [Cl$^-$] into K_{sp} of AgCl gives

$$[Ag^+][1.06 \times 10^{-2}] = 1.8 \times 10^{-10}$$

$$[Ag^+] = \frac{1.8 \times 10^{-10}}{1.06 \times 10^{-2}} = 1.7 \times 10^{-8} \text{ M}$$

37. (a) With what volume of water must a precipitate containing NiCO$_3$ be washed to dissolve 0.100 g of this compound? Assume that the wash water becomes saturated with NiCO$_3$ ($K_{sp} = 1.36 \times 10^{-7}$).
    (b) If the NiCO$_3$ were a contaminant in a sample of CoCO$_3$ ($K_{sp} = 1.0 \times 10^{-12}$), what mass of CoCO$_3$ would have been lost? Keep in mind that both NiCO$_3$ and CoCO$_3$ dissolve in the same solution.

Chapter 18 The Solubility Product Principle   271

Solution

(a) Calculate the molar solubility. Then calculate the number of grams per liter.

$$K_{sp} = 1.36 \times 10^{-7} = [Ni^{2+}][CO_3^{2-}] = x^2$$

$$x = [Ni^{2+}] = [CO_3^{2-}] = 3.688 \times 10^{-4} \text{ M}$$

Molar mass, $NiCO_3 = 118.71$ g mol^{-1}

$$\text{g NiCO}_3 = 118.71 \text{ g/mol} \times 3.688 \times 10^{-4} \text{ mol/L} = 0.0438 \text{ g/L}$$

To contain 0.1 g, $0.1 \text{ g}/0.0438 \text{ g L}^{-1} = 2.28$ L

(b) During the process of removal of $NiCO_3$, some $CoCO_3$ would be lost. The $[CO_3^{2-}]$ is controlled by the amount found in (a). From the solubility product for $CoCO_3$

$$K_{sp} = 1.0 \times 10^{-12} = [Co^{2+}][3.688 \times 10^{-4}]$$

$$[Co^{2+}] = 2.71 \times 10^{-9} \text{ M}$$

The molar mass of $CoCO_3$ is 118.94 g/mol.

$$\text{g CoCO}_3 \text{ in 2.28 L} = 118.94 \text{ g/mol} \times 2.71 \times 10^{-9} \text{ mol/L} \times 2.28 \text{ L}$$

$$= 7.3 \times 10^{-7} \text{ g}$$

FORMATION CONSTANTS OF COMPLEX IONS

41. Calculate the concentration of Ni^{2+} in a 1.0 M solution of $[Ni(NH_3)_6](NO_3)_2$.

Solution

$$Ni^{2+} + 6NH_3 \rightleftharpoons [Ni(NH_3)_6]^{2+} \qquad K_f = 1.8 \times 10^8$$

$$1.8 \times 10^8 = \frac{[Ni(NH_3)_6]^{2+}}{[Ni^{2+}][NH_3]^6} = \frac{(1.0 - x)}{x(6x)^6}$$

$$1.8 \times 10^8 (46656 \, x^7) = 1.0 - x$$

$$8.40 \times 10^{12} \, x^7 = 1.0 - x$$

Since x is small in comparison to 1.0, drop x.

$$8.40 \times 10^{12} \, x^7 = 1.0$$

$$x^7 = 1.19 \times 10^{-13}$$

$$x = 0.014 \text{ M}$$

43. Calculate the silver ion concentration, $[Ag^+]$, of a solution prepared by dissolving 1.00 g of $AgNO_3$ and 10.0 g of KCN in sufficient water to make 1.00 L of solution.

Solution

First calculate the molarity of the two solutions.

Molar mass ($AgNO_3$) = 169.873 g/mol

Molar mass (KCN) = 65.116 g/mol

$$\text{Molarity (AgNO}_3) = \frac{1.00 \text{ g}/169.873 \text{ g mol}^{-1}}{1.00 \text{ L}} = 5.887 \times 10^{-3} \text{ M}$$

$$\text{Molarity (KCN)} = \frac{10.0 \text{ g}/65.11 \text{ g mol}^{-1}}{1.00 \text{ L}} = 0.1536 \text{ M}$$

It is obvious that there is sufficient CN⁻ to cause the equilibrium

$$Ag^+ + 2CN^- \rightleftharpoons [Ag(CN)_2]^- \qquad K_f = 1 \times 10^{20}$$

In aqueous solution the concentration of CN⁻ is controlled by the equilibrium

$$CN^- + H_2O \rightleftharpoons HCN + OH^-$$

$$K_b = K_w/K_a = 1 \times 10^{-14}/4 \times 10^{-10} = 2.5 \times 10^{-5}$$

Since there is so much CN⁻ compared to Ag⁺, the amount of CN⁻ used to complex Ag⁺ may be ignored in the calculation of CN⁻ present from the equilibrium. Let x = concentration of CN⁻ that forms HCN.

$$\frac{[HCN][OH^-]}{[CN^-]} = 2.5 \times 10^{-5} = \frac{x^2}{[0.1536 - x]}$$

Making the assumption that x is small in comparison to 0.1536, we obtain $x = 1.96 \times 10^{-3}$ M = [HCN]; [CN⁻] = 0.1516 M. From the complex formation equation, we have

$$[Ag^+] = \frac{[Ag(CN)_2^-]}{[CN^-]^2 K_f} = \frac{5.887 \times 10^{-3} \text{ M}}{(0.1516)^2 (1 \times 10^{20})} = 3 \times 10^{-21} \text{ M}$$

45. Sometimes equilibria for complex ions are described in terms of dissociation constants, K_d. For the complex ion AlF_6^{3-} the dissociation reaction is

$$AlF_6^{3-} \rightleftharpoons Al^{3+} + 6F^-$$

and

$$K_d = \frac{[Al^{3+}][F^-]^6}{[AlF_6^{3-}]} = 2 \times 10^{-24}$$

(a) Calculate the value of the formation constant, K_f for AlF_6^{3-}.
(b) Using the value of the formation constant for the complex ion $Co(NH_3)_6^{2+}$, calculate the dissociation constant.

Solution

(a) For the formation reaction

$$Al^{3+} + 6F^- \longrightarrow AlF_6^{3-}$$

$$K_f = \frac{[AlF_6^{3-}]}{[Al^{3+}][F^-]^6} = 1/K_d = \frac{1}{2 \times 10^{-24}} = 5 \times 10^{23}$$

(b)

$$K_f = \frac{[Co(NH_3)_6]^{2+}}{[Co^{3+}][NH_3]^6} = 8.3 \times 10^4$$

$$K_d = \frac{1}{K_f} = \frac{1}{8.3 \times 10^4} = 1.2 \times 10^{-5}$$

Chapter 18 The Solubility Product Principle

47. Calculate the cadmium ion concentration [Cd^{2+}], in a solution prepared by mixing 0.100 L of 0.0100 M $Cd(NO_3)_2$ with 0.150 L of 0.100 M $NH_3(aq)$.

Solution

Cadmium ions associate with ammonia molecules in solution to form the complex ion, $[Cd(NH_3)_4]^{2+}$, which is defined by the following equilibrium:

$$Cd^{2+}(aq) + 4NH_3(aq) \rightleftharpoons [Cd(NH_3)_4]^{2+}(aq) \qquad K_f = 4.0 \times 10^6$$

The formation of the complex requires 4 mol of NH_3 for each mol of Cd^{2+}. First calculate the initial amounts of Cd^{2+} and of NH_3 available for association.

$$[Cd^{2+}] = \frac{(0.100 \text{ L})(0.0100 \text{ mol/L})}{0.250 \text{ L}} = 4.00 \times 10^{-3} \text{ M}$$

$$[NH_3] = \frac{(0.150 \text{ L})(0.100 \text{ mol/L})}{0.250 \text{ L}} = 6.00 \times 10^{-2} \text{ M}$$

For the reaction, 4.00×10^{-3} mol/L of Cd^{2+} would require $4(4.00 \times 10^{-3}$ mol/L$)$ of NH_3 or 1.6×10^{-2} mol. Due to the large value of K_f and the substantial excess of NH_3, it can be assumed that the reaction goes to completion with only a small amount of the complex dissociating to form the ions. After reaction, concentrations of the species in solution are

$$[NH_3] = 6.00 \times 10^{-2} \text{ mol/L} - 1.6 \times 10^{-2} \text{ mol/L} = 4.4 \times 10^{-2} \text{ M}$$

$[Cd^{2+}]$ = equilibrium concentration from the dissociation of the complex

$$[Cd(NH_3)_4]^{2+} = 4.00 \times 10^{-3} \text{ M} - x$$

At equilibrium: Let x = moles of $Cd(NH_3)_4^{2+}$ that dissociate.

$$x = [Cd^{2+}]$$

$$[NH_3] = 4.4 \times 10^{-2} + 4x$$

$$K_f = 4.0 \times 10^6 = \frac{[Cd(NH_3)_4^{2+}]}{[Cd^{2+}][NH_3]^4}$$

$$4.0 \times 10^6 = \frac{(4.00 \times 10^{-3} - x)}{(x)(4.4 \times 10^{-2} + 4x)^4}$$

Since x is expected to be about the same size as the number it is subtracted from, the entire expression must be expanded and solved, in this case by successive approximations. We have

$$4.0 \times 10^6 x (4.4 \times 10^{-2} + 4x)^4 = 4.00 \times 10^{-3} - x$$

$$4.0 \times 10^6 x (3.75 \times 10^{-6} + 1.36 \times 10^{-3}x + 0.186x^2 + 9.15x^3 + 64x^4) = 4.00 \times 10^{-3} - x$$

$$16x + 5540x^2 + 7.44 \times 10^5 x^3 + 3.66 \times 10^7 x^4 + 2.56 \times 10^8 x^5 = 4.00 \times 10^{-3}$$

Using 2.50×10^{-4} gives $4.36 \times 10^{-3} - 4.00 \times 10^{-3} = 0$. Using 2.40×10^{-4} gives $3.98 \times 10^{-3} - 4.00 \times 10^{-3} = 0$. Thus 2.40×10^{-4} is close enough to the true value of x. If the approximation to drop $4x$ in comparison to 4.4×10^{-2} is made, the value is 2.35×10^{-4}.

DISSOLUTION OF PRECIPITATES

49. Calculate the minimum concentration of ammonia needed in 1.0 L of solution to dissolve 3.0×10^{-3} mol of silver bromide.

Solution

Two simultaneous equilibria are involved:

$$AgBr(s) + 2NH_3(aq) \rightleftharpoons [Ag(NH_3)_2]^+(aq) + Br^-(aq) \quad (1)$$

$$K_f = \frac{[Ag(NH_3)_2^+]}{[Ag^+][NH_3]^2} = 1.6 \times 10^{-7} \quad (2)$$

and

$$AgBr(s) \rightleftharpoons Ag^+(aq) + Br^-(aq)$$

$$K_{sp} = [Ag^+][Br^-] = 3.3 \times 10^{-13} \quad (3)$$

Express $[Ag^+]$ in terms of Br^- and K_{sp}, then substitute into Equation (2).

$$[Ag^+] = \frac{3.3 \times 10^{-13}}{[Br^-]}$$

$$\frac{[Ag(NH_3)_2^+][Br^-]}{3.3 \times 10^{-13}[NH_3]^2} = 1.6 \times 10^{-7} \quad (4)$$

When 3.0×10^{-3} mol of silver bromide dissolves, $[Br^-] = 3.0 \times 10^{-3}$ mol, and $[Ag(NH_3)_2]^+ = [Br^-]$. Substitution into Equation (4) gives, after rearrangement,

$$[NH_3]^2 = \frac{[Ag(NH_3)_2^+][Br^-]}{(3.3 \times 10^{-13})(1.6 \times 10^7)} = \frac{(3.0 \times 10^{-3})^2}{5.28 \times 10^{-6}} = 1.70 \, M^2$$

$$[NH_3] = 1.3 \, M$$

52. (a) Which of the following slightly soluble compounds will have a solubility greater than that calculated from its solubility product due to reaction of the anion with water? $PbCO_3$, Tl_2S, CuI, $CoSO_3$, $KClO_4$, $PbCl_2$.
(b) For which compound in part (a) will reaction of the anion with water be most extensive?

Solution

(a) When the anion from a weak acid dissociates from a precipitate, it is removed from solution to some extent by its reaction with water, thus allowing the solubility of the precipitate to increase from what is calculated from the solubility product. Therefore, $CoSO_3$, $PbCO_3$, and Tl_2S have increased solubility because SO_3^{2-}, CO_3^{2-}, and S^{2-} are anions from weak acids.

(b) The anion from the weakest acid will react most extensively with water. Therefore, since H_2S is the weakest acid formed, Tl_2S will have the most reaction with water.

56. A roll of 35-mm black and white film contains about 0.27 g of unexposed AgBr before developing. What mass of $Na_2S_2O_3 \cdot 5H_2O$ (hypo) in 1.0 L of developer is required to dissolve the AgBr as $Ag(S_2O_3)_2^{3-}$ ($K_f = 4.7 \times 10^{13}$)?

Chapter 18 The Solubility Product Principle

Solution

The reaction is governed by two equilibria, both of which must be satisfied.

$$AgBr \rightleftharpoons Ag^+ + Br^- \quad K_{sp} = 3.3 \times 10^{-13}$$

$$Ag^+ + 2S_2O_3^{2-} \rightleftharpoons Ag(S_2O_3)_2^{3-} \quad K_f = 4.7 \times 10^{13}$$

The overall equilibrium is obtained by adding the two equations and multiplying their Ks.

$$\frac{[Ag(S_2O_3)_2^{3-}][Br^-]}{[S_2O_3^{2-}]^2} = 15.51$$

If all the Ag is to be dissolved, the concentration of the complex is the molar concentration of AgBr.

mol. wt (AgBr) = 187.772 g/mol

moles present = 0.27 g AgBr/187.772 g mol^{-1} = 1.44×10^{-3} mol

Let $x = [S_2O_3^{2-}]$. Then

$$\frac{(1.44 \times 10^{-3})(1.44 \times 10^{-3})}{x^2} = 15.51$$

$$x^2 = 1.337 \times 10^{-7}$$

$$x = 3.66 \times 10^{-4} \text{ M} = [S_2O_3^{2-}]$$

Mol. wt Na$_2$S$_2$O$_3 \cdot$ 5H$_2$O = 248.13 g/mol. The total [S$_2$O$_3^{2-}$] needed is

$$2(1.44 \times 10^{-3}) + 3.66 \times 10^{-4} = 3.246 \times 10^{-3} \text{ mol}$$

g (hypo) = 3.246×10^{-3} mol $\times$ 248.13 g/mol = 0.81 g

ADDITIONAL EXERCISES

58. Calculate [HgCl$_4^{2-}$] in a solution prepared by adding 8.0×10^{-3} mol of NaCl to 0.100 L of a 0.040 M HgCl$_2$ solution.

Solution

The process is governed by the K_f of [HgCl$_4^{2-}$].

$$K_f = \frac{[HgCl_4^{2-}]}{[Hg^{2+}][Cl^-]^4} = 1.2 \times 10^{15}$$

Because of the very large value of K_f, almost all of the Hg^{2+} will be in the form of [HgCl$_4^{2-}$]. Let x = concentration of Hg^{2+}. Then

$$\frac{0.040 - x}{x(2x + 0.080)^4} = 1.2 \times 10^{15}$$

Since x is small, drop it in comparison to the additive terms in numerator and denominator.

$$\frac{0.040}{x(0.080)^4} = 1.2 \times 10^{15}$$

$$x = \frac{0.040}{(0.080)^4(1.2 \times 10^{15})} = 8.14 \times 10^{-13} = [Hg^{2+}]$$

Therefore, all the Hg^{2+} is in the form $[HgCl_4^{2-}]$, and its concentration is 0.040 M.

60. Even though $Ca(OH)_2$ is an inexpensive base, its limited solubility restricts its use. What is the pH of a saturated solution of $Ca(OH)_2$?

Solution

Let x = concentration of Ca; then

$$K_{sp} = [Ca^{2+}][OH^-]^2 = (x)(2x)^2 = 7.9 \times 10^{-6}$$

$$4x^3 = 7.9 \times 10^{-6}$$

$$x^3 = 1.975 \times 10^{-6}$$

$$x = 0.0125 = [Ca^{2+}]$$

$$[OH^-] = 0.0250 \text{ M}; \quad pOH = 1.60; \quad pH = 12.40$$

61. About 50% of urinary calculi (kidney stones) consist of calcium phosphate, $Ca_3(PO_4)_2$. The normal midrange calcium content excreted in the urine is 0.10 g of Ca^{2+} per day. The normal midrange amount of urine passed may be taken as 1.4 L per day. What is the maximum concentration of phosphate ion possible in urine before a calculus begins to form?

Solution

The dissolution of $Ca_3(PO_4)_2$ yields

$$Ca_3(PO_4)_2(s) \rightleftharpoons 3Ca^{2+}(aq) + 2PO_4^{3-}(aq)$$

Given the concentration of Ca^{2+} in solution, the maximum $[PO_4^{3-}]$ can be calculated by using the K_{sp} expression for $Ca_3(PO_4)_2$.

$$K_{sp} = 1 \times 10^{-25} = [Ca^{2+}]^3[PO_4^{3-}]^2$$

$$[Ca^{2+}]_{urine} = \frac{0.10 \text{ g} \cdot \frac{1 \text{ mol}}{40.08 \text{ g}}}{1.4 \text{ L}} = 1.8 \times 10^{-3} \text{ M}$$

$$[PO_4^{3-}]^2 = \frac{1 \times 10^{-25}}{(1.8 \times 10^{-3})^3} = 1.7 \times 10^{-17}$$

$$[PO_4^{3-}] = 4 \times 10^{-9} \text{ M}$$

63. The calcium ions in human blood serum are necessary for coagulation. In order to prevent coagulation when a blood sample is drawn for laboratory tests, an anticoagulant is added to the sample. Potassium oxalate, $K_2C_2O_4$, can be used as an anticoagulant because it removes the calcium as a precipitate of $CaC_2O_4 \cdot H_2O$. In order to prevent coagulation, it is necessary to remove all but 1.0% of the Ca^{2+} in serum. If normal blood serum with a buffered pH of 7.40 contains 9.5 mg of Ca^{2+} per 100 mL, what mass of $K_2C_2O_4$ is required to prevent the coagulation of a 10-mL blood sample that is 55% serum by volume? [All volumes are accurate to two significant figures. Note that the volume of fluid (serum) in a 10-mL blood sample is 5.5 mL. Assume that the K_{sp} value for CaC_2O_4 in serum is the same as in water.]

Chapter 18 The Solubility Product Principle

Solution

Although oxalic acid is a weak acid, the oxalate ions, $C_2O_4^{2-}$, do not associate significantly with H_3O^+ at the pH of blood. The amount of $K_2C_2O_4$ required equals the equivalent of 99% of the available Ca^{2+} plus the amount needed to maintain the equilibrium defined by the solubility product.

$$K_{sp}(CaC_2O_4) = [Ca^{2+}][C_2O_4^{2-}] = 2.27 \times 10^{-9}$$

The amount of available Ca^{2+} in 5.5 mL of serum is

$$\text{mol } Ca^{2+} = 5.5 \text{ mL} \times \frac{9.5 \text{ mg}}{100 \text{ mL}} \times \frac{1.0 \text{ g}}{1000 \text{ mg}} \times \frac{1 \text{ mol}}{40.08 \text{ g}}$$

$$= 1.304 \times 10^{-5} \text{ mol}$$

$$\text{mol } Ca^{2+} \text{ to be precipitated} = (0.99)(1.304 \times 10^{-5} \text{ mol})$$

$$= 1.291 \times 10^{-5} \text{ mol}$$

At equilibrium the $[Ca^{2+}]$ is 1% of the original:

$$[Ca^{2+}] = \frac{(0.01)(1.304 \times 10^{-5} \text{ mol})}{0.0055 \text{ L}} = 2.371 \times 10^{-5} \text{ mol/L}$$

From the K_{sp} expression,

$$[C_2O_4^{2-}] = \frac{2.27 \times 10^{-9}}{2.371 \times 10^{-5}} = 9.574 \times 10^{-5} \text{ mol/L}$$

$$\text{mol } C_2O_4^{2-} \text{ in 5.5 mL} = 9.574 \times 10^{-5} \text{ mol/L} \times 0.0055 \text{ L}$$

$$= 5.27 \times 10^{-7} \text{ mol}$$

$$\text{Total moles } C_2O_4^{2-} \text{ required} = 5.266 \times 10^{-7} \text{ mol} + 1.291 \times 10^{-5} \text{ mol}$$

$$= 1.34 \times 10^{-5} \text{ mol}$$

$$\text{Mass } K_2C_2O_4 = 1.34 \times 10^{-5} \text{ mol} \times 166.2162 \text{ g/mol} = 2.2 \times 10^{-3} \text{ g}$$

65. The pH of a normal urine sample is 6.30, and the total phosphate concentration

$$[PO_4^{3-}] + [HPO_4^{2-}] + [H_2PO_4^-] + [H_3PO_4] = 0.020 \text{ M}$$

What is the minimum concentration of Ca^{2+} necessary to induce calculus formation? (See exercise 61 for additional information.)

Solution

The concentration of Ca^{2+} depends on the concentration of PO_4^{3-} in solution. But the $[PO_4^{3-}]$ is dependent on pH and the H_3PO_4 equilibrium. Because the total phosphate concentration is 0.020 M, the $[PO_4^{3-}]$ must be calculated from the three expressions involving H_3PO_4. These expressions, solved in terms of $[PO_4^{3-}]$, yield a ratio of $[PO_4^{3-}]$ to the other components.

$$H_3PO_4 \rightleftharpoons H^+ + H_2PO_4^- \qquad K_1 = 7.5 \times 10^{-3} \qquad (1)$$

$$H_2PO_4^- \rightleftharpoons H^+ + HPO_4^{2-} \qquad K_2 = 6.2 \times 10^{-8} \qquad (2)$$

$$HPO_4^{2-} \rightleftharpoons H^+ + PO_4^{3-} \qquad K_3 = 3.6 \times 10^{-13} \qquad (3)$$

At a pH of 6.3, the [H$^+$] = 5.01 × 10^{-7}. For Equation (3),

$$\frac{[H^+][PO_4^{3-}]}{[HPO_4^{2-}]} = 3.6 \times 10^{-13}$$

$$[HPO_4^{2-}] = \frac{5.01 \times 10^{-7}}{3.6 \times 10^{-13}}[PO_4^{3-}] = 1.39 \times 10^6 [PO_4^{3-}]$$

For Equation (2),

$$\frac{[H^+][HPO_4^{2-}]}{[H_2PO_4^-]} = 6.2 \times 10^{-8}$$

$$[H_2PO_4^-] = \frac{5.01 \times 10^{-7}}{6.2 \times 10^{-8}}[HPO_4^{2-}]$$

$$= 8.08(1.39 \times 10^6)[PO_4^{3-}] = 1.12 \times 10^7 [PO_4^{3-}]$$

For Equation (1),

$$\frac{[H^+][H_2PO_4^-]}{[H_3PO_4]} = 7.5 \times 10^{-3}$$

$$[H_3PO_4] = \frac{5.01 \times 10^{-7}}{7.5 \times 10^{-3}}[H_2PO_4^-]$$

$$= 6.7 \times 10^{-5}(1.12 \times 10^7)[PO_4^{3-}] = 750[PO_4^{3-}]$$

[PO$_4^{3-}$] : [HPO$_4^{2-}$] : [H$_2$PO$_4^-$] : [H$_3$PO$_4$]
1 : 1.39 × 10^6 : 1.12 × 10^7 : 750

The fraction of PO$_4^{3-}$ in solution is

$$\frac{[PO_4^{3-}]}{[PO_4^{3-}] + [HPO_4^{2-}] + [H_2PO_4^-] + [H_3PO_4]} = \frac{1}{1.26 \times 10^7} = 7.94 \times 10^{-8}$$

The concentration of PO$_4^{3-}$ in urine at pH 6.3 is

$$[PO_4^{3-}] = 7.94 \times 10^{-8}(0.020) = 1.59 \times 10^{-9}$$

Use the K_{sp} of Ca$_3$(PO$_4$)$_2$ to calculate the minimum [Ca^{2+}] present. For the reaction Ca$_3$(PO$_4$)$_2$ ⇌ 3Ca^{2+} + 2PO$_4^{3-}$,

$$K_{sp} = [Ca^{2+}]^3[PO_4^{3-}]^2$$

$$[Ca^{2+}]^3 = \frac{K_{sp}}{[PO_4^{3-}]^2} = \frac{1 \times 10^{-25}}{(1.59 \times 10^{-9})^2} = 3.96 \times 10^{-8} \text{ M}$$

$$[Ca^{2+}] = (3.96 \times 10^{-8})^{1/3} = 3.4 \times 10^{-3} \text{ M}$$

66. Magnesium metal (a component of alloys used in aircraft and a reducing agent used in the production of uranium, titanium, and other active metals) is isolated from sea water by the following sequence of reactions:

$$Mg^{2+}(aq) + Ca(OH)_2(aq) \rightarrow Mg(OH)_2(s) + Ca^{2+}(aq)$$

$$Mg(OH)_2(s) + 2HCl \rightarrow MgCl_2(aq) + 2H_2O$$

$$MgCl_2(l) \xrightarrow{\text{Electrolysis}} Mg(s) + Cl_2(g)$$

Sea water has a density of 1.026 g/cm^3 and contains 1272 parts per million of magnesium as Mg^{2+}(aq) by mass. What mass, in kilograms, of Ca(OH)$_2$ is required to precipitate 99.9% of the magnesium in 1.00 × 10^3 L of sea water?

Chapter 18 The Solubility Product Principle

Solution

Calculate the amount of Mg^{2+} present in sea water, then use K_{sp} to calculate amount of $Ca(OH)_2$ required to precipitate the magnesium.

$$g\ Mg = 1.00 \times 10^3\ L \times 1000\ cm^3/1\ L \times 1.026\ g/cm^3 \times 1272\ ppm \times 10^{-6}/ppm$$
$$= 1.305 \times 10^3$$

The concentration is 1.305 g/L. If 99.9% is to be recovered, 0.999×1.305 g/L = 1.304 g/L will be obtained. The molar concentration is

$$\frac{1.304\ g/L}{24.305\ g/mol} = 0.05365\ M$$

Since the $Ca(OH)_2$ reacts with Mg^{2+} on a 1:1 mol basis, the amount of $Ca(OH)_2$ required to precipitate 99.9% of the Mg^{2+} in 1 liter is

$$0.05365\ M \times 74.09\ g/mol\ Ca(OH)_2 = 3.97\ g/L$$

For treatment of 1000 L, $1000\ L \times 3.97\ g/L = 3.97 \times 10^3\ g = 3.97$ kg. However, additional $[OH^-]$ must be added to maintain the equilibrium

$$Mg(OH)_2 \rightleftharpoons Mg^2 + 2OH^- = K_{sp} = 1.5 \times 10^{-11}$$

When the initial 1.305 g/L is reduced to 0.001 of the original, the molarity is calculated as

$$[(0.001 \times 1.305\ g/24.305\ g/mol] = 5.369 \times 10^{-5}\ M$$

The added amount of OH^- required is found from the solubility product.

$$[Mg^{2+}][OH^-]^2 = (5.369 \times 10^{-5})[OH^-]^2 = 1.5 \times 10^{-11} = K_{sp}$$
$$[OH^-] = 5.29 \times 10^{-4}$$

Thus an additional $\frac{1}{2} \times 5.29 \times 10^{-4}$ mol (2.65×10^{-4} mol) of $Ca(OH)_2$ per liter is required to supply the OH^-.

$$\text{For 1000 L}\quad 2.65 \times 10^{-4}\ mol/L \times 1.00 \times 10^3\ L \times \frac{74.0946\ g}{mol\ Ca(OH)_2} = 20\ g$$

The total $Ca(OH)_2$ required is 3.97 kg + 0.020 kg = 3.99 kg.

19
Chemical Thermodynamics

INTRODUCTION

The beginnings of modern thermodynamics can be traced to Benjamin Thompson, better known as Count Rumford of the Holy Roman Empire. His observations of the boring of cannon barrels in the early 1800s convinced him that the heat liberated in the machining resulted from the dissipation of mechanical work. Later, in the 1840s, James Joule made quantitative measurements of the conversion of mechanical energy into heat. His value was close to the modern value of the equivalency of heat and work, which is 1 calorie of heat = 4.184 joules of work.

This interconvertibility of heat and work makes it possible to define energy as the ability to produce heat or to do work. The first law of thermodynamics interrelates these two quantities and is basically an outgrowth of the human experience that energy can neither be created nor destroyed. Mathematically stated, the first law is

$$\Delta E = q - w$$

where ΔE is the change in internal energy, q is the heat absorbed by the system, and w is the work done *by* the system

This concept is important in chemistry, because chemists generally are interested in three basic questions that ultimately deal with energy: (1) Will two or more substances react? (2) If a reaction does occur, what is the associated energy change? (3) If the reaction occurs, what will be the equilibrium concentrations of the reactants and products?

The application of the first law of thermodynamics to chemical systems is immediately evident if a reaction is considered involving a phase change, where the work done may be ignored. Thus ice at 0°C may be converted to water at 0°C solely by the addition of heat q. In this reaction the energy content of ice has increased from that initially present by the amount of heat necessary to make this transition. This can be written

$$\Delta E = q - w = q - 0 = q$$

The heat change in this reaction is therefore the change in the internal energy of the system.

In general, during the course of *any* chemical reaction, the internal energy of the system will change by a specific amount because of a change in heat or work. The enthalpy (H), a function related to the internal energy, is introduced to simplify the study of the energy changes that occur in chemical reactions. The study of these heat effects is called *thermochemistry* and constitutes one important aspect of thermodynamics. This field answers the second question above.

Chapter 19 Chemical Thermodynamics

In order to answer the other two questions, two more functions must be introduced. These are the result of the study of fundamental laws in nature, namely: (1) systems tend toward a state of maximum disorder; and (2) systems tend to attain a state of minimum potential energy.

The first of these functions, entropy (S), is historically an outgrowth of the study of steam engines by the Frenchman Sadi Carnot. Carnot's study dates from 1824, but it was Clausius who first introduced the term *entropy* in 1840. Entropy, a measure of the disorder in a system, when properly combined with the enthalpy under the two constraints above, leads to a new function that conveniently allows us to determine whether a reaction will occur. The American J. Willard Gibbs introduced this new function, which was originally known as the *free energy,* but it is now also called the *Gibbs energy* (G).

The free energy or Gibbs energy is an extremely powerful tool, for it not only allows us to determine whether a reaction will occur based only on the knowledge of the state of the system, but it also allows us to predict the concentrations of the reacting substances at equilibrium through the expression

$$\log K = \frac{\Delta G^\circ}{-2.303\ RT}$$

where K is the equilibrium constant. Thus the three questions of interest to chemists posed earlier can be answered through the application of several simple concepts.

The problems in this chapter deal with the calculation of internal energy, enthalpy, entropy, and free energy or Gibbs energy changes as well as the determination of the equilibrium constant.

READY REFERENCE

(Many definitions from Chapter 4 relate to this material but will not be repeated here.)

Chemical thermodynamics (19.1)   That branch of chemistry that studies the energy transformations and transfers that accompany chemical and physical changes.

Entropy (S) (19.7)   Entropy is a measure of the order or randomness of a system. The smaller the value of the entropy, the more ordered the system; the larger the entropy, the greater the disorder or randomness of the system. Entropy is also a function of temperature; it decreases as temperature decreases and increases as temperature increases. The importance of entropy lies in our ability to predict the direction of a chemical process if both the entropy of the system and the entropy of the surroundings are known.

First law of thermodynamics (19.3)   A statement of the law of conservation of energy: The total amount of energy in the universe is constant. Mathematically, $\Delta E = E_2 - E_1 = q - w$. Here ΔE is the internal energy change of the system due to a change in state, q is heat, and w is work.

Free energy change (Gibbs energy change) (ΔG) (19.9)   Perhaps the most useful function of thermodynamics, this is the maximum amount of useful work that can be accomplished by a reaction at constant temperature and pressure. The free energy can also be used to predict the direction of a chemical process using only information about the system. These predictions are possible because reactions tend to proceed to a state of maximum disorder (positive ΔS) and minimum energy (negative ΔH). The sign that accompanies ΔG derived from the free energy expression, $\Delta G = \Delta H - T\Delta S$, indicates whether a reaction is spontaneous or not. A negative sign indicates a spontaneous reaction as written, a positive value indicates a nonspontaneous reaction, and a value of zero indicates a reaction at equilibrium.

Relation of the free energy change ΔG to the equilibrium constant K_e The relation between the standard free energy and the equilibrium constant is given by the equation

$$\Delta G° = -RT \ln K_e$$

where the term $\ln K_e$ is the natural or Naperian logarithm of K_e. Natural logarithms (ln) occur in calculus and are defined in terms of the base e, where $e = 2.71828$. If one is interested in the ln of a number x, x can be expressed as e raised to some power a. Thus,

$$x = e^a$$

Then a is called the natural logarithm of x and

$$a = \ln x$$

Tables are available in handbooks to calculate a. However, it is more customary to work with numbers to the base 10. These are called common or Briggsian logarithms (log) and written

$$x = 10^b$$

where b is called the common logarithm of x. This is expressed as $b = \log x$. (If you are unfamiliar with the fundamental rules for using logs, you should consult a precalculus textbook.)

A relation between these two bases can be derived so that equations written in terms of natural logarithms can be expressed in terms of common logarithms. This is done in the following manner. Since

$$x = e^a \quad \text{and} \quad x = 10^b$$
$$e^a = 10^b$$

Taking logarithms to the base e of both sides gives

$$\ln e^a = \ln 10^b$$

because the ln of e raised to any power is simply that power

$$a = b \ln 10$$

However, as has already been seen, $a = \ln x$; therefore substitution is again possible, since $b = \log x$. Consequently,

$$\ln x = \log x \ln 10$$

or

$$\ln x = 2.303 \log x$$

The free energy expression can, therefore, be written as

$$\Delta G° = -2.303\, RT \log K_e$$

Note that the *standard* free energy change for the reaction is used. The free energy change by itself will not allow the calculation of the equilibrium constant.

Chapter 19 Chemical Thermodynamics   283

Second law of thermodynamics (19.10)   In any spontaneous change the entropy of the universe increases.

Surroundings (19.1)   All of the universe except the system we are studying.

Third law of thermodynamics (19.13)   The entropy of any pure, perfectly crystalline substance at the absolute zero of temperature (0 K) equals zero. Basically, this law allows the establishment of a beginning point, or zero point, for entropy measurements.

Work (w) (19.2)   A positive sign indicates work done by the system; a negative sign corresponds to work done on the system. Work has a pressure-volume equivalent defined as $w = P(V_2 - V_1)$, where P is the pressure restraining the system, V_1 is the initial volume, and V_2 is the final volume.

SOLUTIONS TO CHAPTER 19 EXERCISES

HEAT, WORK, AND INTERNAL ENERGY

2. Calculate the missing value of ΔE, q, or w for a system, given the following data:

    (b) $\Delta E = -7500$ J; $w = 4500$ J
    (d) The system absorbs 2.000 kJ of heat and does 1425 J of work on the surroundings.

    Solution

    (b) The internal energy change ΔE of a system is a balance between the heat and work. This reaction is given by the first law of thermodynamics, $\Delta E = q - w$. Solving for the unknown q,

    $$q = \Delta E + w = -7500 \text{ J} + 4500 \text{ J} = -3000 \text{ J}$$

    (d) Heat absorbed by the system is considered to be positive and work done by the system is considered positive. From the first law,

    $$\Delta E = q - w = 2000 \text{ J} - (1425 \text{ J}) = 575 \text{ J}$$

4. Calculate the work involved in compressing a system consisting of exactly 2 mol of H$_2$O as it changes from a gas at 373 K (volume = 61.2 L) to a liquid at 373 K (volume = 37.8 mL) under a constant pressure of 1.00 atm. Does this work add to or decrease the internal energy of the system?

    Solution

    For pressure-volume work done at constant pressure, the work is given by $P(V_2 - V_1)$. In order to have the work in joules, the pressure must be expressed in pascals and the volume in (meter)^3. Recall that 1 atm = 101,325 Pa, and 1 L = 10^{-3} m^3, so the two volumes may be expressed as

    $$V_1 = 61.2 \text{ L} \times \frac{10^{-3} \text{ m}^3}{\text{L}} = 6.12 \times 10^{-2} \text{ m}^3$$

    and

    $$V_2 = 37.8 \text{ mL} \times \frac{1 \text{ L}}{10^3 \text{ mL}} \times \frac{10^{-3} \text{ m}^3}{1 \text{ L}} = 3.78 \times 10^{-5} \text{ m}^3$$

From $P(V_2 - V_1)$ the work is

$$w = P(V_2 - V_1) = 101{,}325 \text{ Pa}(3.78 \times 10^{-5} - 6.12 \times 10^{-2})\text{m}^3$$
$$= -6.20 \times 10^3 \text{ Pa m}^3 = -6.20 \times 10^3 \text{ J} = -6.20 \text{ kJ}$$

Hence 6.20 kJ of work would be involved. The work done on the system has a negative value. From the first law, $\Delta E = q - w$ and since w enters the equation as a negative quantity, the negative sign in the equation implies an overall increase in the internal energy.

5. What is the change in internal energy of a gas that absorbs 225 J of heat and expands from 10.0 L to 25.0 L against a constant pressure of 0.75 atm?

Solution

$$\Delta E = q - w$$
$$= 225 \text{ J} - 0.75 \text{ atm} \times 101{,}325 \text{ Pa/atm} \times (25.0 \text{ L} - 10.0 \text{ L}) \times 10^{-3} \text{ m}^3/\text{L}$$
$$= 225 \text{ J} - 1140 \text{ J} = -915 \text{ J}$$

ENTHALPY CHANGES: HESS'S LAW

9. What is the difference between ΔE and ΔH for a system undergoing a change at constant pressure?

Solution

The definition is $H = E + PV$. For a change in state at constant pressure, $\Delta H = \Delta E + P\Delta V$. Then the difference is $P\Delta V$.

11. (a) Using the data in Appendix I, calculate the standard enthalpy change, ΔH°_{298}, for each of the following reactions:

   (2) $N_2(g) + O_2(g) \rightarrow 2NO(g)$
   (4) $Fe_2O_3(s) + 3CO(g) \rightarrow 2Fe(s) + 3CO_2(g)$
   (7) $CH_4(g) + N_2(g) \rightarrow HCN(g) + NH_3(g)$

   (b) Which of these reactions are exothermic?

Solution

(a) The solution makes use of Hess's law.

$$\Delta H^\circ = \sum \Delta H^\circ_{f\,products} - \sum \Delta H^\circ_{f\,reactants}$$

(2) $\Delta H^\circ_f = 2\Delta H^\circ_{fNO(g)} - \Delta H^\circ_{fN_2(g)} - \Delta H^\circ_{fO_2(g)}$
$= 2(90.25 \text{ kJ mol}^{-1}) - (0) - (0) = 180.5 \text{ kJ}$

(4) $\Delta H^\circ = 2\Delta H^\circ_{fFe(s)} + 3\Delta H^\circ_{fCO_2(g)} - 3\Delta H^\circ_{fCO(g)} - \Delta H^\circ_{fFe_2O_3(s)}$
$= 2(0) + 3(-393.51) - 3(-110.52) - (-824.2) = -24.8 \text{ kJ}$

(7) $\Delta H^\circ = \Delta H^\circ_{fHCN(g)} + \Delta H^\circ_{fNH_3(g)} - \Delta H^\circ_{fCH_4(g)} - \Delta H^\circ_{fN_2(g)}$
$= (135) + (-46.11) - (-74.81) - (0) = 164 \text{ kJ}$

(b) Reactions are exothermic when ΔH° is negative. Exothermic reactions are 1, 3, 4, 5, and 8.

Chapter 19 Chemical Thermodynamics 285

13. How many kilojoules of heat energy will be liberated when 27.89 g of manganese is burned to form $Mn_3O_4(s)$ at standard state conditions?

Solution

First write the overall reaction

$$3Mn(s) + 2O_2(g) \longrightarrow Mn_3O_4(s)$$

Next calculate the amount of $Mn_3O_4(s)$ in moles that will be formed by the reaction of 27.89 g of Mn. Based on the balanced equation,

$$3 \text{ mol Mn} \longrightarrow 1 \text{ mol Mn}_3O_4$$

Therefore,

$$\text{Amount of Mn}_3O_4 = \frac{\text{mol Mn}}{3} = \frac{\text{mass Mn}}{3 \text{ (at. wt Mn)}}$$

$$= \frac{27.89 \text{ g}}{3(54.9380 \text{ g/mol})} = 0.16922 \text{ mol}$$

Multiply the moles of M_3O_4 produced by the heat released per mole to obtain the heat liberated. Since $\Delta H^\circ_{f298Mn_3O_4} = -1388 \text{ kJ mol}^{-1}$,

$$\text{Heat} = 0.16922 \text{ mol} \times -1388 \text{ kJ mol}^{-1} = |-234.9 \text{ kJ}| = 234.9 \text{ kJ}$$

14. The standard molar heat of formation of $OsO_4(s)$ is -391 kJ mol^{-1}, and the heat of sublimation of $OsO_4(s)$ under standard state is 56.4 kJ mol^{-1}. What is ΔH°_{298} for the process $OsO_4(g) \longrightarrow Os(s) + 2O_2(g)$?

Solution

Sublimation is the process of direct conversion of a solid to a gas. The sublimation process is endothermic and requires heat, as shown by the positive sign for the heat of sublimation.

$$OsO_4(s) \longrightarrow OsO_4(g) \qquad \Delta H^\circ_{sub} = 56.4 \text{ kJ}$$

For the heat of formation

$$Os(s) + 2O_2(g) \longrightarrow OsO_4(s) \qquad \Delta H^\circ_f = -391 \text{ kJ}$$

Reversal of both equations and addition give the desired equation. Addition of ΔH's gives the desired enthalpy change.

$$OsO_4(g) \longrightarrow Os(s) + 2O_2(g) \qquad \Delta H^\circ = 391 - 56.4 = 335 \text{ kJ}$$

16. The white pigment TiO_2 is prepared by the hydrolysis of titanium tetrachloride, $TiCl_4$, in the gas phase.

$$TiCl_4(g) + 2H_2O(g) \longrightarrow TiO_2(s) + 4HCl(g)$$

How much heat is evolved in the production of 1.00 kg of $TiO_2(s)$ under standard state conditions of 25°C and 1 atm?

Solution

Calculate ΔH° for the reaction.

$$\Delta H^\circ = \Delta H^\circ_{fTiO_2(s)} + 4\Delta H^\circ_{fHCl(g)} - \Delta H^\circ_{fTiCl_4(g)} - 2\Delta H^\circ_{fH_2O(g)}$$

$$= -944.7 + 4(-92.307) - (-763.2) - 2(-241.82)$$

$$= -67.088 \text{ kJ mol}^{-1}$$

In 1.00 kg of TiO$_2$(s) there are 1000 g/79.90 g mol^{-1} = 12.52 mol. The heat released is

12.52 mol × 67.088 kJ mol^{-1} = 840 kJ

ENTROPY CHANGES AND ABSOLUTE ENTROPY

19. What is the absolute standard molar entropy of a pure substance?

Solution

According to the third law of thermodynamics, the standard molar entropy is zero for a pure substance in its most stable crystalline state at absolute zero. The entropy change between 0 K and 273.15 K can be measured and the actual entropy of a mole of a substance (the absolute standard molar entropy) can be determined.

20. Arrange the following systems, each of which consists of 1 mol of substance, in order of increasing entropy: H$_2$O(g) at 100°C, N$_2$(s) at −215°C, C$_2$H$_5$OH(g) at 100°C, H$_2$O(l) at 25°C, H$_2$O(s) at −215°C, C$_2$H$_5$OH(s) at 0 K.

Solution

Without thermodynamic data at hand, actual values cannot be obtained but ordering can be done based on three assumptions: The entropy of a gas > entropy of a liquid > entropy of a solid; the entropy will be greatest for the substance at the highest temperature; and for a given temperature, the entropy will be greater for the compound with the more complex structure. The order of increasing entropy is: C$_2$H$_5$OH(s) at 0 K < N$_2$(s) at −215°C < H$_2$O(s) at −215°C < H$_2$O(l) at 25°C < H$_2$O(g) at 100°C < C$_2$H$_5$OH(g) at 100°C.

23. Does the entropy of each of the following systems increase, decrease, or not change in going from the initial to the final state? If the entropy does change, give the sign of ΔS and explain your answers.

INITIAL STATE	FINAL STATE
(a) NaCl(s) at 298 K	NaCl(s) at 0 K
(d) 1 mol CaCO$_3$	1 mol CaO and 1 mol CO$_2$
(f) chicken feed and baby chick	full grown chicken

Solution

When a system becomes more disordered, the entropy increases (ΔS is positive).

(a) Entropy decreases. ΔS is negative. NaCl becomes more ordered at the lower temperature, thus decreasing the entropy.

(d) Entropy increases. ΔS is positive. The gas CO$_2$ has a much higher entropy than a solid. Thus, as the gas is formed from the solid, the entropy increases.

(f) Entropy decreases. ΔS is negative. The state of order of the chicken feed is much less than the order in the full grown chicken. This is a natural process in which the entropy of the system decreases. The overall process, however, in which the surroundings are also considered (the entropy of the products of the biological process, CO$_2$, H$_2$O, etc.), shows a net increase in entropy.

Chapter 19 Chemical Thermodynamics

24. (a) Using the data in Appendix I, calculate $\Delta S°_{298}$, the standard entropy change for each of the following reactions:

    (2) $N_2(g) + O_2(g) \longrightarrow 2NO(g)$
    (4) $Fe_2O_3(s) + 3CO(g) \longrightarrow 2Fe(s) + 3CO_2(g)$
    (7) $CH_4(g) + N_2(g) \longrightarrow HCN(g) + NH_3(g)$

    (b) For which of these reactions are the entropy changes favorable for the reaction to proceed spontaneously?

Solution

(a) The solution makes use of the equation

$$\Delta S° = \Sigma S°_{products} - \Sigma S°_{reactants}$$

(2) $\Delta S°_{298} = 2S°_{NO(g)} - S°_{N_2(g)} - S°_{O_2(g)}$
$\qquad = 2(210.65) - 191.5 - 205.03$
$\qquad = 24.8 \text{ J K}^{-1}$

(4) $\Delta S°_{298} = 2S°_{Fe(s)} + 3S°_{CO_2(g)} - S°_{Fe_2O_3(s)} - 3S°_{CO(g)}$
$\qquad = 2(27.3) + 3(213.6) - 87.40 - 3(197.56)$
$\qquad = 15.3 \text{ J K}^{-1}$

(7) $\Delta S°_{298} = S°_{HCN(g)} + S°_{NH_3(g)} - S°_{CH_4(g)} - S°_{N_2(g)}$
$\qquad = 201.7 + 192.3 - 186.15 - 191.5$
$\qquad = 16.4 \text{ J K}^{-1}$

(b) Spontaneous reactions are favored by a positive value for ΔS. These reactions are 2, 4, 6, and 7.

26. What is $\Delta S°_{298}$ for the following reaction?

    $N_2(g) + 3H_2(g) \longrightarrow 2NH_3(g)$

Solution

$\Delta S°_{298} = 2S°_{NH_3(g)} - S°_{N_2(g)} - 3S°_{H_2(g)}$
$\qquad = 2(192.3) - 191.5 - 3(130.57)$
$\qquad = -198.6 \text{ J K}^{-1}$

FREE ENERGY CHANGES

28. Why is the emphasis in thermodynamics on the free energy change as a system changes rather than on the values of the free energies of the initial and final state?

Solution

Changes in free energy can be determined. The free energy content of the initial and final states cannot be determined.

30. What is a spontaneous reaction?

Solution

A spontaneous reaction is one that proceeds in the direction written under the stated conditions. $\Delta G° < 0$ for a spontaneous reaction.

288 Chapter 19 Chemical Thermodynamics

32. As ammonium nitrate dissolves spontaneously in water at constant pressure, heat is absorbed and the solution gets cold. What is the sign of ΔH for this process? Is it possible to identify the sign of ΔS for this process? Why?

Solution

For the dissolution, ΔH is positive. Since the process is spontaneous, ΔG must be negative. From the relation $\Delta G = \Delta H - T\Delta S$, if ΔG is negative and ΔH is positive, the sign of ΔS must be positive. Since dissolution increases randomness, ΔS increases and so must be positive.

35. Under what conditions is ΔG equal to $\Delta G°$ for the reaction $2H_2(g) + O_2(g) \longrightarrow 2H_2O(l)$?

Solution

$\Delta G = \Delta G°$ when the gases H_2 and O_2 are present at 1 atm of pressure and are converted to liquid water, all at 25°C.

37. (a) Using the data in Appendix I, calculate the standard free energy changes for the following reactions:

   (2) $N_2(g) + O_2(g) \longrightarrow 2NO(g)$
   (4) $Fe_2O_3(s) + 3CO(g) \longrightarrow 2Fe(s) + 3CO_2(g)$
   (7) $CH_4(g) + N_2(g) \longrightarrow HCN(g) + NH_3(g)$

   (b) Which of those reactions are spontaneous?
   (c) For which of these reactions does the value of $\Delta H°$ favor spontaneity (exercise 11)? The value of $\Delta S°$ (exercise 24)?

Solution

(a) The solution makes use of the equation

$$\Delta G° = \Sigma \Delta G°_{products} - \Sigma \Delta G°_{reactants}$$

(2) $\Delta G°_{298} = 2\Delta G°_{fNO(g)} - \Delta G°_{fN_2(g)} - \Delta G°_{fO_2(g)}$
$= 2(86.57) - (0) - (0)$
$= 173.14$ kJ

(4) $\Delta G°_{298} = 2\Delta G°_{fFe(s)} + 3\Delta G°_{fCO_2(g)} - \Delta G°_{fFe_2O_3(s)} - 3\Delta G°_{fCO_2(g)}$
$= (0) + 3(-394.36) - (-742.2) - 3(-137.15)$
$= -29.4$ kJ

(7) $\Delta G°_{298} = \Delta G°_{fHCN(g)} + \Delta G°_{fNH_3(g)} - \Delta G°_{fCH_4(g)} - \Delta G°_{fN_2(g)}$
$= 124.7 + (-16.5) - (-50.75) - 0$
$= 159$ kJ

(b) Reaction 4 is spontaneous.
(c) $\Delta H°$ favors spontaneity in 4.
    $\Delta S°$ favors spontaneity in 2, 4, 7.

39. For a certain process at 300 K, $\Delta G = -17.0$ kJ and $\Delta H = 6.9$ kJ. Find the entropy change for this process at 300 K.

Solution

$$\Delta G = \Delta H - T\Delta S$$

$$\Delta S = -\frac{\Delta G - \Delta H}{T} = -\frac{-17000 - 6900}{300} = 79.7 \text{ J K}^{-1}$$

Chapter 19 Chemical Thermodynamics

FREE ENERGY CHANGES AND EQUILIBRIUM CONSTANTS

41. Explain why equilibrium constants change with temperature.

Solution

Equilibrium constants are calculated from

$$\log K = \frac{\Delta G°}{-2.303\ RT}$$

in which T is the kelvin temperature. Thus, K changes as T changes.

44. Calculate the equilibrium constant at 25°C for the decomposition of solid NH_4Cl to $HCl(g)$ and $NH_3(g)$. $\Delta G°$ for the reaction is 89.7 kJ (exercise 40).

Solution

$$NH_4Cl \longrightarrow HCl(g) + NH_3(g) \quad \Delta G = 89.7\ \text{kJ}$$

$$\Delta G = -RT \ln K_e = -2.303\ RT \log K_e$$

At 25°C,

$$89,700\ \text{J} = -2.303(8.314\ \text{J K}^{-1})(298.15\ \text{K}) \log K_e$$

$$K_e = \text{antilog}\left[\frac{89,700}{-2.303(8.314)(298.15)}\right]$$

$$= \text{antilog}\ (-15.7128)$$

$$= 1.9 \times 10^{-16}$$

46. Consider the reaction

$$2ICl(g) \longrightarrow I_2(g) + Cl_2(g)$$

(a) For this reaction $\Delta H°_{298} = 26.9$ kJ and $\Delta S°_{298} = -11.3$ J K^{-1}. Calculate $\Delta G°_{298}$ for the reaction.
(b) Calculate the equilibrium constant for this reaction at 25°C.

Solution

(a) $\Delta G°_{298} = \Delta H°_{298} - T\Delta S°_{298}$
$= +26.9\ \text{kJ} - 298.15(-11.3\ \text{J K}^{-1})$
$= +26,900 + 3370 = +30,270\ \text{J} = +30.3\ \text{kJ}$

(b) The equilibrium constant for the reaction at 25°C is related to the free energy (Gibbs energy) by the equation

$$\Delta G° = -2.303\ RT \log K$$

Thus, the value of the standard state free energy (Gibbs energy) for a reaction can be used to determine the equilibrium constant K for the reaction. The value of $\Delta G°$ is +30.3 kJ; therefore,

$$+30,300\ \text{J} = (-2.303)(8.314\ \text{J K}^{-1})(298.2\ \text{K}) \log K$$

$$\log K = \frac{30,300}{-(2.303)(8.314)(298.2)} = -5.307$$

$$K = 4.9 \times 10^{-6}$$

47. If the entropy of vaporization of H₂O is equal to 109 J mol⁻¹ K⁻¹ and the enthalpy of vaporization is 40.62 kJ mol⁻¹, calculate the normal boiling temperature of water in °C.

Solution

The boiling point of a liquid represents an equilibrium between liquid and vapor. Consequently, $\Delta G = 0$ and $\Delta G = \Delta H - T\Delta S$ may be written

$$T_b = \frac{\Delta H}{\Delta S} = \frac{40{,}620 \text{ J mol}^{-1}}{109 \text{ J mol}^{-1}\text{ K}^{-1}} = 373 \text{ K}$$

$$T_b = 373 \text{ K} - 273.15 \text{ K} = 100°\text{C}$$

49. The standard molar enthalpy of formation of BaCO₃(s) is −1219 kJ mol⁻¹ and the standard free energy of formation is −1139 kJ mol⁻¹. For the decomposition of BaCO₃(s) into BaO(s) and CO₂(g) at 1 atm,

(a) Estimate the temperature at which the equilibrium pressure of CO₂ would be 1 atm.

(b) Calculate the equilibrium vapor pressure of CO₂(g) above BaCO₃(s) in a closed container at 298 K.

Solution

(a) $\text{BaCO}_3(s) \rightleftharpoons \text{BaO}(s) + \text{CO}_2(g)$

$$\Delta G^\circ_{298} = \Delta G^\circ_{f\text{BaO}(s)} + \Delta G^\circ_{f\text{CO}_2(g)} - \Delta G^\circ_{f\text{BaCO}_3(s)}$$

$$= -528.4 - 394.36 - (-1139)$$

$$\Delta G^\circ_{298} = 216.24 \text{ kJ}$$

$$\Delta H^\circ_{298} = \Delta H^\circ_{f\text{BaO}(s)} + \Delta H^\circ_{f\text{CO}_2(g)} - \Delta H^\circ_{f\text{BaCO}_3(s)}$$

$$= -558.1 - 393.51 - (-1219)$$

$$= 267.39 \text{ kJ}$$

$$\Delta G^\circ = \Delta H^\circ - T\Delta S^\circ$$

$$\Delta S^\circ = -\frac{\Delta G^\circ - \Delta H^\circ}{T} = -\frac{216.24 - 267.39}{298.15} = 0.1716 \text{ kJ mol}^{-1}$$

At 1 atm of CO₂, $\Delta G = 0$ so if we assume ΔH° and ΔS° are constant over the temperature range,

$$T = \frac{\Delta H^\circ}{\Delta S^\circ} = \frac{267.390 \text{ kJ}}{0.1716 \text{ kJ K}^{-1}} = 1560 \text{ K}$$

(b) At 298 K, $\Delta G = 216{,}240 \text{ J} = -2.303\, RT \log K$

$$K_p = \text{antilog}\left[\frac{-216{,}240 \text{ J}}{2.303(8.314 \text{ J K}^{-1})(298.15 \text{ K})}\right] = \text{antilog}(-37.88)$$

$$K_p = 1 \times 10^{-38} \text{ atm} = P_{\text{CO}_2(g)}$$

50. If the enthalpy of vaporization of CH₂Cl₂ is 29.0 kJ mol⁻¹ at 25.0°C and the entropy of vaporization is 92.5 J mol⁻¹ K⁻¹, calculate the normal boiling temperature of CH₂Cl₂, the temperature at which the equilibrium pressure of the vapor is 1 atm.

Chapter 19 Chemical Thermodynamics   291

Solution

ΔG is equal to zero at equilibrium (the boiling point in this case). Thus,

$$\Delta G = \Delta H - T\Delta S = 0$$

$$T = \frac{\Delta H}{\Delta S} = \frac{29{,}000 \text{ J}}{92.5 \text{ J K}^{-1}} = 314 \text{ K} \quad \text{or} \quad 41°C$$

52. The equilibrium constant, K_p, for the reaction $N_2O_4(g) \longrightarrow 2NO_2(g)$ is 0.142 at 298 K. What is $\Delta G°_{298}$ for the reaction?

Solution

$$\Delta G° = -2.303 \, RT \log K_p$$
$$= -2.303(8.314 \text{ J K}^{-1} \text{ mol}^{-1})(298.2 \text{ K})(\log 0.142)$$
$$= -5709.7 \text{ J mol}^{-1}(-0.8477)$$
$$= 4840 \text{ J mol}^{-1} = 4.84 \text{ kJ mol}^{-1}$$

54. (a) Calculate $\Delta G°$ for each of the following reactions from the equilibrium constant at the temperature given.

   (1) $N_2(g) + O_2(g) \longrightarrow 2NO(g)$ $T = 2000°C$
   $K_p = 4.1 \times 10^{-4}$

   (3) $CO_2(g) + H_2(g) \longrightarrow CO(g) + H_2O(g)$ $T = 980°C$
   $K_p = 1.67$

   (b) Assume that $\Delta H°$ does not vary with temperature and calculate $\Delta S°$ for these reactions at the temperature indicated.

Solution

(a) (1) $\Delta G = -RT \ln K_p$
$$= -2.303(8.314 \text{ J K}^{-1})(2273 \text{ K}) \log 4.1 \times 10^{-4} = 147 \text{ kJ}$$
(3) $\Delta G = -RT \ln K_p$
$$= -2.303(8.314 \text{ J K}^{-1})(1253 \text{ K}) \log 1.67 = -5.34 \text{ kJ}$$

(b) ΔH must be calculated from data at 298 K. Over small temperature ranges ΔH may be considered to be constant. This is not true for ranges of 2000° but the value at 25°C may be used as a first approximation.

(1) $\Delta H° = 2\Delta H°_{f NO(g)} - \Delta H°_{f N_2(g)} - \Delta H°_{f O_2(g)}$
$$= 2(90.25)\text{kJ} - 0 - 0 = 180.5 \text{ kJ}$$

$$\Delta G = \Delta H - T\Delta S$$

(3) $\Delta H° = \Delta H°_{CO(g)} + \Delta H°_{H_2O(g)} - \Delta H°_{CO_2(g)} - \Delta H°_{H_2(g)}$
$$= -110.52 \text{ kJ} + (-241.82) \text{ kJ} - (-393.51) \text{ kJ} - 0 = 41.17 \text{ kJ}$$

$$\Delta G = \Delta H° - T\Delta S$$

$$\Delta S = -\left(\frac{\Delta G - \Delta H°}{T}\right) = -\frac{(-5340 - 41{,}170) \text{ J}}{1253 \text{ K}} = \frac{46{,}510 \text{ J}}{1253 \text{ K}} = 37.1 \text{ J K}^{-1}$$

55. Calculate $\Delta G°_{298}$ for the reaction of 1 mol of $H^+(aq)$ with 1 mol of $OH^-(aq)$, using the equilibrium constant for the self-ionization of water at 298 K.

$$H_2O \longrightarrow H^+(aq) + OH^-(aq) \qquad K_w = 1.00 \times 10^{-14}$$

Solution

$$\Delta G = -RT \ln K_w = -2.303 \, RT \log K_w$$
$$= -2.303(8.314 \text{ J K}^{-1})(298 \text{ K}) \log (1.00 \times 10^{-14})$$
$$= 79.9 \text{ kJ}$$

57. Hydrogen sulfide, a pollutant found in some natural gas, is removed by the reaction

$$2H_2S(g) + SO_2(g) \longrightarrow 3S(s) + 2H_2O(g)$$

What is the equilibrium constant for this reaction at 25°C? Is the reaction endothermic or exothermic?

Solution

Find $\Delta G°$ for the reaction, and from $\Delta G°$ find K.

$$\Delta G° = 3\Delta G°_{S(s)} + 2\Delta G°_{H_2O(g)} - 2\Delta G°_{H_2S(g)} - \Delta G°_{SO_2(g)}$$
$$= 3(0) + 2(-228.59) - 2(-33.6) - (-300.19)$$
$$= -89.79 \text{ kJ}$$

$$\Delta G° = -RT \ln K = -2.303 \, RT \log K$$

$$K = \text{antilog} \left(\frac{-\Delta G}{2.303 \, RT} \right) = \text{antilog} \left[\frac{89790}{2.303(8.314)(298.15)} \right]$$
$$= \text{antilog } 15.729 = 5.4 \times 10^{15}$$

ADDITIONAL EXERCISES

60. For the vaporization of bromine liquid to bromine gas, $Br_2(l) \longrightarrow Br_2(g)$, at 25°C.

(a) Calculate the change in enthalpy and the change in entropy at standard state conditions.
(b) Discuss the relative disorder in bromine liquid compared to bromine gas. State what you can about the spontaneity of the vaporization.
(c) Estimate the value of $\Delta G°_{298}$ for the vaporization of bromine from the values of ΔH and ΔS determined in (a).
(d) State what you can about the spontaneity of the process from the value you obtained for ΔG in part (c).
(e) Estimate the temperature at which liquid Br_2 and gaseous Br_2 with a pressure of 1 atm are in equilibrium (assume that $\Delta H°$ and $\Delta S°$ are independent of temperature).
(f) State in which direction the process would be spontaneous at 298 K and at 398 K, using the temperature value obtained in part (e).
(g) Compare $\Delta H°$, $\Delta S°$, and $\Delta G°$ in terms of their usefulness in predicting the spontaneity of the vaporization of bromine.

Chapter 19 Chemical Thermodynamics

Solution

(a) For $Br_2(l) \longrightarrow Br_2(g)$

$$\Delta H^\circ_{298} = \Delta H^\circ_{f\,298\,Br_2(g)} - \Delta H^\circ_{f\,298\,Br_2(l)}$$

$$= 30.91 \text{ kJ} - 0 \text{ kJ} = 30.91 \text{ kJ mol}^{-1}$$

$$\Delta S^\circ_{298} = \Delta S^\circ_{f\,298\,Br_2(g)} - \Delta S^\circ_{f\,298\,Br_2(l)}$$

$$= 245.35 - 152.23 = 93.12 \text{ J mol}^{-1} \text{ K}^{-1}$$

(b) In bromine liquid, weak intermolecular forces hold the liquid together. In view of the fact that the vapor pressure of liquid bromine is high, there is close to maximum disorder in the liquid state consistent with the fact that the molecules must be within a fixed liquid volume. The disorder greatly increases as the molecules enter the vapor state.

(c) For vaporization, $\Delta G^\circ_{298} = \Delta H^\circ_{298} - T \Delta S^\circ_{298}$

$$\Delta G^\circ_{298} = 30{,}910 \text{ J} - 298.15(93.12) \text{ J} = 3146 \text{ J} = 3.146 \text{ kJ mol}^{-1}$$

(d) The vaporization should not be a spontaneous process at 1 atm of pressure and 298 K.

(e) For equilibrium to occur, ΔG°_T must equal zero. Therefore,

$$\Delta G^\circ_T = 0 = \Delta H^\circ - T \Delta S^\circ$$

or

$$\Delta H^\circ = T \Delta S^\circ$$

Assuming that ΔH° and ΔS° do not change significantly with a change in temperature,

$$T = \frac{\Delta H^\circ_{298}}{\Delta S^\circ_{298}} = \frac{30910}{93.12} = 331.9 \text{ K} \quad \text{or} \quad 58.8°C$$

(f) Vapor to liquid is spontaneous at 298 K; liquid to vapor is spontaneous at 398 K.

(g) ΔH with its positive sign indicates that enthalpy will not favor vaporization. ΔG is positive definitively indicating that the process of vaporization is not spontaneous. ΔS with a positive value is a favorable indicator for vaporization. Consequently, ΔS cannot be used as an indicator of spontaneity.

63. Acetic acid, CH_3CO_2H, can form a dimer, $(CH_3CO_2H)_2$, in the gas phase.

$$2CH_3CO_2H(g) \rightleftharpoons (CH_3CO_2H)_2(g)$$

The dimer is held together by two hydrogen bonds

$$CH_3-C\begin{array}{c}O\cdots H-O\\ \diagup \quad \diagdown\\ O-H\cdots O\end{array}C-CH_3$$

with a total strength of 66.5 kJ per mole of dimer. At 25°C the equilibrium constant for the dimerization is 1.3×10^3 (pressure in atm). What is ΔS° for the reaction?

Solution

The equilibrium constant allows the calculation of $\Delta G°_{298}$.

$$\Delta G°_{298} = -2.303 \, RT \log K$$
$$= -2.303(8.314 \text{ J K}^{-1} \text{ mol}^{-1})(298.2 \text{ K})(\log 1.3 \times 10^3)$$
$$= -17.78 \text{ kJ mol}^{-1}$$

The strength of the bond (66.5 kJ) means that it requires 66.5 kJ to pull one mole of bonds apart. In other words, $\Delta H°_{298} = -66.5$ kJ. The values of $\Delta G°_{298}$ and $\Delta H°_{298}$ allow us to calculate $\Delta S°_{298}$ by use of the equation

$$\Delta G°_{298} = \Delta H°_{298} - T\Delta S°_{298}$$

$$\Delta S°_{298} = -\frac{\Delta G°_{298} - \Delta H°_{298}}{T}$$

$$\Delta S°_{298} = -\frac{(-17,780) - (-66,500)}{298.2} = -163 \text{ J K}^{-1}$$

64. At 1000 K the equilibrium constant for the decomposition of bromine molecules, $Br_2(g) \rightleftharpoons 2Br(g)$, is 2.8×10^4 (pressure in atm). What is $\Delta G°$ for the reaction? Assume that the bond energy of Br_2 does not change between 298 K and 1000 K and calculate the approximate value of $\Delta S°$ for the reaction at 1000 K.

Solution

$$\Delta G° = -2.303 \, RT \log K$$
$$\Delta G°_{1000} = -2.303(8.314)(1000)(\log 2.8 \times 10^4)$$
$$= -19,147(4.447) = -85 \text{ kJ mol}^{-1}$$

The value of the change in enthalpy of the reaction $Br_2(g) \rightleftharpoons 2Br(g)$ may be calculated from the ΔH_f's at 298 K on the assumption that the value of ΔH will not change with temperature. Over such a large temperature range, this is a first approximation. Techniques are developed in a course in physical chemistry to treat this situation. As an approximation, then,

$$\Delta H = 2\Delta H°_{fBr(g)} - \Delta H°_{fBr_2(g)}$$
$$= 2(111.88) - 30.91 \quad (\Delta H°_f \text{ in kJ})$$
$$= 192.85 \text{ kJ}$$

Then, from

$$\Delta G = \Delta H - T \Delta S$$

$$\Delta S = -\frac{\Delta G - \Delta H}{T}$$

$$= -\frac{-85,000 - 192,850}{1000 \text{ K}} = \frac{277,860 \text{ J}}{1000 \text{ K}}$$

$$= 278 \text{ J K}^{-1}$$

20

A Survey of the Nonmetals

INTRODUCTION

We now cover the general behavior of nonmetals, reserving detailed treatment of specific elements for later chapters. The nonmetals covered are hydrogen, carbon, nitrogen, oxygen, phosphorus, sulfur, arsenic, selenium, and all members of groups VIIA and VIIIA, the halogens and noble gases.

Aside from hydrogen, these elements do not form monatomic positive ions because of their large ionization energies. However, when combined with a more electronegative element, a nonmetal can exhibit a positive oxidation number as in IF where I has a +1 value.

Most nonmetal oxides are acidic. In water they react to form oxyacids, either with no change in oxidation number or, in the case of NO_2 and ClO_2, with a change in oxidation number producing both an oxidized and a reduced species in a disproportionation reaction. The strength of the oxyacids so produced increases as the electronegativity of the central atom in the acid increases and as the oxidation number of the central atom increases.

Many other guidelines are given in this chapter to provide a general introduction to the behavior of nonmetals. Chapters 22 and 23 complete this study, except for carbon containing compounds, which are covered in Chapters 30 and 31.

READY REFERENCE

Allotropes (20.2) Two forms of the same element in the same state but with different structures.

Hydride (20.10) Hydrogen compounds of the metals of Group IA and of Ca, Sr, and Ba are ionic compounds that contain the hydride anion, H^-.

Metallic hydride (20.10) Hydrogen can dissolve in many transition-type metals to produce metallic hydrides. These are so named because they possess metallic properties. They do not contain the hydride ion because the hydrogen atoms are merely located in the voids of the metal lattice.

SOLUTIONS TO CHAPTER 20 EXERCISES

1. Both hydrogen and oxygen are colorless, odorless, and tasteless gases. How may we distinguish between them using chemical properties?

Solution

Simple chemical tests involve combustion. Hydrogen burns in air. Oxygen will cause glowing wood splints to burst into flame.

3. Describe five chemical properties of sulfur, or of compounds of sulfur, that characterize it as a nonmetal.

    Solution

    1. It readily forms negative ions.
    2. It functions as an oxidizer.
    3. It forms oxides that are acidic in water.
    4. H_2S is a weak acid.
    5. It readily forms compounds through covalent bonding with other nonmetals.

5. With what elements can chlorine form binary compounds in which it exhibits a positive oxidation number?

    Solution

    Only oxygen and fluorine are more electronegative than chlorine.

7. With which elements will phosphorus form compounds in which it has a positive oxidation number?

    Solution

    The halogens and oxygen all oxidize phosphorus—they are more electronegative. Nitrogen is more electronegative, but is not very reactive.

9. From the positions of the elements in the Periodic Table, predict which of the following pairs will:

    (a) reduce S, Cl_2 or C
    (b) neutralize H_2SO_4, Li_2O or ClO_2
    (c) oxidize S, P_4 or Cl_2
    (d) oxidize P_4, Al or Cl_2
    (e) reduce Se, Zn or O_2
    (f) neutralize $NH_3(aq)$, Li_2O or SO_2

    Solution

    (a) C
    (b) Li_2O
    (c) Cl_2
    (d) Cl_2
    (e) Zn
    (f) SO_2

11. Write a complete and balanced equation for the reaction of each of the following with water.

    (a) CO_2
    (b) N_2O_3
    (c) N_2O_5

Chapter 20 A Survey of the Nonmetals

(d) P_4O_6
(e) P_4O_{10}
(f) SO_2
(g) SO_3
(h) Cl_2O_7
(i) I_2O_5

Solution

(a) $CO_2 + H_2O \longrightarrow H_2CO_3$
(b) $N_2O_3 + H_2O \longrightarrow 2HNO_2$
(c) $N_2O_5 + H_2O \longrightarrow 2HNO_3$
(d) $P_4O_6 + 6H_2O \longrightarrow 4H_3PO_3$
(e) $P_4O_{10} + 6H_2O \longrightarrow 4H_3PO_4$
(f) $SO_2 + H_2O \longrightarrow H_2SO_3$
(g) $SO_3 + H_2O \longrightarrow H_2SO_4$
(h) $Cl_2O_7 + H_2O \longrightarrow 2HClO_4$
(i) $I_2O_5 + H_2O \longrightarrow 2HIO_3$

13. Write a complete and balanced equation for the reaction of each of the following. In some cases there may be more than one correct answer, depending on the amounts of reactants used.

(a) $NaF + H_2SO_4$
(b) $Li_3N + H_2O$
(c) $Mg_3P_2 + HCl$
(d) $FeS + HCl$
(e) $NiO + H_2SO_4$
(f) $HC \equiv C^- + H_2O$
(g) $Na_2O + H_2O$
(h) $CaO + HF$

Solution

(a) $2NaF + H_2SO_4 \longrightarrow Na_2SO_4 + 2HF$
(b) $2Li_3N + 3H_2O \longrightarrow 3Li_2O + 2NH_3$
(c) $Mg_3P_2 + 6HCl \longrightarrow 3MgCl_2 + 2PH_3$
(d) $FeS + 2HCl \longrightarrow FeCl_2 + H_2S$
(e) $NiO + H_2SO_4 \longrightarrow NiSO_4 + H_2O$
(f) $HC \equiv C^- + H_2O \longrightarrow HC \equiv CH + OH^-$
(g) $Na_2O + H_2O \longrightarrow 2NaOH$
(h) $CaO + 2HF \longrightarrow CaF_2 + H_2O$

HYDROGEN

15. Why does hydrogen not exhibit an oxidation number of -1 when bonded to nonmetals?

Solution

Nonmetals have high ionization potentials and electronegativity and thus tend not to form positive ions. Hydrogen is the least electronegative nonmetal and it does not bond to any less electronegative element in compounds with other nonmetals. Hence, hydrogen forms covalent bonds in which hydrogen becomes partially positive.

17. Which of the binary hydrides of the nonmetals are strong acids? Which is the weakest acid?

Solution

Strong acids: HCl, HBr, HI
Weak acids: H_2O, H_2S, HF
NH_3 and PH_3 are extremely weak acids and are generally considered bases. Hydrides of carbon have little or no tendency to function as acids.

19. What mass of hydrogen would result from the reaction of 27.2 g of BaH_2 with water?

Solution

$$\underset{\underset{1\text{ mol}}{27.2\text{ g}}}{BaH_2} + 2H_2O \longrightarrow Ba(OH)_2 + \underset{\underset{2\text{ mol}}{\text{Mass?}}}{2H_2}$$

g-form. wts.: $BaH_2 = 139.3458$; $H_2 = 2.0158$

$$\text{Mass}(H_2) = 27.2 \text{ g } BaH_2 \times \frac{1 \text{ mol}}{139.3458 \text{ g}} \times \frac{2 \text{ mol } H_2}{1 \text{ mol } BaH_2} \times 2.0158 \frac{g}{\text{mol}} = 0.787 \text{ g}$$

CARBON

21. Why does carbon form a maximum of four single covalent bonds?

Solution

Carbon has only four orbitals in its valence shell and therefore can share four electrons through four single bonds.

23. (a) Compare and contrast the bonding in diamond and graphite.

Solution

(a) Both are covalent, but in diamond, carbon forms four single bonds. The carbon atoms bond to each other and form repeating tetrahedra. In graphite the carbon atoms form 2 single bonds and a double bond. The unhybridized p orbitals extend above and below the plates to form the double bonds

$$\overset{\|}{\underset{/\,\backslash}{C}}$$

These layers easily move, slide, over each other. Graphite can function as a lubricant. The layers also enable graphite to be a good electrical conductor.

25. Write the chemical equation for the reaction of CO_2 with water. With an aqueous solution of NaOH.

Solution

$$CO_2 + H_2O \longrightarrow H_2CO_3$$

$$H_2CO_3 + NaOH \longrightarrow NaHCO_3 + H_2O$$

Chapter 20 A Survey of the Nonmetals 299

27. How could the heat of combustion of graphite and diamond be used to show that the conversion

$$C(s, graphite) \longrightarrow C(s, diamond)$$

is endothermic?

Solution

Use Hess's law to express the heat of combustion of both diamond and graphite. Rearrange the equation to obtain the conversion energy.

$$C(s, graphite) + O_2(g) \longrightarrow CO_2(g)$$

$$\Delta H_{comb.} = \Delta H_{fCO_2} - \Delta H_{fC_{(graph)}} - \Delta H_{fO_2}$$

$$= -393.51 - 0 - 0 = -393.51 \text{ kJ}$$

$$C(s, diamond) + O_2(g) \longrightarrow CO_2(g)$$

$$\Delta H_{comb.} = \Delta H_{fCO_2} - \Delta H_{fdiamond} - \Delta H_{fO_2}$$

$$= -393.51 - 1.897 - 0 = -395.407 \text{ kJ}$$

Rearrange and add the equations:

$$C(graphite) + O_2 \longrightarrow CO_2 \quad \Delta H = -393.51 \text{ kJ}$$

$$CO_2 \longrightarrow C(diamond) + O_2 \quad \Delta H = +395.407 \text{ kJ}$$

$$\overline{C(graphite) \longrightarrow C(diamond) \quad \Delta H = \quad 1.897 \text{ kJ}}$$

NITROGEN, PHOSPHORUS, AND ARSENIC

29. Although PF_5 and AsF_5 are stable, nitrogen does not form NF_5 molecules. Explain this difference between members of the same group.

Solution

The valence shell of nitrogen does not contain d orbitals as do P and As.

31. What are the products of the reaction between aluminum nitride and water?

Solution

$$AlN + 3H_2O \longrightarrow Al(OH)_3 + NH_3$$

33. The bond energy of a phosphorus-phosphorus single bond is 215 kJ per mole of bonds and of a phosphorus-phosphorus triple bond, 490 kJ per mole of bonds. What is the approximate value of $\Delta H°$ for the following reaction?

$$P_4(g) \longrightarrow 2P_2(g)$$

Solution

The tetrahedral bonding arrangment in P_4 (each atom bonded to three others) requires that six bonds be broken to release four atoms. Using bond energies, ΔH can be estimated by:

$$P_4(g) \longrightarrow 2P_2(g)$$
$$\Delta H = 6(215 \text{ kJ}) \quad \Delta H = 2(490 \text{ kJ})$$
$$\Delta H = 1290 \text{ kJ} \quad \Delta H = -980 \text{ kJ}$$
$$\text{Endothermic} \quad \text{Exothermic}$$

$$\Delta H = 310 \text{ kJ}$$

300   Chapter 20   A Survey of the Nonmetals

35. Is an As—As triple bond stronger or weaker than three As—As single bonds. Explain your answer.

   Solution

   Triple bonds typically involve σ and π bonds and have at least one strong σ bond. The triple bond will be weaker than three single bonds because of the weaker π bonds. However, experimental evidence shows that the structure of As consists of sheets of As atoms forming single bonds rather than of As≡As bonded molecules.

OXYGEN, SULFUR, SELENIUM

37. Why does oxygen form diatomic molecules, whereas sulfur forms eight membered rings?

   Solution

   Sulfur is able to form double bonds only under substantial endothermic conditions, which is not the case for oxygen.

39. What are the molecular structures of sulfur found in the liquid, solid, and gaseous states?

   Solution

   Sulfur exists in eight-membered puckered rings in which each atom is linked to two of its neighbors by single covalent bonds. The normal state of sulfur is the yellow rhombic structure that is characteristic of the element. The form changes as sulfur is heated, because the rings tend to open forming "plastic" chains. In the gas phase, sulfur vapor contains S_2, S_6 and S_8 molecules.

41. Write Lewis structures for the following:

   (a) O_2 and S_2
   (b) H_2O, H_2S, and H_2Se
   (c) OF_2, SF_2, and SeF_2
   (d) O_2^{2-}, S_2^{2-}, and Se_2^{2-}
   (e) O_3 and SO_2
   (f) OH^-, SH^-, and SeH^-

   Solution

   (a) Ö=Ö   S̈=S̈

   (b) H–Ö–H   H–S̈–H   H–S̈e–H

   (c) :F̈–Ö–F̈:   :F̈–S̈–F̈:   :F̈–S̈e–F̈:

   (d) $[:\ddot{O}-\ddot{O}:]^{2-}$ $[:\ddot{S}-\ddot{S}:]^{2-}$ $[:\ddot{S}e-\ddot{S}e:]^{2-}$

   (e) O=Ö–Ö   O=S̈–Ö

   (f) $[H-\ddot{O}:]^-$ $[H-\ddot{S}:]^-$ $[H-\ddot{S}e:]^-$

Chapter 20 A Survey of the Nonmetals

43. Write a balanced equation for the reaction that occurs when each of the following is burned in air.

(a) C
(b) Mg
(c) Se
(d) Ge
(e) Al
(f) As
(g) S (two reactions)

Solution

(a) $C + O_2 \longrightarrow CO_2$
(b) $2Mg + O_2 \longrightarrow 2MgO$
(c) $Se + O_2 \longrightarrow SeO_2$
(d) $Ge + O_2 \longrightarrow GeO_2$
(e) $4Al + 3O_2 \longrightarrow 2Al_2O_3$
(f) $4As + 3O_2 \longrightarrow 2As_2O_3$
(g) $S + O_2 \longrightarrow SO_2$
$2S + 3O_2 \longrightarrow 2SO_3$

45. Elements generally exhibit their highest oxidation numbers in compounds containing oxygen or fluorine. What is the oxidation number of osmium in 0.3789 g of an osmium compound prepared by the reaction of 0.2827 g of Os with O_2? Write the equation for the reaction that gives this compound.

Solution

Use the mass data to determine the empirical formula of the compound.

Os: $\dfrac{0.2827 \text{ g}}{190.2 \text{ g/mol}} = 0.001486 \text{ mol}; \quad \dfrac{0.001486 \text{ mol}}{0.001486} = 1$

O: $\dfrac{0.0962 \text{ g}}{15.9994 \text{ g/mol}} = 0.00601 \text{ mol}; \quad \dfrac{0.00601 \text{ mol}}{0.001486} = 4$

The oxidation number of Os in OsO_4 is $+8$.

$Os + 2O_2 \longrightarrow OsO_4$

47. Write chemical equations describing three chemical reactions in which elemental sulfur acts as an oxidizing agent. Three in which it acts as a reducing agent.

Solution

Oxidizing Agent:

1. $H_2 + S \longrightarrow H_2S$
2. $Fe + S \longrightarrow FeS$
3. $C + 2S \longrightarrow CS_2$

Reducing Agent:

1. $S + O_2 \longrightarrow SO_2$
2. $S + Cl_2 \longrightarrow SCl_2$
3. $S + 3F_2 \longrightarrow SF_6$

49. Why are solutions of oxides or of sulfides basic? Write equations.

Solution

Metallic oxides and sulfides are basic. Both ions are stable and electron rich which makes them good proton acceptors.

$S^{2-} + H_2O \longrightarrow HS^- + OH^-$

$O^{2-} + H_2O \longrightarrow 2OH^-$

HALOGENS

51. Why does fluorine not form allotropes like oxygen does?

Solution

Fluorine forms a stable molecule through the formation of a single covalent bond. No other bond can be formed, a requirement for variability.

53. Write Lewis structures for the following:

(a) HF, HCl, HBr, and HI
(b) ClF, Cl₂, ClBr, and ClI
(c) PF₃, PCl₃, PBr₃, and PI₃
(d) CF₄, CCl₄, CBr₄, and CI₄
(e) KF, KCl, KBr, and KI
(f) HOCl, HOBr, and HOI
(g) HOClO, HOBrO, and HOIO
(h) ClF₃, BrF₃, and IF₃

Solution

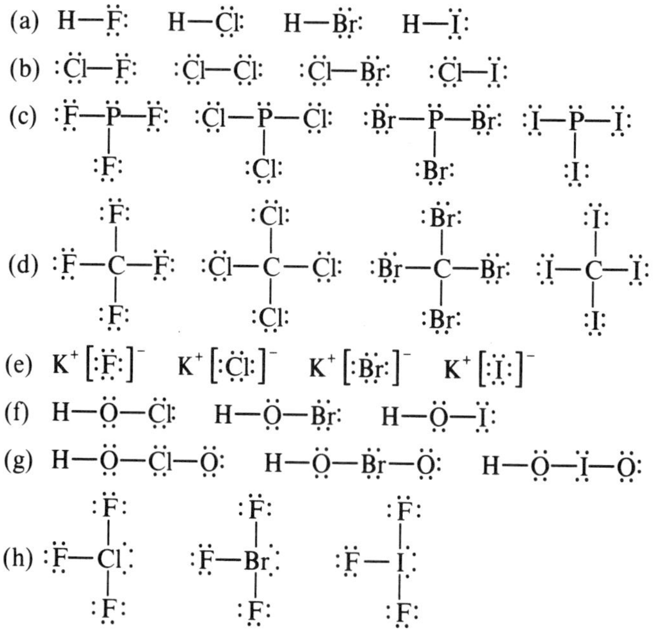

(a) H—F̈:   H—C̈l: H—B̈r: H—Ï:

(b) :C̈l—F̈:   :C̈l—C̈l: :C̈l—B̈r: :C̈l—Ï:

(c) :F̈—P̈—F̈:   :C̈l—P̈—C̈l: :B̈r—P̈—B̈r: :Ï—P̈—Ï:
     | | | |
     :F̈:   :C̈l: :B̈r: :Ï:

(d)   :F̈:   :C̈l: :B̈r: :Ï:
         | | | |
     :F̈—C—F̈:   :C̈l—C—C̈l: :B̈r—C—B̈r: :Ï—C—Ï:
         | | | |
         :F̈:   :C̈l: :B̈r: :Ï:

(e) K⁺[:F̈:]⁻   K⁺[:C̈l:]⁻   K⁺[:B̈r:]⁻   K⁺[:Ï:]⁻

(f) H—Ö—C̈l: H—Ö—B̈r: H—Ö—Ï:

(g) H—Ö—C̈l—Ö: H—Ö—B̈r—Ö: H—Ö—Ï—Ö:

(h)   :F̈:   :F̈:   :F̈:
         | | |
     :F̈—Cl: :F̈—Br: :F̈—I:
         | | |
         :F̈:   :F̈:   :F̈:

Cl, Br, and I have 2 lone pairs of electrons.

Chapter 20 A Survey of the Nonmetals   303

55. Using the reactions of the halogens with sulfur, show that the oxidizing abilities of the halogens decrease as their size increases.

   Solution

   Sulfur forms multiple compounds with fluorine and chlorine, only one with bromine and none directly with iodine.

   $S + 3F_2 \longrightarrow SF_6$
   $S + 2Cl_2 \longrightarrow SCl_4$
   $2S + Br_2 \longrightarrow S_2Br_2$
   $S + I_2 \longrightarrow$ no reaction

57. Write a balanced chemical equation describing the reaction that occurs in each of the following cases.

   (a) Calcium is added to hydrobromic acid.
   (b) Potassium hydroxide is added to hydrofluoric acid.
   (c) Ammonia is bubbled through hydrofluoric acid.
   (d) Silver oxide is added to hydrobromic acid.
   (e) Chlorine is bubbled through hydrobromic acid.

   Solution

   (a) $Ca + 2HBr \longrightarrow CaBr_2 + H_2$
   (b) $KOH + HF \longrightarrow KF + H_2O$
   (c) $NH_3 + HF \longrightarrow NH_4F$
   (d) $Ag_2O + 2HBr \longrightarrow 2AgBr + H_2O$
   (e) $Cl_2 + 2HBr \longrightarrow Br_2 + 2HCl$

59. What mass of PCl_3 can be prepared by the reaction of chlorine with 13.55 g of phosphorus?

   Solution

   $$\begin{array}{cc} 13.55 \text{ g} & \text{Mass?} \\ 2P + 3Cl_2 & \longrightarrow 2PCl_3 \\ 2 \text{ mol} & 2 \text{ mol} \end{array}$$

   g-form. wts.: $PCl_3 = 137.33276$; $P = 30.97376$

   $$\text{Mass } PCl_3 = 13.55 \text{ g P} \times \frac{1 \text{ mol P}}{30.97376 \text{ g}} \times \frac{2 \text{ mol } PCl_3}{2 \text{ mol P}} \times \frac{137.33276 \text{ g}}{\text{mol } PCl_3} = 60.08 \text{ g}$$

61. Calculate the bond energies of the F—F, Cl—Cl, Br—Br, and I—I bonds from the data in Appendix I, and arrange these molecules in increasing order of their X—X bond energy.

   Solution

   $F_2(g) \longrightarrow 2F$ $\Delta H = 2 \times 78.99 \text{ kJ} = 157.98 \text{ kJ}$

   $Cl_2(g) \longrightarrow 2Cl$ $\Delta H = 2 \times 121.68 \text{ kJ} = 243.36 \text{ kJ}$

   $Br_2(l) \longrightarrow Br_2(g) \longrightarrow 2Br$ $\Delta H = 2 \times 111.88 \text{ kJ} - 30.91 \text{ kJ} = 192.85 \text{ kJ}$

   $I_2(s) \longrightarrow I_2(g) \longrightarrow 2I$ $\Delta H = 2 \times 106.84 \text{ kJ} - 62.438 \text{ kJ} = 151.24 \text{ kJ}$

   $I_2 < F_2 < Br_2 < Cl_2$

304   Chapter 20   A Survey of the Nonmetals

NOBLE GASES

63. Write the Lewis structures for and describe the molecular structures of XeF_2 and XeF_4.

    Solution

    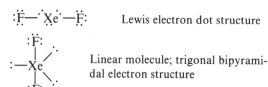
    Lewis electron dot structure

    Linear molecule; trigonal bipyramidal electron structure

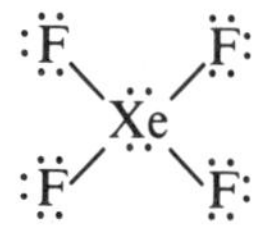

    Lewis electron dot structure

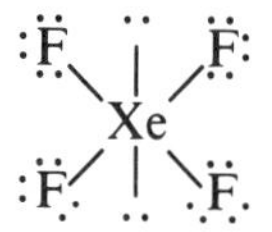

    Square planar molecule: octahedral electron structure

65. A 0.492-g sample of the XeF_4 described in exercise 64 exerted a pressure of 47.8 torr at 127°C in a bulb with a volume of 1238 mL. Determine the molecular formula of xenon tetrafluoride from these data.

    Solution

    Use the ideal gas equation to calculate the molecular weight of the compound.

    $$n = \frac{PV}{RT} = \frac{(47.8/760 \text{ atm})(1.238 \text{ L})}{(.08206 \text{ L atm/mol K})(400.15 \text{ K})} = 0.0023713 \text{ mol}$$

    0.0023713 mol = 0.492 g

    1 mol = 207.48 g

    This molecular weight can only contain one atom of xenon, therefore the difference in mass must be due to fluorine.

    Mass (F) = 207.48 − 131.30 = 76.18

    No. F atoms = 76.18/18.9984 = 4

    XeF_4

ADDITIONAL EXERCISES

67. Which of the following compounds will react with water to give an acid solution? Which will give a basic solution? Write a balanced equation for the reaction of each with water.

    (a) NaH
    (b) H_2S
    (c) HCl
    (d) SO_3

Chapter 20 A Survey of the Nonmetals 305

(e) CaO
(f) Na₃P
(g) PCl₃
(h) N₂O₅
(i) SF₄

Solution

(a) $NaH + H_2O \longrightarrow NaOH + H_2$ Basic
(b) $H_2S + H_2O \longrightarrow H_3O^+ + HS^-$ Acidic
(c) $HCl + H_2O \longrightarrow H_3O^+ + Cl^-$ Acidic
(d) $SO_3 + H_2O \longrightarrow H_2SO_4$ Acidic
(e) $CaO + H_2O \longrightarrow Ca(OH)_2$ Basic
(f) $Na_3P + 3H_2O \longrightarrow 3NaOH + PH_3$ Basic
(g) $PCl_3 + 3H_2O \longrightarrow 3HCl + H_3PO_3$ Acidic
(h) $N_2O_5 + H_2O \longrightarrow 2HNO_3$ Acidic
(i) $SF_4 + 2H_2O \longrightarrow 4HF + SO_2$ Acidic
$SO_2 + H_2O \longrightarrow H_2SO_3$

69. Air contains 20.99% oxygen by volume. What volume of air in cubic meters at 27.0°C and 0.9868 atm is required to oxidize 1.00 metric ton (1000 kg) of sulfur to sulfur dioxide, assuming that all of the oxygen reacts with sulfur?

Solution

Calculate the number of moles of oxygen required from the balanced equation. Use the ideal gas equation to calculate volume at the stated conditions. Finally express the oxygen volume in terms of its air equivalent

$$\begin{array}{ccc} 1 \times 10^6 \text{ g} & & \text{mol?} \\ S & + \quad O_2 & \longrightarrow SO_2 \\ 1 \text{ mol} & 1 \text{ mol} & \end{array}$$

No. mol $(O_2) = 1 \times 10^6$ g S $\times \dfrac{1 \text{ mol}}{32.06 \text{ g}} \times \dfrac{1 \text{ mol } O_2}{1 \text{ mol S}} = 3.11915 \times 10^4$ mol

$PV = nRT \quad V = nRT/P$

$V = (3.11915 \times 10^4 \text{ mol})(0.08206 \text{ l atm/mol K})(300.15 \text{ K})/0.9868$ atm

$= 7.7853 \times 10^5$ L

$V \text{ (air)} = \dfrac{7.7853 \times 10^5 \text{ L}}{0.2099} = 3.71 \times 10^6$ L or 3.71×10^3 m³

71. Iodine reacts with liquid chlorine at −40°C to give an orange compound containing 54.3% iodine and 45.7% chlorine. Write the balanced equation for the reaction.

Solution

Determine the empirical formula for the compound. From the formula, write the equation.

I: $54.3/126.9045 = 0.427881$ $0.427881/1.28903 = 0.33$
Cl: $45.7/35.453 = 1.28903$ $1.28903/1.28903 = 1$

$I_{0.33} Cl_1$ or ICl_3

$I_2 + 3Cl_2 \longrightarrow 2ICl_3$

73. The heat of formation of $CS_2(l)$ is $+21.4$ kcal mol^{-1}. How much heat is required for the reaction of 10.00 g of S_8 with carbon according to the following equation?

$$4C(s) + S_8(s) \longrightarrow 4CS_2(l)$$

Is the reaction exothermic or endothermic?

Solution

$$\underset{1 \text{ mol}}{4C(s)} + \underset{1 \text{ mol}}{\overset{10.00 \text{ g}}{S_8(s)}} \longrightarrow \underset{4 \text{ mol}}{\overset{\text{mol?}}{4CS_2(l)}}$$

No. mol CS_2 = 10.00 g $S_8 \times \dfrac{1 \text{ mol}}{256.48 \text{ g}} \times \dfrac{4 \text{ mol } CS_2}{1 \text{ mol } S_8}$ = 0.15596 mol

ΔH = 0.15596 mol $\times$ 21.4 kcal/mol = 3.34 kcal

ΔH is positive, so the reaction is endothermic.

75. At 422 K the partial pressures of gaseous PCl_5, PCl_3, and Cl_2 in a closed system were 0.453 atm, 6.08×10^{-2} atm, and 6.08×10^{-2} atm, respectively. Determine K_p for the oxidation.

$$PCl_3(g) + Cl_2(g) \longrightarrow PCl_5(g)$$

Solution

Write the expression for K_p, substitute the partial pressures accordingly and compute K_p.

$$PCl_3(g) + Cl_2(g) \longrightarrow PCl_5(g)$$

$$K_p = \dfrac{P_{PCl_5}}{(P_{PCl_3})(P_{Cl_2})} = \dfrac{0.453}{(6.08 \times 10^{-2})(6.08 \times 10^{-2})} = 1.23 \times 10^2$$

21

Electrochemistry and Oxidation-Reduction

INTRODUCTION

Historically, the terms *oxidation* and *reduction* came from the field of metallurgy. Oxidation referred to reactions involving the addition of oxygen to a metal. Reduction, on the other hand, referred to the removal of oxygen from a reacting metal oxide, often by the passage of hydrogen over or through it. Today, these terms are used in chemistry in a much broader sense to describe reactions involving the transfer of electrons from one substance to another.

Reactions involving oxidation and reduction are called *oxidation-reduction* or *redox* reactions; they are quite common. For example, the energy that is required for you to assimilate your thoughts while you read this page is derived from foods that have undergone a variety of complex biochemical oxidation-reduction reactions. Another example may be drawn from industry. The extraction of aluminum from bauxite ore is accomplished by a redox reaction in which an electric current is passed through a molten mixture of the ore. These examples by no means exhaust the diverse forms possible in redox reactions.

The first evidence of a relationship between chemical reactions and electricity was the result of work done by Luigi Galvani on the twitching of frog legs caused by an electrical shock. The results of further work by Galvani set the stage in 1800 for the development by Allesandro Volta of a practical battery based on a silver-zinc couple. This discovery made possible for the first time a continuous source of electric current. In 1807, Sir Humphrey Davy isolated the elements sodium and potassium by electrolyzing fused mixtures of their solid hydroxides. Bear in mind that all this work was accomplished prior to the discovery of subatomic particles — protons, electrons, and neutrons.

The development of this chapter treats the quantitative relationships between chemical change and electrical phenomena. The fundamental laws of electrochemical work were discovered by Michael Faraday during the years 1832 and 1833. Faraday's work showed that the electrical decomposition that occurs in a quantity of substance is proportional to the electrical current passing through that substance and to the time the current is applied. He further extended his research to show that the masses of substances deposited at the electrodes in electrolysis cells are proportional to their chemical equivalent weights. Faraday's laws are now widely applied in research and industry.

READY REFERENCE

Anions (21.2) Negatively charged ions.

Anode (21.2)   The electrode toward which negatively charged ions are attracted; electrons are withdrawn from the electrolytic liquid causing oxidation.

Cathode (21.2)   The electrode toward which positively charged ions are attracted; electrons enter the electrolytic liquid causing reduction.

Cations (21.2)   Positively charged ions.

Electrolysis (21.2)   An oxidation-reduction reaction taking place in an electrolytic cell.

Electrolytic cell (21.2)   A chemical reaction system in which electrical energy is consumed to bring about desired chemical changes. Such chemical changes are by definition nonspontaneous. In the process, electrons are forced from an outside source onto the cathode, making it negatively charged, and electrons are withdrawn from the anode, making it positively charged.

Electromotive series (21.8)   An ordering of the elements according to their tendency to form positive ions. In terms of reduction potentials, potassium has the largest negative value, -2.925 volts for the reaction $K^+ + e^- \longrightarrow K$.

emf (21.7)   An acronym for electromotive force, a force or potential causing an electron flow. It is normally measured in volts.

Faraday's law (21.5)   During electrolysis, 96,485 coulombs (1 faraday) of electricity reduce one gram-equivalent of the oxidizing agent and oxidize one gram-equivalent of the reducing agent. In other words, the amount of substance reacted at each electrode during electrolysis is directly proportional to the quantity of electricity passed through the electrolytic cell.

Nernst equation (21.11)   The equation is defined for reactions and for half-reactions having the general form

$$a\text{A} + b\text{B} \rightleftharpoons c\text{C} + d\text{D} \tag{1}$$

and

$$x\text{M} + n e^- \rightleftharpoons y\text{N} \tag{2}$$

as

$$E = E° - \frac{0.05915}{n} \log Q$$

where

$$Q = \frac{[\text{C}]^c [\text{D}]^d}{[\text{A}]^a [\text{B}]^b}$$

for equation (1) and

$$Q = \frac{[\text{N}]^y}{[\text{M}]^x}$$

for equation (2).

E = emf for the reaction or half-reaction.

$E°$ = standard electrode potential for the cell reaction or the half-reaction.

n = number of electrons required in the redox transfer process according to the balanced equation or half-reactions.

Oxidation (21.2) Oxidation involves the removal or loss of electrons from a substance. The elements undergoing oxidation are determined from changes in oxidation states or numbers in the following way. The oxidation number 0 is assigned to neutral species. For each electron in excess of the number in the neutral species, the state is decreased by 1, or for each electron less than that occurring in the neutral species, the state is increased by 1. Thus Cl^{-1} indicates a gain of one electron; its oxidation number is -1. On the other hand, Cu^{2+} indicates a loss of two electrons; its oxidation number is $+2$. By use of this convention, the substance undergoing oxidation shows a net increase in oxidation number. Examples include

$$Sn^{2+} - 2e^- \longrightarrow Sn^{4+}$$
$$Fe^0 - 2e^- \longrightarrow Fe^{2+}$$
$$O^{2-} - 2e^- \longrightarrow O^0$$

Oxidizing agent (21.2) The substance in a reaction causing the oxidation of a substance.

Reducing agent (21.2) The substance in a reaction causing the reduction of a substance. In the following reaction,

$$Fe^0 + Sn^{2+} \longrightarrow Fe^{2+} + Sn^0$$

iron (0) is oxidized to iron (2+) and tin (2+) is reduced to tin (0). Tin (2+) is the oxidizing agent and iron (0) is the reducing agent.

Reduction (21.2) The reduction of a substance involves the absorption or gain of electrons. By the convention used to assign oxidation numbers to reactants, the substance undergoing reduction shows a net decrease in oxidation number. Examples include

$$Fe^{3+} + e^- \longrightarrow Fe^{2+}$$
$$Cl^0 + e^- \longrightarrow Cl^-$$
$$Ag^+ + e^- \longrightarrow Ag^0$$

Standard hydrogen electrode (21.6) Prepared by bubbling hydrogen gas at 25°C and a pressure of 1 atm around a platinized platinum electrode immersed in a solution in which hydrogen ions are at unit activity (approximately 1 M).

Standard potential (21.7) Potential of the electrode measured at 25°C when the concentration of the ions in the solution are at unit activity ($\approx$ 1 M) and the pressure of any gas involved is 1 atm.

Thermodynamic functions and their relation to $E°$ (21.12) Several relations are possible:

$$\Delta G° = -nFE°$$

$$E° = \frac{RT}{nF} \ln K_e = \frac{0.05915}{n} \log K_e$$

where n is the number of electrons transferred and F is the faraday, a constant in units of kilojoules per volt ($F = 96.485$ kJ/V), and the temperature is defined at 298.15 K.

Voltaic cells (21.7) Commonly thought of as batteries, voltaic cells have as their negative terminal the anode, where oxidation occurs. Reduction occurs at the cathode, but in contrast to the situation in electrolytic cells, the cathode is the positive terminal.

BALANCING EQUATIONS: TECHNIQUES FOR OXIDATION-REDUCTION REACTIONS

Several methods are commonly used to balance equations involving oxidation and reduction. Regardless of the method used to balance equations, electrons lost through oxidation must be absorbed through reduction; no net gain or net loss of electrons can occur in a reaction. Further, for an equation to be balanced, the total of the ion charges on both sides of the equation must be equal. The two methods most used for balancing equations are presented here for your study.

Change in Oxidation Number Method: This is based on the concept that in a redox reaction the total increase in units of positive oxidation number must equal the total decrease in units of negative oxidation number. Consider the following reaction as an example:

$$MnO_4^- + C_2O_4^{2-} + H^+ \longrightarrow Mn^{2+} + CO_2 + H_2O$$

Assign oxidation numbers to each element, determine those elements that undergo oxidation or reduction, and then write half-reactions that indicate the oxidation number changes (sometimes these are half-reactions for the overall reaction). In this case, the half-reactions are

$$C_2^{3+} \longrightarrow 2C^{4+} + 2e^- \qquad \uparrow 2 \quad \text{(gain in positive oxidation number = 2)}$$

$$Mn^{7+} \longrightarrow Mn^{2+} - 5e^- \qquad \downarrow 5 \quad \text{(decrease in positive oxidation number = 5)}$$

To balance the equation, the proper number of manganese-containing and carbon-containing ions must be selected, so that the total increase of electrons will equal the total decrease. This can be done by choosing the smallest factor common to both 2 and 5 and multiplying each half-reaction by the multiple of each in the factor. For these half-reactions, 10 is the common factor, and the half-reactions must be multiplied by 5 and 2, respectively.

$$5[C_2^{3+} \longrightarrow 2C^{4+} + 2e^-] \qquad \uparrow 2 \quad 5 \times 2e^- = 10e^-$$

$$2[Mn^{7+} \longrightarrow Mn^{2+} - 5e^-] \qquad \downarrow 5 \quad 2 \times 5e^- = 10e^-$$

The proper coefficients for those substances changing oxidation state become

$$2MnO_4^- + 5C_2O_4^{2-} + ?H^+ \longrightarrow 2Mn^{2+} + 10CO_2 + ?H_2O$$

The remaining factors in the equation can be balanced by inspection, yielding

$$2MnO_4^- + 5C_2O_4^{2-} + 16H^+ \longrightarrow 2Mn^{2+} + 10CO_2 + 8H_2O$$

Ion-Electron Method: Based on combining half-reactions, one-half of which represents the oxidation step and the other, the reduction step. This method is very useful in electrochemical cells, for which, generally, the half-reactions are known. The same reaction used in the change in oxidation number method will be used to illustrate this method.

The oxalate ion is oxidized to carbon dioxide

$$C_2O_4^{2-} \longrightarrow 2CO_2 + 2e^- \tag{1}$$

and is balanced by inspection. The oxidizing agent in this reaction, permanganate, is reduced in acid solution to manganous ion and water:

Chapter 21 Electrochemistry and Oxidation-Reduction 311

$$MnO_4^- + 8H^+ + 5e^- \rightarrow Mn^{2+} + 4H_2O \qquad (2)$$

Again, the balancing is done by inspection.

It is now obvious that multiplying Equation (1) by 5 and Equation (2) by 2 will make the number of electrons lost by oxalate the same as the number gained by permanganate. The two equations are added to give the following:

$$5C_2O_4^{2-} \rightarrow 10CO_2 + 10e^-$$
$$\underline{2MnO_4^- + 16H^+ + 10e^- \rightarrow 2Mn^{2+} + 8H_2O}$$
$$2MnO_4^- + 5C_2O_4^{2-} + 16H^+ \rightarrow 2Mn^{2+} + 10CO_2 + 8H_2O$$

Both methods give the same result, as should be expected.

SOLUTIONS TO CHAPTER 21 EXERCISES

2. How does a voltaic cell differ from an electrolytic cell?

   Solution

   A voltaic cell spontaneously delivers current at a particular voltage. The electrolytic cell requires an outside source of electricity such as a battery or rectified line current in order for the cell to produce the desired products.

4. Complete and balance the following half-reactions. (In each case indicate whether oxidation or reduction occurs.)

   (b) $I_2 \rightarrow I^-$
   (c) $AgCl \rightarrow Ag + Cl^-$
   (e) $H_2O_2 \rightarrow O_2$ (in acid)
   (g) $MnO_4^- \rightarrow MnO_2$ (in base)

   Solution

   (b) $I_2 + 2e^- \rightarrow 2I^-$; the iodine gains two electrons, so the process is a reduction.

   (c) $AgCl + e^- \rightarrow Ag + Cl^-$; the Ag goes from an oxidation state of $+1$ to 0 in elemental silver. This is a gain of electrons and is a reduction.

   (e) $H_2O_2 \rightarrow O_2 + 2H^+ + 2e^-$; the oxygen goes from a -1 oxidation state to 0, a loss of electrons. The hydrogen peroxide is oxidized; it is a reducing agent.

   (g) $MnO_4^- + 2H_2O + 3e^- \rightarrow MnO_2 + 4OH^-$; the Mn goes from a $+7$ to a $+4$ oxidation state, a gain of electrons or reduction. Permanganate ion is an oxidizing agent in base or neutral solution.

ELECTROLYTIC CELLS AND FARADAY'S LAW OF ELECTROLYSIS

5. Describe the electrolytic purification of aluminum, the Hoopes process, using half-reactions to describe the changes at the anode and cathode.

   Solution

   In the Hoopes process, the bottom molten layer serves as the anode and consists of a molten alloy of copper and impure aluminum. In the bottom layer, the anode is

   $$Al \text{ (alloy)} \rightarrow Al^{3+} + 3e^- \quad \text{(oxidation)}$$

312   Chapter 21   Electrochemistry and Oxidation-Reduction

In the top layer, the cathode is

$$Al^{3+} + 3e^- \longrightarrow Al \text{ (pure)} \quad \text{(reduction)}$$

7. Using a diagram of the type

$$\xrightarrow{e^-}$$

Anode | anode soln ‖ cathode soln | cathode

diagram the electrolytic cell for each of the following cell reactions:

(b) $2NaCl(aq) + 2H_2O \longrightarrow 2NaOH(aq) + H_2(g) + Cl_2(g)$
(using inert electrodes of titanium, Ti)

(c) $Cu + 2Ag^+ \longrightarrow Cu^{2+} + 2Ag$

Solution

See exercise 29 for a discussion of cell conventions.

(b) The anodic reaction is the oxidation of chloride ions: $2Cl^- \longrightarrow Cl_2 + 2e^-$. The cathodic reaction is the reduction of water: $2H_2O + 2e^- \longrightarrow H_2 + 2OH^-$ giving an overall reaction of

$$2H_2O(l) + 2Cl^-(aq) \longrightarrow H_2(g) + Cl_2(g) + 2OH^-$$

Using the convention of the anode on the left, we have

$$Ti \mid Cl^- \mid Cl_2 \parallel H_2O, OH^- \mid H_2 \mid Ti$$

The sodium does not appear because it is unchanged in the reaction.

(c) $Cu + 2Ag^+ \longrightarrow Cu^{2+} + 2Ag$

anodic reaction: $Cu \longrightarrow Cu^{2+} + 2e^-$
cathodic reaction: $Ag^+ + e^- \longrightarrow Ag$

$$Cu \mid Cu^{2+} \parallel Ag^+ \mid Ag$$

8. Write the anode half-reaction, the cathode half-reaction, and the cell reaction of the following electrolytic cells:

$$\xrightarrow{e^-}$$

(a) $C \mid NaCl(l) \mid Cl_2 \parallel NaCl(l) \mid Na(l) \mid Fe$

$$\xrightarrow{e^-}$$

(d) $Pb \mid PbSO_4 \mid H_2SO_4(aq) \parallel H_2SO_4(aq) \mid PbO_2 \mid PbSO_4 \mid Pb$

Solution

(a) The anodic half-reaction is

$$2Cl^- \longrightarrow Cl_2 + 2e^- \quad \text{(anodic oxidation)}$$

The cathodic half-reaction is

$$Na^+ + e^- \longrightarrow Na \quad \text{(cathodic reduction)}$$

The overall cell reaction is obtained by adding the electrode reactions.

anode: $2Cl^- \longrightarrow Cl_2 + 2e^-$
cathode: $\underline{2(Na^+ + e^- \longrightarrow Na)}$
               $2Na^+ + 2Cl^- \longrightarrow Cl_2 + 2Na$

(d) This is the diagram for the charging of a lead-acid battery. The anodic half-reaction is

Chapter 21 Electrochemistry and Oxidation-Reduction

$$PbSO_4(s) + 2H_2O \longrightarrow PbO_2 + 4H^+ + SO_4^{2-} + 2e^- \quad \text{(anodic oxidation)}$$

The cathodic half-reaction is

$$PbSO_4(s) + 2e^- \longrightarrow Pb + SO_4^{2-} \quad \text{(cathodic reduction)}$$

The overall cell reaction is obtained by adding the two electrode reactions.

$$2PbSO_4(s) + 2H_2O \longrightarrow Pb + PbO_2 + 4H^+ + 2SO_4^{2-}$$

9. Tarnished silverware is coated with Ag_2S. The tarnish can be removed by placing the silverware in an aluminum pan and covering it with a solution of an inert electrolyte such as NaCl. Explain the electrochemical basis for this procedure.

Solution

Since Al is higher in the electromotive series than is Ag, Al will replace Ag from its compounds if the two are in contact.

$$2Al(s) + 3Ag_2S(s) \xrightarrow[\text{electrolyte}]{\text{NaCl}} 6Ag(s) + Al_2S_3(s)$$

12. Calculate the value of the Faraday constant, F, from the charge on a single electron, 1.6021×10^{-19} C.

Solution

The Faraday constant is the charge on one mole of electrons.

$$F = 1.6021 \times 10^{-19} \frac{C}{e^-} \times 6.022 \times 10^{23} \frac{e^-}{mol} = 9.648 \times 10^4 \text{ C}$$

13. How many moles of electrons are involved in the following electrochemical changes?

   (a) 0.800 mol of I^- is converted to I_2
   (c) 27.6 g of SO_3 is converted to SO_3^{2-}
   (g) The Mn^{2+} in 12.58 mL of 0.1145 M $Mn(NO_3)_2$ is converted to MnO_4^-

Solution

(a) The reaction is

$$2I^- \longrightarrow I_2 + 2e^-$$

$$\text{mol } e^- = \frac{1 \text{ mol } e^-}{\text{mol } I^-} \times 0.800 \text{ mol} = 0.800 \text{ mol}$$

(c) The reaction is

$$SO_3 + 2e^- \longrightarrow SO_3^{2-}$$

One mol SO_3 requires 2 mol e^- for reduction.

$$\text{mol }(SO_3) = 27.6 \text{ g} \times \frac{1 \text{ mol}}{80.06 \text{ g}} = 0.345 \text{ mol}$$

$$\text{mol } e^- = \frac{2 \text{ mol } e^-}{\text{mol } SO_3} \times 0.345 \text{ mol} \quad SO_3 = 0.690 \text{ mol}$$

314   Chapter 21 Electrochemistry and Oxidation-Reduction

(g) The reaction is

$$Mn^{2+} + 4H_2O \longrightarrow MnO_4^- + 8H^+ + 5e^-$$

The amount of Mn^{2+} is 0.01258 L × 0.1145 M = 1.4404 × 10^{-3} mol

$$\text{mol } e^- = \frac{5 \text{ mol } e^-}{1 \text{ mol } Mn^{2+}} \times 1.4404 \times 10^{-3} \text{ mol} = 7.202 \times 10^{-3} \text{ mol}$$

14. How many faradays of electricity are involved in the electrochemical changes described in exercise 13?

Solution

One faraday of charge provides 1 mole of electrons.

(a) $\text{faradays} = 0.800 \text{ mol } e^- \times \dfrac{1 \text{ faraday}}{\text{mol } e^-} = 0.800 \text{ F}$

(c) $\text{faradays} = 0.690 \text{ mol } e^- \times \dfrac{1 \text{ faraday}}{\text{mol } e^-} = 0.690 \text{ F}$

(g) $\text{faradays} = 7.202 \times 10^{-3} \text{ mol} \times \dfrac{1 \text{ faraday}}{\text{mol } e^-} = 7.202 \times 10^{-3} \text{ F}$

15. How many coulombs of electricity are involved in the electrochemical changes described in exercise 13?

Solution

(a) $\text{coulombs} = 0.800 \text{ mol } e^- \times \dfrac{1 \text{ faraday}}{\text{mol } e^-} \times \dfrac{96,485 \text{ C}}{\text{faraday}} = 7.72 \times 10^4 \text{ C}$

(c) $\text{coulombs} = 0.690 \text{ mol } e^- \times \dfrac{1 \text{ faraday}}{\text{mol } e^-} \times \dfrac{96,485 \text{ C}}{\text{faraday}} = 6.66 \times 10^4 \text{ C}$

(g) $\text{coulombs} = 7.202 \times 10^{-3} \text{ mol } e^- \times \dfrac{1 \text{ faraday}}{\text{mol } e^-} \times \dfrac{96,485 \text{ C}}{\text{faraday}} = 6.95 \times 10^2 \text{ C}$

18. Ammonium perchlorate, NH_4ClO_4, used in the solid fuel in the booster rockets on the space shuttle, is prepared from sodium perchlorate, $NaClO_4$, which is produced commercially by the electrolysis of a hot, stirred solution of sodium chloride.

$$NaCl + 4H_2O \longrightarrow NaClO_4 + 4H_2$$

How many moles of electrons are required to produce 1.00 kg of sodium perchlorate? How many faradays? How many coulombs?

Solution

The reaction at the anode is an oxidation

$$Cl^- \longrightarrow Cl^{7+} + 8e^-$$

This equation tells us that 1 mol of Cl^{7+} is produced for each 8 mol of electrons passing through the cell. There are

$$1000 \text{ g} \times \frac{1 \text{ mol}}{122.440 \text{ g}} = 8.17 \text{ mol } NaClO_4$$

in 1.00 kg. This requires

$$8.17 \text{ mol NaClO}_4 \times \frac{8 \text{ mol e}^-}{\text{mol NaClO}_4} = 65.3 \text{ mol e}^-$$

$$65.3 \text{ mol e}^- \times \frac{1 \text{ faraday}}{1 \text{ mol e}^-} = 65.3 \text{ faraday}$$

$$65.3 \text{ faraday} \times \frac{96,485 \text{ C}}{1 \text{ faraday}} = 6.30 \times 10^6 \text{ C}$$

20. How many grams of tin will be deposited from a solution of tin(II) nitrate by 3.40 F of electricity?

Solution

The reduction of 1 mol of tin(II) ions requires 2 mol of electrons.

$$\text{Sn}^{2+} + 2\text{e}^- \longrightarrow \text{Sn}$$

And 1 faraday of charge furnishes 1 mol of electrons for the reaction. Hence, 2 faraday of charge reduce 1 mol of tin(II) ions.

$$\text{Mass Sn} = 3.40 \text{ faraday} \times \frac{1 \text{ mol Sn}}{2 \text{ faraday}} \times \frac{118.69 \text{ g}}{\text{mol}} = 202 \text{ g}$$

21. An experiment is conducted using the apparatus depicted in Figure 21.8. How many grams of gold would be plated out of solution by the current required to plate out 4.97 g of silver? How many moles of hydrogen gas simultaneously are released from the hydrochloric acid solution? How many moles of oxygen gas are freed at each anode in the copper sulfate and silver nitrate solutions?

Solution

The reactions are $\text{Ag}^+ + \text{e}^- \longrightarrow \text{Ag}$ and $\text{Au}^{3+} + 3\text{e}^- \longrightarrow \text{Au}$. The moles of Ag plated will be 3 times the amount of Au plated, because Au requires 3 times the amount of electrons for reduction.

$$\text{Mol Ag} = 4.97 \text{ g}/107.868 \text{ g mol}^{-1} = 0.0461 \text{ mol}$$

$$\text{g Au/mol Ag} = \frac{1 \text{ mol Au}}{3 \text{ mol Ag}} \times 0.0461 \text{ mol Ag} \times 196.9665 \text{ g mol}^{-1} \text{ Au} = 3.03 \text{ g}$$

The amount of H$_2$ released is found from the reaction: $2\text{H}^+ + 2\text{e}^- \rightleftharpoons \text{H}_2(g)$. The reduction requires 2 mol e$^-$. Therefore,

$$\text{Mol H}_2 = 0.0461 \text{ mol Ag} \times \frac{1 \text{ mol H}_2}{2 \text{ mol Ag}} = 2.30 \times 10^{-2} \text{ mol}$$

The amount of O$_2$ generated in Au electrolysis = amount of O$_2$ generated in Ag electrolysis = $\frac{1}{2}$(moles of H$_2$ produced). Therefore,

$$\text{Mol O}_2 = \tfrac{1}{2}(2.30 \times 10^{-2} \text{ mol}) = 1.15 \times 10^{-2} \text{ mol}$$

23. How many grams of cobalt will be deposited from a solution of cobalt(II) chloride electrolyzed with a current of 10.0 A for 30.0 min?

Solution

$$\text{g Co} = 10.0 \text{ A} \times 30.0 \text{ min} \times \frac{60 \text{ s}}{\text{min}} \times \frac{1 \text{ mol e}^-}{96,485 \text{ C}} \times \frac{1 \text{ mol Co}}{2 \text{ mol e}^-} \times \frac{58.9332 \text{ g}}{\text{mol}} = 5.50 \text{ g}$$

25. Chromium metal can be plated electrochemically from an acidic aqueous solution of CrO_4^{2-}.

   (a) What is the half-reaction for the process?
   (b) What mass of chromium, in grams, will be deposited by a current of 250 A passing for 20.0 min?
   (c) How long will it take to deposit 1.0 g of chromium using a current of 10.0 A?

   Solution

   (a) $CrO_4^{2-} + 8H^+ + 6e^- \rightarrow Cr + 4H_2O$
   (b) amps × time → coulombs → moles of electrons → moles of Cr → mass of copper

   $$g\ Cr = 250\ A \times 20.0\ min \times \frac{60\ s}{min} \times \frac{1\ mol\ e^-}{96,485\ C} \times \frac{1\ mol\ Cr}{6\ mol\ e^-} \times \frac{51.996\ g\ Cr}{mol\ Cr}$$

   $$= 26.9\ g$$

   (c) $t = \dfrac{1.0\ g \times \dfrac{96485\ C}{1\ mol\ e} \times \dfrac{6\ mol\ e^-}{1\ mol\ Cr} \times \dfrac{1\ mol\ Cr}{51.996\ g\ Cr}}{10.0\ A} = 1100\ s$

27. Which metals in Table 21.1 could be purified by an electrolysis similar to that used to purify copper?

   Solution

   In aqueous solution, as long as the oxidation of the metal is easier than the oxidation of water, the anode metal dissolves as the metal ion. In molten metal, impurities that do not oxidize as easily as the metal of interest will not dissolve. Those impurity metals which do dissolve can be prevented from being reduced at the cathode by careful regulation of voltage. From aqueous solution the metals that can be purified are those with $E°$ greater than -0.8277 V for the reaction

   $$2H_2O + 2e^- \rightleftharpoons H_2 + 2OH^-$$

   namely, Zn, Cr, Fe, Cd, Co, Ni, Sn, Pb, Cu, Ag, Hg, Pt, Au.

STANDARD REDUCTION POTENTIALS AND ELECTROMOTIVE FORCE

29. Diagram voltaic cells having the following net reactions:

   (b) $Mn + 2Ag^+ \rightarrow Mn^{2+} + Ag$
   (d) $Cr_2O_7^{2-} + 14H^+ + 6I^- \rightarrow 3I_2 + 2Cr^{3+} + 7H_2O$

   Solution

   The anode reaction, by convention, is placed on the left. Single vertical lines represent a phase separation, and the double vertical lines represent a salt bridge or separation of the two electrolytic solutions. The oxidized form of the anode is placed adjacent to and to the right of the anode substance. The cathode always appears to the extreme right with its oxidized form to its left separated by a single vertical line.

Chapter 21 Electrochemistry and Oxidation-Reduction 317

(b) anodic oxidation: $Mn \longrightarrow Mn^{2+} + 2e^-$
 cathodic reduction: $Ag^+ + e^- \longrightarrow Ag$

   $Mn \mid Mn^{2+} \parallel Ag^+ \mid Ag$

(d) anodic oxidation: $2I^- \longrightarrow I_2 + 2e^-$
 cathodic reduction: $Cr_2O_7^{2-} + 14H^+ + 6e^- \longrightarrow 2Cr^{3+} + 7H_2O$

   $I^- \mid I_2 \parallel H^+, Cr_2O_7^{2-}, H_2O \mid Cr^{3+}$

31. Why is the potential for the standard hydrogen electrode listed at 0.00?

Solution

Single electrode potentials are impossible to determine. Consequently, they can be measured only by reference to some other electrode. By convention the hydrogen electrode is picked. For convenience, the value is chosen as 0.00.

32. Which is the better oxidizing agent in each of the following pairs at standard conditions?

(a) Al^{3+} or Cu^{2+}
(c) MnO_4^- or $Cr_2O_7^{2-}$ (in acid solution)
(e) MnO_4^- in acid or MnO_4^- in base

Solution

In each of the following, the half-reaction with the higher $E°$ is the better oxidizing agent.

(a) Cu^{2+}

  $Cu^{2+} + 2e^- \longrightarrow Cu$ $E° = +0.337$ V

  $Al^{3+} + 3e^- \longrightarrow Al$ $E° = -1.66$ V

(c) MnO_4^-

  $MnO_4^- + 8H^+ + 5e^- \longrightarrow Mn^{2+} + 4H_2O$ $E° = +1.51$ V

  $Cr_2O_7^{2-} + 14H^+ + 6e^- \longrightarrow 2Cr^{3+} + 7H_2O$ $E° = +1.33$ V

(e) In acid: $MnO_4^- + 8H^+ + 5e^- \rightleftharpoons Mn^{2+} + 4H_2O$ $E° = 1.51$ V
 In base: $MnO_4^- + 2H_2O + 3e^- \rightleftharpoons MnO_2 + 4OH^-$ $E° = 0.595$ V
 It is a better oxidizing agent in acid.

33. Which is the better reducing agent in each of the following pairs at standard state conditions?

(a) F^- or Cl^-
(c) Sn^{2+} or Co^{2+}

Solution

The better reducing agent is itself oxidized. The species with the smallest reduction potential is the best reducing agent in its reduced form.

(a) Cl^-

  $F_2 + 2e^- \rightleftharpoons 2F^-$ $E° = +2.87$ V

  $Cl_2 + 2e^- \rightleftharpoons 2Cl^-$ $E° = +1.36$ V

(c) Sn^{2+}

$$Co^{3+} + e^- \rightarrow Co^{2+} \quad E° = +1.81 \text{ V}$$
$$Sn^{4+} + 2e^- \rightarrow Sn^{2+} \quad E° = +0.15 \text{ V}$$

34. Calculate the emf of each cell at standard conditions based on the following reactions:

   (a) $Co + Cl_2 \rightarrow Co^{2+} + 2Cl^-$
   (c) $Pt^{2+} + 2Cl^- \rightarrow Pt + Cl_2$
   (e) $Mn + 2Hg_2Cl_2 \rightarrow Mn^{2+} + 2Cl^- + 2Hg$

Solution

In order to determine the emf, combine the half-cell reactions so that the electrons cancel in the equation. The emf is the sum of the potentials.

(a)
$$Co \rightarrow Co^{2+} + 2e^- \quad E° = +0.277 \text{ V}$$
$$\underline{Cl_2 + 2e^- \rightarrow 2Cl^- \quad E° = +1.3595 \text{ V}}$$
$$Co + Cl_2 \rightarrow Co^{2+} + 2Cl^- \quad E° = +1.636 \text{ V}$$

(c)
$$Pt^{2+} + 2e^- \rightarrow Pt \quad E° = +1.2 \text{ V}$$
$$\underline{2Cl^- \rightarrow Cl_2 + 2e^- \quad E° = -1.3595 \text{ V}}$$
$$Pt^{2+} + 2Cl^- \rightarrow Pt + Cl_2 \quad E° = -0.2 \text{ V}$$

(e)
$$Mn \rightarrow Mn^{2+} + 2e^- \quad E° = +1.18$$
$$\underline{Hg_2Cl_2 + 2e^- \rightarrow 2Hg + 2Cl^- \quad E° = +0.27 \text{ V}}$$
$$Mn + Hg_2Cl_2 \rightarrow Mn^2 + 2Cl^- + 2Hg \quad E° = +1.45 \text{ V}$$

35. Determine the standard emf for each of the following cells:

   (a) $Co \mid Co^{2+}(1M) \xrightarrow{e^-} \parallel Cr^{3+}(1\,M) \mid Cr$

   (c) $Pb \mid PbSO_4(s) \mid SO_4^{2-}(1\,M) \xrightarrow{e^-} \parallel H^+(1\,M) \mid H_2(1 \text{ atm}) \mid Pt$

Solution

(a) Cr lies higher than Co in the electromotive series and will reduce cobalt ion. However, in this cell Co is oxidized to Co^{2+}, and Cr^{3+} is reduced to Cr. This reversal from normal conditions would indicate that an overall negative sign is expected. To obtain the net cell reaction, the half-reactions are added as follows:

$$3(Co \rightleftharpoons Co^{2+} + 2e^-) \quad E° = +0.277 \text{ V}$$
$$\underline{2(Cr^{3+} + 3e^- \rightleftharpoons Cr) \quad E° = -0.74 \text{ V}}$$
$$3Co + 2Cr^{3+} \rightleftharpoons 3Co^{2+} + 2Cr \quad E° = -0.46 \text{ V}$$

(c)
$$Pb(s) + SO_4^{2-} \rightarrow PbSO_4(s) + 2e^- \quad E° = +0.356 \text{ V}$$
$$\underline{2H^+ + 2e^- \rightarrow H_2(g) \quad E° = +0.000 \text{ V}}$$
$$Pb + H_2SO_4 \rightarrow H_2(g) + PbSO_4 \quad E° = +0.356 \text{ V}$$

Chapter 21 Electrochemistry and Oxidation-Reduction

36. Write the cell reaction for a voltaic cell based on each of the following pairs of half-reactions, and calculate the emf of the cell under standard conditions.

    (b) $S + 2e^- \longrightarrow S^{2-}$
    $I_2 + 2e^- \longrightarrow 2I^-$

    (e) $HClO_2 + 2H^+ + 2e^- \longrightarrow HClO + H_2O$
    $ClO_3^- + 3H^+ + 2e^- \longrightarrow HClO_2 + H_2O$

Solution

(b) $\quad S + 2e^- \rightleftharpoons S^{2-} \quad E° = -0.47627$ V (1)

$\quad\quad I_2 + 2e^- \rightleftharpoons 2I^- \quad E° = +0.5355$ V (2)

Reversing the sense of (1) and adding to (2) gives

$\quad\quad \underline{S^{2-} \longrightarrow S + 2e^- \quad\quad E° = +0.47627}$

$\quad\quad I_2 + S^{2-} \longrightarrow 2I^- + S \quad E° = +1.0118$ V

(e) $\quad HClO_2 + 2H^+ + 2e^- \rightleftharpoons HClO + H_2O \quad E° = 1.64$ V

$\quad\quad ClO_3^- + 3H^+ + 2e^- \rightleftharpoons HClO_2 + H_2O \quad E° = 1.21$ V

The first reaction is more favored to occur in the direction written than the second one because of the larger positive sign. Therefore, reverse the sense of the second equation for the anode reaction and add.

Anode: $\quad HClO_2 + H_2O \longrightarrow ClO_3^- + 3H^+ + 2e^- \quad E° = -1.21$ V

Cathode: $\quad \underline{HClO_2 + 2H^+ + 2e^- \longrightarrow HClO + H_2O \quad E° = +1.64\ V}$

Cell reaction: $\quad 2HClO_2 \longrightarrow HClO + ClO_3^- + H^+ \quad E° = +0.43$ V

37. Rechargeable nickel-cadmium cells are used in calculators and other battery-powered devices. The Telstar communication satellite also uses these cells. The cell reaction is

    $NiO_2 + Cd + 2H_2O \longrightarrow Ni(OH)_2 + Cd(OH)_2$

    Calculate the emf of such a cell using the following half-cell potentials:

    $NiO_2 + 2H_2O + 2e^- \longrightarrow Ni(OH)_2 + 2OH^- \quad E° = +0.49$ V

    $Cd(OH)_2 + 2e^- \longrightarrow Cd + 2OH^- \quad E° = -0.81$ V

Solution

The tendency of Cd to go to $Cd(OH)_2$ is greater than that of NiO_2 to go to $Ni(OH)_2$. Write the $Cd(OH)_2$ half-cell in the reverse sense and change the sign of its $E°$. The two half-cells are

$\quad NiO_2 + 2H_2O + 2e^- \longrightarrow Ni(OH)_2 + 2OH^- \quad E° = +0.49$ V

$\quad Cd + 2OH^- \longrightarrow Cd(OH)_2 + 2e^- \quad E° = +0.81$ V

The reactions are added and give an $E°$ of 0.49 V + 0.81 V = +1.30 V for the cell reaction.

39. Under standard conditions the potential of the cell diagrammed below is +0.74 V. What is the metal M? Show your calculations.

    $M \mid M^{n+} \parallel Cu^{2+} \mid Cu$

Solution

Copper acts as the cathode in this process because of its position in the diagram. Reduction occurs at that electrode. The process is

$$Cu^{2+} + 2e^- \rightleftharpoons Cu \qquad E° = 0.337$$

At the anode, the metal must go into solution. The reaction is

$$M \longrightarrow M^{n+} + ne^- \qquad E° = +?\ V$$

The value must be positive since ? V + 0.337 V = +0.74 V. The unknown voltage must be a +0.403 V. In a table of reduction values, the listing will be

$$M^{n+} + ne^- \longrightarrow M \qquad E° = -0.403\ V$$

This corresponds to the metal Cd.

THE NERNST EQUATION

42. Why does the potential of a single electrode change as the concentrations of the species involved in the half-reaction change?

Solution

The potential changes because the ratio of oxidized to reduced forms influences the ability of one form to convert to the other. The "driving force" for reaction differs for example, when there is none of the second form present from the situation in which there is a large amount of the second form present. The expected voltage is controlled by the Nernst equation.

44. Calculate the emf for each of the following half-reactions:

(a) $Sn^{2+}(0.0100\ M) + 2e^- \longrightarrow Sn$

(c) $O_2(0.0010\ atm) + 4H^+(0.100\ M) + 4e^- \longrightarrow 2H_2O(l)$

(d) $Cr_2O_7^{2-}(0.150\ M) + 14H^+(0.100\ M) + 6e^- \longrightarrow 2Cr^{3+}(0.000100\ M)$
$$+ 7H_2O(l)$$

Solution

(a) $Sn^{2+}(0.0100\ M) + 2e^- \longrightarrow Sn$. The emf for this half-cell is calculated from the Nernst equation by assuming that the activity or concentration of the free metal has a value of 1. At unit concentration

$$E = E° - \frac{0.05915\ V}{n} \log Q$$

the emf, $E°$, is -0.136 V, but at a lower concentration of Sn^{2+}, 0.0100 M, the cell potential is

$$E = -0.136\ V - \frac{0.05915\ V}{2} \log \frac{[Sn]}{[Sn^{2+}]}$$

$$= -0.136\ V - \frac{0.05915}{2} \left[\log \frac{1}{0.01}\right] V$$

$$= -0.136\ V = \frac{0.05915}{2}(2)\ V = -0.195\ V$$

(c) $O_2(0.0010 \text{ atm}) + 4H^+(0.100 \text{ M}) + 4e^- \longrightarrow 2H_2O(l)$

$$E = E° - \frac{0.05915}{n} \log Q$$

$$= 1.23 \text{ V} - \frac{0.05915}{4} \log \frac{[H_2O]^2}{P_{O_2}[H^+]^4}$$

The water produced is pure and its concentration or activity is considered unity.

$$E = 1.23 \text{ V} - \frac{0.05915}{4} \log \frac{(1)^2}{(10^{-3})(10^{-1})^4}$$

$$= 1.23 \text{ V} - \frac{0.05915}{4} (7) \text{ V} = 1.13 \text{ V}$$

(d) $Cr_2O_7^{2-}(0.150 \text{ M}) + 14H^+(0.100 \text{ M}) + 6e^- \longrightarrow 2Cr^{3+}(0.000100 \text{ M}) + 7H_2O(l)$

$$E = E° - \frac{0.05915}{n} (\log Q) \text{ V}$$

$$= 1.33 \text{ V} - \frac{0.05915}{6} \left(\log \frac{[Cr^{3+}]^2}{[Cr_2O_7^{2-}][H^+]^{14}} \right) \text{ V}$$

$$= 1.33 \text{ V} - \frac{0.05915}{6} \left(\log \frac{(1.00 \times 10^{-4})^2}{(0.150)(10^{-1})^{14}} \right) \text{ V}$$

$$= 1.33 \text{ V} - \frac{0.05915}{6} (\log 6.67 \times 10^6) \text{ V}$$

$$= 1.33 \text{ V} - 0.0673 \text{ V} = 1.26 \text{ V}$$

45. Hypochlorous acid, HOCl, is a stronger oxidizing agent in acidic solution than in neutral solution. Calculate the potential for the reduction of HOCl to Cl$^-$ in a solution with a pH of 7.00 in which [HOCl] and [Cl$^-$] both equal 1.00 M.

Solution

In neutral solution

$$HClO + H^+ + 2e^- \longrightarrow Cl^- + H_2O \qquad E° = +1.49 \text{ V}$$

From the Nernst equation

$$E = E° - \frac{0.05915}{n} \log Q$$

$$= 1.49 \text{ V} - \frac{0.05915}{2} \left(\log \frac{[Cl^-]}{[HClO][H^+]} \right) \text{ V} \qquad (1)$$

$$= 1.49 \text{ V} - \frac{0.05915}{2} \left(\log \frac{1.00}{(1.00)(10^{-7})} \right) \text{ V} \qquad (2)$$

$$= 1.49 \text{ V} - \frac{0.05915}{2} \log 10^7 \text{ V} \qquad (3)$$

$$= 1.49 \text{ V} - \frac{0.05915}{2} (7) \text{ V} \qquad (4)$$

$$= 1.49 \text{ V} - 0.20702 \text{ V} = 1.28 \text{ V} \tag{5}$$

At stage (2):

KEYSTROKES: $1 \div 1 \boxed{EE} \boxed{+/-} 7 = \boxed{2nd} \boxed{\log} \times .05912 \div 2 = 2.07025 - 01$

47. The Nernst equation for the net reaction of an electrochemical cell is

$$E = E° - \frac{0.05915}{n} \log Q$$

where Q is the reaction quotient for the reaction and n is the total number of moles of electrons transferred in the net reaction. Show that the Nernst equation for the net reaction

$$2\text{Au} + 3\text{Cl}_2 \longrightarrow 2\text{Au}^{3+} + 6\text{Cl}^-$$

is identical to that derived by addition of the Nernst equations for the two half-reactions

$$\text{Au} \longrightarrow \text{Au}^{3+} + 3e^- \qquad E° = -1.50 \text{ V}$$
$$\text{Cl}_2 + 2e^- \longrightarrow 2\text{Cl}^- \qquad E° = +1.3595 \text{ V}$$

Solution

The reaction quotient is written in the same way as an equilibrium constant. Thus, for the net reaction

$$Q = \frac{[\text{Au}^{3+}]^2[\text{Cl}^-]^6}{[\text{Au}]^2[\text{Cl}_2]^3}$$

The concentration under standard conditions are 1 M for the ions, 1 for solids since the concentration is a constant and 1 atm for gases. Consequently, $Q = 1$ and the log of 1 is zero. Therefore, the second term in the Nernst equation reduces to 0. $E = E°$.

48. Calculate the voltage produced by each of the following cells:

(a) Zn | Zn^{2+}(0.0100 M) || Cu^{2+}(1.00 M) | Cu
(c) Pt | Br^-(0.450 M) | $\text{Br}_2(l)$ || $\text{Cl}_2(g)$(0.900 atm) | Cl^-(0.0500 M) | Pt

Solution

(a) First calculate $E°$ for the reaction. The half-cells are from Table 21.1.

$$\text{Zn}^{2+} + 2e^- \rightleftharpoons \text{Zn} \qquad E° = -0.763 \text{ V}$$

Cathode half-reaction:

$$\text{Cu}^{2+} + 2e^- \rightleftharpoons \text{Cu} \qquad E° = +0.337 \text{ V}$$

The first of these is reversed along with the sign of its $E°$.

Anode half-reaction:

$$\text{Zn} \rightleftharpoons \text{Zn}^{2+} + 2e^- \qquad E° = +0.763$$

This and the cathode reaction are added.

Net cell reaction:

$$\text{Zn} + \text{Cu}^{2+} \longrightarrow \text{Zn}^{2+} + \text{Cu} \qquad E° = 1.100 \text{ V}$$

Chapter 21 Electrochemistry and Oxidation-Reduction

This value can now be used in the Nernst equation

$$E = E° - \frac{0.05915}{n} \log Q$$

$$E = 1.100 \text{ V} - \frac{0.05915 \text{ V}}{2} \log \frac{[Zn^{2+}]}{[Cu^{2+}]}$$

where $n = 2$ since two electrons are transferred in each reaction.

$$E = 1.100 - 0.029575 \log\left(\frac{0.01}{1.00}\right) = 1.100 + 0.05915 = 1.16 \text{ V}$$

(c) The $E°$ is determined from the half-cells in Table 21.1, where the bromine reaction is reversed from the way it is written in the table.

Anode half-reaction:	$2Br^- \longrightarrow Br_2(l) + 2e^-$	$E° = -1.0652$
Cathode half-reaction:	$Cl_2(g) + 2e^- \longrightarrow 2Cl^-$	$E° = +1.3595$
Net reaction:	$2Br^- + Cl_2(g) \longrightarrow Br_2(l) + 2Cl^-$	$E° = +0.2943$

The Nernst equation is

$$E = E° - \frac{0.05915}{n} \log Q = E° - \frac{0.05915}{n} \log \frac{[Cl^-]^2}{[Br^-]^2[0.9]}$$

$$= 0.2943 - \frac{0.05915}{2} \log \frac{(0.0500)^2}{(0.450)^2(0.9)}$$

$$= 0.2943 - \frac{0.05915}{2} \log\left(\frac{(0.0025)}{(0.2025)(0.9)}\right)$$

$$= 0.2943 + 0.0551 = +0.3494 \text{ V}$$

FREE ENERGY CHANGES AND EQUILIBRIUM CONSTANTS

51. For a cell based on each of the following reactions run at standard state conditions, calculate the emf of the cell, the standard free energy change of the reaction, and the equilibrium constant of the reaction:

(a) $Mn(s) + Cd^{2+}(aq) \longrightarrow Mn^{2+}(aq) + Cd(s)$
(c) $2Br^-(aq) + I_2(s) \longrightarrow Br_2(l) + 2I^-(aq)$

Solution

(a) The reaction represents the oxidation of Mn to Mn^{2+} and reduction of Cd^{2+} to Cd. Under standard conditions the emf of the cell is the sum of the half-cell potentials.

Anode half-cell:	$Mn \longrightarrow Mn^{2+} + 2e^-$	$E° = +1.18 \text{ V}$
Cathode half-cell:	$Cd^{2+} + 2e^- \longrightarrow Cd$	$E° = -0.40 \text{ V}$
Net cell reaction:	$Mn + Cd^{2+} \longrightarrow Mn^{2+} + Cd$	$E° = +0.78 \text{ V}$

The change in free energy for this reaction at standard conditions is calculated from

$$\Delta G° = -nFE°$$

in which 2 mol of electrons are transferred from Mn to Cd per mole of reactant

$$\Delta G° = -(2)\left(96.485 \frac{\text{kJ}}{\text{V}}\right)(0.78 \text{ V}) = -150 \text{ kJ}$$

The equilibrium constant can be calculated directly from $\Delta G°$ or from $E°$ equated to K. This latter method is preferred if a value for $E°$ is available, because less chance for rounding error is involved. In this equation,

$$E° = \frac{0.05915 \text{ V}}{n} \log K$$

the unit of $E°$ is the volt, but the unit is conventionally deleted and the equation left as

$$E° = \frac{0.05915}{n} \log K$$

with volts understood as the unit of measurement. In this problem the value of K is calculated as

$$0.78 = \frac{0.05915}{2} \log K$$

$$\log K = \frac{2(0.78)}{0.05915} = 26.3736$$

$$K = 10^{26.3736} = 2.4 \times 10^{26}$$

KEYSTROKES: 26.3736 |2nd| |10ˣ| = 2.3638852 26

(c) The emf of the cell is the sum of the half-cell potentials.

Anode half-cell:	$2\text{Br}^- \longrightarrow \text{Br}_2(l) + 2e^-$	$E° = -1.0652$ V
Cathode half-cell:	$\text{I}_2 + 2e^- \longrightarrow 2\text{I}^-$	$E° = +0.5355$ V
Net cell reaction:	$2\text{Br}^- + \text{I}_2 \longrightarrow \text{Br}_2 + 2\text{I}^-$	$E° = -0.5297$ V

$$\Delta G° = -nFE° = -(2)\left(96.485 \frac{\text{kJ}}{\text{V}}\right)(-0.5297 \text{ V}) = +102.2 \text{ kJ}$$

$$E° = \frac{0.05915}{n} \log K$$

$$-0.5297 = \frac{0.05915}{2} \log K$$

$$\log K = -17.9104$$

$$K = 10^{-17.9104} = 1.229 \times 10^{-18}$$

52. Calculate the standard free energy change and equilibrium constant for the reaction

$$2\text{Br}^- + \text{Cl}_2 \longrightarrow 2\text{Cl}^- + \text{Br}_2$$

Solution

The change in free energy for this reaction can be calculated from the emf of the cell:

Chapter 21 Electrochemistry and Oxidation-Reduction

$\Delta G° = -nFE°$

Anode half cell:	$2Br^- - 2e^- \longrightarrow Br_2$	$E° = -1.0652$ V
Cathode half cell:	$Cl_2 + 2e^- \longrightarrow 2Cl^-$	$E° = +1.3595$ V
Net cell reaction:	$2Br^- + Cl_2 \longrightarrow Br_2 + 2Cl^-$	$E°_{cell} = 0.2943$ V

$$\Delta G° = -2\left(96.485 \frac{kJ}{V}\right)(0.2943 \text{ V}) = -56.79 \text{ kJ}$$

$$E° = \frac{0.05915}{n} \log K$$

$$0.2943 = \frac{0.05915}{2} \log K$$

$\log K = 9.951 \qquad K = 8.93 \times 10^9$

54. Copper(I) salts disproportionate in water to form copper(II) salts and copper metal:

$$2Cu^+ \longrightarrow Cu^{2+} + Cu$$

What concentration of Cu^+ remains at equilibrium in 1.00 L of a solution prepared from 1.00 mol of Cu_2SO_4?

Solution

The concentration of Cu^+ remaining in equilibrium is <u>determined from the equilibrium constant</u>, which can be found from the emf. The half-reactions and calculations of the emf of the cell are

$$Cu^{2+} + 2e^- \longrightarrow Cu \qquad E_1 = 0.337 \text{ V} \qquad (1)$$

$$Cu^{2+} + e^- \longrightarrow Cu^+ \qquad E_2 = 0.153 \text{ V} \qquad (2)$$

If reaction (2) is subtracted from reaction (1),

$$2Cu^+ \longrightarrow Cu^{2+} + Cu \qquad E = E_1 - E_2 = 0.184 \text{ V}$$

$$\log K = \frac{nE}{0.05915} = \frac{2 \times 0.184}{0.05915} = 6.22$$

$$K = 1.66 \times 10^6$$

Let

$x = [Cu^{2+}]$ formed

$2x = [Cu^+]$ consumed

We start with 1 M Cu_2SO_4, so the initial concentration of Cu^+ is 2 M. The concentration remaining is $(2.00 - 2x)$ M
From the equilibrium constant,

$$K = \frac{[Cu^{2+}]}{[Cu^+]^2} = 1.66 \times 10^6 = \frac{x}{(2.00 - 2x)^2}$$

Since the K is so large, almost all of the Cu^+ has been converted to Cu^{2+} and Cu. The denominator is, therefore, very small and goes to zero. Therefore, assume that x in the numerator is 1 and solve for the small difference $(2.00 - 2x)$. Call it y. Then

$$\frac{1}{y^2} = 1.66 \times 10^6$$

$$y^2 = \frac{1}{1.66 \times 10^6} = 6.02 \times 10^{-7}$$

$$y = 7.8 \times 10^{-4}$$

This is the remaining concentration of Cu^+.

55. Use the emf values of the following half-cells and show that hydrogen peroxide, H_2O_2, is unstable with respect to decomposition into oxygen and water:

$$H_2O_2 + 2H^+ + 2e^- \longrightarrow 2H_2O \qquad E° = +1.77 \text{ V}$$

$$O_2 + 2H^+ + 2e^- \longrightarrow H_2O_2 \qquad E° = +0.68 \text{ V}$$

Solution

If the emf for the decomposition into O_2 and H_2O is positive, hydrogen peroxide will tend to decompose. Reverse the sense of the second equation and add.

$$H_2O_2 + 2H^+ + 2e^- \longrightarrow 2H_2O \qquad E° = +1.77 \text{ V}$$
$$\underline{H_2O_2 \longrightarrow O_2 + 2H^+ + 2e^- \qquad E° = -0.68 \text{ V}}$$
$$2H_2O_2 \longrightarrow O_2 + 2H_2O \qquad E° = +1.09 \text{ V}$$

OXIDATION-REDUCTION REACTIONS

57. Balance the following redox equations.

(c) $MnO_4^- + S^{2-} + H_2O \longrightarrow MnO_2 + S + OH^-$
(e) $Cu + H^+ + NO_3^- \longrightarrow Cu^{2+} + NO_2 + H_2O$
(g) $Cu + H^+ + NO_3^- \longrightarrow Cu^{2+} + NO + H_2O$
(j) $MnO_4^- + NO_2^- + H_2O \longrightarrow MnO_2 + NO_3^- + OH^-$
(n) $Br_2 + SO_2 + H_2O \longrightarrow H^+ + Br^- + SO_4^{2-}$

Solution

(c) $\quad 3[S^{2-} \longrightarrow S^0 + 2e^-] \qquad \uparrow 2 \quad 2 \times 3 = 6$
$\quad \underline{2[Mn^{7+} \longrightarrow Mn^{4+} - 3e^-]} \qquad \downarrow 3 \quad 3 \times 2 = 6$
$\quad 3S^{2-} + 2Mn^{7+} \longrightarrow 3S^0 + 2Mn^{4+}$
$\quad 2MnO_4^- + 3S^{2-} + 4H_2O \longrightarrow 2MnO_2 + 3S + 8OH^-$

(e) $\quad [Cu^0 \longrightarrow Cu^{2+} + 2e^-] \qquad \uparrow 2 \quad 2 \times 1 = 2$
$\quad \underline{2[N^{5+} \longrightarrow N^{4+} - 1e^-]} \qquad \downarrow 1 \quad 1 \times 2 = 2$
$\quad Cu^0 + 2N^{5+} \longrightarrow Cu^{2+} + 2N^{4+}$
$\quad Cu + 4H^+ + 2NO_3^- \longrightarrow Cu^{2+} + 2NO_2 + 2H_2O$

(g) $\quad 3[Cu^0 \longrightarrow Cu^{2+} + 2e^-] \qquad \uparrow 2 \quad 2 \times 3 = 6$
$\quad \underline{2[N^{5+} \longrightarrow N^{2+} - 3e^-]} \qquad \downarrow 3 \quad 3 \times 2 = 6$
$\quad 3Cu^0 + 2N^{5+} \longrightarrow 3Cu^{2+} + 2N^{2+}$
$\quad 3Cu + 8H^+ + 2NO_3^- \longrightarrow 3Cu^{2+} + 2NO + 4H_2O$

Chapter 21 Electrochemistry and Oxidation-Reduction

(j) $\quad 3[N^{3+} \longrightarrow N^{5+} + 2e^-]$ $\qquad \uparrow 2 \quad 2 \times 3 = 6$

$\quad\underline{2[Mn^{7+} \longrightarrow Mn^{4+} - 3e^-]}$ $\qquad \downarrow 3 \quad 3 \times 2 = 6$

$\quad 2Mn^{7+} + 3N^{3+} \longrightarrow 2Mn^{4+} + 3N^{5+}$

$\quad 2MnO_4^- + 3NO_2^- + H_2O \longrightarrow 2MnO_2 + 3NO_3^- + 2OH^-$

(n) $\quad S^{4+} \longrightarrow S^{6+} + 2e^-$ $\qquad \uparrow 2 \quad 2 \times 1 = 2$

$\quad\underline{Br_2^0 \longrightarrow 2Br^- - 2e^-}$ $\qquad \downarrow 2 \quad 2 \times 1 = 2$

$\quad Br_2 + S^{4+} \longrightarrow 2Br^- + S^{6+}$

$\quad Br_2 + SO_2 + 2H_2O \longrightarrow 4H^+ + 2Br^- + SO_4^{2-}$

58. Balance the following redox equations.

    (b) $OH^- + NO_2 \longrightarrow NO_3^- + NO_2^- + H_2O$
    (g) $C + HNO_3 \longrightarrow NO_2 + H_2O + CO_2$
    (k) $H_2S + H_2O_2 \longrightarrow S + H_2O$

    Solution

    (b) $\quad N^{4+} \longrightarrow N^{5+} + e^-$ $\qquad \uparrow 1 \quad 1 \times 1 = 1$

    $\quad\underline{N^{4+} \longrightarrow N^{3+} - e^-}$ $\qquad \downarrow 1 \quad 1 \times 1 = 1$

    $\quad 2N^{4+} \longrightarrow N^{5+} + N^{3+}$

    $\quad 2OH^- + 2NO_2 \longrightarrow NO_3^- + NO_2^- + H_2O$

    (g) $\quad C^0 \longrightarrow C^{4+} + 4e^-$ $\qquad \uparrow 4 \quad 4 \times 1 = 4$

    $\quad\underline{4[N^{5+} \longrightarrow N^{4+} - 1e^-]}$ $\qquad \downarrow 1 \quad 1 \times 4 = 4$

    $\quad C^0 + 4N^{5+} \longrightarrow C^{4+} + 4N^{4+}$

    $\quad C + 4HNO_3 \longrightarrow 4NO_2 + 2H_2O + CO_2$

    (k) $\quad S^{2-} \longrightarrow S^0 + 2e^-$ $\qquad \uparrow 2 \quad 2 \times 1 = 2$

    $\quad\underline{O_2^{2-} \longrightarrow 2O^{2-} - 2e^-}$ $\qquad \downarrow 2 \quad 2 \times 1 = 2$

    $\quad S^{2-} + O_2^{2-} \longrightarrow S^0 + 2O^{2-}$

    $\quad H_2S + H_2O_2 \longrightarrow S + 2H_2O$

59. Complete and balance the following equations. When the reaction occurs in acidic solution H^+ and/or H_2O may be added on either side of the equation, as necessary, to balance the equation properly; when the reaction occurs in basic solution, OH^- and/or H_2O may be added, as necessary, on either side of the equation. No indication of the acidity of the solution is given if neither H^+ nor OH^- is involved as a reactant or product.

    (a) $Ag + NO_3^- \longrightarrow Ag^+ + NO$ (acidic solution)
    (c) $H_2S + I_2 \longrightarrow S + I^-$ (acidic solution)
    (f) $PbO_2 + Cl^- \longrightarrow Pb^{2+} + Cl_2$ (acidic solution)
    (i) $H_2O_2 + ClO_2 \longrightarrow ClO_2^- + O_2$ (basic solution)

Solution

(a) $3[Ag^0 \longrightarrow Ag^+ + 1e^-]$ $\uparrow 1$ $1 \times 3 = 3$

$\underline{1[N^{5+} \longrightarrow N^{2+} - 3e^-]}$ $\downarrow 3$ $3 \times 1 = 3$

$3Ag + NO_3^- \longrightarrow 3Ag^+ + NO$

$3Ag + NO_3^- + 4H^+ \longrightarrow 3Ag^+ + NO + 2H_2O$

(c) $1[S^{2-} \longrightarrow S^0 + 2e^-]$ $\uparrow 2$

$\underline{1[I_2 \longrightarrow 2I^- - 2e^-]}$ $\downarrow 2$

$H_2S + I_2 \longrightarrow S + 2I^- + 2H^+$

(f) $1[2Cl^- \longrightarrow Cl_2 + 2e^-]$ $\uparrow 2$

$\underline{1[Pb^{4+} \longrightarrow Pb^{2+} - 2e^-]}$ $\downarrow 2$

$PbO_2 + 2Cl^- + 4H^+ \longrightarrow Pb^{2+} + Cl_2 + 2H_2O$

(i) $[O_2^{2-} \longrightarrow O_2 + 2e^-]$ $\uparrow 2$ $2 \times 1 = 2$

$\underline{2[Cl^{+4} \longrightarrow Cl^{+3} - 1e^-]}$ $\downarrow 1$ $1 \times 2 = 2$

$H_2O_2 + 2ClO_2 + 2OH^- \longrightarrow O_2 + 2ClO_2^- + 2H_2O$

60. Complete and balance the following equations (see instruction for exercise 59).

(a) $Fe^{2+} + MnO_4^- \longrightarrow Fe^{3+} + Mn^{2+}$ (acidic solution)
(d) $Fe^{3+} + I^- \longrightarrow Fe^{2+} + I_2$
(g) $Fe^{2+} + Cr_2O_7^{2-} \longrightarrow Fe^{3+} + Cr^{3+}$ (acidic solution)
(j) $CrO_4^{2-} + HSnO_2^- \longrightarrow HSnO_3^- + CrO_2^-$ (basic solution)

Solution

(a) $5[Fe^{2+} \longrightarrow Fe^{3+} + 1e^-]$ $\uparrow 1$ $1 \times 5 = 5$

$\underline{[Mn^{7+} \longrightarrow Mn^{2+} - 5e^-]}$ $\downarrow 5$ $5 \times 1 = 5$

$5Fe^{2+} + MnO_4^- + 8H^+ \longrightarrow 5Fe^{3+} + Mn^{2+} + 4H_2O$

(d) $[2I^- \longrightarrow I_2 + 2e^-]$ $\uparrow 2$ $2 \times 1 = 2$

$\underline{2[Fe^{3+} \longrightarrow Fe^{2+} - 1e^-]}$ $\downarrow 1$ $1 \times 2 = 2$

$2I^- + 2Fe^{3+} \longrightarrow 2Fe^{2+} + I_2$

(g) $6[Fe^{2+} \longrightarrow Fe^{3+} + 1e^-]$ $\uparrow 1$ $1 \times 6 = 6$

$\underline{[2Cr^{6+} \longrightarrow 2Cr^{3+} + 6e^-]}$ $\downarrow 6$ $6 \times 1 = 6$

$6Fe^{2+} + Cr_2O_7^{2-} + 14H^+ \longrightarrow 6Fe^{3+} + 7H_2O + 2Cr^{3+}$

(j) $3[Sn^{2+} \longrightarrow Sn^{4+} + 2e^-]$ $\uparrow 2$ $2 \times 3 = 6$

$\underline{2[Cr^{+6} \longrightarrow Cr^{+3} - 3e^-]}$ $\downarrow 3$ $3 \times 2 = 6$

$3HSnO_2^- + 2CrO_4^{2-} + H_2O \longrightarrow 3HSnO_3^- + 2CrO_2^- + 2OH^-$

Chapter 21 Electrochemistry and Oxidation-Reduction

ADDITIONAL EXERCISES

62. When gold is plated electrochemically from a basic solution of $[Au(CN)_4]^-$, O_2 forms at one electrode and Au is deposited at the other. Write the half-reactions occurring at each electrode and the net reaction for the electrochemical cell. (The cyanide ion, CN^-, is not oxidized or reduced under these conditions.)

Solution

Cathodic reduction:

$$[Au(CN)_4]^- + 3e^- \longrightarrow Au + 4(CN)^-$$

Anodic oxidation:

$$4OH^- \longrightarrow O_2 + 2H_2O + 4e^-$$

Cell reaction:

$$4[Au(CN)_4]^- + 12OH^- \longrightarrow 4Au + 3O_2 + 6H_2O + 16CN^-$$

64. A current of 15.0 A flowed for 25.0 min through water containing a small quantity of sodium hydroxide. How many liters of gas were formed at the anode at 27.0°C and 1.03 atm pressure.

Solution

$$\text{Amperes} \times \text{time} \longrightarrow \text{coulombs} \longrightarrow \text{faradays} \longrightarrow \text{mol of } e^-$$

$$15.0 \text{ A} \times 25.0 \text{ min} \times \frac{60 \text{ s}}{\text{min}} \times \frac{1 \text{ F}}{96,485} = 0.233 \text{ mol}$$

The reaction producing gas is

$$4OH^- \longrightarrow O_2(g) + 2H_2O + 4e^-$$

The production of 0.233 mol e^- generates $0.233/4 = 0.058$ mol O_2.
From the ideal gas law;

$$PV = nRT, \quad V = \frac{nRT}{P} = \frac{0.058 \text{ mol} \times 0.08206 \frac{\text{L atm}}{\text{K mol}} \times 300.1 \text{ K}}{1.03 \text{ atm}} = 1.39 \text{ L}$$

65. A lead storage battery has 9.00 kg of lead and 9.00 kg of PbO_2, plus excess H_2SO_4. Theoretically, how long could this cell deliver a current of 50.0 A, without recharging, if it were possible to operate it so that the reaction goes to completion?

Solution

The overall equation for discharge is

$$Pb + PbO_2 + 4H^+ + 2SO_4^{2-} \longrightarrow 2PbSO_4(s) + 2H_2O$$

Both anode and cathode processes require 2 mol of e^- to occur. The ratio of Pb to PbO_2 is 1 to 1. First find the limiting reagent.

$$\text{mol Pb} = \frac{9000 \text{ g}}{207.2 \text{ g/mol}} = 43.4; \quad \text{mol PbO}_2 = \frac{9000 \text{ g}}{239.2 \text{ g/mol}} = 37.6$$

330   Chapter 21 Electrochemistry and Oxidation-Reduction

PbO₂ is the limiting agent
Next calculate the time to dissipate 1 mol of e⁻ at 50.0 A. From

$$\text{mol e}^- = 50.0 \text{ A} \times t \times \frac{1 \text{ F}}{96{,}485 \text{ coulombs}} \times \frac{1 \text{ mol e}^-}{\text{F}}$$

$t = 1929.7$ s/mol e⁻

For 37.6 mol of PbO₂ and Pb to react requires 2 mol e⁻ each and 1929.7 s for each mol e⁻.

1929.7 s/mol e⁻ × 2 mol e⁻/mol PbO₂ × 37.6 mol PbO₂

$= 1.45 \times 10^5$ s or 1.45×10^5 s/3600 s h⁻¹ = 40.3 h

67. A current of 20.0 A is applied for 0.50 h to 1.0 L of a solution containing 1.0 mol of HCl. Calculate the pH of the solution after this time.

Solution

The reduction of H⁺ to produce H₂ at the cathode will produce a change in pH as the reaction proceeds.

Cathode half-reaction:

$2\text{H}^+ + 2e^- \longrightarrow \text{H}_2$

Anode half-reaction:

$2\text{Cl}^- \longrightarrow \text{Cl}_2 + 2e^-$

The number of moles of H⁺ consumed by the reaction is calculated as follows

mol H⁺ reduced = faradays passed

$$\text{faradays} = (20.0 \text{ A})(0.50 \text{ h} \times 3600 \text{ s/h}) \frac{1.00 \text{ faraday}}{96{,}485 \text{ C}} = 0.373$$

moles H⁺ reduced = 0.373

[H⁺] after reduction = 1.0 mol/L − 0.373 mol/L = 0.63 mol/L

pH = −log 0.63 = 0.20

69. Describe briefly how you could determine the solubility product of AgCl ($K_{sp} \approx 1.8 \times 10^{-10}$), using an electrochemical measurement.

Solution

In order to solve problems of this type in the general case, pick two reactions that, when combined, give the ions as reactants and the final precipitate as the product. For example, see exercise 71. Measure concentrations and the voltage. With this data and $E°$s, use the Nernst equation to determine Q, from which the solubility can be found.

71. The standard reduction potentials for the reactions

$\text{Ag}^+ + e^- \longrightarrow \text{Ag}$ and $\text{AgCl} + e^- \longrightarrow \text{Ag} + \text{Cl}^-$

are +0.7991 V and +0.222 V, respectively. From these data and the Nernst equation, calculate a value for the solubility product (K_{sp}) for AgCl. Compare your answer with the value given in Appendix D.

Chapter 21 Electrochemistry and Oxidation-Reduction 331

Solution

The solubility product constant for the dissolution of AgCl, is

$$AgCl(s) \longrightarrow Ag^+(aq) + Cl^-(aq)$$

$$K_{sp} = [Ag^+][Cl^-]$$

The two half-cells given in the problem can be rearranged to give the desired equilibrium.

$$\begin{array}{ll} Ag \longrightarrow Ag^+ + e^- & E° = -0.7991 \text{ V} \\ AgCl + e^- \longrightarrow Ag + Cl^- & E° = 0.222 \text{ V} \\ \hline AgCl \longrightarrow Ag^+ + Cl^- & E°_{cell} = -0.557 \text{ V} \end{array}$$

This value can be substituted into the equation, relating $E°$ to log K. We solve that equation to obtain

$$E° = \frac{0.05915}{n} \log K$$

$$-0.577 = \frac{0.05915}{1} \log K$$

$$\log K = \frac{-0.577}{0.05915} = -9.755$$

$$K = K_{sp} = 10^{-9.755} = 1.76 \times 10^{-10}$$

This compares to 1.8×10^{-10} as given in Appendix D.

73. The standard reduction potentials for the reactions

$$Ag^+ + e^- \longrightarrow Ag$$

and

$$[Ag(NH_3)_2]^+ + e^- \longrightarrow Ag + 2NH_3$$

are +0.7991 V and +0.373 V, repectively. From these values and the Nernst equation, determine K_f for the $[Ag(NH_3)_2]^+$ ion. Compare your answer with the value given in Appendix E.

Solution

The formation constant for $[Ag(NH_3)_2]^+$ is

$$K_f = \frac{[Ag(NH_3)_2]^+}{[Ag^+][NH_3]^2}$$

By combining the half-cells given,

$$\begin{array}{ll} Ag + 2NH_3 \longrightarrow [Ag(NH_3)_2]^+ + e^- & E° = -0.373 \text{ V} \\ Ag^+ + e^- \longrightarrow Ag & E° = 0.7991 \text{ V} \\ \hline Ag^+ + 2NH_3 \longrightarrow [Ag(NH_3)_2]^+ & E° = 0.426 \text{ V} \end{array}$$

This value is substituted into the equation relating $E°$ to log K.

$$E° = \frac{0.05915}{n} \log K$$

$$\log K = \frac{0.426}{0.05915} = 7.20$$

$$K_f = 10^{7.20} = 1.6 \times 10^7$$

22

The Nonmetals, Part 1: Hydrogen, Oxygen, Sulfur, and the Halogens

INTRODUCTION

In Chapter 20 the general behavior of the nonmetals was described. We now look at hydrogen, oxygen, sulfur, and the halogens. Aside from hydrogen, which is normally a reducing agent in reactions, the others, especially the lighter members of the halogen family, are oxidizing agents. This explains, to a large extent, the chemical behavior of these elements in both the elemental and combined states.

READY REFERENCE

Disproportionation (22.19)   Disproportionation refers to a situation in which one compound forms two different products. In the process, the oxidation state of the central ion of the reactant both increases and decreases to form the final products.

Oxyacids (22.23)   The oxyacids are acids formed from the halogens in union with varying amounts of oxygen. Not all acids exist for each halogen. The formula and naming follow where X represents any halogen.

HXO   hypohalous acid
HXO_2 halous acid
HXO_3 halic acid
HXO_4 perhalic acid

Photosynthesis (22.8)   A process in which carbon dioxide and water are transformed into glucose and oxygen through the action of chlorophyll (a green plant substance) and sunlight (the source of energy).

SOLUTIONS TO CHAPTER 22 EXERCISES

1. Both hydrogen and oxygen are colorless, odorless and tasteless gasses. How may we distinguish between them using physical properties? Using chemical properties?

Chapter 22 Nonmetals, Part 1: Hydrogen, Oxygen, Sulfur, and the Halogens

Solution

Using physical properties: The Dumas method of filling an evacuated bulb with the gas and determining the density of the gas can be used. The densities are quite different — H_2, 0.089 g/L; O_2, 1.43 g/L.

Using chemical properties: The gas could be passed over hot CuO. If it is H_2, the CuO will be reduced to Cu. No reaction will occur if it is O_2.

3. Describe five chemical properties of sulfur or of compounds of sulfur that characterize it as a nonmetal.

Solution

Sulfur oxidizes metals. Elemental sulfur oxidizes less electronegative nonmetals. Sulfur commonly exhibits an oxidation state of -2. Most reactive metals will displace H_2 from H_2S, a characteristic of a nonmetal united with hydrogen. Sulfur forms polysulfide complex ions, S_n^{2-}, that contain covalent S—S bonds.

5. Arrange the halogens in order of increasing
   (a) atomic radius
   (b) electronegativity
   (c) boiling point
   (d) oxidizing activity

Solution

(a) $F < Cl < Br < I$
(b) $I < Br < Cl < F$
(c) $F < Cl < Br < I$
(d) $I < Br < Cl < F$

HYDROGEN

7. Write the equation that describes the preparation of hydrogen when steam is passed through a red-hot gun barrel.

Solution

$$3Fe(s) + 4H_2O(g) \xrightarrow{\Delta} Fe_3O_4(s) + 4H_2(g)$$

9. Write balanced equations for and name the compounds formed in the reaction, if any, of each of the following metals with water and with hydrobromic acid: K, Mg, Ga, Bi, Fe, Pt.

Solution

$2K(s) + 2H_2O(l) \longrightarrow H_2(g) + 2K^+(aq) + 2OH^-(aq)$

$2K(s) + 2HBr(aq) \longrightarrow H_2(g) + 2K^+(aq) + 2Br^-(aq)$

$Mg(s) + H_2O(g) \xrightarrow{\Delta} H_2(g) + MgO(s)$ (with no heat, no reaction)

$Mg(s) + 2HBr(aq) \longrightarrow H_2(g) + Mg^{2+}(aq) + 2Br^-(aq)$

$2Ga(s) + 3H_2O(g) \xrightarrow{\Delta} 3H_2(g) + Ga_2O_3(s)$

$2Ga(s) + 6HBr(aq) \longrightarrow 3H_2(g) + 2Ga^{3+}(aq) + 6Br^-(aq)$

$$2Bi(s) + 3H_2O(g) \xrightarrow{\Delta} 3H_2(g) + Bi_2O_3(s)$$

$$2Bi(s) + 6HBr(aq) \longrightarrow 3H_2(g) + 2Bi^{3+}(aq) + 6Br^-(aq)$$

$$4Fe(s) + 6H_2O(l) \xrightarrow{\Delta} 6H_2(g) + 2Fe_2O_3(s)$$

$$Fe(s) + 2HBr(aq) \longrightarrow H_2(g) + Fe^{2+}(aq) + 2Br^-(aq)$$

$$Pt(s) + H_2O(l) \longrightarrow \text{No reaction}$$

$$Pt(s) + HBr(aq) \longrightarrow \text{No reaction}$$

11. Explain: "Metal hydrides are convenient and portable sources of hydrogen."

Solution

The metal hydrides react to form 1 to 2 mol H_2 per mol of hydride. They are easily stored and transported. The hydrides readily react with water to form hydrogen.

13. Hydrogen is not very soluble. At 25°C a saturated solution of H_2 in water is 7.86×10^{-5} M. How many milliliters of H_2 at 25°C and 1.00 atm are dissolved in a liter of solution?

Solution

There are 7.86×10^{-5} mol H_2 in 1.00 L of water at 25°C or 298.1 K. From the ideal gas law

$$PV = nRT$$

$$V = \frac{nRT}{P} = \frac{7.86 \times 10^{-5} \text{ mol} \left(82.1 \frac{\text{mL atm}}{\text{K mol}}\right)(298.1 \text{ K})}{1.00 \text{ atm}} = 1.92 \text{ mL}$$

15. What mass of CaH_2 is required to provide enough hydrogen by reaction with water to fill a balloon at STP with a volume of 3.50 L?

Solution

The reaction is

$$CaH_2(s) + 2H_2O(l) \longrightarrow Ca(OH)_2(aq) + 2H_2(g)$$

The number of mol H_2 required is obtained from the ideal gas law.

$$PV = nRT$$

$$n = \frac{PV}{RT} = \frac{1.00 \text{ atm }(3.50 \text{ L})}{0.08206 \frac{\text{L atm}}{\text{K mol}}(273.1 \text{ K})} = 0.1562 \text{ mol}$$

$$\text{g CaH}_2 = 0.1562 \text{ mol H}_2 \times \frac{1 \text{ mol CaH}_2}{2 \text{ mol H}_2} \times \frac{42.10 \text{ g CaH}_2}{\text{mol CaH}_2} = 3.29 \text{ g}$$

17. Deuterium, ^2_1H, usually indicated as D, may be separated from ordinary hydrogen by repeated distillation of water. Ultimately, pure D_2O can be isolated. What is the molecular weight of D_2O to three significant figures? (The atomic weight of D may be found in Table 5.2).

Solution

Mol. wt $D_2O = 2(2.0166) + 15.9994 = 20.0$

OXYGEN

19. Write the reaction describing the preparation of oxygen by using a lens to focus the sun's rays on a sample of mercury(II) oxide.

    ### Solution

    $$2HgO(s) \longrightarrow 2Hg(s) + O_2(g)$$

21. Write balanced equations for the preparation of oxygen from

    (a) $Cu(NO_3)_2$
    (b) $NaNO_3$
    (c) H_2O
    (d) BaO_2

    ### Solution

    (a) $2Cu(NO_3)_2 \xrightarrow{\Delta} 2CuO + 4NO_2 + O_2$
    (b) $2NaNO_3 \xrightarrow{\Delta} 2NaNO_2 + O_2$
    (c) $2H_2O(l) \xrightarrow{electrolysis} 2H_2(g) + O_2(g)$
    (d) $2BaO_2 \xrightarrow{\Delta} 2BaO + O_2$

23. Which of the following materials will burn in O_2: $SiH_4(g)$, SiO_2, CO, CO_2, Mg, CaO? Why won't some of these materials burn in oxygen?

    ### Solution

    Substances that will burn (oxidize) in O_2 are $SiH_4(g)$, CO, Mg. The elements Si, C, and Ca in SiO_2, CO_2 and CaO already have reached their greatest positive oxidation state or ability to combine with oxygen.

25. Write a balanced equation for the reaction of an excess of oxygen with each of the following. (Keep in mind that an element tends to reach its highest oxidation number when it combines with an excess of oxygen.)

    (a) Ca
    (b) Cs
    (c) As
    (d) Ge
    (e) Na_2SO_3
    (f) AlP
    (g) C_2H_6
    (h) CO

    ### Solution

    (a) $Ca + O_2 \longrightarrow CaO_2$ (a peroxide)
    (b) $Cs + O_2 \longrightarrow CsO_2$ (a superoxide)
    (c) $4As + 5O_2 \longrightarrow 2As_2O_5$
    (d) $Ge + O_2 \longrightarrow GeO_2$
    (e) $2Na_2SO_3 + O_2 \longrightarrow 2Na_2SO_4$

336 Chapter 22 Nonmetals, Part 1: Hydrogen, Oxygen, Sulfur, and the Halogens

(f) $4AlP + 8O_2 \longrightarrow 2Al_2O_3 + P_4O_{10}$
(g) $C_2H_6 + 5O_2 \longrightarrow 2CO_2 + 3H_2O$
(h) $2CO + O_2 \longrightarrow 2CO_2$

27. Assume that when O_2 dissolves in water, the volume of the solution is the same as the volume of water from which the solution was made. What is the molar concentration of O_2 in a solution at 0°C that is saturated with O_2?

Solution

The solubility is

$$\frac{0.0489 \text{ L } O_2}{\text{L } H_2O}$$

$$n = \frac{PV}{RT} = \frac{1.0 \text{ atm} \times 0.0489 \text{ L}}{0.08206 \frac{\text{L atm}}{\text{K mol}} (273.1 \text{ K})} = 2.2 \times 10^{-3} \text{ M}$$

29. What mass of oxygen can be obtained by the thermal decomposition of 48.19 g of $Cu(NO_3)_2$? What is the volume of O_2 produced at STP?

Solution

$$2Cu(NO_3)_2 \longrightarrow 2CuO + 4NO_2 + O_2$$

$$\text{g } O_2 = 48.19 \text{ g } Cu(NO_3)_2 \times \frac{1 \text{ mol } Cu(NO_3)_2}{187.556 \text{ } Cu(NO_3)_2} \times \frac{1 \text{ mol } O_2}{2 \text{ mol } Cu(NO_3)_2}$$

$$\times \frac{2(15.9994 \text{ g})}{1 \text{ mol } O_2} = 4.111 \text{ g}$$

$$V = \frac{nRT}{P} = \frac{\frac{4.111 \text{ g}}{2(15.9994)} \left(0.08206 \frac{\text{L atm}}{\text{K mol}}\right) 273. \text{K}}{1.000 \text{ atm}} = 2.879 \text{ L}$$

SULFUR

31. Based on the location of sulfur in the Periodic Table, predict the products of and write a balanced equation for each of the following. (There may be more than one correct answer, depending on your choice of stoichiometries.)

(a) $Ca + S \longrightarrow$
(b) $Al + S \longrightarrow$
(c) $F_2 + S \longrightarrow$
(d) $H_2 + S \longrightarrow$
(e) $C + S \longrightarrow$

Solution

(a) $Ca + S \longrightarrow CaS$
(b) $2Al + 3S \longrightarrow Al_2S_3$
(c) $3F_2 + S \longrightarrow SF_6$
(d) $H_2 + S \longrightarrow H_2S$
(e) $C + 2S \longrightarrow CS_2$

Chapter 22 Nonmetals, Part 1: Hydrogen, Oxygen, Sulfur, and the Halogens   337

33. Write chemical equations describing three chemical reactions in which elemental sulfur acts as an oxidizing agent; three in which it acts as a reducing agent.

    Solution

    As oxidizing agent:

    $Fe + S \longrightarrow FeS$

    $C + 2S \longrightarrow CS_2$

    $Pb + S \longrightarrow PbS$

    As reducing agent:

    $S + 6HNO_3 \longrightarrow 2H^+ + SO_4^{2-} + 2H_2O + 6NO_2$

    $S + 2H_2SO_4 \longrightarrow 3SO_2 + 2H_2O$

    $S + O_2 \longrightarrow SO_2$

35. Explain the fact that hydrogen sulfide is a gas at room temperature, whereas water, which has a smaller molecular weight, is a liquid.

    Solution

    The hydrogen bonds in water are very strong because of the higher electronegativity of oxygen. These are sufficient to keep water in the liquid state. In spite of the fact that H_2S has almost twice the mol. wt of H_2, it remains a gas because it does not hydrogen bond due to sulfur's lower electronegativity and larger size.

37. Write an equation for the reaction of hydrogen sulfide in acidic solution with each of the following:

    (a) Fe^{3+}
    (b) Br_2
    (c) MnO_4^-
    (d) $Cr_2O_7^{2-}$

    Solution

    (a) $2Fe^{3+} + 3H_2S \longrightarrow 2FeS + 6H^+ + S$
    (b) $Br_2 + H_2S \longrightarrow 2Br^- + S + 2H^+$
    (c) $8MnO_4^- + 5H_2S + 14H^+ \longrightarrow 8Mn^{2+} + 5SO_4^{2-} + 12H_2O$
    (d) $4Cr_2O_7^{2-} + 3H_2S + 26H^+ \longrightarrow 8Cr^{3+} + 3SO_4^{2-} + 16H_2O$

39. Why do the two sulfur-oxygen bonds in sulfur dioxide have the same length?

    Solution

    The two sulfur-oxygen bonds in sulfur dioxide have the same length because of resonance. There is no way to justify two different lengths as initially predicted by the Lewis electron dot picture. The "real" structure is somewhere between a single and a full double bond.

41. How does the hybridization of the sulfur atom change when gaseous SO_3 condenses to give solid SO_3?

Solution

SO₃ exhibits sp^2 hybridization in the gas state. In the solid state the sulfur forms tetrahedra and is sp^3 hybridized.

43. Which is the stronger acid, NaHSO₃ or NaHSO₄?

Solution

NaHSO₄ is the stronger acid. An oxyacid is stronger when the oxidation number of the central atom is higher.

45. Write the Lewis structure of each of the following:

   (a) S^{2-}
   (b) H_2S_2
   (c) SO_2
   (d) H_2SO_3
   (e) SO_3
   (f) Na_2SO_4
   (g) H_2SO_5
   (h) $H_2S_2O_8$

Solution

(a)  $[\ddot{\underset{\cdot\cdot}{S}}]^{2-}$

(b)  $H:\ddot{\underset{\cdot\cdot}{S}}:\ddot{\underset{\cdot\cdot}{S}}:H$

(c) [resonance structures of SO₂]

(d) [Lewis structure of H₂SO₃]

(e) [three resonance structures of SO₃]

(f) $2Na^+ + \left[\begin{array}{c}:\ddot{O}:\\:\ddot{O}:S:\ddot{O}:\\:\ddot{O}:\end{array}\right]^{2-}$

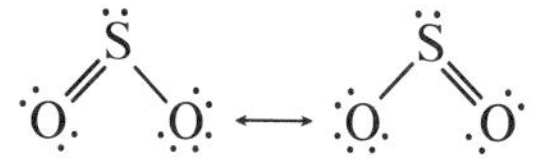

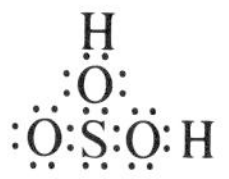

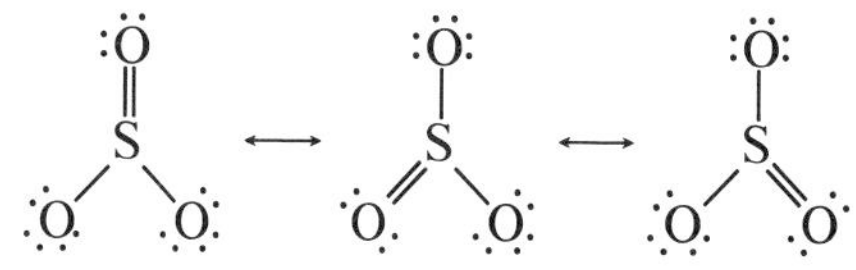

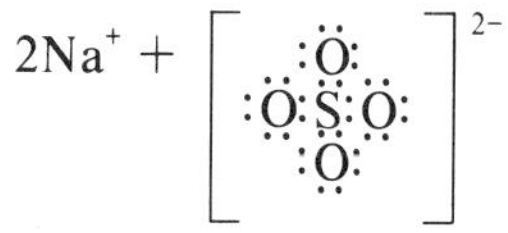

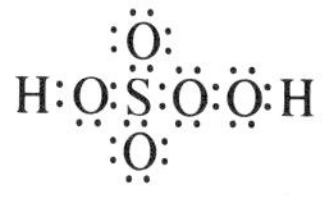

(g) [Lewis structure of H₂SO₅]

(h) [Lewis structure of H₂S₂O₈]

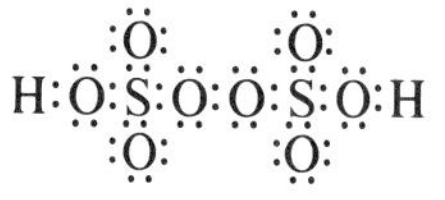

47. A volume of 22.85 mL of a 0.1023 M standard sodium thiosulfate solution is required to titrate a 25.00-mL sample of a solution containing iodine. What is the iodine concentration in the 25.00-mL sample?

Solution

The reaction is

$$2S_2O_3^{2-}(aq) + I_2 \longrightarrow S_4O_6^{2-}(aq) + 2I^-$$

Each $S_2O_3^{2-}$ ion produces one I^- ion. The final molarity of iodide ion is twice the molarity of the I_2 from which it came. Therefore, find the molarity from $M_iV_i = M_fV_f$ and divide M_f by 2.

$$22.85 \text{ mL} \times 0.1023 \text{ M} = 25.00 \text{ mL} \times M_f$$

$$M_f = \frac{22.85 \times 0.1023 \text{ M}}{25.00} = 0.09350 \text{ M}$$

Molarity of iodine = 0.09350 M/2 = 0.04675 M

49. What volume of hydrogen sulfide at STP can be produced by the reaction of 308 g of aluminum sulfide, Al_2S_3, with an excess of phosphoric acid?

    Solution

    $$Al_2S_3 + 2H_3PO_4 \longrightarrow 3H_2S + 2AlPO_4$$

    From the stoichiometry first find the moles of H_2S produced, then its volume.

    $$\text{g } H_2S = 308 \text{ g } Al_2S_3 \times \frac{1 \text{ mol } Al_2S_3}{150.14 \text{ g}} \times \frac{3 \text{ mol } H_2S}{1 \text{ mol } Al_2S_3} = 6.154 \text{ mol}$$

    $$PV = nRT$$

    $$V = \frac{nRT}{P} = \frac{6.154 \text{ mol} \left(0.08206 \dfrac{\text{L atm}}{\text{K mol}}\right) 273.1 \text{ K}}{1.00 \text{ atm}} = 138 \text{ L}$$

HALOGENS

51. Why can fluorine, which reacts with all of the metals, be stored in certain metal cylinders?

    Solution

    An adherent film of the metal fluoride protects the surfaces of Cu, Fe, Mg, and Ni permitting their use for handling fluorine.

53. Suggest two reasons why fluorine is not used in the extraction of bromine from brine.

    Solution

    Fluorine is not used to extract bromine from brine, because it must be prepared by electrolysis so its cost is comparatively high, and because fluorine reacts with water, producing O_2, OF_2, H_2O_2, and HF.

55. Write balanced chemical equations that describe the reaction of chlorine with each of the following:

    (a) lithium
    (b) magnesium
    (c) hydrogen
    (d) an excess of phosphorus
    (e) iodine
    (f) sulfur

340   Chapter 22   Nonmetals, Part 1: Hydrogen, Oxygen, Sulfur, and the Halogens

Solution

(a) $2Li(s) + Cl_2(g) \longrightarrow 2LiCl(s)$
(b) $Mg(s) + Cl_2(g) \longrightarrow MgCl_2(s)$
(c) $H_2(g) + Cl_2(g) \longrightarrow 2HCl(g)$
(d) $2P(s) + 3Cl_2(g) \longrightarrow 2PCl_3(s)$
(e) $I_2(s) + 3Cl_2(g) \longrightarrow 2ICl_3(s)$
(f) $S_8 + 8Cl_2(g) \longrightarrow 8SCl_2(s)$

57. Show that the reaction of bromine with water is a disproportionation reaction by considering the changes in oxidation number that occur.

Solution

The reaction of bromine with water is

$$Br_2(l) + H_2O(l) \longrightarrow H^+(aq) + Br^-(aq) + HOBr(aq)$$

This is a disproportionation reaction since Br_2 has an oxidation state of zero, whereas in Br^- it is -1, and in HOBr it is $+1$. Thus $Br_2^0 \longrightarrow Br^{+1} + Br^{-1}$

59. Why is iodine monofluoride more polar than iodine monochloride?

Solution

Fluorine is more electronegative than chlorine. Consequently, in union with iodine, the negative charge is much more concentrated on fluorine than on chlorine, making the monofluoride a more polar compound compared to the monochloride.

61. Write the Lewis structure for each of the iodine flourides.

Solution

Iodine and fluorine form IF, IF_3, IF_5, and IF_7.

IF:

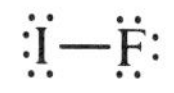

:Ï—F̈:

IF_3:

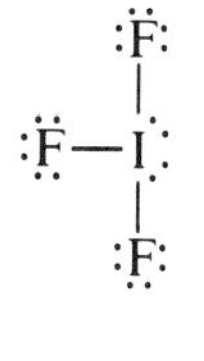

:F̈:
|
:F̈—Ï:
|
:F̈:

IF_5:

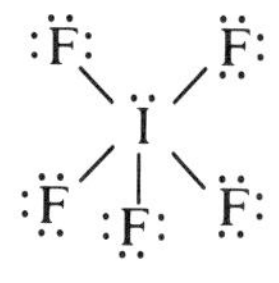

:F̈:   :F̈:
  \ /
   Ï
  /|\
:F̈ :F̈: :F̈:

IF_7:

:F̈: :F̈: :F̈:
  \ | /
   I—F̈:
  / | \
:F̈: :F̈: :F̈:

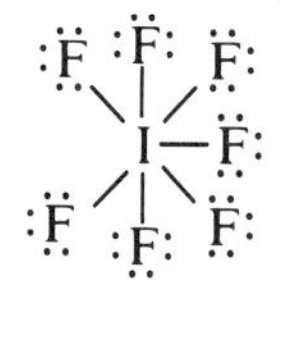

63. Write an equation describing a convenient laboratory preparation for each of the hydrogen halides.

Solution

F: $CaF_2 + H_2SO_4 \longrightarrow CaSO_4 + 2HF(g)$
Cl: $2NaCl + H_2SO_4 \longrightarrow Na_2SO_4 + 2HCl(g)$
Br: $PBr_3 + 3H_2O \longrightarrow 3HBr(g) + H_3PO_3$
I: $PI_3 + 3H_2O \longrightarrow 2HI(g) + H_3PO_3$

65. Write a balanced chemical equation that describes the reaction that occurs in each of the following cases.

   (a) Calcium is added to hydrobromic acid.
   (b) Potassium hydroxide is added to hydrofluoric acid.
   (c) Ammonia is bubbled through hydrofluoric acid.
   (d) Sodium acetate is added to hydroiodic acid.
   (e) Silver oxide is added to hydrobromic acid.
   (f) Chlorine is bubbled through hydrobromic acid.

Solution

(a) $Ca + 2HBr \longrightarrow H_2(g) + CaBr_2$
(b) $2K + 2HF \longrightarrow H_2(g) + 2KF$
(c) $NH_3(g) + HF(aq) \longrightarrow NH_4F$
(d) $NaCH_3CO_2 + HI \rightleftharpoons CH_3CO_2H + NaI$
(e) $Ag_2O + 2HI \longrightarrow 2AgI + H_2O$
(f) $Cl_2 + 2HBr \longrightarrow 2H^+ + 2Cl^- + Br_2$

67. Write the Lewis structure of each of the oxyacids of chlorine. Calculate the oxidation number of chlorine in each of the acids.

Solution

hypochlorous acid, HOCl (+1): $H^+[:\ddot{O}:\ddot{C}l:]^-$

chlorous acid, $HClO_2$ (+3): $H^+[:\ddot{O}:\ddot{C}l:\ddot{O}:]^-$

chloric acid, $HClO_3$ (+5): $H^+\begin{bmatrix}:\ddot{O}:\ddot{C}l:\ddot{O}:\\:\ddot{O}:\end{bmatrix}^-$

perchloric acid, $HClO_4$ (+7): $H^+\begin{bmatrix}:\ddot{O}:\\:\ddot{O}:\ddot{C}l:\ddot{O}:\\:\ddot{O}:\end{bmatrix}^-$

69. Which is the stronger acid, $HClO_2$ or $HClO_3$? Why?

Solution

$HClO_3$ is the stronger acid, because the Cl with the higher +5 oxidation number exerts a strong pull for the electrons in the O—H bond, thus reducing the strength of the bond. A weaker bond allows the hydrogen ion to dissociate more thus by making $HClO_3$ the stronger acid.

342 Chapter 22 Nonmetals, Part 1: Hydrogen, Oxygen, Sulfur, and the Halogens

71. Predict the products of each of the following.

   (a) Aluminum reacts with oxygen difluoride.
   (b) A solution of hypochlorous acid is heated.
   (c) Hydrochloric acid is added to a solution of hypochlorous acid.

   Solution

   (a) $4Al + 3OF_2 \longrightarrow 2AlF_3 + Al_2O_3$
   (b) $3HClO_2 \longrightarrow HCl + 2HClO_3$
   (c) $HCl + HClO \longrightarrow Cl_2 + H_2O$

   ADDITIONAL EXERCISES

73. Write the balanced equations or equations necessary to carry out the following transformations (H_2O, H_2, and/or O_2 may be used as needed):

   (a) Na_2O_2 from Na
   (b) NaOH from Na and O_3
   (c) NaCl from Na_2O_2 and Cl_2
   (d) $ZnSO_4$ from Zn and H_2S
   (e) Fe from Fe_3O_4

   Solution

   (a) $2Na + O_2 \longrightarrow Na_2O_2$
   (b) $6Na + O_3 \longrightarrow 3Na_2O$
   $Na_2O + H_2O \longrightarrow 2NaOH$
   (c) $Cl_2 + H_2 \longrightarrow 2HCl$
   $Na_2O_2 + 2HCl \longrightarrow 2NaCl + H_2O_2$
   (d) $H_2S + O_2 \longrightarrow SO_2 + H_2O$
   $2SO_2 + O_2 \longrightarrow 2SO_3$
   $SO_3 + H_2O \longrightarrow H_2SO_4$
   $Zn + H_2SO_4 \longrightarrow ZnSO_4 + H_2$
   (e) $Fe_3O_4 + 4H_2 \longrightarrow 3Fe + 4H_2O$

75. Write the chemical reaction for the preparation of ethylene dibromide, $C_2H_4Br_2$, a constituent of leaded gasoline, from ethylene and bromine. Use Lewis structures instead of chemical formulas in your equation.

   Solution

$$\underset{H}{\overset{H}{\diagdown}}C=C\underset{H}{\overset{H}{\diagup}} + :\ddot{Br}-\ddot{Br}: \longrightarrow :\ddot{Br}-\underset{H}{\overset{H}{C}}-\underset{H}{\overset{H}{C}}-\ddot{Br}:$$

77. The reaction of titanium metal with F_2 yields a titanium fluoride that contains 38.7% titanium. Write the chemical equation that describes the reaction.

   Solution

   Find the empirical formula of the Ti compound. Assume 100 g of fluoride, then

Chapter 22 Nonmetals, Part 1: Hydrogen, Oxygen, Sulfur, and the Halogens   343

$$\frac{38.7 \text{ g Ti}}{47.90 \text{ g mol}^{-1} \text{ Ti}} = 0.8079 \text{ mol Ti} \qquad \text{LOWEST RATIO} \quad \frac{0.8079}{0.8079} = 1$$

$$\frac{61.3 \text{ g F}}{18.998403 \text{ g mol}^{-1} \text{ F}} = 3.2266 \text{ mol F} \qquad \frac{3.2266}{0.8079} = 3.99$$

The formula of the product is TiF_4. Therefore, the reaction is

$$Ti(s) + 2F_2(g) \longrightarrow TiF_4$$

79. The reaction of V_2O_3 with Cl_2 gives a yellow liquid that contains 29.42% vanadium, 61.3% chlorine, and the remainder oxygen. At 19°C a sample of the liquid with a mass of 0.433 g vaporized in a 115-mL flask, giving a gas with a pressure of 390 torr. What is the molecular formula of this vanadium oxychloride?

    Solution

    Find the empirical formula; then on the basis of the mol. wt, determined from the ideal gas law, determine how many formula units comprise the molecular weight. Assume 100 g of sample.

$$\frac{29.42 \text{ g V}}{50.9415 \text{ g mol}^{-1}} = 0.5775 \qquad \text{LOWEST RATIO} \quad \frac{0.5775}{0.5775} = 1$$

$$\frac{61.3 \text{ g Cl}}{35.453 \text{ g mol}^{-1}} = 1.729 \qquad \frac{1.729}{0.5775} = 3$$

$$\frac{9.28 \text{ g O}}{15.9994 \text{ g mol}^{-1}} = 0.5800 \qquad \frac{0.5800}{0.5775} = 1$$

The empirical formula is $VOCl_3$.

$$PV = \frac{g\, RT}{\text{mol. wt}}; \qquad \text{mol. wt} = \frac{g\, RT}{PV} = \frac{0.433 \text{ g} \left(0.08206 \frac{\text{L atm}}{\text{K mol}}\right) 292.1 \text{ K}}{\frac{390}{760} \text{ atm} \times 0.115 \text{ L}}$$

$$= 176 \text{ g mol}^{-1}$$

The mol. wt of $VOCl_3$ is 173.300 g. Therefore, the molecular formula is $VOCl_3$.

81. The bond length in the O_2 molecule is 1.209 Å, and that in the O_3 molecule is 1.278 Å. Why does ozone have a longer bond?

    Solution

    In O_3 In O_2

    The bond has an order of $1\frac{1}{2}$, the average of a single and a double bond. In O_2, the Lewis electron dot picture shows a double bond between the two oxygens, and thus a shorter bond length. Of course, the dot picture does not show that O_2 is paramagnetic.

83. The average oxidation number of sulfur is not one of its common ones in Na$_2$S$_2$, H$_2$S$_2$, K$_2$S$_5$, and Na$_2$S$_2$O$_3$. Calculate the average oxidation number of sulfur in these compounds. What is the common structural feature in these compounds? In view of your answer to the preceding question, write a Lewis structure of S$_2$F$_{10}$. What is the oxidation number of sulfur in S$_2$F$_{10}$?

Solution

$$\begin{array}{c}
\ddot{\mathrm{F}}\!:\;\;\ddot{\mathrm{F}}\!:\\
\ddot{\mathrm{F}}\!:\;\;\;|\;\;\;|\;\;\;\ddot{\mathrm{F}}\!:\\
:\ddot{\mathrm{F}}\!-\!\mathrm{S}\!-\!\mathrm{S}\!-\!\ddot{\mathrm{F}}\!:\\
\ddot{\mathrm{F}}\!:\;\;\;|\;\;\;|\;\;\;\ddot{\mathrm{F}}\!:\\
\ddot{\mathrm{F}}\!:\;\;\ddot{\mathrm{F}}\!:
\end{array}$$

Oxidation number of S	−1	−1	−2/5	+2
Compound	Na$_2$S$_2$	H$_2$S$_2$	K$_2$S$_5$	Na$_2$S$_2$O$_3$

The common structural feature is that there are S to S bonds present. For S$_2$F$_{10}$, the oxidation number of S is +5.

85. Iodine reacts with chlorine to give a compound containing 78.2% iodine. Write the balanced equation for the reaction.

Solution

Find the empirical formula; then balance the equation. Assume 100 g of compound.

$$\frac{78.2 \text{ g I}}{2(126.9045 \text{ g mol}^{-1})} = 0.308 \text{ mol I}_2$$

$$\frac{21.8 \text{ g Cl}}{70.906 \text{ g mol}^{-1}} = 0.307 \text{ mol Cl}_2$$

The ratio is 1I to 1Cl.

The formula is ICl.

The reaction is

$$\mathrm{I}_2 + \mathrm{Cl}_2 \longrightarrow 2\mathrm{ICl}$$

87. A molten mixture of rubidium fluoride and uranium (IV) fluoride is oxidized with fluorine to produce a uranium compound in which most but not all of the uranium has an oxidation number of +5. The product is found to contain 54.43% uranium. A 1.0357-g solution reacts according to the following equation:

$$2\mathrm{I}^- + 2\mathrm{UF}_6^- \longrightarrow 2\mathrm{UF}_4 + \mathrm{I}_2 + 4\mathrm{F}^-$$

The iodine produced is titrated with 14.80 mL of 0.1494 M sodium thiosulfate solution. What percentage of the original uranium was oxidized to the +5 oxidation number?

Solution

On the basis of the molarity of I$_2$, calculate the mass of UF$_6^-$ present. Then calculate the grams of U present in UF$_4$; from that information the final percentage occurring as UF$_4$ can be calculated. The reaction of I$_2$ is the same as in exercise 47.

$$2S_2O_3^{2-}(aq) + I_2 \longrightarrow S_4O_6^{2-}(aq) + 2I^-$$

$$M_iV_i = \text{mol } I_2$$

$$\frac{0.01480 \text{ L} \times 0.1494 \text{ M}}{2} = 1.10556 \times 10^{-3} \text{ mol } I_2$$

The mol I_2 corresponds to 2 mol UF_4 (mol. wt = 314.023) or

$$\frac{2 \text{ mol } UF_4}{\text{mol } I_2} \times 1.10556 \times 10^{-3} \text{ mol } I_2 \times 314.023 \text{ g } \frac{UF_4}{\text{mol } UF_4} = 0.6943 \text{ g } UF_4$$

Total amount of U present is

$$1.0357 \text{ g} \times 54.43\% = 0.56373 \text{ g U}$$

$$\text{Percentage U in } UF_4 = \frac{238.029}{314.023} \times 100 = 75.799\%$$

$$\text{Grams U in } UF_4 = 0.6943 \text{ g } UF_4 \times 0.75799 \frac{U}{UF_4} = 0.5263 \text{ g U}$$

$$\text{Percent } UF_4 \text{ in sample} = \frac{0.5263}{0.56373} \times 100 = 93.36\%$$

89. The reactions involved in the preparation of sulfuric acid are highly exothermic. Using the data in Appendix I, calculate the enthalpy changes for the following reactions:

    (a) $S(s) + O_2(g) \longrightarrow SO_2(g)$
    (b) $2SO_2(g) + O_2(g) \longrightarrow 2SO_3(g)$
    (c) $SO_3(g) + H_2O(l) \longrightarrow H_2SO_4(l)$

Solution

(a) $\Delta H = \Delta H_{fSO_2} - \Delta H_{fS} - \Delta H_{fO_2} = -296.83 \text{ kJ} - 0 - 0 = -296.83 \text{ kJ}$
(b) $\Delta H = 2\Delta H_{fSO_3} - 2\Delta H_{fSO_2} - 2\Delta H_{fO_2} = 2(-395.7 \text{ kJ}) - 2(-296.83 \text{ kJ}) = -197.8 \text{ kJ}$
(c) $\Delta H = \Delta H_{fH_2SO_4} - \Delta H_{fSO_3} - \Delta H_{fH_2O} = -813.989 \text{ kJ} - (-395.7 \text{ kJ}) - (-285.83 \text{ kJ}) = -132.5 \text{ kJ}$

91. From the data in Appendix I, calculate the free energy change for the reaction of hydrogen with each of the halogens. Determine which of these reactions is not spontaneous at 25°C.

Solution

$H_2 + F_2 \longrightarrow 2HF \quad \Delta G_f^\circ = -273 \text{ kJ mol}^{-1}$

$H_2 + Cl_2 \longrightarrow 2HCl \quad \Delta G_f^\circ = -95.299 \text{ kJ mol}^{-1}$

$H_2 + Br_2 \longrightarrow 2HBr \quad \Delta G_f^\circ = -53.43 \text{ kJ mol}^{-1}$

$H_2 + I_2 \longrightarrow 2HI \quad \Delta G_f^\circ = +1.7 \text{ kJ mol}^{-1}$

The formation of HI is not spontaneous because of the positive ΔG°.

93. Thallium(I) chloride is one of the few insoluble chlorides. What is the molar solubility of TlCl, whose solubility product is 1.8×10^{-4}?

Solution

$$K_{sp} = [Tl^+][Cl^-] = 1.8 \times 10^{-4} = x^2$$

$$x = [Tl^+] = [Cl^-] = 0.0134 \text{ mol/L} = 1.3 \times 10^{-2} \text{ M}$$

23

The Nonmetals, Part 2: Carbon, Nitrogen, Phosphorus, and the Noble Gases

INTRODUCTION

We now expand our study of nitrogen, phosphorus, carbon, and the noble gases. Nitrogen and phosphorus are important to agriculture. The tonnages of fertilizer used in North America are huge. Nitrogen fixation by bacteria is essential for growth of plants, and farmers supply additional nitrogen in the form of ammonia or ammonium salts. Phosphates, made soluble by chemical processes, are no less important for the production of the outstanding yields given by American and Canadian fields.

Carbon compounds in which carbon is not bonded to hydrogen are treated here; the industrial use of "inorganic carbon," especially in the form of CO_2, is very large. Only the rare gas compounds are in their infancy as industrial chemicals. Compounds in which carbon is bonded directly to hydrogen are examined in Chapters 30 and 31.

READY REFERENCE

Aqua regia (23.13)   A strong oxidizing acid made by mixing 1 part concentrated nitric acid and 3 parts concentrated hydrochloric acid.

Azide (23.14)   A salt of hydroazoic acid, HN_3. Azides are unstable.

Ostwald process (23.13)   Nitric acid is produced by oxidation of ammonia to nitric oxide and then to nitrogen dioxide. Nitrogen dioxide is dissolved in water to form the acid.

SOLUTIONS TO CHAPTER 23 EXERCISES

1. Compare and contrast the chemical properties of elemental phosphorus and nitrogen; phosphoric acid and nitric acid; phosphine and ammonia.

Chapter 23 Nonmetals, Part 2: Carbon, Nitrogen, Phosphorus, and the Noble Gases

Solution

A major difference between nitrogen and phosphorus is in their reactivity at room temperature to a variety of reagents. Nitrogen is practically unreactive while phosphorus burns in air; white phosphorus ignites spontaneously on contact with air. Both elements react with active metals at elevated temperatures. In addition, their oxides are acidic in water.

Nitric acid is a strong, oxidizing acid while phosphoric acid, although triprotic, is a relatively weak acid with little oxidizing power. Phosphates are sufficiently mild as acids and bases (HPO_4^{2-}) that they are widely used as buffers in foods and drinks.

Phosphine, unlike ammonia, is only slightly soluble in water and is a much weaker base. Phosphonium salts are unstable in water and decompose to release phosphine. Phosphine cannot be prepared from the elements as is ammonia, and it is extremely poisonous.

3. With which elements will phosphorus form compounds in which it has a positive oxidation number?

Solution

Phosphorus has a positive oxidation number when it combines with the major, abundant nonmetals — oxygen, sulfur and halogens.

NITROGEN

5. Write electronic structures for the following:

   (a) N_2
   (b) NH_3
   (c) NH_4^+
   (d) N_2H_4
   (e) HN_3
   (f) NH_2OH

Solution

(a) :N≡N:

(b) 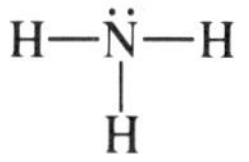 H—N̈—H
　　　|
　　　H

(c) 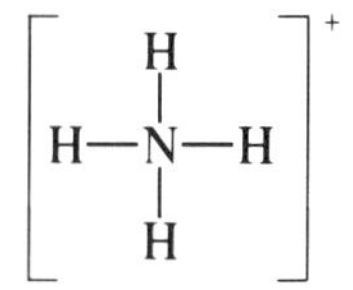 $\left[\begin{array}{c} H \\ | \\ H-N-H \\ | \\ H \end{array}\right]^+$

(d) 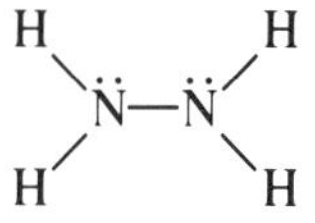 H　　　　H
　　 \\..　../
　　　N—N
　　 /　　　\\
　　H　　　　H

(e) 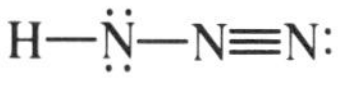 H—N̈—N≡N:

(f) 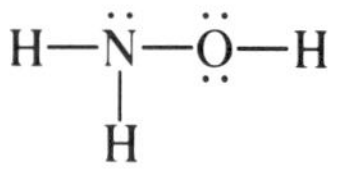 H—N̈—Ö—H
　　　|
　　　H

348 Chapter 23 Nonmetals, Part 2: Carbon, Nitrogen, Phosphorus, and the Noble Gases

7. Explain the effects of temperature, pressure, and a catalyst on the direct synthesis of ammonia from hydrogen and nitrogen.

Solution

$$N_2(g) + 3H_2(g) \underset{}{\overset{\text{catalyst}}{\rightleftharpoons}} 2NH_3(g) \quad \Delta H° = -92 \text{ kJ}$$

Because the reaction is exothermic, an increase in temperature will decrease the yield. A pressure increase will force the equilibrium in favor of ammonia. A catalyst is necessary to produce ammonia at practical industrial rates.

9. What mass of ammonia is produced by adding water to the product of the reaction of 93.0 g of lithium with elemental nitrogen?

Solution

Lithium reacts with nitrogen to form lithium nitride, Li_3N, which in turn reacts with water to produce ammonia.

$$6Li + N_2 \longrightarrow 2Li_3N \quad \quad (1)$$

$$Li_3N + 3H_2O \longrightarrow 3LiOH + NH_3 \quad \quad (2)$$

Equation (2) shows that one mole of Li_3N produces one mole of NH_3. Use equation (1) to determine the moles of Li_3N produced from 93.0 g Li.

$$\text{No. mol } Li_3N = 93.0 \text{ g Li} \times \frac{1 \text{ mol}}{6.941 \text{ g}} \times \frac{2 \text{ mol } Li_3N}{6 \text{ mol Li}} = 4.4662 \text{ mol}$$

$$\text{Mass } NH_3 = 4.4662 \text{ mol } Li_3N \times \frac{1 \text{ mol } NH_3}{1 \text{ mol } Li_3N} \times 17.0304 \frac{\text{g}}{\text{mol}} = 76.1 \text{ g}$$

11. What is the maximum volume of ammonia that can be collected at 27°C and 750 torr by the treatment of 2.00 g of ammonium chloride with 0.500 L of 0.500 M sodium hydroxide?

Solution

$$\underset{\underset{1 \text{ mol}}{2.00 \text{ g}}}{NH_4Cl(aq)} + \underset{\underset{1 \text{ mol}}{0.5 \text{ L of } 0.5 \text{ M}}}{NaOH(aq)} \longrightarrow \underset{\underset{1 \text{ mol}}{V}}{NH_3(g)} + NaCl(aq)$$

Determine the limiting reagent, and from that, moles of NH_3. Then convert moles to volume at the stated conditions.

$$\text{mol } NH_4Cl = 2.00 \text{ g} \times \frac{1 \text{ mol}}{53.4913 \text{ g}} = 0.03738 \text{ mol}$$

$$\text{mol } NaOH = .500 \text{ L} \times .500 \text{ M} = 0.25 \text{ mol}$$

NH_4Cl is the limiting reagent.

$$\text{mol } NH_3 = 0.03738; \quad V = nRT/P$$

$$V = \frac{(0.03738 \text{ mol})(0.08206 \text{ L atm/mol K})(300.15 \text{ K})}{(750/760 \text{ atm})} = 0.933 \text{ L}$$

Chapter 23 Nonmetals, Part 2: Carbon, Nitrogen, Phosphorus, and the Noble Gases   349

13. Write equations for the preparation of each of the oxides of nitrogen.

    Solution

    $$4NH_3 + 5O_2 \xrightarrow{\Delta} 4NO + 6H_2O$$
    $$NH_4NO_3 \xrightarrow{\Delta} N_2O + 2H_2O$$
    $$2NO + O_2 \xrightarrow{\Delta} 2NO_2$$
    $$NO + NO_2 \xrightarrow{\Delta} N_2O_3$$
    $$2NO_2 \xrightarrow{\Delta} N_2O_4$$
    $$4HNO_3 + P_4O_{10} \xrightarrow{\Delta} 4HPO_3 + 2N_2O_5$$

15. Outline the chemistry of the production of nitric acid from ammonia.

    Solution

    $$4NH_3(g) + 5O_2 \longrightarrow 4NO(g) + 6H_2O(g)$$
    $$2NO(g) + O_2(g) \longrightarrow 2NO_2(g)$$
    $$3NO_2 + H_2O \longrightarrow 2HNO_3 + NO$$

17. Write electronic structures for the following:

    (a) HNO_3
    (b) NO_3^-
    (c) HNO_2
    (d) N_2O
    (e) N_2O_4
    (f) NO_2^-
    (g) N_2O_3
    (h) N_2O_5
    (i) NCl_3
    (j) ClNO

    Solution

    (a) 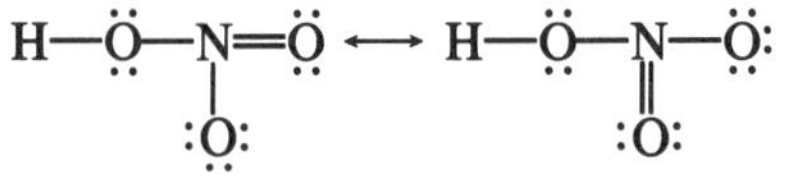 H—Ö—N=Ö ⟷ H—Ö—N—Ö:
                  ‖   ‖
                 :Ö: :O:

    (b) 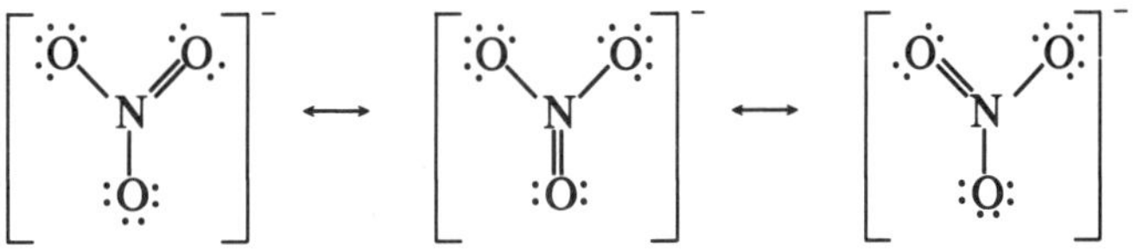 [:Ö⟍N⟋Ö:]⁻ ⟷ [:Ö⟍N⟋Ö:]⁻ ⟷ [·Ö⟍N⟋Ö·]⁻
         ‖ | | ‖   ‖   ‖
        :Ö: :Ö: :O:

    (c) 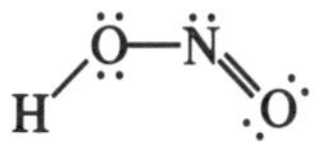 H—Ö—N=Ö

    (d)  N̈=N=Ö ⇌ :N≡N—Ö:

    (e) 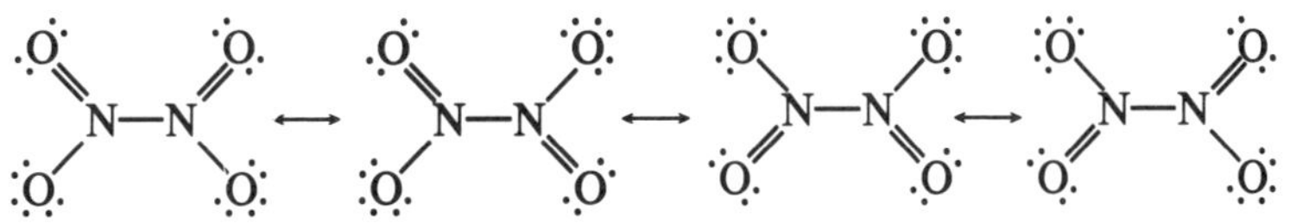 ·Ö⟍N—N⟋Ö· ⟷ ·Ö⟍N—N⟋Ö: ⟷ :Ö⟍N—N⟋Ö: ⟷ :Ö⟍N—N⟋Ö·
        :Ö⟋   ⟍Ö: :Ö⟋   ⟍Ö: ·Ö⟋   ⟍Ö: ·Ö⟋   ⟍Ö:

(f) $\left[:\overset{..}{\underset{..}{N}}\overset{\overset{..}{O}:}{\underset{\underset{..}{O}:}{}} \right]^{-} \longleftrightarrow \left[:\overset{..}{\underset{..}{N}}\overset{\overset{..}{O}:}{\underset{\underset{..}{O}:}{}} \right]^{-}$

(g) (two resonance structures of N_2O_4)

(h) (four resonance structures of N_2O_5)

(i) $:\overset{..}{\underset{..}{Cl}}-\overset{.}{N}(-\overset{..}{\underset{..}{Cl}}:)-\overset{..}{\underset{..}{Cl}}:$

(j) $:\overset{..}{\underset{..}{Cl}}-\overset{..}{N}-\overset{..}{\underset{..}{O}}: \longleftrightarrow :\overset{..}{\underset{..}{Cl}}-\overset{..}{N}=\overset{..}{O}$

19. The oxidation of ammonia and of nitric oxide are exothermic processes. Using the data in Appendix I, calculate ΔH°_{298} for these reactions.

Solution

(a) $4NH_3(g) + 5O_2(g) \longrightarrow 4NO(g) + 6H_2O(g)$

$\Delta H_{Rx} = 6\Delta H_{fH_2O(g)} + 4\Delta H_{fNO(g)} - 4\Delta H_{fNH_3(g)}$

$= 6(-241.82) + 4(90.25) - 4(-46.11) = -905.48$ kJ

(b) $2NO(g) + O_2(g) \longrightarrow 2NO_2(g)$

$\Delta H_{Rx} = 2\Delta H_{fNO_2(g)} - 2\Delta H_{fNO(g)}$

$= 2(33.2) - 2(90.25) = -114.1$ kJ

PHOSPHORUS

21. Complete and balance the following equations (*xs* indicates that the reactant is present in excess):

(a) $P_4 + xsO_2 \longrightarrow$
(b) $P_4 + K \longrightarrow$
(c) $P_4 + xsS \longrightarrow$
(d) $P_4 + Ca \longrightarrow$
(e) $P_4 + xsF_2 \longrightarrow$
(f) $xsP_4 + Cl_2 \longrightarrow$
(g) $P_4 + I_2 \longrightarrow$

Solution

(a) $P_4 + 5O_2 \longrightarrow P_4O_{10}$
(b) $P_4 + 12K \longrightarrow 4K_3P$
(c) $P_4 + 10S \longrightarrow P_4S_{10}$
(d) $P_4 + 6Ca \longrightarrow 2Ca_3P_2$
(e) $P_4 + 10F_2 \longrightarrow 4PF_5$
(f) $P_4 + 6Cl_2 \longrightarrow 4PCl_3$
(g) $P_4 + 6I_2 \longrightarrow 4PI_3$

23. Write a Lewis structure for each of the following:

   (a) PH_3
   (b) P_2H_4
   (c) H_3PO_2
   (d) H_3PO_3
   (e) H_3PO_4

   Solution

   (a)
   $$H-\ddot{P}-H$$
   $$\quad\quad |$$
   $$\quad\quad H$$

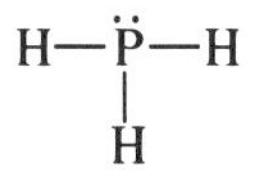

   (b)
   $$H\diagdown\quad\quad\diagup H$$
   $$\quad\ddot{P}-\ddot{P}$$
   $$H\diagup\quad\quad\diagdown H$$

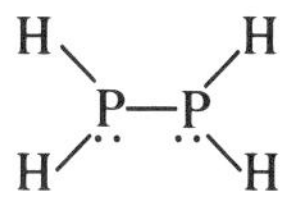

   (c)
   $$\quad\quad H$$
   $$\quad\quad |$$
   $$:\ddot{O}-P-\ddot{O}-H$$
   $$\quad\quad |$$
   $$\quad\quad H$$

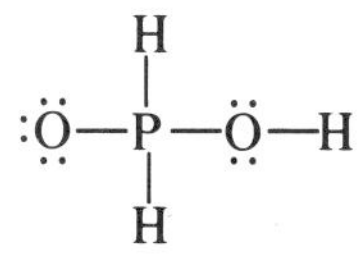

   (d)
   $$\quad\quad H$$
   $$\quad\quad |$$
   $$\quad\quad :\ddot{O}:$$
   $$\quad\quad |$$
   $$H-P-\ddot{O}:$$
   $$\quad\quad |$$
   $$\quad\quad :\ddot{O}:$$
   $$\quad\quad |$$
   $$\quad\quad H$$

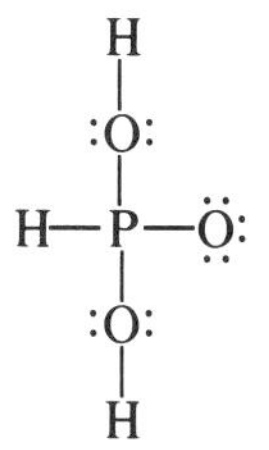

   (e)
   $$\quad\quad H$$
   $$\quad\quad |$$
   $$\quad\quad :\ddot{O}:$$
   $$\quad\quad |$$
   $$:\ddot{O}-P-\ddot{O}-H$$
   $$\quad\quad |$$
   $$\quad\quad :\ddot{O}:$$
   $$\quad\quad |$$
   $$\quad\quad H$$

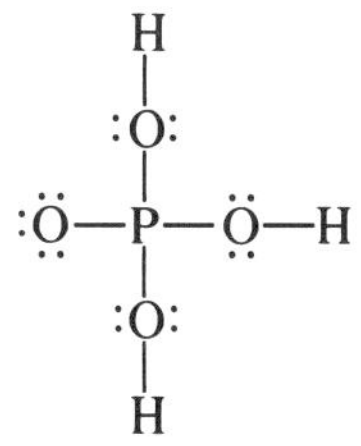

25. Write equations for the preparation and hydrolysis of sodium phosphide. Compare the hydrolysis of sodium phosphide to that of sodium nitride.

   Solution

   $$P_4 + 12Na \longrightarrow 4Na_3P$$

   $$Na_3P + 3H_2O \longrightarrow 3NaOH + PH_3$$

   $$Na_3N + 3H_2O \longrightarrow 3NaOH + NH_3$$

27. Write the equation to show the effect of heating H_3PO_4. The effect of heating NaH_2PO_4.

   Solution

   $$2H_3PO_4 \xrightarrow{\Delta} H_4P_2O_7 + H_2O$$

   $$nNaH_2PO_4 \xrightarrow{\Delta} (NaPO_3)_n + nH_2O$$

352 Chapter 23 Nonmetals, Part 2: Carbon, Nitrogen, Phosphorus, and the Noble Gases

29. Write equations for the preparation of hypophosphorous acid starting with white phosphorus.

   Solution

   $$2P_4 + 3Ba(OH)_2 \longrightarrow 2PH_3 + 3Ba(H_2PO_2)_2$$
   $$Ba(H_2PO_2)_2 + H_2SO_4 \longrightarrow BaSO_4 + 2H_3PO_2$$

31. Write equations for each of the following preparations:

   (a) P_4 from $Ca_3(PO_4)_2$
   (b) P_4O_{10} from P_4
   (c) H_3PO_4 from P_4O_{10}
   (d) Na_2HPO_4 from H_3PO_4
   (e) $Na_4P_2O_7$ from H_3PO_4

   Solution

   (a) $2Ca_3(PO_4)_2 + 6SiO_2 + 10C \longrightarrow 6CaSiO_3 + 10CO + P_4$
   (b) $P_4 + 5O_2 \longrightarrow P_4O_{10}$
   (c) $P_4O_{10} + 6H_2O \longrightarrow 4H_3PO_4$
   (d) $H_3PO_4 + 2NaOH \longrightarrow Na_2HPO_4 + 2H_2O$
   (e) $2H_3PO_4 \xrightarrow{\Delta} H_4P_2O_7 + H_2O$
   $H_4P_2O_7 + 4NaOH \longrightarrow Na$

33. Complete and balance the following equations (*xs* indicates that the reactant is present in excess):

   (a) $NaH_2PO_4 + NH_3 \longrightarrow$
   (b) $PF_5 + KF \longrightarrow$
   (c) $P_4O_{10} + K_2O \longrightarrow$
   (d) $xsK_3PO_4 + HCl \longrightarrow$
   (e) $Na_3P + xsH_2O \longrightarrow$
   (f) $Na_4P_2O_7 + xsH_2O \longrightarrow$
   (g) $PCl_3 + CH_3OH \longrightarrow$
      (treat CH_3OH as a derivative of water in which one H is replaced by the $-CH_3$ group.)

   Solution

   (a) $NaH_2PO_4 + NH_3 \longrightarrow Na(NH_4)HPO_4$
   (b) $PF_5 + KF \longrightarrow PF_6^- + K^+$
   (c) $P_4O_{10} + 6K_2O \longrightarrow 4K_3PO_4$
   (d) $K_3PO_4 + HCl \longrightarrow K_2HPO_4 + KCl$
   (e) $Na_3P + 3H_2O \longrightarrow 3NaOH + PH_3$
   (f) $Na_4P_2O_7 + H_2O \longrightarrow 2Na_2HPO_4$
   (g) $PCl_3 + 3CH_3OH \longrightarrow P(OCH_3)_3 + 3HCl$

35. What volume of 0.100 M NaOH will be required to neutralize the solution produced by dissolving 1.00 g of PCl_3 in an excess of water? Note that when H_3PO_3 is titrated under these conditions, only one proton of the phosphorous acid molecule reacts.

Solution

$$PCl_3 + 3H_2O \longrightarrow H_3PO_3 + 3HCl$$
1.0 g
1 mol   1 mol   3 mol

$$H_3PO_3 + NaOH \longrightarrow NaH_2PO_4 + H_2O \quad (1\ H^+/H_3PO_3)$$

$$HCl + NaOH \longrightarrow NaCl + H_2O \quad (1\ H^+/HCl)$$

$$\text{mol } H^+ = 4(\text{mol } PCl_3) = 4(1.00\ g\ PCl_3 \times \frac{1\ mol}{137.33\ g}) = 0.02913\ mol$$

$$V(0.1\ M\ NaOH) = 0.02913\ mol \times \frac{1\ L}{0.100\ mol} = 0.291\ L \text{ or } 291\ mL$$

CARBON

37. Write the equation describing the chemical reaction that occurs when carbon dioxide is bubbled through sodium hydroxide solution.

    Solution

    $$NaOH(aq) + CO_2(g) \longrightarrow NaHCO_3(aq)$$

39. Determine the heat of combustion of carbon monoxide and of carbon when carbon dioxide is the product.

    Solution

    (a) $2CO(g) + O_2(g) \longrightarrow 2CO_2(g)$

    $$\Delta H_{Rx} = 2\Delta H_{fCO_2(g)} - 2\Delta H_{fCO(g)}$$
    $$= 2(-393.51) - 2(-110.52) = -565.98\ kJ \text{ or } -282.99\ kJ/mol$$

    (b) $C(s) + O_2(g) \longrightarrow CO_2(g)$

    $$\Delta H_{Rx} = \Delta H_{fCO_2(g)} - \Delta H_{fC(g)} - \Delta H_{fO_2(g)}$$
    $$= -393.51\ kJ/mol - 0 - 0$$

41. Write equations for the production of the following:

    (a) carbon disulfide
    (b) carbon tetrachloride
    (c) calcium carbide
    (d) acetylene
    (e) sodium cyanide
    (f) hydrogen cyanide
    (g) calcium cyanamide

    Solution

    (a) $C + 2S \xrightarrow{\Delta} CS_2$
    (b) $CH_4 + 4Cl_2 \xrightarrow{\Delta} CCl_4 + 4HCl$
    (c) $CaO + 3C \xrightarrow{\Delta} CaC_2 + CO$
    (d) $CaC_2 + 2H_2O \longrightarrow Ca(OH)_2 + C_2H_2$
    (e) $Ca(CN)_2 + C + Na_2CO_3 \longrightarrow CaCO_3 + 2NaCN$
    (f) $NaCN + HCl \longrightarrow HCN + NaCl$
    (g) $CaC_2 + N_2 \xrightarrow{\Delta} CaCN_2 + C$

THE NOBLE GASES

43. Based on the periodic properties of the elements, write balanced chemical equations for the reactions, if any, between the components of air given in Table 29.1 and hot magnesium.

Solution

$$3Mg + N_2 \xrightarrow{\Delta} Mg_3N_2$$

$$2Mg + O_2 \xrightarrow{\Delta} 2MgO$$

$$Mg + \text{noble gases} \longrightarrow \text{No reaction (Ar, Ne, He, Kr, Xe)}$$

$$Mg + CO_2 \xrightarrow{\Delta} 2MgO + C$$

$$6Mg + 2O_3 \longrightarrow 6MgO$$

$$Mg + H_2 \longrightarrow MgH_2 \text{ (under pressure)}$$

45. Write complete and balanced equations for the following reactions:

(a) $XeO_3 + Ba(OH)_2$ (in water) $\longrightarrow$
(b) $XeF_2 + HClO_4 \longrightarrow$
(c) $XeF_2 + 2HClO_4 \longrightarrow$
(d) $XeF_6 + Na \longrightarrow$
(e) $xsXeF_6 + P \longrightarrow$

Solution

(a) $2XeO_3 + 2Ba(OH)_2 \longrightarrow Xe + O_2 + Ba_2XeO_6 + 2H_2O$
(b) $XeF_2 + HClO_4 \longrightarrow F—Xe—O—ClO_3 + HF$
(c) $XeF_2 + 2HClO_4 \longrightarrow ClO_3—O—Xe—O—ClO_3 + 2HF$
(d) $XeF_6 + 6Na \longrightarrow 6NaF + Xe$
(e) $10XeF_6 + 3P_4 \longrightarrow 12PF_5 + 10Xe$

47. Basic solutions of Na_4XeO_6 are powerful oxidants. What mass of $Mn(NO_3)_2 \cdot 6H_2O$ will react with 125.0 mL of a 0.1717 M basic solution of Na_4XeO_6 if the products include Xe and a solution of sodium permanganate?

Solution

This is a redox reaction and must be treated accordingly.

$$\underset{}{4NaOH} + \underset{\substack{\text{125 mL of 0.1717 M} \\ \text{5 mol}}}{5Na_4XeO_6} + \underset{\substack{\text{Mass?} \\ \text{8 mol}}}{8Mn(NO_3)_2 \cdot 6H_2O} \longrightarrow$$

$$8NaMnO_4 + 5Xe + 16NaNO_3 + 48H_2O$$

mol (Xe compound) = 0.125 L × 0.1717 M = 0.021462 mol

Mass (Mn compound) = 0.021462 mol Xe × $\dfrac{8 \text{ mol Mn}}{5 \text{ mol Xe}}$ × $\dfrac{287.039 \text{ g}}{\text{mol}}$ = 9.857 g

ADDITIONAL EXERCISES

49. Which of the following compounds will react with water to give an acid solution? Which will give a basic solution? Write a balanced equation for the reaction of each with water.

Chapter 23 Nonmetals, Part 2: Carbon, Nitrogen, Phosphorus, and the Noble Gases

(a) Na_3N
(b) N_2O_3
(c) Na_2CO_3
(d) CO_2
(e) Ca_3P_2
(f) Na_2NH
(g) CaC_2
(h) $LiNO_2$

Solution

(a) $Na_3N + 3H_2O \longrightarrow NH_3 + 3NaOH$ basic
(b) $N_2O_3 + H_2O \longrightarrow 2HNO_2$ acidic
(c) $Na_2CO_3 + H_2O \longrightarrow NaHCO_3 + NaOH$ basic
(d) $CO_2 + H_2O \longrightarrow H_2CO_3$ acidic
(e) $Ca_3P_2 + 6H_2O \longrightarrow 3Ca(OH)_2 + 2PH_3$ basic
(f) $Na_2NH + 2H_2O \longrightarrow NH_3 + 2NaOH$ basic
(g) $CaC_2 + 2H_2O \longrightarrow Ca(OH)_2 + C_2H_2$ basic
(h) $LiNO_2 + H_2O \longrightarrow LiOH(aq) + HNO_2(aq)$ basic

51. Increases in the price of petroleum result in increases in the price of ammonia. Why?

Solution

The industrial preparation of ammonia utilizes hydrogen from the decomposition of natural gas and other hydrocarbons.

53. What evidence is there that a coordinate-covalent bond between the proton and an ammonia molecule is stronger than that between the proton and water?

Solution

In water, ammonia acts as a base in which water donates a proton to ammonia forming the ammonium ion.

$$NH_3 + H_2O \longrightarrow NH_4^+ + OH^- \qquad K_b = 1.8 \times 10^{-5}$$

57. What mass of phosphorus is required to prepare 60.08 g of PCl_3 by the reaction of chlorine with phosphorus?

Solution

Mass?   60.08 g
$2P + 3Cl_2 \longrightarrow 2PCl_3$
2 mol   2 mol

$$\text{Mass (P)} = 60.08 \text{ g } PCl_3 \times \frac{1 \text{ mol}}{137.33 \text{ g}} \times \frac{2 \text{ mol P}}{2 \text{ mol } PCl_3} \times 30.97376 \text{ g} = 13.55 \text{ g}$$

59. At 25°C and 1.0 atm a mixture of $N_2O_4(g)$ and $NO_2(g)$ contains 30.0% $NO_2(g)$ by volume. Calculate the partial pressures of the two gases when they are at equilibrium at 25°C with a total pressure of 9.0 atm.

Solution

This equilibrium is based on the reaction:

$$2NO_2(g) \longrightarrow N_2O_4(g)$$

Given that the NO$_2$ concentration is 30% by volume and that the total pressure is 1 atm, the partial pressures of NO$_2$ and N$_2$O$_4$ are 0.30 atm and 0.70 atm, respectively. The equilibrium constant at 1 atm is:

$$K_p = \frac{P_{N_2O_4}}{(P_{NO_2})^2} = \frac{0.70 \text{ atm}}{(0.30 \text{ atm})^2} = 7.7778$$

At constant temperature, K_p is valid at the new pressure. Therefore, let x equal the pressure of NO$_2$ remaining at equilibrium and $9.0 - 0.5x$ equal the pressure of N$_2$O$_4$.

$$K_p = 7.7778 = \frac{9.0 - 0.5x}{(x)^2}$$

Rearranging and solving the quadratic equation yields:

$$7.7778x^2 + 0.5x - 9 = 0; \quad x = 1.0$$

$$P_{NO_2} = 1.0 \text{ atm}$$

$$P_{N_2O_4} = 9.0 - x = 8.0 \text{ atm}$$

61. The solubility product of Ca$_3$(PO$_4$)$_2$ is 1×10^{-25}. What is the concentration of Ca$_3$(PO$_4$)$_2$ in a saturated solution of Ca$_3$(PO$_4$)$_2$?

Solution

$$\text{Ca}_3(\text{PO}_4)_2 \rightleftharpoons 3\text{Ca}^{2+} + 2\text{PO}_4^{3-} \quad K_{sp} = 1 \times 10^{-25}$$

$$K_{sp} = [\text{Ca}^{2+}]^3[\text{PO}_4^{3-}]^2 = 1 \times 10^{-25}$$

Let x = moles of Ca$_3$(PO$_4$)$_2$ dissolving

$$3x = \text{Ca}^{2+}; \quad 2x = \text{PO}_4^{3-}$$

$$(3x)^3(2x)^2 = 1 \times 10^{-25}$$

$$108x^5 = 1 \times 10^{-25}$$

$$x = (9.259 \times 10^{-28})^{1/5} = 4 \times 10^{-6} \text{ mol}$$

The concentration of Ca$_3$(PO$_4$)$_2$ in a saturated solution is 4×10^{-6} M.

24

Nuclear Chemistry

INTRODUCTION

In 1896, Henri Becquerel observed that ores containing the compound potassium uranyl sulfate, $K_2SO_4 \cdot UO_2SO_4 \cdot 2H_2O$, emitted high-energy rays similar to X rays. One of his students, Marie Sklodowska Curie, named this phenomenon *radioactivity*. Two years later, Marie and her husband Pierre were able to identify two new elements, polonium (Po, named after her native Poland) and radium (Ra), both radioactive. The Nobel Prize was awarded to the Curies and Becquerel in 1903 for their work.

Their discovery led to the understanding that certain atomic nuclei are unstable and spontaneously disintegrate with the release of energy and penetrating radiations to produce new nuclei. These may be either nuclei of an element different from the original element or an isotope of the original element. About three-fourths of the more than 1000 known isotopes are unstable or radioactive and can exhibit the phenomena described. All isotopes with atomic number Z greater than 83 are radioactive, and all elements beyond uranium ($Z = 92$) in the Periodic Table are artificially created as well as radioactive.

READY REFERENCE

Binding energy (B) (24.2)   The mass of a nucleus is always less than the combined mass of its constituent particles. The mass difference, or mass defect, is related to energy through the equation $E = mc^2$. This energy, called the binding energy of the nucleus, is the energy required to break up the nucleus into its constituent particles. To calculate the binding energy, calculate the mass difference between the mass of a nuclide and the mass of its components. This difference, the mass defect, is substituted into $E = mc^2$.

Energy equivalences (24.2)   The energy produced by conversion of a mass equivalent to 1 amu, or $1.6605655 \times 10^{-27}$ kg, to energy through the Einstein equation, $E = mc^2$, is

$$E = (1.6605 \times 10^{-27} \text{ kg})(2.99792468 \times 10^8 \text{ m/s})^2 = 1.492442 \times 10^{-10} \text{ J}$$

$$1 \text{ joule} = 6.24146 \times 10^{18} \text{ electron volts} = 6.24146 \times 10^{12} \text{ MeV}$$

or

$$1 \text{ MeV} = 1.602189 \times 10^{-13} \text{ J}$$

$$E = (1.49244 \times 10^{-10} \text{ J})(6.24146 \times 10^{12} \text{ MeV/J}) = 931.450 \text{ MeV}$$

Half-life ($t_{1/2}$) (24.4)   The amount of time required for one-half the number of nuclei in a radioactive sample to decay to new nuclei. Half-life values range from microseconds to billions of years.

Mass number (*A*) (24.1) The sum of the number of neutrons and protons in the nucleus of a specific nuclide.

Nuclear reaction rate (24.7) The rate of a nuclear reaction is kinetically of the first order. That is, the number of nuclei of a specific nuclide remaining in a sample after an elapsed time, t, is a function of the initial number of nuclei, N_0, and the half-life of the nuclide. The value N_t, or the number of nuclei at time, t, is related to N_0 by the equation $N_t = N_0 e^{-kt}$, where k is a proportionality constant called the ***decay constant***. This equation may be rearranged as

$$\log \frac{N_0}{N_t} = \frac{kt}{2.303} \quad \text{or} \quad k = \frac{0.693}{t_{\frac{1}{2}}}$$

SOLUTIONS TO CHAPTER 24 EXERCISES

2. Indicate the number of protons and neutrons in each of the following nuclei:

   (a) $^{16}_{8}O$
   (c) lead-208

   Solution

Symbol	Protons	Neutrons
(a) $^{16}_{8}O$	8	$16 - 8 = 8$
(c) lead-208 ($^{208}_{82}Pb$)	82	$208 - 82 = 126$

   NUCLEAR STABILITY

3. Which of the following nuclei lie within the band of stability shown in Figure 24.2?

   (a) $^{5}_{3}Li$
   (d) $^{60}_{30}Zn$
   (e) radon-210

   Solution

   None.

6. The mass of the atom $^{19}_{9}F$ is 18.99840 amu.

   (a) Calculate its binding energy per atom in millions of electron-volts.
   (b) Calculate its binding energy per nucleon. (See Appendix C.)

   Solution

   (a) The binding energy per atom of $^{19}_{9}F$ is calculated from the mass defect, the difference between the actual mass of the nuclide and its theoretical mass. First, determine the theoretical mass of $^{19}_{9}F$, which contains 9 protons, 9 electrons, and $19 - 9$ or 10 neutrons.

Neutrons	10×1.0087 amu	$= 10.0870$ amu
Protons	9×1.0073 amu	$= 9.0657$ amu
Electrons	9×0.00055 amu	$= 0.00495$ amu
Theoretical Mass		19.15765 amu

   Mass defect = 19.15765 ammu − 18.99840 amu = 0.15925 amu

Chapter 24 Nuclear Chemistry

To use the Einstein conversion, the mass must be expressed in kg.

$$\text{mass defect} = 0.15925 \text{ amu} \times \frac{1.6605 \times 10^{-27} \text{ kg}}{1 \text{ amu}} = 2.644 \times 10^{-28} \text{ kg}$$

$$E = mc^2 = (2.644 \times 10^{-28} \text{ kg})(2.9979 \times 10^8 \text{ m s}^{-1})^2 = 2.3766 \times 10^{-11} \text{ kg m}^2 \text{ s}^{-2}$$

$$= 2.3766 \times 10^{-11} \text{ J/nucleus}$$

In terms of MeV, use the conversion factor

$$1 \text{ MeV} = 1.602189 \times 10^{-13} \text{ J}$$

which gives

$$2.376 \times 10^{-11} \text{ J/nucleus} \times \frac{1 \text{ MeV}}{1.602189 \times 10^{-13} \text{ J}} = 148.3 \text{ MeV/nucleus}$$

(b) $\quad 148.3 \dfrac{\text{MeV}}{\text{nucleus}} \times \dfrac{1 \text{ nucleus}}{19 \text{ nucleons}} = 7.805 \dfrac{\text{MeV}}{\text{nucleon}}$

8. The mass of a hydrogen atom (^1_1H) is 1.007825 amu; that of a tritium atom (^3_1H), 3.01605 amu; and that of an α particle, 4.00150 amu. How much energy in kilojoules per mole of ^4_2He produced is released by the following reaction?

$$^1_1H + ^3_1H \longrightarrow ^4_2He$$

Solution

Calculate the mass change that occurs; then convert this mass to energy.

$$\text{Mass defect} = \text{mass } ^1_1H + \text{mass } ^3_1H - \text{mass } ^4_2He$$

$$= 1.007825 \text{ amu} + 3.01605 \text{ amu} - 4.00150 \text{ amu} = 0.022375 \text{ amu}$$

Convert amu to kg:

$$\text{Mass defect} = 0.022375 \text{ amu} \times 1.6605 \times 10^{-27} \frac{\text{kg}}{\text{amu}} = 3.7154 \times 10^{-29} \text{ kg}$$

$$E = mc^2 = (3.7154 \times 10^{-29} \text{ kg})(2.9979 \times 10^8 \text{ m s}^{-1})^2$$

$$= 3.339 \times 10^{-12} \text{ kg m}^2 \text{ s}^{-2} = 3.339 \times 10^{-12} \text{ J/nucleus}$$

On a per mole basis,

$$E = \frac{E}{\text{nucleus}} \times \frac{6.022 \times 10^{23}}{\text{mol}} = 3.339 \times 10^{-12} \text{ J/nucleus} \times 6.022 \times 10^{23} \frac{\text{nuclei}}{\text{mol}}$$

$$= 2.011 \times 10^{12} \text{ J mol}^{-1} = 2.011 \times 10^9 \text{ kJ mol}^{-1}$$

10. What percentage of $^{254}_{102}No$ remains of a 0.100-g sample, 5.0 min after it is formed (half-life of 55 s)? 1.0 h after it is formed?

Solution

Determine the rate constant from the half-life equation.

$$k = \frac{0.693}{t_{1/2}} = \frac{0.693}{55 \text{ s}} = 0.0126 \text{ s}^{-1}$$

From the integrated first-order rate equation,

$$\log \frac{C_0}{C} = \frac{kt}{2.303}$$

$$\log \frac{0.100 \text{ g}}{C} = \frac{0.0126 \text{ s}^{-1} \times 5.0 \text{ min} \times 60 \text{ s min}^{-1}}{2.303}$$

$$-1.00 - \log C = \frac{0.0126 \times 5.0 \times 60}{2.303}$$

$$\log C = -1.6413 - 1.00$$

$$\log C = -2.6413$$

$$C = 2.28 \times 10^{-3} \text{ g}$$

The percentage remaining after 5.0 min is

$$\frac{2.28 \times 10^{-3} \text{ g}}{0.100 \text{ g}} \times 100 = 2.3\%$$

After 1.0 h,

$$\log \frac{0.100 \text{ g}}{C} = \frac{0.0126 \text{ s}^{-1} \times 60 \text{ min} \times 60 \text{ s min}^{-1}}{2.303}$$

$$\log C = -19.696 - 1.00 = -20.696$$

$$C = 2.013 \times 10^{-21}$$

The percentage remaining is

$$\frac{2.013 \times 10^{-21} \text{ g}}{0.100 \text{ g}} \times 100 = 2.0 \times 10^{-18}\%$$

11. The isotope ^{208}Tl undergoes β decay with a half-life of 3.1 min.

    (a) What isotope is the product of the decay?
    (b) Is ^{208}Tl more stable or less stable than an isotope with a half-life of 54.5 s?
    (c) How long will it take for 99.0% of a sample of pure ^{208}Tl to decay?
    (d) What percentage of a sample of pure ^{208}Tl will remain undecayed after 1.0 h?

Solution

(a) $^{208}_{81}\text{Tl} \longrightarrow {}^{0}_{-1}\text{e} + {}^{208}_{82}\text{Pb}$
(b) Tl-208, with a half-life of 3.1 min, is considered to be more stable, because its half-life is longer.
(c) Determine the decay constant, k, then use the integrated rate equation.

$$k = \frac{0.693}{t_{1/2}} = \frac{0.693}{3.1 \text{ min}} = 0.224 \text{ min}^{-1}$$

The percentage of sample remaining after a period of time is independent of the mass. If 99% of the sample has decayed, 1% of the original mass of 100% remains.

$$\log \frac{C_0}{C} = \frac{kt}{2.303} \qquad \log \frac{100\%}{1\%} = \frac{0.224 \text{ min}^{-1}}{2.303} t$$

$$2.000 = (0.09726 \text{ min}^{-1})t \qquad t = 20.6 \text{ min}$$

(d) Let $C_0 = 1.00$; upon substitution,

$$\log \frac{C_0}{C} = \frac{kt}{2.303} \qquad \log \frac{100\%}{C\%} = \frac{0.224 \text{ min}^{-1} \times 60 \text{ min}}{2.303}$$

$$\log C\% = -3.836 \qquad C = 1.5 \times 10^{-4}\%$$

Chapter 24 Nuclear Chemistry

NUCLEAR DECAY

16. What is the change in the nucleus that gives rise to a β particle? A β^+ particle?

Solution

High-energy electrons or β^- are emitted from the nucleus by the disintegration of a neutron.

$${}^1_0n \longrightarrow {}^1_1p + {}^{0}_{-1}e$$

A β^+ or positron has the same mass as an electron and the same magnitude positive charge as the electron has negative charge. When a nucleus has a neutron-to-proton ratio that is too low, a proton is converted to a neutron with the emission of a positron.

$${}^1_1p \longrightarrow {}^1_0n + {}^{0}_{+1}e$$

17. The loss of an α particle by a nucleus causes what change in the atomic number and the mass of the nucleus? What is the change in the atomic number and mass when a β particle is emitted?

Solution

Loss of an α particle by a nucleus reduces the mass number by 4 and the atomic number by 2. Emission of a β produces a proton from a neutron, thereby increasing the atomic number by one; the mass number does not change.

19. Many nuclides with atomic numbers greater than 83 decay by processes such as electron emission. Rationalize the observation that the emissions from these unstable nuclides normally include α particles also.

Solution

Large isotopes tend to undergo decay, emitting alpha particles and other large masses in order to decrease the n/p ratio faster.

21. Write a balanced equation for each of the following nuclear reactions.

 (a) Uranium-230 undergoes α decay.
 (c) Beryllium-8 and a positron are produced by the decay of an unstable nucleus.

Solution

(a) ${}^{230}_{92}U \longrightarrow {}^4_2He + {}^{226}_{90}Th$

(c) ${}^8_5B \longrightarrow {}^8_4Be + {}^{0}_{+1}e$

23. Complete the following equations:

 (b) ${}^7_3Li + ? \longrightarrow 2\,{}^4_2He$
 (d) ${}^{14}_{6}C \longrightarrow {}^{14}_{7}N + ?$

Solution

(b) ${}^7_3Li + {}^1_1p \longrightarrow 2\,{}^4_2He$

(d) ${}^{14}_{6}C \longrightarrow {}^{14}_{7}N + {}^{0}_{-1}e$

24. Fill in the atomic number of the initial nucleus and write out the complete nuclear symbol for the product of each of the following nuclear reactions:

   (a) ^{9}Be (α, n)
   (c) ^{33}S (n, p)
   (e) ^{10}B (α, p)
   (g) ^{63}Cu (p, n)

   Solution

   (a) $^9_4$Be $+ ^4_2$He $\longrightarrow ^1_0$n $+ ^{12}_6$C
   (c) $^{33}_{16}$S $+ ^1_0$n $\longrightarrow ^1_1$H $+ ^{33}_{15}$P
   (e) $^{10}_5$B $+ ^4_2$He $\longrightarrow ^1_1$H $+ ^{13}_6$C
   (g) $^{63}_{29}$Cu $+ ^1_1$H $\longrightarrow ^1_0$n $+ ^{63}_{30}$Zn

25. Complete the following notations by filling in the missing parts:

   (b) $(\alpha, n)^{30}$P
   (e) ^{232}Th $(\ ,n)^{235}$U
   (h) ^{238}U $(^{12}_6$C, 4n$)$

   Solution

   (b) $^{27}_{13}$Al $(\alpha, n)^{30}_{15}$P
   (e) $^{232}_{90}$Th $(\alpha, n)^{235}_{92}$U
   (h) $^{238}_{92}$U $(^{12}_6$C, 4n$)^{246}_{98}$Cf

26. Use the abbreviated system of notation, as in exercise 25, to describe:

   (a) the production of ^{17}O from ^{14}N by α-particle bombardment
   (c) the production of ^{233}Th from ^{232}Th by neutron bombardment

   Solution

   (a) $^{14}_7$N $+ ^4_2$He $\longrightarrow ^{17}_8$O $+ ^1_1$H

   ^{14}N $(\alpha, p)^{17}$O

   (c) $^{232}_{90}$Th $+ ^1_0$n $\longrightarrow ^{233}_{90}$Th $+ \gamma$

   ^{232}Th $(n, \gamma)^{233}$Th

27. For each of the following unstable isotopes, predict by what mode(s) spontaneous radioactive decay might proceed:

   (a) $^6_2$He (n/p ratio too large)
   (b) $^{60}_{30}$Zn (n/p ratio too small)
   (e) ^{18}F

   Solution

   (a) $^6_2$He (n/p ratio too large): β decay would cause a decrease in the number of neutrons and an increase in the number of protons.
   (b) $^{60}_{30}$Zn (n/p ratio too small): β^+ emission decreases the number of protons and, to a much smaller extent, electron capture increases the number of neutrons.
   (e) ^{18}F: the only stable isotope is fluorine-19. β^+ decay or electron capture will increase the neutron/proton ratio.

Chapter 24 Nuclear Chemistry

28. Which of the following nuclei is most likely to decay by positron emission: chromium-53, manganese-51, or iron-59? Explain your choice.

Solution

$^{53}_{24}$Cr is stable; $^{51}_{25}$Mn could use β^+ decay to raise the n/p ratio to $^{53}_{23}$V, which is 2 mass numbers from a stable isotope. $^{59}_{26}$Fe decays by β^-. Decay by β^+ would form $^{59}_{25}$Mn, which is far from being stable. Therefore, Mn-51 is more probable for β^+ decay.

30. Technetium is used in nuclear medicine because it is absorbed by certain damaged tissues. The location of the technetium (and the tissue) can be detected by the γ ray that an excited Tc nucleus emits. Technetium is prepared from ^{98}Mo. Molybdenum-98 combines with a neutron to give molybdenum-99, an unstable isotope which decays by β^- emission to give an excited form of ^{99}Tc. This excited nucleus relaxes to the ground state by emission of a γ ray. The ground state of ^{99}Tc decays by β emission. Write the equations for each of these nuclear reactions.

Solution

$^{98}_{42}$Mo + $^{1}_{0}$n $\longrightarrow$ $^{99}_{42}$Mo

$^{99}_{42}$Mo $\longrightarrow$ $^{99}_{43}$Tc* + $^{0}_{-1}$e

$^{99}_{43}$Tc* $\longrightarrow$ $^{99}_{43}$Tc + γ

$^{99}_{43}$Tc $\longrightarrow$ $^{99}_{44}$Ru + $^{0}_{-1}$e

NUCLEAR POWER

32. How are atomic bombs and hydrogen bombs detonated?

Solution

Atomic bombs are ignited using conventional explosives that quickly compress the fissionable material into a critical mass so that enough neutrons can be captured to create and sustain a chain reaction. A hydrogen bomb or fusion bomb requires the heat of an exploding atomic bomb to trigger it so that the chain reaction can be sustained.

35. What is a breeder reactor?

Solution

A breeder reactor produces more fissionable material through various decay schemes than it consumes.

37. Discuss and compare the problems of radioactive wastes for radioactive substances of short half-life and those of long half-life.

Solution

The intensity of the radiation is the major problem involving short-lived isotopes, whereas long-term storage of long-lived isotopes is a problem.

25

The Semi-Metals

INTRODUCTION

The semi-metals or metalloids have properties that are characteristic of both metals and nonmetals. Boron brings to our study an encounter with the three-centered bond. Boron can form some compounds such as B_2H_6, in which the normal two-electron bond cannot be accommodated because boron is electron deficient. Instead, a single electron from hydrogen is shared between two boron atoms by overlap of the s orbital of hydrogen with an sp^3-type orbital based on each boron. In this arrangement only one electron is provided from both borons in the bond. Compounds containing such bonds have been carefully studied because of the novelty of this bonding arrangement.

Silicon is of considerable interest in two respects. Because it is below carbon in the periodic table, it might be expected that silicon can form the wide variety of compounds that carbon does. This expectation has not been fulfilled, although chains with Si-Si bonds are known. The main interest has been in silicon's ability, in extremely high-purity crystals, to convert sunlight into electricity. More recently, very thin layers of amorphous crystals of silicon have shown this phenomenon.

Perhaps what we should learn from the behavior of boron and silicon is that as we gain confidence in predicting and explaining phenomenon, all the more we should not be surprised when we find new examples of behavior that do not fit into our preconceived framework.

READY REFERENCE

Band theory (25.3)   By considering the molecular orbitals of the covalently or metallic bonded crystals, band theory explains the variation in electrical conductivity among metals, semiconductors, and insulators.

Semi-metals (25.1)   Elements that have chemical behavior intermediate between that of metals and nonmetals — namely, boron, silicon, germanium, antimony, and tellurium.

Three-centered-bond (25.2)   A bond formed by using two electrons shared by three atoms.

SOLUTIONS TO CHAPTER 25 EXERCISES

1. What are the common oxidation numbers exhibited by each of the semi-metals?

Solution

Boron, 3
Silicon, 4
Germanium, 2, 4
Antimony, 3, 5
Tellurium, 2, 4, 6

3. Predict the products of the reaction of antimony with an excess of each of the following: F_2, S, Cl_2, H_2SO_4.

Solution

$$2Sb + 5F_2 \longrightarrow 2SbF_5$$
$$2Sb + 5S \longrightarrow Sb_2S_5$$
$$2Sb + 5Cl_2 \longrightarrow 2SbCl_5$$
$$2Sb + 6H_2SO_4 \longrightarrow Sb_2(SO_4)_3 + 3SO_2 + 6H_2O$$

5. Why are boron compounds generally good Lewis acids?

Solution

The lone, unhybridized p orbital in sp^2 hybridized compounds readily accepts an electron pair and functions as a strong Lewis acid.

7. The bonding in crystalline boron is more like that of a metal than a non-metal, whereas the bonding in crystalline tellurium is more like that of a non-metal than a metal. Explain this statement.

Solution

Boron atoms form regular, crystalline geometries similar to closest packing of spheres common to metals. Tellurium forms chains of atoms similar to sulfur. Each atom in the chain is bound to two others.

9. Explain why the partially filled band in lithium is exactly half filled.

Solution

The electron configuration of lithium, $1s^2 2s^1$, shows one electron in the $n = 2$ energy level. The energy band arising from the $2s$ orbital is only half filled because it contains one electron per s orbital. Each energy level can hold two electrons.

BORON

11. What is the electron configuration of a boron atom?

Solution

$1s^2 2s^2 2p^1$

13. Describe the molecular structure and the hybridization of boron in each of the following molecules or ions:

   (a) BCl_3
   (b) BH_4^-

(c) B_2H_6
(d) $B(OH)_3$
(e) $B(OH)_4^-$
(f) H_3BCO

Solution

(a) 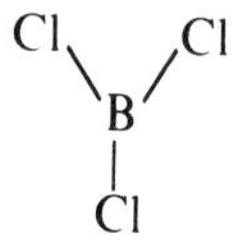
```
Cl    Cl
  \  /
   B
   |
   Cl
```
trigonal planar sp^2

(b) 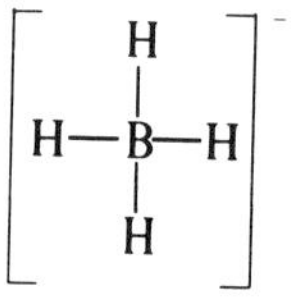
$$\left[\begin{array}{c} H \\ | \\ H-B-H \\ | \\ H \end{array} \right]^-$$
tetrahedral sp^3

(c) 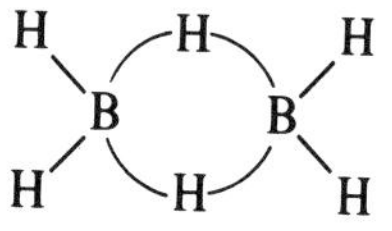
```
   H    H    H
    \  / \  /
     B    B
    /  \ /  \
   H    H    H
```
The borons and 4 hydrogens are planar using sp^2 hybrids but with one hydrogen above and one below the plane.

(d) 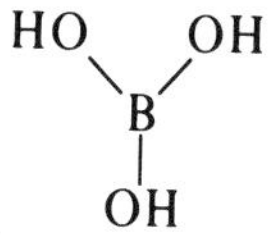
```
HO    OH
  \  /
   B
   |
   OH
```
trigonal planar sp^2

(e) 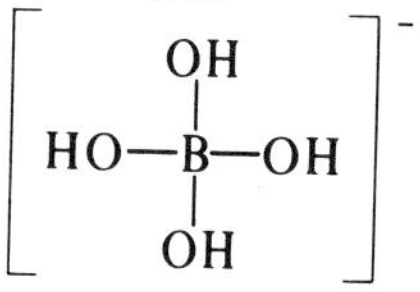
$$\left[\begin{array}{c} OH \\ | \\ HO-B-OH \\ | \\ OH \end{array} \right]^-$$
tetrahedral sp^3

(f) 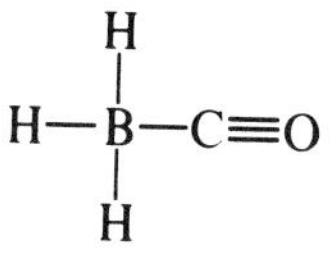
```
    H
    |
H - B - C≡O
    |
    H
```
tetrahedral sp^3

15. Write two equations for reactions in which boron exhibits metallic behavior and two equations for reactions in which boron exhibits nonmetallic behavior.

Solution

Metallic:

$$B_2H_6 + 6H_2O \longrightarrow 2B(OH)_3 + 6H_2$$

$$B_2O_3 + 3H_2O \longrightarrow 2B(OH)_3$$

Nonmetallic:

$$H_3BO_3 + 2H_2O \longrightarrow H_3O^+ + B(OH)_4^-$$

$$2B + 3Cl_2 \longrightarrow 2BCl_3(g)$$

17. Why is B_2H_6 said to be an electron-deficient compound?

Solution

Each boron atom contains only three valence electrons. The compound B_2H_6 would require a total of 14 electrons for normal covalent bonds to form an

Chapter 25 The Semi-Metals   367

sp^3 hybridized structure. Only 12 electrons are available; hence the name electron-deficient.

19. Write an equation to show the formation of fluoroboric acid from boron trifluoride.

Solution

$$BF_3 + HF \longrightarrow HBF_4$$

21. Why does an aqueous solution of borax turn red litmus blue?

Solution

Borax is a salt of a strong base and a weak acid. The salt hydrolyzes to form excess OH^-.

$$B_4O_7^{2-} + 7H_2O \longrightarrow 4H_3BO_3 + 2OH^-$$

23. Why is boron nitride sometimes referred to as inorganic graphite?

Solution

It is iso-electronic with graphite and forms hexagonal layers similar to graphite.

25. Complete and balance the following equations. (In some instances, two or more reactions may be correct, depending on the stoichiometry.)
   (a) $B + Cl_2 \longrightarrow$
   (b) $B + S_8 \longrightarrow$
   (c) $B + SO_2 \longrightarrow$
   (d) $B_2H_6 + NaH \longrightarrow$
   (e) $BCl_3 + C_2H_5OH \longrightarrow$
   (f) $HBO_2 + H_2O \longrightarrow$
   (g) $B_2O_3 + CuO \longrightarrow$

Solution

(a) $2B + 3Cl_2 \longrightarrow 2BCl_3$
(b) $16B + 3S_8 \longrightarrow 8B_2S_3$
(c) $4B + 3SO_2 \longrightarrow 2B_2O_3 + 3S$
(d) $B_2H_6 + 2NaH \longrightarrow 2NaBH_4$
(e) $BCl_3 + 3C_2H_5OH \longrightarrow B(OC_2H_5)_3 + 3HCl$
(f) $HBO_2 + H_2O \longrightarrow H_3BO_3$
(g) $B_2O_3 + CuO \longrightarrow Cu(BO_2)_2$

27. From the data given in Appendix I, determine the standard enthalpy change and standard free energy change for each of the following reactions:
   (a) $BF_3(g) + 3H_2O(l) \longrightarrow B(OH)_3(s) + 3HF(g)$
   (b) $BCl_3(g) + 3H_2O(l) \longrightarrow B(OH)_3(s) + 3HCl(g)$
   (c) $B_2H_6(g) + 6H_2O(l) \longrightarrow 2B(OH)_3(s) + 6H_2(g)$

Solution

(a) $BF_3(g) + 3H_2O(l) \longrightarrow B(OH)_3(s) + 3HF(g)$

$$\Delta H° = \Delta H°_{fB(OH)_3} + 3\Delta H°_{f(HF)} - \Delta H°_{fBF_3} - 3\Delta H°_{fH_2O}$$

$$= -1094.3 + 3(-271) - (-1137.3) - 3(-285.83) = 87 \text{ kJ}$$

$$\Delta G° = -969.01 + 3(-273) - (-1120.3) - 3(-237.18) = 44 \text{ kJ}$$

(b) $BCl_3(g) + 3H_2O(l) \longrightarrow B(OH)_3(s) + 3HCl(g)$

$$\Delta H° = \Delta H°_{B(OH)_3(s)} + 3\Delta H°_{HCl(g)} - \Delta H°_{BCl_3(g)} - 3\Delta H°_{H_2O(l)}$$
$$= -1094.3 + 3(-92.307) - (-403.8) - 3(-285.83) = -109.9 \text{ kJ}$$
$$\Delta G° = -969.01 + 3(-95.299) - (-388.7) - 3(-237.18) = -154.7 \text{ kJ}$$

(c) $B_2H_6(g) + 6H_2O(l) \longrightarrow 2B(OH)_3(s) + 6H_2(g)$

$$\Delta H° = 2\Delta H°_{B(OH)_3(s)} + 6\Delta H°_{H_2(g)} - \Delta H°_{B_2H_6(g)} - 6\Delta H°_{H_2O(l)}$$
$$= 2(-1094.3) + 0 - 36 - 6(-285.83) = -510 \text{ kJ}$$
$$\Delta G° = 2(-969.01) + 0 - 86.6 - 6(-237.18) = -601.5 \text{ kJ}$$

SILICON

29. Describe the hybridization and the bonding of a silicon atom in elemental silicon.

Solution

Each atom is bonded to four other atoms at the corners of a regular tetrahedron. Its hybridization is sp^3.

31. Write a Lewis structure for each of the following molecules or ions:

(a) $SiCl_4$
(b) SiO_4^{4-}
(c) $(CH_3)_2SiH_2$
(d) SiF_6^{2-}
(e) Si_3H_8 (contains Si—Si single bonds)

Solution

(a) tetrahedral sp^3

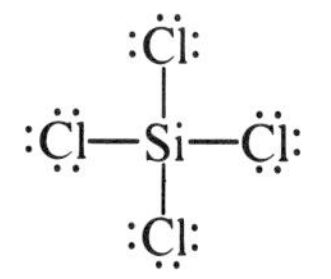

(b) tetrahedral sp^3

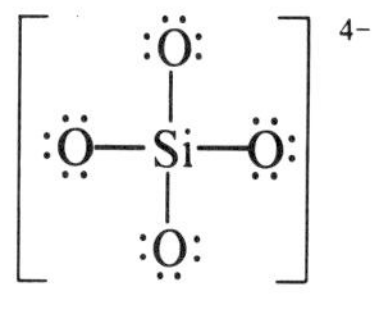

(c) tetrahedral sp^3

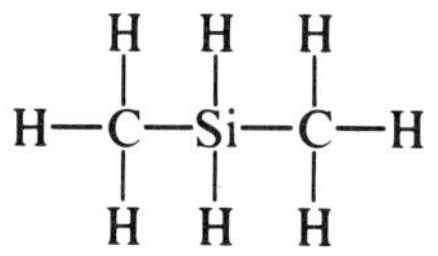

(d) octahedral d^2sp^3

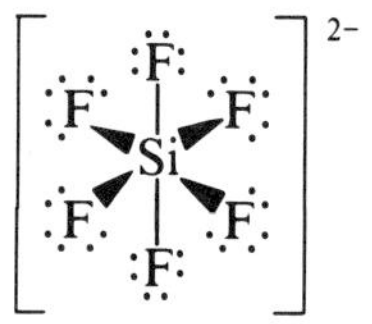

(e) 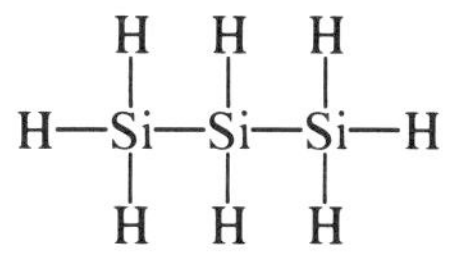
```
      H   H   H
      |   |   |
  H — Si— Si— Si — H      tetrahedral  $sp^3$
      |   |   |
      H   H   H
```

33. Heating a mixture of sand and coke in an electric furnace produces either silicon or silicon carbide. What determines which will be formed?

   Solution

   Heating large excesses of coke (C) tends to produce the carbide. Lesser stoichiometric amounts of coke tend to produce the element.

35. How does the following reaction show the acidic character of SiO_2? Which acid-base theory applies?

   $$CaO + SiO_2 \xrightarrow{\Delta} CaSiO_3$$

   Solution

   It functions as a Lewis acid incorporating Ca and O into a tetrahedral structure.

37. Account for the existence of the great variety of silicates in terms of how SiO_4 tetrahedra are linked.

   Solution

   The tetrahedra may be joined corner to corner or edge to edge, thereby giving silicates of many different forms. The number of tetrahedra linked together is arbitrary.

39. How does the internal structure of silica glass differ from that of quartz?

   Solution

   Silica glass is an amorphous material, while quartz (several types) is a crystalline substance.

41. Compare and contrast the chemistry of carbon and that of silicon.

   Solution

   A main difference in carbon and silicon is that carbon readily forms strong single, double, and triple covalent bonds. Silicon-silicon bonds are weak. Silicon is rather unreactive to air and water and its halides are very reactive to water — very different behavior than that of carbon.

43. Write equations to contrast the hydrolysis of SiF_4, $SiCl_4$, and CCl_4.

   Solution

   $$3SiF_4 + 4H_2O \longrightarrow H_4SiO_4 + 4H^+ + 2SiF_6^{2-}$$

   $$SiCl_4 + 2H_2O \longrightarrow SiO_2 + 4HCl$$

   $$CCl_4 + H_2O \longrightarrow \text{No reaction}$$

ADDITIONAL EXERCISES

45. The experimentally determined ratio of hydrogen atoms to boron atoms in a gas is 3 to 1. How could it be proved that the gas actually consists of B_2H_6 molecules?

Solution

B_2H_6 is a gas, so a molecular weight determination utilizing the ideal gas equation could be made. The molecular weight of B_2H_6 would be twice that of BH_3.

47. Write balanced equations describing at least two different reactions in which solid silica exhibits acidic character.

Solution

$$SiO_2 + 4NaOH \longrightarrow Na_4SiO_4 + 2H_2O$$

$$SiO_2 + Na_2CO_3 \longrightarrow Na_2SiO_3 + CO_2$$

49. In what ways does the chemistry of silicates resemble that of phosphates?

Solution

The silicates and phosphates of most metals have low solubility in water. Their common compounds have tetrahedral structures: SiO_4^{4-}, PO_4^{3-} and these condense to give corner-sharing chains. Both silicates and phosphates are formed by single covalent bonds.

51. (a) From the table of bond energies given in Section 6.9 calculate the approximate enthalpy change for the following reaction:

$$SiH_4(g) + 4HF(g) \longrightarrow SiF_4(g) + 4H_2(g)$$

(b) Using bond energy data, calculate the approximate enthalpy change for the corresponding reaction with methane:

$$CH_4(g) + 4HF(g) \longrightarrow CF_4(g) + 4H_2(g)$$

(c) Why do the enthalpy changes differ so greatly?
(d) Which reaction is more likely to proceed spontaneously?

Solution

(a) $SiH_4(g) + 4HF(g) \longrightarrow SiF_4(g) + 4H_2(g)$

Break:

4Si—H × 395 = 1580
4H—F × 569 = 2276
$\Delta H = +3856$ kJ

Form:

4Si—F × 540 = 2160
4H—H × 436 = 1744
$\Delta H = -3904$ kJ

$\Delta H_{RX} = -48$ kJ

(b) $CH_4(g) + 4HF(g) \rightarrow CF_4(g) + 4H_2(g)$

Break:

4C—H × 415 = 1660
4H—F × 569 = 2276
$\Delta H = +3936$ kJ

Form:

4C—F × 439 = 1756
4H—H × 436 = 1744
$\Delta H = -3500$ kJ

$\Delta H_{RX} = 436$ kJ

(c) The C—H bond is much stronger than the Si—H bond.
(d) The SiH$_4$ reaction is exothermic in contrast to the CH$_4$ reaction and is more likely spontaneous.

53. A hydride of silicon prepared by the reaction of Mg$_2$Si with acid exerted a pressure of 306 torr at 26°C in a bulb with a volume of 57.0 mL. If the mass of the hydride was 0.0861 g, what is its molecular weight? What is the molecular formula of the hydride?

Solution

Use the ideal gas equation to calculate the molecular weight, then convert to the mass of one mole.

$$n = PV/RT = \frac{(306/760 \text{ atm})(0.057 \text{ L})}{(0.08206 \text{ L atm/mol K})(299.15 \text{ K})} = 0.00093489 \text{ mol}$$

$$\text{mol. wt} = \frac{0.0861 \text{ g}}{0.00093489 \text{ mol}} = 92$$

Molecular formula: Si$_x$H$_y$

$x(28.0855) + y(1.0079) = 92$

The hydride is Si$_3$H$_8$.

26

Coordination Compounds

INTRODUCTION

Coordination compounds have been used by human beings for hundreds of years. They also are important in nature. Chlorophyll and hemoglobin of blood are coordination compounds. What distinguishes coordination compounds or complexes is that ions and/or neutral molecules are attached to the central metal atom. To accomplish this, the ion or molecule must have at least one pair of electrons available for coordinate covalent bonding, and the central metal ion or atom must have sufficient attraction to form a bond with the electron pair.

The work on the elucidation of the structure of these compounds reads like a detective story, because the donor ions or molecules can take different orientations about the central atom. That is, structural isomerism is possible. Learning the naming system devised originally by Alfred Werner is only the beginning of what can be a fascinating study that has already taken us to one drug that has been used in the fight against cancer.

READY REFERENCE

Coordination number (26.1)   The number of coordinate covalent bonds formed with ligands or the number of donor atoms attached to the central metal ion.

Donor atom (26.1)   Within a ligand, the atom that attaches directly to the central atom through a coordinate covalent bond is called the donor atom.

e_g orbital (26.7)   This is a name for the d_{z^2} and $d_{x^2-y^2}$ orbitals in an octahedral complex. The symbol refers to the symmetry of the orbitals.

High-spin complex (26.8)   This is a complex formed with ligands of low field strength in which the spins of the central atom's d electrons are not paired. It is also called a spin-free complex.

Ligand field theory (26.7, 26.8)   In this theory the central atom and its d-orbital electrons are considered fixed in space. As different ligands approach, the electron interaction is considered between the donor and the central atom. This causes the d-orbitals directed toward the ligands to be higher in energy and the orbitals directed away from them to be lower in energy.

Ligand (26.1)   An ion or molecule that can provide a pair of electrons to the central atom or ion.

Chapter 26 Coordination Compounds

Low-spin complex (26.8) A complex in which a strong ligand can cause the electrons of the central ion to pair within the *d* orbitals, thereby lowering the spin of the system.

t_{2g}-orbitals (26.7) This description refers to the set of orbitals d_{xy}, d_{xz}, d_{yz} in an octahedral complex.

SOLUTIONS TO CHAPTER 26 EXERCISES

STRUCTURE AND NOMENCLATURE OF COMPLEXES

1. Indicate the coordination number for the central metal atom in each of the following compounds or ions:

    (a) $[Cu(NH_3)_4]^{2+}$
    (b) $[Cr(NH_3)_2(H_2O)_2Br_2]^+$
    (c) $[Co(NH_3)_4Br_2]_2SO_4$
    (d) $[Co(CO_3)_3]^{3-}$ (CO_3^{2-} is bidentate in this complex.)
    (e) $[Cr(en)_3]^{3+}$ (en = ethylenediamine)
    (f) $[Co(en)_2(NO_2)Cl]^+$
    (g) $[Pt(NH_3)(py)(Cl)(Br)]$ (py = pyridine, C_5H_5N)
    (h) $[Zn(NH_3)_2Cl_2]$
    (i) $[Co(C_2O_4)_2Cl_2]^{3-}$
    (j) $[Zn(NH_3)(py)(Cl)(Br)]$

    Solution

    (a) There are four donor atoms; the coordination number is 4.
    (b) Six donor atoms; coordination number 6.
    (c) Six donor atoms; coordination number 6.
    (d) Three bidentate ions act as six donors; coordination number 6.
    (e) Ethylenediamine is a bidentate, equivalent to two donors each; coordination number 6.
    (f) Two bidentates equivalent to four donors plus two additional donors; coordination number 6.
    (g) Four donor atoms; coordination number 4.
    (h) Four donor atoms; coordination number 4.
    (i) Six donor atoms; two are from bidentates; coordination number 6.
    (j) Four donor atoms; coordination number 4.

3. Explain what is meant by

    (a) coordination sphere,
    (b) ligand,
    (c) donor atom,
    (d) coordination number,
    (e) coordination compound,
    (f) complex, and
    (g) chelating ligand.

    Solution

    (a) The coordination sphere includes the central metal ion and the attached ligands.
    (b) A ligand is an ion or molecule that can donate an electron pair.

(c) The donor atom is the specific atom in the ligand that donates the electron pair.

(d) The coordination number is the number of coordinate bonds formed with the central atom.

(e) This refers to any compound that contains a coordination complex.

(f) This is a compound in which ions or neutral molecules are attached directly to the metal ion.

(g) A chelating ligand is capable of attaching to a metal ion by forming two or more bonds from its donor atoms.

5. Give the coordination numbers and write the formulas for each of the following, including all isomers where appropriate:

   (a) diamminedibromoplatinum(II)
   (b) dichlorobis(ethylenediamine)chromium(III) nitrate
   (c) hexacyanopalladate(IV) ion
   (d) potassium diamminetetrabromocobaltate(III)
   (e) hexaamminecobalt(III) hexacyanochromate(III)
   (f) tetrahydroxonickelate(II) ion
   (g) dibromoaurate(I) ion (*aurum* is Latin for gold)

Solution

   (a) $[Pt(NH_3)_2Br_2]$; coordination number 4.
   (b) *cis* and *trans* $[Cr(en)_2(Cl)_2]NO_3$; coordination number 6. An optical isomer of the *cis* form also exists.
   (c) $[Pd(CN)_6]^{2-}$; coordination number 6.
   (d) *cis* and *trans* $K[Co(NH_3)_2Br_4]$; coordination number 6.
   (e) $[Co(NH_3)_6][Cr(CN)_6]$; coordination number 6.
   (f) $[Ni(OH)_4]^{2-}$; coordination number 4.
   (g) $[AuBr_2]^{1-}$; coordination number 2.

7. Name and sketch the structures of the following complexes. Indicate any *cis*, *trans*, and optical isomers.

   (a) $[Zn(NH_3)_2Cl_2]$
   (b) $[Zn(NH_3)(py)(Cl)(Br)]$ (py = pyridine, C_5H_5N)
   (c) $[Pt(H_2O)_2Cl_2]$
   (d) $[Pt(NH_3)(py)(Cl)(Br)]$
   (e) $[Co(C_2O_4)_2Cl_2]^{3-}$
   (f) $[Ni(H_2O)_4Cl_2]$
   (g) $[Pt(en)(NH_3)_2Br_2]^{2+}$

Solution

   (a) diamminedichlorozinc(II)

   (b) amminebromochloropyridinezinc(II); there are four different ligands bound to the central ion, therefore, there is an optical isomer.

(c) diaquadichloroplatinum(II)

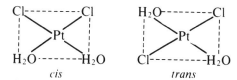

(d) amminebromochloropyridineplatinum(II); there are four different ligands bonded to the central ion in a square planar arrangement.

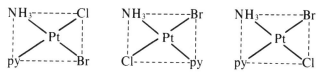

(e) dichlorobisoxalatocobaltate(III) ion; this is octahedral and has *cis* and *trans* positions for the Cl ligands. The *cis* form has an optical isomer.

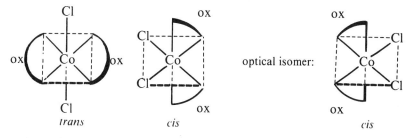

(f) tetraaquadichloronickel(II)

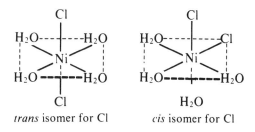

(g) diamminedibromoethylenediamineplatinum(IV) ion

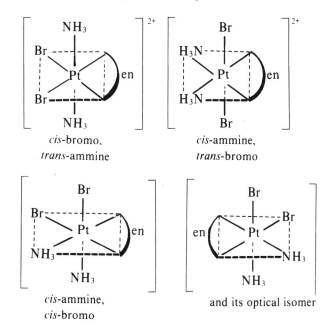

9. Using NH_3 and NO_2^- as ligands, (a) give the formula of a six-coordinated cobalt(III) complex that would not give ions when dissolved in water; (b) give the formula of a complex containing one six-coordinated cobalt(III) ion that would give two ions when dissolved in water; and (c) sketch the isomers of the compounds formulated in (a) and (b).

Solution

(a) A neutral complex will not ionize. Such a complex could be $[Co(NH_3)_3(NO_2)_3]$.

(b) $[Co(NH_3)_4(NO_2)_2]NO_2$

(c)

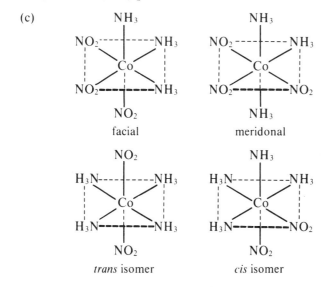

BONDING IN COMPLEXES

11. Explain how the diphosphate ion, $P_2O_7^{4-}$ (Section 23.23) can function as a water softener by complexing Fe^{2+}.

Solution

Because there are nonbonded pairs of electrons on the oxygen atoms, the diphosphate ion can function as an excellent bidentate ligand. Three of the diphosphate ions can strongly bind Fe^{2+} in an octahedral complex, making it a highly useful water softening material.

13. Show by means of orbital diagrams the hybridization for each of the following examples given in Table 26.1.

Solution

Linear; $[Cu(CN)_2]^-$; Cu^{1+} = sp hybridization for 2 ligands

⑪	⑪	⑪⑪⑪	⑪	⑪⑪⑪	○	⑪⑪⑪⑪⑪	○○○
1s	2s	2p	3s	3p	4s	3d	4p

Trigonal planar; $[HgCl_3]^-$; Hg^{2+} = sp^2 hybridization for 3 ligands

⑪	⑪	⑪⑪⑪	⑪	⑪⑪⑪	○	⑪⑪⑪⑪⑪	○○○
1s	2s	2p	3s	3p	4s	3d	4p

Tetrahedral; $[Co(Br)_4]^{2-}$; Co^{2+} = sp^3 hybridization for 4 ligands

⑪	⑪	⑪⑪⑪	⑪	⑪⑪⑪	○	⑪⑪○○○	○○○
1s	2s	2p	3s	3p	4s	3d	4p

Chapter 26 Coordination Compounds   377

Square planar; [Ni(CN)₄]²⁻; Ni²⁺ = dsp^2 hybridization for 4 ligands. The CN⁻ is a low-field ligand, i.e., it forms low-field complexes.

⑪ ⑪ ⑪⑪⑪ ⑪ ⑪⑪⑪ ◯ ⑪⑪⑪⑪◯ ◯◯◯
1s 2s   2p   3s   3p   4s   3d   4p

Square pyramidal; [VOCl₄]²⁻; V⁴⁺ = d^4s hybridization for 4 ligands

⑪ ⑪ ⑪⑪⑪ ⑪ ⑪⑪⑪ ◯ ⑪◯◯◯◯ ◯◯◯
1s 2s   2p   3s   3p   4s   3d   4p

Trigonal bipyramidal; [Fe(CO)₅]; Fe⁰ = dsp^3 hybridization for 5 ligands

⑪ ⑪ ⑪⑪⑪ ⑪ ⑪⑪⑪ ◯ ⑪⑪⑪⑪◯ ◯◯◯
1s 2s   2p   3s   3p   4s   3d   4p

Octahedral; [Co(NH₃)₆]³⁺; Co³⁺ = sp^3d^2 hybridization for 6 ligands

⑪ ⑪ ⑪⑪⑪ ⑪ ⑪⑪⑪ ◯ ⑪⑪⑪◯◯ ◯◯◯
1s 2s   2p   3s   3p   4s   3d   4p

15. How many unpaired electrons will be present in each of the following?

  (a) [MnCl₆]⁴⁻   (high-spin)
  (b) [Mn(CN)₆]³⁻   (low-spin)
  (c) [Mn(CN₆]⁴⁻   (low-spin)
  (d) [CoF₆]³⁻   (high-spin)
  (e) [RhCl₆]³⁻   (low-spin)
  (f) [Co(en)₃]³⁻   (low-spin)

Solution

(a) Mn⁺²   ◯   ⑪⑪⑪⑪⑪   high spin = 5 unpaired
           4s   3d

(b) Mn⁺³   ◯   ⑪⑪⑪◯◯   low spin = 0 unpaired
           4s   3d

(c) Mn⁺²   ◯   ⑪⑪⑪◯◯◯   low spin = 1 unpaired
           4s   3d

(d) Co⁺³   ◯   ⑪⑪⑪⑪⑪   high spin = 4 unpaired
           4s   3d

(e) Rh⁺³   ◯   ⑪⑪⑪◯◯   low spin = 0 unpaired
           5s   4d

(f) Co⁺³   ◯   ⑪⑪⑪◯◯   low spin = 0 unpaired
           4s   3d

17. Assume that you have a complex of a transition metal ion with a d^6 configuration. Can you tell whether the complex is octahedral or tetrahedral if a measurement of the magnetic moment establishes that it has no unpaired electrons? (See exercise 16 for the ligand field splitting in a tetrahedral complex.)

Solution

Octahedral ligand field splitting for d^6:

Strong ligand   ◯◯   Weak ligand   ⑪⑪
= paired   ⑪⑪⑪ = 10Dq   = 4 unpaired   ⑪⑪⑪ = 10 Dq

Tetrahedral ligand field splitting for d^6:

Strong ligand = 2 unpaired: ①①○ / ①①①① = 10 Dq

Weak ligand = 4 unpaired: ①①① / ①①① = 10 Dq

Since only the octahedral complex has no unpaired electrons, the complex in question can only be octahedral.

19. For complexes of the same metal ion with no change in oxidation number, the stability increases as the LFSE of the complex increases. Which complex in each of the following pairs of complexes is the more stable? $[Fe(H_2O)_6]^{2+}$ or $[Fe(CN)_6]^{4-}$; $[Co(NH_3)_6]^{3+}$ or $[CoF_6]^{3-}$; $[Mn(H_2O)_6]^{2+}$ or $[MnCl_6]^{4-}$; $[Co(NH_3)_6]^{3+}$ or $[Co(en)_3]^{3+}$.

Solution

The stability depends on the strength of the ligands.

$CN^- > H_2O$, so $[Fe(CN)_6]^{4-}$ is more stable.
$NH_3 > F^-$, so $[Co(NH_3)_6]^{3+}$ is more stable.
$H_2O > Cl^-$, so $[Mn(H_2O)_6]^{2+}$ is more stable.
en $> NH_3$, so $[Co(en)_3]^{3+}$ is more stable.

ADDITIONAL EXERCISES

21. What type of tetrahedral complexes can form optical isomers?

Solution

Tetrahedral complexes that have four different ligands can form optical isomers.

For example,

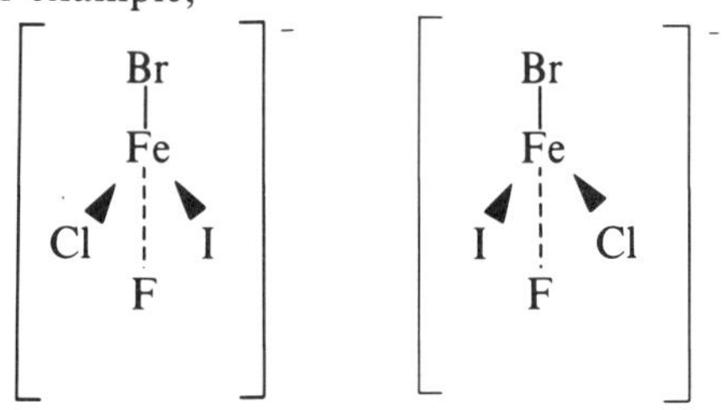

$$\begin{bmatrix} \text{Br} \\ | \\ \text{Fe} \\ \text{Cl} \quad \text{I} \\ \text{F} \end{bmatrix}^- \quad \begin{bmatrix} \text{Br} \\ | \\ \text{Fe} \\ \text{I} \quad \text{Cl} \\ \text{F} \end{bmatrix}$$

23. Determine the ligand field splitting of the d orbitals in a two-coordinated complex as the two ligands approach the metal along the z axis.

Solution

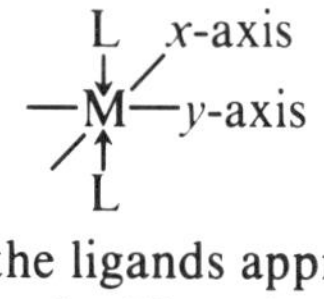

L, x-axis
—M—y-axis
L

As the ligands approach along the z axis, the d_{z^2} orbital will interact most strongly. The other d orbitals with z components (d_{xz}, d_{yz}) will have the next

Chapter 26 Coordination Compounds

highest interaction and energy. The $d_{x^2-y^2}$ and the d_{xy} will have minimal interaction and energy stabilization. The splitting will be

$$\begin{array}{c} — d_{z^2} \\ d_{yz} \; — — \; d_{xz} \\ d_{xy} \; — — \; d_{x^2-y^2} \end{array}$$

25. Calculate the concentration of free nickel ion that is present in equilibrium with 1.5×10^{-3} M $[Ni(NH_3)_6]^{2+}$ and 0.10 M NH_3.

Solution

$$Ni^{2+} + 6NH_3 \rightleftharpoons [Ni(NH_3)_6]^{2+}$$

$$K_f = \frac{[Ni(NH_3)_6]^{2+}}{[Ni^{2+}][NH_3]^6}$$

$$1.8 \times 10^8 = \frac{1.5 \times 10^{-3}}{[Ni^{2+}][1 \times 10^{-1}]^6}$$

$$[Ni^{2+}] = 8.3 \times 10^{-6}$$

27. Trimethylphosphine, :$P(CH_3)_3$, can act as a ligand by donating the free pair of electrons on the phosphorus atom. If trimethylphosphine is added to a solution of nickel(II) chloride in acetone, a blue compound with a molecular weight of approximately 270 and containing 21.5% Ni, 26.0% Cl, and 52.5% $P(CH_3)_3$ can be isolated. This blue compound does not have any isomeric forms. What is the geometry and molecular formula of the blue compound?

Solution

First, determine the empirical formula from the percentage composition, assuming 100 g of complex.

21.5% Ni = 21.5 g Ni × 1 mol Ni/58.70 g = 0.366 mol Ni

26.0% Cl = 26.0 g Cl × 1 mol Cl/35.453 g = 0.733 mol Cl

52.5% $P(CH_3)_3$ = 52.5 g P × 1 mol $P(CH_3)_3$/76.078 g = 0.690 mol $P(CH_3)_3$

Divide each amount by the smallest mole value to find the simplest ratio.

Ni: 0.366 mol/0.366 mol = 1.00 = 1

Cl: 0.733 mol/0.366 mol = 2.00 = 2

$P(CH_3)_3$: 0.690 mol/0.366 mol = 1.89 ≅ 2

The empirical formula is therefore $NiCl_2(P(CH_3)_3)_2$ which has a mol. wt of 281.76 g/mol. A square planar geometry could have *cis* and *trans* isomers. Since there are no isomers, the geometry must be tetrahedral.

29. Using the radius ratio rule and the ionic radii given inside the back cover, predict whether complexes of Ni^{2+}, Mn^{2+}, and Sc^{3+} with Cl^- will be tetrahedral or octahedral. Write the formulas for these complexes.

Solution

The limiting values for the radius ratio (r^+ is radius of the cation; r^- is the radius of the anion) are:

Coordination number	Complex shape	Limiting value of r^+/r^-
6	octahedral	0.414 to 0.732
4	tetrahedral	0.225 to 0.414

The required ratios are

$$\frac{r^+(\text{Ni}^{2+})}{r^-(\text{Cl}^-)} = \frac{0.70}{1.81} = 0.387; \quad \frac{r^+(\text{Mn}^{2+})}{r^-(\text{Cl}^-)} = \frac{0.80}{1.81} = 0.442; \quad \frac{r^+(\text{Sc}^{3+})}{r^-(\text{Cl}^-)} = \frac{0.81}{1.81} = 0.448$$

The Ni^{2+} complex is tetrahedral; those of Mn^{2+} and Sc^{3+} are octahedral. The formulas of the complexes are $[NiCl_4]^{2-}$, $[MnCl_6]^{4-}$, $[ScCl_6]^{3-}$.

27

The Transition Elements

INTRODUCTION

The transition elements include many elements that people generally think of as metals, namely, iron, copper, chromium, nickel, silver, and gold. The amount of these used yearly is high, except for gold because of its scarcity. Actually the percentage use of gold by industry is high compared to use in jewelry.

These metals, as do all the transition elements, span several subclassifications. Chromium, iron, nickel, and copper belong to the first transition series. Silver belongs to the second transition series whose other members are important in many applications. For example, molybdenum, "moly", is used to harden steel. Gold lies directly under Ag in the periodic table and belongs to the third transition series. The fourth transition series is all radioactive and presently incomplete.

In each of the above, d-shell electrons are filling. Two other series occur in which inner f orbitals are filling; these are the lanthanide and actinide series. Because of a partially filled d subshell, many of the transition metal compounds are highly colored and/or have interesting chemical and electrical properties.

READY REFERENCE

Actinide contraction (27.8) This is a reduction in size as the atomic number increases from number 89 to 103.

Basic oxygen process (27.22) This process is the principal method for the production of steel. It uses a cylindrical furnace with a basic lining, such as magnesium oxides and calcium oxides. The charge consists of scrap iron, new molten iron, and limestone (to form slag). High-purity oxygen directed into the white-hot molten charge oxidizes the charge's impurities.

Cementite (27.22) Cementite is a compound of carbon and iron, Fe_3C, and is stable only at high temperature.

Hydrometallurgy (27.2) Hydrometallurgy is the separation of an element from other elements, using aqueous solutions.

Mond process (27.24) Carbon monoxide is passed over a mixture of nickel and other metals, forming nickel carbonyl. The nickel carbonyl is separated, carried by the excess carbon monoxide. It is decomposed at 200°C, the metal depositing as a fine dust.

Parkes process (27.2) An extraction process to remove silver from lead is based on the differential preference of silver to dissolve in zinc rather than lead. The zinc-silver alloy is removed from the lead and the zinc distilled off.

Superconductor (27.28) This is a substance that conducts electricity without any loss due to resistance to flow of the electrons. In addition, the substance must exhibit the Meissner effect in which the superconductor excludes the magnetic lines of force of a magnet. Recent interest in this field is great because of the discovery of copper oxide ceramics that are superconductors at or above liquid nitrogen temperature. This phenomenon was previously restricted to below 36 K.

SOLUTIONS TO CHAPTER 27 EXERCISES

1. Write the electron configurations for the following elements: Sc, Ti, Cr, Fe, Mo, Ru.

   Solution

   Sc $1s^22s^22p^63s^23p^64s^23d^1$
   Ti $1s^22s^22p^63s^23p^64s^23d^2$
   Cr $1s^22s^22p^63s^23p^64s^13d^5$
   Fe $1s^22s^22p^63s^23p^64s^23d^6$
   Mo $1s^22s^22p^63s^23p^63d^{10}4s^24p^65s4d^55s^1$
   Ru $1s^22s^22p^63s^23p^63d^{10}4s^24p^65s^24d^6$

3. Write the electron configurations for the following elements and their +3 ions: La, Sm, and Lu.

   Solution

		atom	+3 ion
La	$1s^22s^22p^63s^23p^63d^{10}4s^24p^64d^{10}$	$4f^05s^25p^65d^16s^2$	$4f^05s^25p^6$
Sm	$1s^22s^22p^63s^23p^63d^{10}4s^24p^64d^{10}$	$4f^65s^25p^65d^06s^2$	$4f^65s^25p^5$
Lu	$1s^22s^22p^63s^23p^63d^{10}4s^24p^64d^{10}$	$4f^{14}5s^25p^65d^16s^2$	$4f^{14}5s^25p^6$

5. What are the various reducing agents used to isolate the elements of the first transition series?

   Solution

   Al, C, Ca, CO, H_2, Mg, Cu_2S

   Co: conversion to oxide and reduction with aluminum or H_2
   Cr: reduction of the oxide with aluminum metal
   Cu: conversion to oxide and reduction with copper(I) sulfide
   Fe: reduction with carbon or carbon monoxide
   Mn: reduction of oxide with aluminum metal
   Ni: conversion to oxide and reduction with carbon
   Sc: reduction of the fluoride with calcium metal
   Ti: conversion of oxide to chloride with C and Cl_2 followed by reduction with magnesium metal
   V: reduction of oxide by calcium

Chapter 27 The Transition Elements   383

7. Describe the electrolytic process for refining copper.

   Solution

   Impure copper is used as anodes in the electrolytic refining of copper. Thin sheets of pure copper serve as the cathode. Copper sulfate in sulfuric acid serves as the electrolyte. The impure copper dissolves from the anodes and, under the influence of the voltage, passes to the cathode where it is plated out in high purity. Inert metals drop out of solution, whereas elements more active than Cu that are in solution are prevented from plating out by fine control of the voltage.

9. Why are the rare earths found associated with one another in nature?

   Solution

   The similarity in size and oxidation number of the rare earths allows for their substitution one for another in a wide variety of oxides in which they are found.

11. Which of the following elements is most likely to be used to prepare La by the reduction of La_2O_3: Al, C, or Fe? Why?

    Solution

    Al is used because Al_2O_3 has the highest negative Gibbs energy per oxygen atom of any compound.

13. Which of the following ions are not stable in water: Sc^{3+}, Ti^{2+}, Cr^{2+}, Fe^{4+}, Co^{3+}, Ni^{2+}, Cu^+?

    Solution

    The ions that are unstable are Ti^{2+}, Fe^{4+}, Co^{3+}, and Cu^+.

15. Describe the lanthanide contraction.

    Solution

    The decrease in size with increasing atomic number in the rare earth elements is known as the lanthanide contraction. This causes a regular graduation of properties. It also causes the elements filling d orbitals after the lanthanide series to be closer than expected in size to members of the same family one period earlier.

17. Predict the products of the following reactions.

    (a) $V + VCl_3 \xrightarrow{\Delta}$
    (b) $Ti + xsF_2 \xrightarrow{\Delta}$
    (c) $Co + xsF_2 \longrightarrow$
    (d) $Mn(OH)_2 + HBr(aq) \longrightarrow$
    (e) $CuCO_3 + HI(aq) \longrightarrow$
    (f) $Mn_2O_3 + HCl(aq) \longrightarrow$
    (g) $TiBr_4 + H_2O \longrightarrow$
    (h) $Cr + O_2 \xrightarrow{\Delta}$
    (i) $CoO + O_2 \xrightarrow{\Delta}$

(j) $La + O_2 \xrightarrow{\Delta}$
(k) $W + O_2 \xrightarrow{\Delta}$
(l) $CrO_3 + CsOH(aq) \longrightarrow$
(m) $Fe + H_2SO_4 \longrightarrow$
(n) $LaCl_3(aq) + NaOH(aq) \longrightarrow$

Solution

(a) $V(s) + 2VCl_3(s) \xrightarrow{\Delta} 3VCl_2(s)$
(b) $Ti(s) + 2F_2(g) \longrightarrow TiF_4(g)$
(c) $2Co(s) + 3F_2(g) \longrightarrow 2CoF_3(s)$
(d) $Mn(OH)_2(s) + 2H^+(aq) + 2Br^-(aq) \longrightarrow Mn^{2+}(aq) + 2Br^- + 2H_2O(l)$
(e) $CuCO_3 + 2H^+(aq) + 2I^-(aq) \longrightarrow Cu^{2+}(aq) + 2I^-(aq) + H_2O(l) + CO_2(g)$
(f) $Mn_2O_3(s) + 6H^+(aq) + 6Cl^-(aq) \longrightarrow 2MnCl_3(s) + 3H_2O(l)$
(g) $TiBr_4(l) + H_2O(l) \longrightarrow TiO^{2+}(aq) + 2H^+(aq) + 4Br^-(aq)$
(h) $2Cr + \frac{3}{2}O_2(g) \xrightarrow{\Delta} Cr_2O_3(s)$
(i) $3CoO(s) + \frac{1}{2}O_2 \xrightarrow{\Delta} Co_3O_4(s) \xrightarrow[\text{above}]{\Delta} 3CoO + \frac{1}{2}O_2$
(j) $2La(s) + \frac{3}{2}O_2(g) \xrightarrow{\Delta} La_2O_3(s)$ 950°C
(k) $W(s) + \frac{3}{2}O_2 \xrightarrow{\Delta} WO_3(s)$
(l) $CrO_3(s) + 2Cs^+(aq) + 2OH^-(aq) \longrightarrow 2Cs^+(aq) + CrO_4^{2-}(aq) + H_2O(l)$
(m) $Fe(s) + 2H^+(aq) + SO_4^{2-} \longrightarrow Fe^{2+}(aq) + SO_4^{2-}(aq) + H_2(g)$
(n) $LaCl_3(aq) + 3Na^+(aq) + 3OH^-(aq) \longrightarrow La(OH)_3 + 3Na^+(aq) + 3Cl^-(aq)$

19. In what ways are the Group VIIIB metals Fe, Co, and Ni typical of transition metals in general?

    Solution

    The Group VIIIB metals are typical of transition elements in general because they all have d orbitals that are filling; they show a variety of oxidation states; they form many colored species; and they form stable coordination compounds.

CHROMIUM

21. Write an equation for the decomposition of ammonium dichromate by ignition.

    Solution

    $$(NH_4)_2Cr_2O_7(s) \xrightarrow{\Delta} N_2(g) + 4H_2O(g) + Cr_2O_3(s)$$

23. Reduction of chromite ore by carbon yields an alloy of iron and chromium. How would you separate the two components of this alloy into solutions of their respective cations, if necessary?

    Solution

    To prepare chromium from the chromium and iron ore, chromite, $FeCr_2O_4$, treat with molten alkali and oxygen to convert Cr(III) to chromate(VI). This material is dissolved in water and precipitated as sodium dichromate. The oxide is reduced to Cr(III) oxide with carbon; the oxide then is reduced with aluminum to form pure chromium. If starting from the alloy, dissolve the alloy in hydrochloric acid, then add hydrogen peroxide and NaOH. The iron as Fe^{2+} is oxidized to Fe^{3+} and precipitated as $Fe(OH)_3$; the chromium is converted to soluble CrO_4^{2-}. The solution is separated from the precipitate of $Fe(OH)_3$. The solution containing CrO_4^{2-} is neutralized with HCl and made alkaline with NH_4OH. The chromate can be precipitated by the addition of barium acetate $Ba(C_2H_3O_2)_2$. A yellow precipitate is confirmation of the presence of chromium.

25. Write the Lewis structure of chromic acid.

Solution

Chromic acid is H_2CrO_4 and has tetrahedral symmetry about Cr with coordination number 4. The Lewis structure is

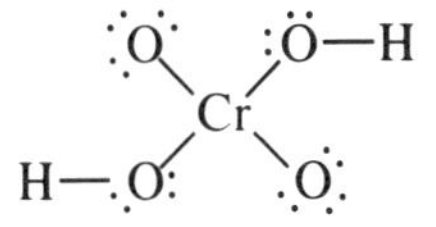

$$\begin{array}{c} :\ddot{O}: \quad :\ddot{O}-H \\ \diagdown\!\!\!\!\diagup \\ Cr \\ \diagup\!\!\!\!\diagdown \\ H-\!:\ddot{O}: \quad :\ddot{O}: \end{array}$$

27. Predict the products of the following reactions.
    (a) $Cr^{2+} + CrO_4^{2-}$ in acid solution.
    (b) $CrO + 2HNO_3$ in water.
    (c) $Cr_2(SO_4)_3 + Zn$ in acid.
    (d) $CrCl_3$ is added to an aqueous solution of NaOH.
    (e) $TiCl_2$ is added to a solution containing an excess of CrO_4^{2-} ion.
    (f) Mn is heated with CrO_3.

Solution

(a) In acid solution between pH 2 and pH 6, CrO_4^{2-} forms $HCrO_4^-$ which is in equilibrium with dichromate ion. The reaction is:

$$2HCrO_4^- \rightleftharpoons Cr_2O_7^{2-} + H_2O$$

$$3Cr^{2+} + CrO_4^{2-} + 8H^+ \longrightarrow 4Cr^{3+} + 4H_2O$$

(b) $CrO(s) + 2H^+(aq) + 2NO_3^-(aq) \longrightarrow Cr^{2+}(aq) + 2NO_3^-(aq) + H_2O(l)$
(c) $Cr_2(SO_4)_3(aq) + 2Zn(s) + 2H^+(aq) \longrightarrow 2Zn^{2+}(aq) + H_2(g)$
$ + 2Cr^{2+}(aq) + 3SO_4^{2-}(aq)$
(d) $CrCl_3(aq) + 3NaOH(aq) \longrightarrow Cr(OH)_3(s) + 3Na^+(aq) + 3Cl^-(aq)$
(e) $3TiCl_2(s) + 2CrO_4^{2-}(aq) + 8H_2O \longrightarrow 3Ti^{4+} + 10OH^- + 2Cr(OH)_3 + 6Cl^-$
(f) $16CrO_3(s) + 18Mn(s) \xrightarrow{\Delta} 8Cr_2O_3(s) + 6Mn_3O_4(s)$

MANGANESE

29. Write a balanced equation for the reaction of manganese dioxide and hydrochloric acid.

Solution

$$MnO_2 + 4HCl \longrightarrow MnCl_2 + Cl_2(g) + 2H_2O$$

This reaction can be used for the small-scale generation of chlorine in the laboratory.

31. Identify two uses of manganese dioxide.

Solution

Manganese dioxide is used as an oxidizing agent to correct the green color in glass produced by iron(II) compounds, and as a depolarizer in dry cells.

33. How is potassium permanganate prepared commercially?

Solution

Potassium permanganate is prepared by oxidizing potassium manganate in alkaline solution with chlorine.

$$2MnO_4^{2-} + Cl_2 \longrightarrow 2MnO_4^- + 2Cl^-$$

35. How can the insolubility of barium sulfate be applied to the preparation of free permanganic acid?

Solution

$$Ba^{2+} + 2MnO_4^- + 2H^+ + SO_4^{2-} \longrightarrow BaSO_4(s) + 2H^+ + 2MnO_4^-$$

IRON

37. Write the names and formulas of the principal ores of iron.

Solution

The major iron ores are *hematite*, Fe_2O_3, *lodestone* or *magnetite*, Fe_3O_4, *limonite*, $FeO(OH)$, and *siderite*, $FeCO_3$.

39. What is the composition of iron rust?

Solution

Iron rust is partially hydrated iron(III) oxide, $Fe_2O_3 \cdot nH_2O$.

41. What is the gas produced when iron(II) sulfide is treated with a nonoxidizing acid?

Solution

The gas formed is H_2S.

43. Iron(II) can be titrated to iron(III) by dichromate ion, which is reduced to chromium(III) in acid solution. A 2.500-g sample of iron ore is dissolved and the iron converted to iron(II). Exactly 19.17 mL of 0.0100 M $Na_2Cr_2O_7$ is required in the titration. What percentage of iron was the ore sample?

Solution

The reaction is

$$6Fe^{2+} + Cr_2O_7^{2-} + 14H^+ \longrightarrow 6Fe^{3+} + 2Cr^{3+} + 7H_2O$$

Moles of $Cr_2O_7^{2-}$ = L × M = 0.01917 L × 0.0100 M = 0.0001917 mol

Since 6Fe^{2+} mol react for each mole of dichromate ion,

$$\text{Moles of Fe}^{2+} = 6 \times 0.001917 \text{ mol} = 1.1502 \times 10^{-3} \text{ mol}$$

$$\text{g Fe}^{2+} = 1.1502 \times 10^{-3} \text{ mol} \times 55.847 \text{ g/mol} = 0.06424 \text{ g}$$

$$\text{Percent Fe} = \frac{0.06424 \text{ g}}{2.500 \text{ g}} \times 100 = 2.57\%$$

45. Predict the products of the following reactions.

    (a) Fe is heated in air.
    (b) Fe is added to a dilute solution of H$_2$SO$_4$.
    (c) A solution of Fe(NO$_3$)$_2$ and HNO$_3$ is allowed to stand in air.
    (d) Fe is heated in an atmosphere of steam.
    (e) FeCO$_3$ is added to a solution of HClO$_4$.
    (f) NaOH is added to a solution of Fe(NO$_3$)$_3$.
    (g) FeSO$_4$ is added to an acidic solution of K$_2$CrO$_4$.

    Solution

    (a) $3\text{Fe}(s) + 2\text{O}_2(g) \xrightarrow{\Delta} \text{Fe}_3\text{O}_4(s)$
    (b) $\text{Fe}(s) + 2\text{H}^+(aq) + \text{SO}_4^{2-}(aq) \longrightarrow \text{Fe}^{2+}(aq) + \text{SO}_4^{2-}(aq) + \text{H}_2(g)$
    (c) $4\text{Fe}^{2+} + \text{O}_2 + 4\text{HNO}_3 \longrightarrow 4\text{Fe}^{3+} + 2\text{H}_2\text{O} + 4\text{NO}_3^-$
    (d) $3\text{Fe} + 4\text{H}_2\text{O} \longrightarrow \text{Fe}_3\text{O}_4 + 4\text{H}_2$
    (e) $\text{FeCO}_3 + \text{HClO}_4 \longrightarrow \text{Fe}(\text{ClO}_4)_2 + \text{H}_2\text{O} + \text{CO}_2(g)$
    (f) $6\text{NaOH} + \text{Fe(NO}_3)_3 \xrightarrow{\text{H}_2\text{O}} \text{Fe(OH)}_6^{3-} + 6\text{Na}^+ + 3\text{NO}_3^-$
    (g) $2\text{CrO}_4^{2-} + 2\text{H}^+ \longrightarrow 2\text{HCrO}_4^- \rightleftharpoons \text{Cr}_2\text{O}_7^{2-} + \text{H}_2\text{O}$
    $6\text{Fe}^{2+} + \text{Cr}_2\text{O}_7^{2-} + 14\text{H}^+ \longrightarrow 6\text{Fe}^{3+} + 2\text{Cr}^{3+} + 7\text{H}_2\text{O}$

COBALT

47. Balance the following equation by oxidation-reduction methods:

    $$\text{Co(NO}_3)_2 \longrightarrow \text{Co}_2\text{O}_3 + \text{NO}_2 + \text{O}_2$$

    Solution

    $$\text{Co}^{2+} \longrightarrow \text{Co}^{3+} + e^- \qquad \text{N}^{5+} + e^- \longrightarrow \text{N}^{4+} \qquad 2\text{O}^{2-} \longrightarrow \text{O}_2 + 4e^-$$

    First establish that there is a ratio of 4NO$_3^-$ for each O$_2$ produced. At the same time one additional NO$_3^-$ is required for each Co^{2+} converted. The lowest even number of Co(NO$_3$)$_2$ units provides 4NO$_3^-$, only enough for the production of one O$_2$. Then

    $$4\text{Co(NO}_3)_2 \longrightarrow 2\text{Co}_2\text{O}_3 + 8\text{NO}_2 + \text{O}_2$$

49. What is the percent by mass of cobalt in sodium hexanitrocobaltate(III)?

    Solution

    Mol. wt Na$_3$[Co(NO$_2$)$_6$] = 403.9355

    $$\text{Percent Co} = \frac{58.9332}{403.9355} \times 100 = 14.59\%$$

51. Predict the products of the following reactions.

   (a) Co is heated in air.
   (b) Co is added to a dilute solution of H_2SO_4.
   (c) NaOH is added to a solution of $Co(NO_3)_2$.
   (d) Co is heated with I_2.
   (e) $CoCO_3$ is added to a solution of $HClO_4$.
   (f) NaOH and Na_2S are added to a solution of $Co(NO_3)_2$.
   (g) $CoSO_4$ is added to solution of NH_3.

   Solution

   (a) $3Co(s) + 2O_2(g) \longrightarrow Co_3O_4$
   (b) $Co + H_2SO_4 \longrightarrow Co^{2+}(aq) + SO_4^{2-}(aq) + H_2(g)$
   (c) $2Na^+(aq) + 2OH^-(aq) + Co^{2+}(aq) + 2NO_3^-(aq) \longrightarrow Co(OH)_2(aq) + 2Na^+ + 2NO_3^-(aq)$
   (d) $Co + I_2 \xrightarrow{\Delta} CoI_2$
   (e) $CoCO_3 + 2HClO_4(aq) \longrightarrow Co^{2+}(aq) + H_2O(l) + CO_2(g) + 2ClO_4^-(aq)$
   (f) $Na_2S(aq) + Co(NO_3)_2(aq) \xrightarrow{OH^-} CoS + 2Na^+ + 2NO_3^-$
   (g) $CoSO_4 + 6NH_3(aq) \longrightarrow [Co(NH_3)_6]SO_4$

NICKEL

53. How is nickel rendered passive?

   Solution

   Nickel can be rendered passive by treatment with concentrated nitric acid.

55. Write the equation for the dissolution of nickel(II) hydroxide in aqueous ammonia.

   Solution

   $$Ni(OH)_2(aq) + 2H_2O(l) + 4NH_3(aq) \longrightarrow [Ni(H_2O)_2(NH_3)_4]^{2+} 2OH^-$$

57. What is the potential of the following electrochemical cell:

   $Cd \mid Cd^{2+}, M = 0.10 \parallel Ni^{2+}, M = 0.50 \mid Ni$

   Solution

   Cathode: $Ni^{2+} + 2e^- \longrightarrow Ni$ $\qquad -0.250$ V
   Anode: $Cd \longrightarrow Cd^{2+} + 2e^-$ $\qquad +0.403$ V

   $Cd + Ni^{2+} \longrightarrow Cd^{2+} + Ni$ $\quad E = +0.153$ V

   Using the Nernst equation, we have

   $$E = E° - \frac{0.059}{2} \log \frac{0.10}{0.50} = 0.153 + 0.021 = 0.17 \text{ V}$$

COPPER

59. Copper plate makes up the surface of the Statue of Liberty in New York harbor. Why does it appear green?

Solution

Copper, when exposed to moisture and air, forms a basic carbonate, $Cu_2(OH)_2CO_3$. This and/or the hydroxysulfate is responsible for the green color of weathered copper.

61. Explain the use of copper(II) sulfate in urine analysis.

Solution

A reducing sugar, such as glucose, in the urine indicates that a problem such as diabetes may exist. Copper(II) sulfate in Benedict's solution will react with any reducing agent to precipitate Cu_2O giving a proof of abnormality.

63. A white precipitate forms when copper metal is added to a solution of copper(II) chloride and hydrochloric acid. The white precipitate dissolves in excess concentrated hydrochloric acid. Dilution with water causes the white precipitate to reappear. Write equations for the reactions involved.

Solution

$$Cu + CuCl_2 \longrightarrow 2CuCl(s)$$

$$CuCl + Cl^- \rightleftharpoons [CuCl_2]^-$$

$$[CuCl_2]^- + H_2O \longrightarrow CuCl + H_2O + Cl^-$$

65. Why are most slightly soluble copper salts readily dissolved by aqueous ammonia?

Solution

Ammonia forms a soluble complex with the slightly soluble copper salts. The reaction is

$$[Cu(H_2O)_4]^{2+} + 4NH_3 \rightleftharpoons [Cu(NH_3)_4]^{2+} + 4H_2O$$

67. Sketch the structure of $Cu(H_2O)_4SO_4 \cdot H_2O$.

Solution

69. A 1.008-g sample of a silver-copper alloy is dissolved and treated with excess iodide ion. The liberated I_2 is titrated with a 0.1052 M $S_2O_3^{2-}$ solution. If 29.84 mL of this solution is required, what is the percentage of Cu in the alloy?

Solution

This method is based on the fact that I^- will be oxidized to I_2 by Cu^+ as it goes to Cu. Silver ion will not react. The reaction is

$$Cu^{2+} + 2I^- \longrightarrow Cu + I_2$$

The thiosulfate reacts with I_2 on a 1:1 basis so that each mol of thiosulfate reacted is equivalent to one mole of Cu^{2+} ion.

mol Cu^{2+} = mol I_2 = mol thiosulfate = 0.1052 M $S_2O_3^{2-}$ × 0.02984 L

$$= 3.13916 \times 10^{-3} \text{ mol}$$

g Cu = 63.546 g Cu/mol × 3.13916×10^{-3} mol = 0.19948 g

$$\text{Percent Cu} = \frac{0.19948 \text{ g}}{1.008 \text{ g}} \times 100 = 19.79\%$$

SILVER

71. Explain the tarnishing of silver in the presence of sulfur-containing substances.

Solution

In the presence of air and $H_2S(g)$ generated from sulfur-containing compounds, silver forms a sulfide

$$4Ag(s) + 2H_2S(g) + O_2(g) \longrightarrow 2Ag_2S(s) + 2H_2O(l)$$

73. Why does silver dissolve in nitric acid but not in hydrochloric acid?

Solution

Nitric acid is an oxidizing acid that causes silver to form ions whereas the NO_3^- goes to NO(g) to accept the electron from silver metal.

$$3Ag + 4H^+ + NO_3^- \longrightarrow 3Ag^+ + NO(g) + 2H_2O$$

On the other hand, silver lies below hydrogen in the activity series, and therefore H^+ cannot be reduced to H_2 by silver. HCl is not an oxidizing acid.

75. Dilute sodium cyanide solution is slowly dripped into a slowly stirred silver nitrate solution. A white precipitate forms temporarily but dissolves as the addition of sodium cyanide continues. Use chemical equations to explain this observation.

Solution

As CN^- is added:

$$Ag^+ + CN^- \longrightarrow AgCN(s)$$

As more CN^- is added:

$$Ag^+ + 2CN^- \longrightarrow [Ag(CN)_2]^-$$

$$AgCN(s) + CN^- \longrightarrow [Ag(CN)_2]^-$$

77. The formation constant of $[Ag(CN)_2]^-$ is 1.0×10^{20}. What will be the equilibrium concentration of Ag^+ if 1.0 g of Ag is oxidized and put into 1 L of 1.0×10^{-1} M CN^- solution?

Solution

The reaction is

$$Ag + 2CN^- \longrightarrow Ag(CN)_2^-$$

Let x = amount of silver:

$$1.0 \text{ g Ag}/107.868 \text{ g mol}^{-1} = 9.27 \times 10^{-3} \text{ mol Ag}^+$$

$$K = 1 \times 10^{20} = \frac{[Ag(CN)_2^-]}{[Ag^+][CN^-]^2} = \frac{(9.27 \times 10^{-3} - x)}{(x)[0.1 - 2(9.27 \times 10^{-3})]^2}$$

Dropping x in comparison to 9.27×10^{-3} M,

$$1 \times 10^{20} = \frac{9.27 \times 10^{-3}}{x(0.08146)^2}$$

$$1 \times 10^{20} \times 6.636 \times 10^{-3} x = 9.27 \times 10^{-3}$$

$$x = [Ag^+] = 1.4 \times 10^{-20} \text{ M}$$

GOLD

79. Account for the variable oxidation number of gold in terms of electronic structure.

Solution

Gold has 1 electron beyond a closed d-shell configuration. Removal of that electron gives Au^+. Energetics are such that 2 other electrons can easily be removed from the full d orbital because of the energy differential of the $6s$ orbital compared to that of the $5d$'s. With the loss of these 2 additional electrons, the Au^{3+} ion is formed.

81. Write balanced equations for the following changes, which occur during the recovery of gold from an ore:

$$Au \longrightarrow [Au(CN)_2]^- \longrightarrow Au$$

Solution

$$4Au + 8CN^- + O_2 + 2H_2O \longrightarrow 4[Ag(CN)_2]^- + 4OH^-$$

$$2[Au(CN)_2]^- + Zn \longrightarrow 2Au + [Zn(CN)_4]^{2-}$$

28

The Post-Transition Metals

INTRODUCTION

The post-transition metals have often been included along with the transition metals immediately preceeding them. However, their filled d shells put them into a class by themselves. Several of these elements, lead, mercury, and tin, were known from ancient times. These, along with bismuth, cadmium, and zinc, are very commonly used in a wide variety of compounds in modern society. Gallium, indium, thallium are important but are used in much smaller tonnage. Radioactive polonium is of lesser importance.

READY REFERENCE

Betts process (28.2) This is an electrolytic process in which thin sheets of pure lead are used as cathode, and plates of impure lead are used as anode. The electrolyte contains a solution of lead hexafluorosilicate and hexafluorosilicic acid.

Inert pair effect (28.1) This refers to the stability of an oxidation number that is two lower than the group oxidation number. It is a result of the pair of s electrons in the valence shell behaving as an inert pair, resisting their removal. The effect is stronger the lower the position of the element in the Periodic Table.

SOLUTIONS TO CHAPTER 28 EXERCISES

PERIODIC PROPERTIES

1. How do the properties of the post-transition metals tin and lead differ from the properties of the elements at the top of Group IVA in the Periodic Table?

    Solution

    The lighter members of Group IVA are primarily nonmetallic in character, whereas tin and lead are metallic in their behavior. Tin and lead form stable dipositive ions which are two below the oxidation state of the lighter members of the group.

3. Which post-transition metals form common oxides and halides in which the metal does not exhibit the group oxidation number?

Chapter 28 The Post-Transition Metals

Solution

Thallium and bismuth exhibit the inert pair effect and consequently do not exhibit the normal group oxidation number except under special conditions.

5. Describe the inert pair effect.

Solution

In some cases, the s electrons in the valence shell resist removal in the formation of ions. These two s electrons behave as an inert pair and the oxidation number of the resulting ion has a value two lower than the group oxidation number.

7. Write the electron configurations for an Sn atom, an Sn^{2+} ion, and Sn^{4+} ion.

Solution

Sn   [Ar] $4s^2 4p^6 4d^{10} 5s^2 5p^2$
Sn^{2+} [Ar] $4s^2 4p^6 4d^{10} 5s^2$
Sn^{4+} [Ar] $4s^2 4p^6 4d^{10}$

9. Which of the post-transition metals do not dissolve readily in acids, with the evolution of hydrogen?

Solution

Pb; Hg

ZINC

11. What is the common ore of zinc and how is zinc extracted from it?

Solution

Zinc is obtained from sphalerite, or zinc blende, ZnS. The ore is concentrated by flotation using organic surfactants and foaming agents. It is then roasted (heated in air) to convert zinc sulfide to the oxide. The oxide is then heated with coal in a fire-clay retort from which zinc is distilled out and then condensed.

13. What is galvanized iron, how is it produced, and what useful properties does it have?

Solution

Galvanized iron can be made by a hot-dip process in which iron is dipped into molten zinc. When the iron is removed, a thin coating of zinc remains on the iron. It can also be electroplated producing a more uniform coating than in the hot-dip process. Galvanizing protects iron from rusting.

15. What is lithopone and how is it produced? In what way is lithopone superior to white lead as a paint pigment?

Solution

Lithophone is a mixture of barium sulfate and zinc sulfide in a vehicle to spread it in paint. Its major advantage is that it is not toxic if ingested by children.

CADMIUM

17. What is the common source of cadmium and how is cadmium extracted from it?

Solution

Most cadmium comes as a by-product from zinc smelters and from the sludge obtained from the electrolytic refining of zinc. In smelting cadmium-containing zinc ores, both metals are reduced but cadmium can be separated by fractional distillation because of its lower boiling point. In the electrolytic process cadmium can be obtained by selective deposition at a lower voltage than zinc.

19. What properties make cadmium effective as a protective coating for other metals?

Solution

Cadium can protect other base metals because it is comparatively inactive yet can still function as an anode and the base metal acts as the cathode.

21. Write balanced chemical equations for the following reactions:

   (a) Cadmium is burned in air.
   (b) An aqueous solution of ammonia is added dropwise to a solution of cadmium chloride until the mixture becomes clear again.
   (c) Cadmium is heated with sulfur.
   (d) Cadmium is added to a solution of hydrochloric acid.
   (e) Cadmium hydroxide is added to a solution of acetic acid, CH_3CO_2H.
   (f) Cadmium hydroxide is heated until weight loss stops.
   (g) Cadmium metal is added to a solution of mercury(II) nitrate.
   (h) Cadmium metal is added to a solution of lead(II) nitrate.
   (i) Cadmium sulfide is heated in a stream of oxygen gas.

Solution

(a) $2Cd(s) + O_2(g) \longrightarrow 2CdO(s)$
(b) $CdCl_2(aq) + 4NH_3(aq) \longrightarrow [Cd(NH_3)_4]^{2+}(aq) + 2Cl^-(aq)$
(c) $Cd(s) + S(s) \longrightarrow CdS(s)$
(d) $Cd(s) + 2HCl(aq) \longrightarrow CdCl_2(aq) + H_2(g)$
(e) $Cd(OH)_2(aq) + 2CH_3CO_2H(aq) \longrightarrow Cd(CH_3CO_2)_2(aq) + 2H_2O(l)$
(f) $Cd(OH)_2(aq) \xrightarrow{\Delta} CdO(s) + H_2O(g)$
(g) $Cd(s) + Hg(NO_3)_2(aq) \longrightarrow Cd(NO_3)_2(aq) + Hg(s)$
(h) $Cd(s) + Pb(NO_3)_2(aq) \longrightarrow Pb(s) + Cd(NO_3)_2(aq)$
(i) $CdS(s) + O_2(g) \longrightarrow Cd(s) + SO_2(g)$

23. Calculate the emf of the following electrochemical cell:

   $Cd \mid Cd^{2+}, M = 0.10 \parallel Ni^{2+}, M = 0.50 \mid Ni$

Chapter 28 The Post-Transition Metals

Solution

Method: Use the Nernst equation to calculate the desired value.

$E°_{Cd^{2+},Cd} = -0.40$ V, $E°_{Ni^{2+},Ni} = -0.250$

The reaction is

$$Cd + Ni^{2+} (M = 0.50) \longrightarrow Cd^{2+} (M = 0.10) + Ni$$

$$E° = E°_{Ni^{2+},Ni} - E°_{Cd^{2+},Cd} = -0.25 - (-0.40) = 0.15 \text{ V}$$

$$E = E° - \frac{0.05915}{n} \log Q$$

$$E = 0.15 - \frac{0.05915}{2} \log \frac{0.10}{0.50} = 0.17 \text{ V}$$

MERCURY

25. What is the common ore of mercury and how is mercury separated from it?

Solution

Mercury occurs as cinnabar and is separated by roasting in air.

27. The roasting of an ore of a metal usually results in the conversion of the metal to the oxide. Why does the roasting of cinnabar produce metallic mercury rather than an oxide of mercury?

Solution

The mercury oxides are thermally unstable and decompose to the elements in the furnace.

29. What does it mean to say that mercury(II) halides are weak electrolytes?

Solution

Weak electrolytes do not fully dissociate in solution. Therefore to say mercury(II) halides are weak electrolytes means that they are mostly in the form of HgX_2 in solution where X represents a halide.

31. Write Lewis structures for $HgCl_2$ and Hg_2Cl_2.

Solution

:Cl̈—Hg—C̈l: :Cl̈—Hg—Hg—C̈l:

33. How many moles of ionic species would be present in a solution marked 1.0 M mercury(I) nitrate? How would you demonstrate the accuracy of your prediction?

Solution

There would be three moles of ionic species in 1.0 M mercury(I) nitrate: 1 mol Hg_2^{2+} and 2 mol NO_3^-. To test the accuracy of the prediction, measure either boiling point elevation or freezing point depression.

35. How many pounds of mercury can be obtained from 10 tons of ore that is 7.4% cinnabar?

Solution

$$HgS + O_2 \longrightarrow Hg + SO_2$$

$$\text{Mass Hg} = 10 \text{ ton} \times \frac{2000 \text{ lb}}{\text{ton}} \times \frac{1 \text{ lb-mol HgS}}{232.64 \text{ lb}} \times \frac{1 \text{ lb-mol Hg}}{1 \text{ lb-mol H}_2\text{S}}$$

$$\times \frac{200.59 \text{ lb Hg}}{1 \text{ lb-mol Hg}} \times 0.074 = 1.3 \times 10^3 \text{ lb}$$

TIN

37. What is the common ore of tin and how is tin separated from it?

Solution

Roasting of cassiterite or tinstone, (SnO$_2$), removes arsenic and sulfur as volatile oxides. Hydrochloric acid removes other metal impurities. The purified ore is reduced by carbon.

39. Mercury(I) solutions can be protected from air oxidation to mercury(II) by keeping them in contact with metallic mercury. Can tin(II) solutions be similarly stabilized against oxidation to tin(IV) by keeping them in contact with metallic tin? (The standard reduction potential for tin(IV) to tin(II) is +0.15 V.)

Solution

Mercury is added to force the equilibrium to the left.

$$Hg_2Cl_2 \longrightarrow HgCl_2 + Hg$$

The electropotentials are favorable for this process. However, tin lies above mercury in the table and when considered with the potentials for the +2 state, the potential still favors oxidation of the Sn^{2+} state.

41. What is tin pest, also known as tin disease?

Solution

Tin pest or tin disease occurs because at temperatures below 13.2° an allotropic form of tin, grey tin, is the stable form. White tin will slowly convert to grey tin that is powdery. The transformation will cause articles of tin to disintegrate at low temperatures.

43. Write balanced chemical equations describing:
   (a) the burning of tin metal
   (b) the dissolution of tin in a solution of hydrochloric acid
   (c) the preparation of sodium metastannate
   (d) the purification of tin by electrolysis
   (e) the reaction of dry bromine with tin (write equations for both possible products)
   (f) the formation of thiostannate ion

(g) the dissolution of tin oxides by basic solution (two equations)
(h) the reactions involved when an excess of aqueous NaOH is slowly added to a solution of tin(II) chloride
(i) the thermal decomposition of tin(II) oxalate
(j) the reaction of tin with sulfur (write equations for both posssible products)

Solution

(a) $Sn(s) + O_2(g) \longrightarrow SnO_2(s)$
(b) $Sn(s) + 2HCl(aq) \longrightarrow SnCl_2(aq) + H_2(g)$
(c) $SnO_2 \cdot H_2O(s) + 2NaOH(s) \longrightarrow Na_2SnO_3(s) + 2H_2O(l)$
(d) $Sn^{2+} + 2e^- \longrightarrow Sn$
(e) $Sn(s) + Br_2(l) \longrightarrow SnBr_2(s)$
    $Sn(s) + 2Br_2(l) \longrightarrow SnBr_4(s)$
(f) $SnS_2 + Na_2S \longrightarrow Na_2SnS_3$
(g) $SnO_2 + 2H_2O + 2OH^- \longrightarrow Sn(OH)_6^{2-}$
    $SnO + H_2O + 4OH^- \longrightarrow Sn(OH)_6^{2-} + 2e^-$
(h) $SnCl_2(aq) + 2OH^-(aq) \longrightarrow Sn(OH)_2(s) + 2Cl^-(aq)$
(i) $SnC_2O_4(s) \longrightarrow SnO(s) + CO_2(g) + CO(g)$
(j) $Sn(s) + S(s) \longrightarrow SnS$
    $Sn(s) + 2S(s) \longrightarrow SnS_2(s)$

45. Why must aqueous solutions of SnCl$_2$ be protected from the air?

Solution

Aqueous solutions of SnCl$_2$ are air oxidized to Sn(IV) compounds.

LEAD

47. What is the common ore of lead and how is lead separated from it?

Solution

Lead is obtained from galena, PbS. The ore is concentrated by flotation and is then roasted in air to convert the sulfide to the oxide. The oxide is then reduced in a blast furnace with coke and scrap iron to the metal.

49. Describe the Betts process for the refining of lead.

Solution

The Betts process is an electrolytic process using pure lead sheets as the cathode and plates of impure lead as the anodes. The electrolyte is a solution of lead hexafluorosilicate PbSiF$_6$, and hexafluorosilicic acid, H$_2$SiF$_6$.

51. Show by suitable equations that lead(II) hydroxide is amphoteric.

Solution

$$Pb(OH)_2 + 2H^+ \longrightarrow Pb^{2+} + 2H_2O$$
$$Pb(OH)_2 + OH^- \longrightarrow Pb(OH)_3^-$$

53. Why should water to be used for human consumption not be conveyed in lead pipes?

Solution

The water in lead pipe will leach enough lead that consumption of the water will cause lead posioning in humans. Indeed, a recent concern in environmental fields is that of the use of lead-containing solder to join copper pipe used for conveying water. The water sitting in contact with even that small amount of lead will extract measurable amounts of lead overnight, thus making the consumption of that water dangerous. Two methods exist to stop the problem. The first is to use silver solder in place of lead on pipes carrying water, the second is to flush the system before use after periods during which the water has stood for a long time.

55. When elements in the first transition metal series form dipositive ions, they usually give up their *s* electrons. Does this apply to lead when it forms Pb^{2+} ion?

Solution

No. The electrons are lost from the $6p$ level.

BISMUTH

57. What are the properties of metallic bismuth that make it commercially useful?

Solution

Bismuth has the property that it expands upon solidification. It also forms low melting alloys with tin and lead. The first leads to uses in casting alloys, the second leads to uses in fuses.

59. Write equations for the reaction of $BiCl_3$ with water.

Solution

$$BiCl_3(s) + 3H_2O(l) \longrightarrow Bi(OH)_3(s) + 3HCl(l)$$

ADDITIONAL EXERCISES

61. Write balanced equations for the following reactions:
    (a) Gallium is heated in air.
    (b) An aqueous solution of sodium hydroxide is added dropwise to a solution of gallium chloride until the solution becomes clear again.
    (c) Indium is heated with an excess of sulfur.
    (d) Indium is added to a solution of hydrobromic acid.
    (e) Gallium hydroxide is added to a solution of nitric acid.
    (f) Indium hydroxide is heated until weight loss stops.
    (g) Gallium metal is added to a solution of indium nitrate.
    (h) Indium metal is added to a solution of lead(II) nitrate.
    (i) Indium sulfide is heated in a stream of oxygen gas.

Solution

(a) $4Ga(s) + 3O_2(g) \xrightarrow{\Delta} 2Ga_2O_3(s)$

(b) $GaCl_3(aq) + 4NaOH(aq) \longrightarrow [Ga(OH)_4]^-(s) + 4Na^+(aq) + 3Cl^-(aq)$

(c) $2\text{In}(s) + (3xs)\text{S}(s) \xrightarrow{\Delta} \text{In}_2\text{S}_3(s)$

(d) $2\text{In}(s) + 6\text{HBr}(aq) \longrightarrow 2\text{InBr}_3(aq) + 3\text{H}_2(g)$

(e) $\text{Ga(OH)}_3(aq) + 3\text{HNO}_3(aq) \longrightarrow \text{Ga(NO}_3)_3(aq) + 3\text{H}_2\text{O}(l)$

(f) $2\text{In(OH)}_3 \xrightarrow{\Delta} \text{In}_2\text{O}_3(s) + 3\text{H}_2\text{O}(g)$

(g) $\text{Ga}(s) + \text{In(NO}_3)_3(aq) \longrightarrow \text{In}(s) + \text{Ga(NO}_3)_3(aq)$

(h) $2\text{In}(s) + 3\text{Pb(NO}_3)_2(aq) \longrightarrow 3\text{Pb}(s) + 2\text{In(NO}_3)_3(aq)$

(i) $2\text{In}_2\text{S}_3(s) + 9\text{O}_2(g) \xrightarrow{\Delta} 2\text{In}_2\text{O}_3(s) + 6\text{SO}_2(g)$

63. Calculate the molar solubilities of AgBr, Hg$_2$Cl$_2$, and PbSO$_4$.

Solution

From the solubility product of each substance calculate the molar solubility.

AgBr:

$$K_{sp} = 3.3 \times 10^{-13} = [\text{Ag}^+][\text{Br}^-] = x^2$$
$$x = 5.7 \times 10^{-7} \text{ M}$$

Hg$_2$Cl$_2$:

$$K_{sp} = 1.1 \times 10^{-18} = [\text{Hg}_2^{2+}][\text{Cl}^-]^2$$

Let $x = [\text{Hg}_2^{2+}]$. Then $[\text{Cl}^-] = 2x$.

$$x(2x)^2 = 1.1 \times 10^{-18} = 4x^3$$
$$x^3 = 2.75 \times 10^{-19}$$
$$x = 6.5 \times 10^{-7} \text{ M}$$

PbSO$_4$:

$$K_{sp} = 1.8 \times 10^{-8} = [\text{Pb}^{2+}][\text{SO}_4^{2-}] = x^2$$
$$x = 1.3 \times 10^{-4} \text{ M}$$

65. What is the maximum [S^{2-}] possible in a solution that is 0.0010 M in Cd^{2+}?

Solution

CdS:

$$K_{sp} = 2.8 \times 10^{-35} = [\text{Cd}^{2+}][\text{S}^{2-}]$$
$$[\text{S}^{2-}] = \frac{2.8 \times 10^{-35}}{0.0010} = 2.8 \times 10^{-32}$$

29

The Atmosphere and Natural Waters

INTRODUCTION

The Greek philosopher Aristotle viewed matter as being made from air, water, earth, and fire. The importance of air and water is no less paramount today than it was some twenty-three hundred years ago. Indeed, the importance of the atmosphere and naturally occurring waters is more clear to us today. Our very existence requires air and water. Industrially, large amounts of the atmosphere are liquefied or otherwise collected and used as is or separated into components. In the United States alone, some 400 billion cubic feet of oxygen were produced in 1986 for uses as far-ranging as medicine to steel making. (*Chemical and Engineering News,* July 13, 1987, p. 12.) Fresh water needs soar as consumer products require ever-increasing amounts of water for processing. Farm irrigation systems add to the need for pure water.

Unfortunately, pollution of both the atmosphere and natural waters (lakes, streams, and even oceans) threaten to cause drastic changes in our lives. Destruction of aquatic life in our lakes and streams because of acid rain, contamination of our ground waters by improper waste disposal, the destruction of our beaches by garbage swept up from the oceans (where both municipal and chemical wastes are being dumped), the thinning of the ozone layer by fluorocarbons, and the increase in temperature of the atmosphere brought about by the greenhouse effect (carbon dioxide from fossil fuel fires prevents the reradiation of heat to space once heat has been absorbed from the sun) — all indicate how fragile our ecosystem is. It should be important to us to learn the chemistry involved in the atmosphere and oceans and what can be done to combat pollution of our world.

READY REFERENCE

Causes of air pollution (29.3–29.5)   Sulfur oxides, particulates, carbon monoxide, nitrogen oxides, hydrocarbons, and photochemical oxidants all constitute problems for air quality.

Causes of water pollution (29.9)   Water quality is degraded by: phosphates and other nutrients for algae, bacteria, and aquatic plants; sewage and other waste materials; toxic pollutants; acid rain; and thermal pollution.

… Chapter 29 The Atmosphere and Natural Waters

SOLUTIONS TO CHAPTER 29 EXERCISES

THE ATMOSPHERE

1. What evidence can you cite to show that air is a mixture rather than a compound?

   Solution

   A normal sample of air will contain moisture. When a sample of air is brought into contact with a cold surface, water will condense out as frost. If air is liquefied and then allowed to boil, different boiling points will be observed for each gaseous component: oxygen at $-183.0°C$, nitrogen at $-195.8°C$. A pure compound would boil at only one temperature.

3. Write balanced equations for each of the following processes:
   (a) the formation of sulfur dioxide from zinc sulfide present in burning coal
   (b) the formation of nitrogen dioxide in an internal combustion engine
   (c) the formation of sulfuric acid mist from sulfur dioxide
   (d) the removal of carbon monoxide from an automobile's exhaust using a catalytic converter.

   Solution

   (a) $2ZnS(s) + 3O_2(g) \rightarrow 2ZnO(s) + 2SO_2(g)$
   (b) $N_2(g) + O_2(g) \rightleftharpoons 2NO(g)$
   $2NO(g) + O_2(g) \rightleftharpoons 2NO_2(g)$
   (c) $SO_2(g) + NO_2(g) \rightleftharpoons NO(g) + SO_3(g)$
   $SO_3(g) + H_2O(l) \rightleftharpoons H_2SO_4(l)$
   (d) $2CO(g) + O_2(g) \rightleftharpoons 2CO_2(g)$

5. Based on the periodic properties of the elements, write balanced chemical equations for the reactions, if any, between the components of air given in Table 29.1 and hot magnesium.

   Solution

   $3Mg(s) + N_2(g) \rightarrow Mg_3N_2(s)$
   $2Mg(s) + O_2(g) \rightarrow 2MgO$
   $Mg(s) + Ar(g) \rightarrow$ no reaction
   $2Mg(s) + CO_2(g) \rightarrow 2MgO(s) + C(s)$
   $Mg(s) + Ne(g) \rightarrow$ no reaction
   $Mg(s) + He(g) \rightarrow$ no reaction
   $Mg(s) + Kr(g) \rightarrow$ no reaction
   $Mg(s) + O_3(g) \rightarrow MgO(s) + O_2(g)$
   $Mg(s) + H_2(g) \rightarrow$ no reaction (presence of slight amount of oxygen could trigger an explosion)
   $Mg(s) + Xe(g) \rightarrow$ no reaction

NATURAL WATERS

7. What naturally occurring impurities are typically present in unpolluted rain water? In unpolluted river water?

Solution

Impurities in rain water include particulate matter such as $CaCO_3$ dust, airborne soil, and salts such as $NaCl$ and Na_2CO_3. In addition, CO_2 is dissolved in rain, forming H_2CO_3. In addition to the impurities listed for rain water, more impurities will be present in river water, which comes into contact with river banks and consequently leaches out various salts and minerals. Furthermore, decay of plants, fish, and other organisms puts large numbers of other compounds into the water. Suspended materials that can pollute include clays and silts.

9. Define the following terms:
   (a) BOD
   (b) eutrophication
   (c) aerobic decomposition
   (d) anaerobic decomposition
   (e) hard water

Solution

(a) BOD stands for the Biological Oxygen Demand which is the volume of oxygen reacted by a given amount of water at 20°C in five days.
(b) Eutrophication is the process by which a lake grows rich in nutrients causing growth of algae and plants that die and decay, gradually filling the lake with organic sediment.
(c) Aerobic decomposition is the decay of organic material by bacteria in the presence of oxygen.
(d) Anaerobic decomposition is the decay of organic matter in the absence of oxygen.
(e) Hard water is water containing dissolved calcium, magnesium and/or iron salts.

11. What is thermal pollution of water? Is it necessarily undesirable?

Solution

Thermal pollution is the artificial raising of the temperature of a body of water by the dissipation of heat from some human endeavor. A slight increase of temperature (even a few °C) can be fatal to some species of fish. Other species might thrive at the higher temperature. It has been proposed to use the heat generated in nuclear power plants to foster the decomposition of sewage in lagoons located near such plants in northern latitudes; in general, however, thermal pollution is harmful.

13. What metal ions are commonly responsible for the hardness of water?

Solution

The three common metal ions responsible for hard water are calcium, magnesium, and iron ions.

15. Distinguish between softening and demineralizing of water. Does demineralization soften water?

Solution

Water softening removes dissolved calcium, magnesium, and iron ions. Demineralization removes all metal ions and their counter ions, leaving essentially pure water; it is the most complete form of water softening.

17. What are the molarities of Na^+ and Cl^- in sea water (see Table 29.2)?

Solution

$[Na^+]$ = 10.56 g/1.000 L H_2O × 1 mol Na/22.98977 g = 0.4593 M

$[Cl^-]$ = 18.98 g/1.000 L H_2O × 1 mol Cl/35.453 g = 0.5354 M

30

Organic Chemistry

INTRODUCTION

Organic chemistry deals with the wide variety of compounds composed principally of carbon and hydrogen that are made possible by the peculiar ability of carbon to bond with itself through single, double, and triple bonds.

Few new concepts are needed here, application being made of the ideas developed in earlier chapters. Nomenclature, concepts of isomerism, and the reactions and properties that particular functional groups have are the focuses of this chapter.

READY REFERENCE

Asymmetric carbon (28.6)   A carbon atom that is bonded to four different groups is asymmetric. Such atoms give compounds the property of optical activity.

Functional group (28.1)   Particular groups, such as carbonyl, carbonylic acid, and ester, are found to behave in a particular manner throughout a wide variety of compounds and are known as functional groups.

Geometrical isomers (28.3)   Where two or more compounds have the same number and kind of atom attached to each other and differ only in the arrangement (geometry) of the atoms, the different forms are geometrical isomers.

Optical isomers (28.6)   Optical isomers occur when a compound, with an asymmetric carbon atom, and its mirror image are not superimposable.

Structural isomers (28.1)   These isomers have the same molecular formula but different physical and chemical properties, because the arrangement of the atoms in their molecules is different.

SOLUTIONS TO CHAPTER 30 EXERCISES

HYDROCARBONS

1. Write the chemical formula and Lewis structure of an alkane, alkene, an alkyne, and an aromatic hydrocarbon, each of which contains six carbon atoms.

Solution

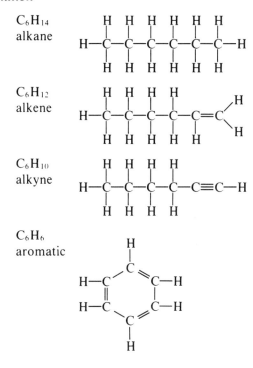

3. What is the difference between the electronic structures of saturated and unsaturated hydrocarbons?

Solution

In saturated hydrocarbons each carbon is bonded to the next with sp^3-type bonds. In unsaturated hydrocarbons either sp- or sp^2-type bonds are used to join the carbons. Electronically, some electrons remain in p orbitals in the unsaturated compounds. The p orbitals overlap above and below the line joining adjacently bonded carbon atoms, thus forming π orbitals.

5. Draw a three-dimensional structure for each of the following, using solid and striped wedge-shaped bonds where appropriate:

   (a) CH_4
   (b) C_2H_6
   (c) $n\text{-}C_4H_{10}$
   (d) C_3H_6
   (e) C_3H_4

Solution

(a)

(b)

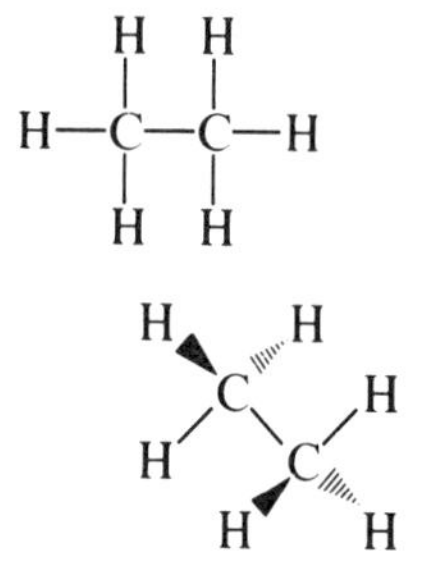

```
    H H
    | |
H—C—C—H
    | |
    H H
```

(c)

```
    H H H H
    | | | |
H—C—C—C—C—H
    | | | |
    H H H H
```

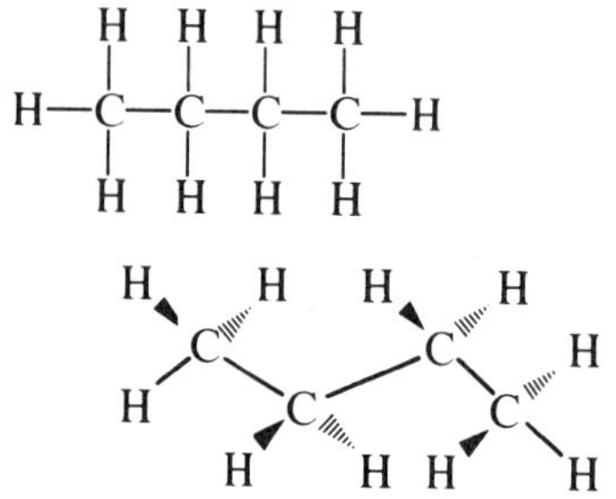

(d)

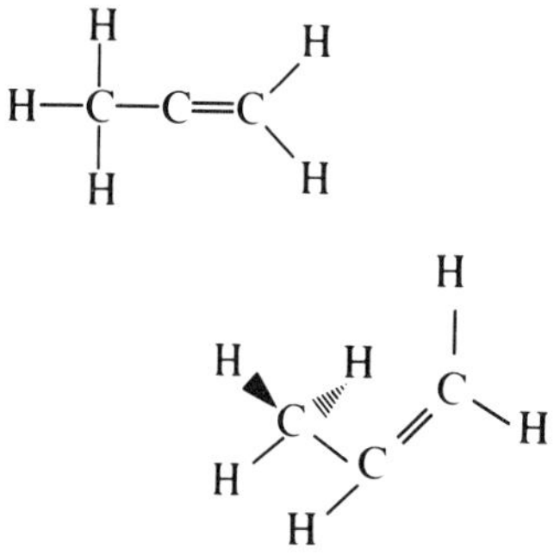

```
    H       H
    |      /
H—C—C=C
    |      \
    H       H
```

(e)

```
 H          H
  \        /
   C=C=C
  /        \
 H          H
```

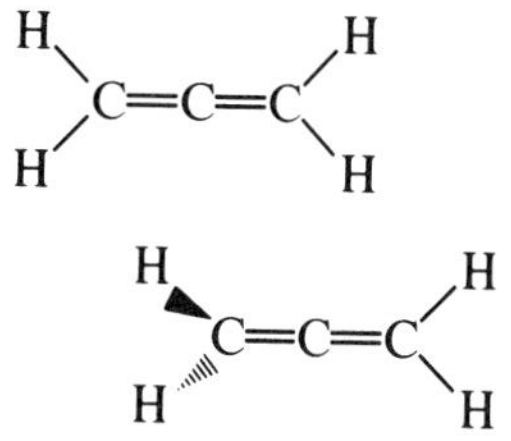

7. Write Lewis structures for all of the isomers of the alkyne C_5H_8.

Solution

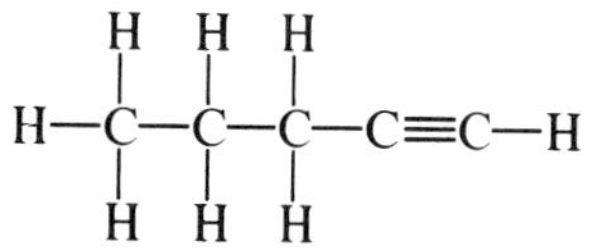

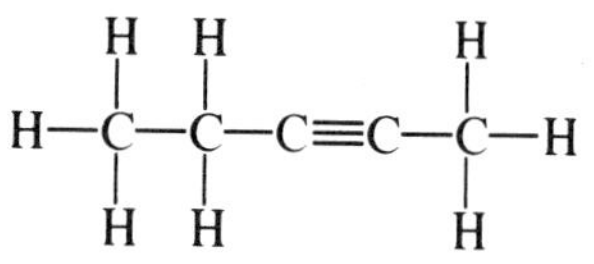

```
    H H H
    | | |
H—C—C—C—C≡C—H
    | | |
    H H H
```

```
    H H        H
    | |        |
H—C—C—C≡C—C—H
    | |        |
    H H        H
```

Chapter 30 Organic Chemistry   407

```
        H
        |
    H—C—H
        |
    H—C—C≡C—H
        |
    H—C—H
        |
        H
```

9. Write Lewis structures and the names of all isomers of the alkyl groups
    —C₃H₇ and —C₄H₉.

    Solution

    ```
        H H H                    H H H
        | | |                    | | |
      —C—C—C—H              H—C—C—C—H
        | | |                    |   |
        H H H                    H   H
    ```
        n-propyl group   *sec*-propyl group

    ```
        H H H H                  H H H H
        | | | |                  | | | |
      —C—C—C—C—H            H—C—C—C—C—H
        | | | |                  |   | |
        H H H H                  H   H H
    ```
        n-butyl group   *sec*-butyl group

    ```
              H                    H   H   H
         HH—C—HH                   |   |   |
          |  |                   H—C———C———C—H
      H—C——C——C—H                    |
          |  |                   HH—C—HH
          H  H                       |
                                     H
    ```
        tert-butyl group   2-methyl-*n*-propyl group

11. Indicate which of the following molecules can form optical isomers:

    (a) *trans*-2-pentene
    (b) dichlorodifluoromethane
    (c) 3-ethyloctane
    (d) 4-ethyloctane
    (e) 2-butanethiol
    (f) 1-cyanoethanol, [CH₃CH(CN)OH]

    Solution

    (a)
    ```
         H H H
         | | |
       H—C—C—C    H
         | | \\   |
         H H  C—C—H
              |  |
              H  H
    ```
    No optical isomers

    (b)
    ```
       Cl   F
         \\ ◢
          C
         ╱ ⦀
       Cl   F
    ```
    No optical isomers

(c)

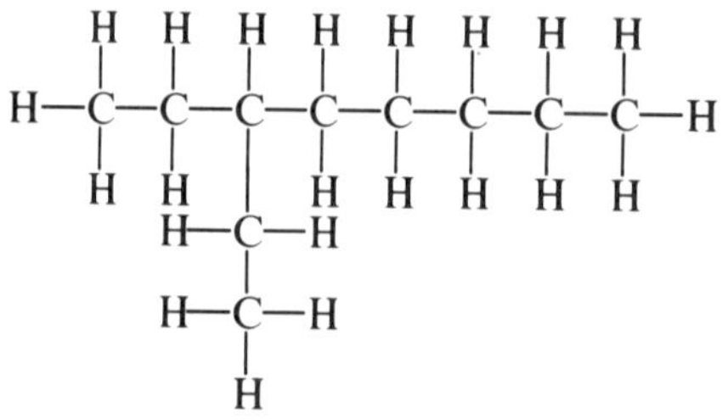

        H H H   H H H H H
        | | | | | | | |
    H—C—C—C—C—C—C—C—C—H
        | | | | | | | |
        H H H   H H H H H
              H—C—H
                 |
              H—C—H
                 |
                 H

No optical isomers

(d)

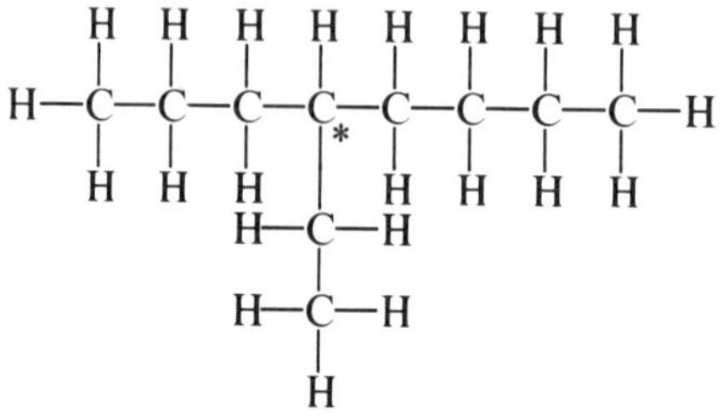

        H H H   H H H H H
        | | | | | | | |
    H—C—C—C—C*—C—C—C—C—H
        | | | | | | | |
        H H H   H H H H H
              H—C—H
                 |
              H—C—H
                 |
                 H

Optical isomers exist
*asymmetric carbon atom

(e)

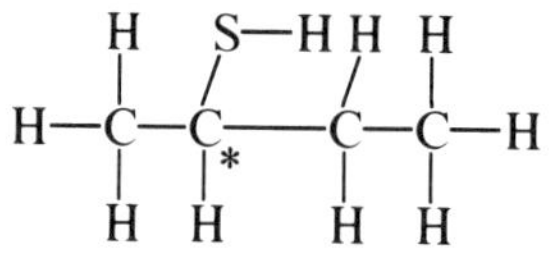

        H   S—H H   H
        | | | |
    H—C—C*—C—C—H
        | | | |
        H   H   H   H

Optical isomers exist
*asymmetric carbon atom

(f)

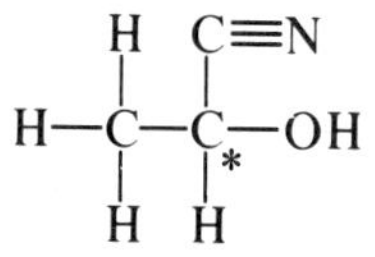

        H   C≡N
        | |
    H—C—C*—OH
        | |
        H   H

Optical isomers exist
*asymmetric carbon atom

13. Does the following molecule exist as *cis-trans* isomers? Optical isomers?

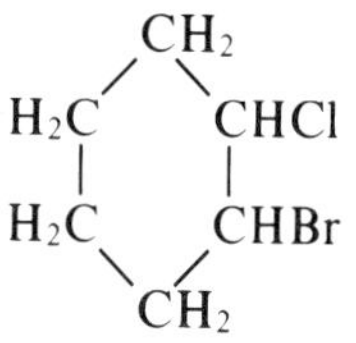

$$\begin{array}{c} CH_2 \\ H_2C \quad CHCl \\ | \quad\quad | \\ H_2C \quad CHBr \\ CH_2 \end{array}$$

Solution

There are no *cis-trans* isomers because there is free rotation about each carbon. However, there are optical isomers because there are 2 asymmetric carbon atoms.

HYDROCARBON DERIVATIVES

15. Identify each of the following hydrocarbon derivatives, in which R is an alkyl group:

(a) ROH
(b) RNH$_2$
(c) RCO$_2$H
(d) ROR
(e) RCHO
(f) RC(O)R

Solution

(a) alkyl alcohol
(b) alkyl amine

(c) carboxylic acid
(d) dialkyl ether
(e) alkyl aldehyde
(f) dialkyl ketone

17. Draw a three-dimensional structure, using solid and striped wedge-shaped bonds where appropriate, for each of the following molecules:

    (a) CH_3OCH_3
    (b) CH_3CO_2H
    (c) C_2H_5CHO
    (d) CH_3OH
    (e) $C_2H_5NH_2$

Solution

(a) [three-dimensional structure of CH_3OCH_3]

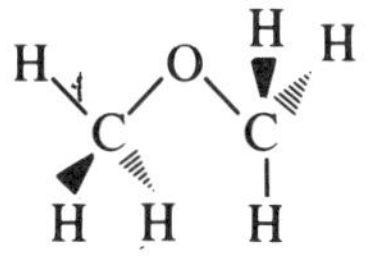

(b) [three-dimensional structure of CH_3CO_2H]

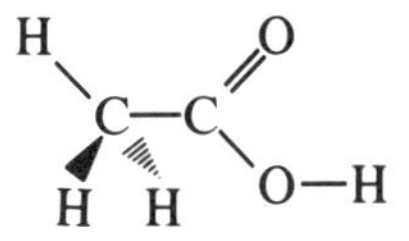

(c) [three-dimensional structure of C_2H_5CHO]

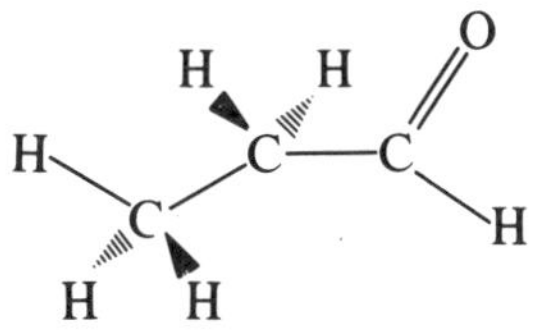

(d) [three-dimensional structure of CH_3OH]

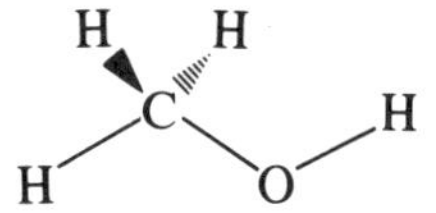

(e) [three-dimensional structure of $C_2H_5NH_2$]

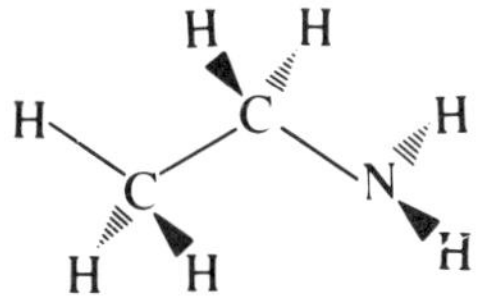

19. What are the products formed by the stepwise oxidation of methanol? Of ethanol?

Solution

For methanol:

$$CH_3OH \xrightarrow{\text{oxidation}} CH_2O \xrightarrow{\text{oxidation}} HCOOH \xrightarrow{\text{oxidation}} CO_2$$

For ethanol:

$$CH_3CH_2OH \xrightarrow{\text{oxidation}} CH_3CHO \xrightarrow{\text{oxidation}} CH_3COOH \xrightarrow{\text{oxidation}} CO_2$$

21. What alcohol can be oxidized to give the acid $(CH_3)_2CHCH_2CO_2H$? Which aldehyde?

Solution

$$(CH_3)_2CHCH_2CH_2OH \xrightarrow{\text{oxidation}} (CH_3)_2CHCH_2CO_2H$$

$$(CH_3)_2CHCH_2CH_2O \xrightarrow{\text{oxidation}} (CH_3)_2CHCH_2CO_2H$$

ADDITIONAL EXERCISES

23. Write a complete balanced equation for each of the following reactions:
  (a) $CH_3Li + H_2O \longrightarrow$
  (b) $(CH_3)_2NH + Li \longrightarrow$
  (c) $(CH_3)_2CHCH_2OH + Na \longrightarrow$
  (d) $(CH_3)_3CCH_2SH + NaOH \longrightarrow$
  (e) $CH_3OLi + CS_2 \longrightarrow$

Solution

  (a) $CH_3Li + H_2O \longrightarrow CH_4 + LiOH$
  (b) $(CH_3)_2NH + Li \longrightarrow [(CH_3)_2NH]^- Li^+$
  (c) $(CH_3)_2CHCH_2OH + Na \longrightarrow \frac{1}{2}H_2(g) + (CH_3)_2CHCH_2O^-Na^+$
  (d) $(CH_3)_3CCH_2SH + NaOH \longrightarrow (CH_3)_3CCH_2S^-Na^+ + H_2O$
  (e) $CH_3OLi + CS_2 \xrightarrow{\Delta} [CH_3OCS-S]^- + Li^+$ (a xanthate)

25. If all of the 30.55 billion lb of ethylene produced in the United States during 1985 was produced by the pyrolysis of ethane

$$C_2H_6 \longrightarrow C_2H_4 + H_2$$

what mass of hydrogen, in pounds, would have been produced?

Solution

The ratio of ethane to ethylene to hydrogen is 1:1:1

$$\text{lb }(H_2) = 30.55 \text{ billion lb ethylene} \times \frac{\text{lb mol ethylene}}{28.0532 \text{ lb}} \times \frac{2.0158 \text{ lb}}{\text{lb mol } H_2}$$

$$= 2.195 \text{ billion lb}$$

27. (a) From the reactions

$$C_2H_4(g) + 2H_2(g) \longrightarrow 2CH_4(g)$$

$$C(g) + 2H_2(g) \longrightarrow CH_4(g)$$

and by considering the bonds that must be broken and formed to get from reactants to products, show that the C=C bond energy, $D_{C=C}$, is given by

$$D_{C=C} = \Delta H°_{f C(g)} + \Delta H°_{f CH_4(g)} - \Delta H°_{f C_2H_4(g)}$$

Calculate, using data in Appendix I, a value for $D_{C=C}$.
  (b) By considerations similar to those in part (a), show that the reaction

$$C_2H_4(g) + H_2(g) \longrightarrow C_2H_6(g)$$

Chapter 30 Organic Chemistry

leads to the conclusion that

$$D_{C=C} = D_{C-C} + 2D_{C-H} - D_{H-H} + \Delta H°_{fC_2H_6(g)} - \Delta H°_{fC_2H_4(g)}$$

and calculate, using data in Table 6.3 and Appendix I, a value for $D_{C=C}$.

Solution

(a)

$$\underset{H}{\overset{H}{>}}C=C\underset{H}{\overset{H}{<}} + 2(H-H) \longrightarrow 2\left(H-\underset{\underset{H}{|}}{\overset{\overset{H}{|}}{C}}-H\right)$$

$$C(g) + 2(H-H) \longrightarrow H-\underset{\underset{H}{|}}{\overset{\overset{H}{|}}{C}}-H$$

Subtraction of the second equation from the first gives

$$\underset{H}{\overset{H}{>}}C=C\underset{H}{\overset{H}{<}} \longrightarrow H-\underset{\underset{H}{|}}{\overset{\overset{H}{|}}{C}}-H + C(g)$$

Reactants	Products	Change
1(C=C)		−1(C=C)
4(C—H)	4(C—H)	

$$D_{C=C} = \Delta H°_{fC} + \Delta H°_{fCH_4(g)} - \Delta H°_{fC_2H_4(g)}$$

From Appendix I.

$$D_{C=C} = 716.681 - 74.81 - (+52.26) = 589.6 \text{ kJ}$$

(b) For the reaction $C_2H_4(g) + H_2(g) \longrightarrow C_2H_6(g)$, the number and type of bonds are:

Reactants	Products	Change in number of bonds
4(C—H)	6(C—H)	2(C—H)
1(C=C)	1(C—C)	1(C—C)
1(H—H)		−1(H—H)
		−1(C=C)

$$D_{C=C} = D_{C-C} + 2D_{C-H} - D_{H-H} + \Delta H°_{fC_2H_6(g)} - \Delta H°_{fC_2H_4(g)}$$
$$= 345 + (415) - 436 - 84.68 - 52.26 = 602 \text{ kJ}$$

29. On the basis of hybridization of orbitals of the carbon atom, explain why cyclohexane exists in the form of a "puckered" (bent) ring and benzene in the form of a planar ring.

Solution

In cyclohexane the carbons are bonded through sp^3-hybrid bonds which angle at 109°. Consequently, from the side the ring is "puckered" with the first carbon fixed, the next down and the next up. In benzene, sp^2-hybrid orbitals are used. Since the sp^2 orbital set is coplanar, that is flat, the benzene ring is flat.

31

Biochemistry

INTRODUCTION

Biochemistry is the study of the chemical composition and structure of living organisms and the reactions that take place within them. Biochemistry is an outgrowth of the study of organic chemistry as applied to living systems and is related to biology. Progress in such interdisciplinary fields is rapid today. Great strides have been made, for instance, in recombinant DNA technology. In another important area, much biochemical work has been done but much more is needed to find a chemical solution to the new terror facing the world, acquired immune deficiency syndrome, AIDS.

READY REFERENCE

Codon (31.12)   A sequence of three adjacent nucleotides in a DNA sequence (codon) is the basis for coding the sequence of amino acids in proteins. Based on the sequence of nucleotides on the strand of DNA and reading from the 5′ to 3′ end, an m-RNA segment forms in the reverse order but replaces the thymine in DNA with uracil; the base pairs used during transcription are A-U and G-C. The three letter codon then gives the amino acid as determined from Table 31.1 in the main text.

Replication (31.12)   This is a term applied to the copying or reproduction of the DNA molecule. The information coded within the DNA double helix must be capable of being transcribed or rewritten to messenger RNA. This code is translated from one "language" to another during the assembly of the protein, from the language of nucleic acids to the amino acid sequence of the protein.

SOLUTIONS TO CHAPTER 31 EXERCISES

1. List the four principal chemical constituents of living cells. Identify the specific functions of each of these constituents.

    Solution

    The four constituents of eukaryotic cells are *cytosol* (water-soluble contents of the cells), *ribosomes* (site of genetic coding), *mitochondria* (site of energy production and storage), and *chloroplasts* (converters of light energy to chemical energy).

Chapter 31 Biochemistry

PROTEINS

3. Define the following: amino acid, peptide, and protein.

Solution

An *amino acid* contains both an amine group, —NH$_2$, and a carboxylic acid group, —CO$_2$H.

A *peptide* is a polymer formed by linking two or more amino acids through covalent amide linkages.

A *protein* is a complex polymer containing 40–10,000 or more amino acid units.

5. Define what is meant by *denaturation*.

Solution

Denaturation is the disruption of the natural conformation of the protein without chemically altering the primary structure.

7. Describe what happens to the rate of an enzyme-catalyzed reaction from the instant that the enzyme and substrate are mixed until the substance is consumed.

Solution

When the enzyme and substrate are mixed, the substrate binds to the enzyme, and the reaction rate increases, in some cases by many orders of magnitude. The maximum rate is reached quickly and then, depending on a number of factors, may decrease by first-order kinetics or some other scheme until the substrate is consumed or the enzyme loses its ability to catalyze the reaction through inhibition.

9. Write the structure of the tripeptide CYS-LYS-GLU at neutral pH.

Solution

Peptides are written and named starting with the N-terminal amino acid on the left. The three-letter code refers to the specific amino acids: CYS cysteine, LYS lysine, GLU glutamic acid. Using peptide bonds to link these together, we have:

$$H_3^+N-\underset{\underset{\underset{SH}{|}}{\underset{CH_2}{|}}}{\overset{\overset{H}{|}}{C}}-\overset{\overset{O}{\|}}{C}-\underset{H}{\overset{\text{peptide bond}}{N}}-\underset{\underset{\underset{\underset{NH_2}{|}}{\underset{CH_2}{|}}}{\underset{CH_2}{|}}}{\overset{\overset{H}{|}}{C}}-\overset{\overset{O}{\|}}{C}-\underset{H}{\overset{\text{peptide bond}}{N}}-\underset{\underset{\underset{CO_2H}{|}}{\underset{CH_2}{|}}}{\overset{\overset{H}{|}}{C}}-CO_2^-$$

CYS — cysteine ; LYS — lysine ; GLU — glutamic acid

11. Describe the process of protein synthesis by the body and indicate its significance.

    Solution

    The double helix of DNA serves to control the sequence of incorporation of amino acids into the proteins that are produced in the body. The DNA double helix replicates the information by transcribing the sequence to another nucleic acid called messenger RNA. From m-RNA the information is translated from the DNA code of nucleic acids to the amino acid sequence. The significance of this process is that an error in the sequence can lead to genetic diseases.

CARBOHYDRATES

13. What is the literal meaning of carbohydrate, and how did this name arise?

    Solution

    The name *carbohydrate* originally meant a compound with the empirical formula CH_2O. In other words, it is a hydrate of carbon.

15. How does a pyranose differ from a furanose?

    Solution

    A pyranose is a six-membered ring formed from an aldose (a six-carbon chain with an aldehyde function on the end carbon). A furanose is a five-membered ring formed from a ketose (a six-carbon chain with a ketone function on the second carbon atom).

17. What is the difference between the α and β isomers of glucopyranose?

    Solution

    Glucopyranose is formed from a six-membered carbon chain terminating with an aldehyde function on the number 1 carbon atom. When the glucopyranose ring is formed the joining occurs at carbon atom number 1, where the aldehyde function forms a hydroxy group on the ring. In α-glucopyranose the hydroxy function is placed down relative to the ring and in β-glucopyranose, the —OH group is pointed up.

LIPIDS

19. Write the general formula for a neutral fat. How do fats and oils differ in composition?

    Solution

    The formula for a typical neutral fat is that of tristearylglycerol.

    $$\begin{array}{l} \mathrm{CH_2OC(CH_2)_{16}CH_3} \\ \quad \parallel \\ \quad \mathrm{O} \\ \mathrm{HCOC(CH_2)_{16}CH_3} \\ \quad \parallel \\ \quad \mathrm{O} \\ \mathrm{CH_2OC(CH_2)_{16}CH_3} \\ \quad \parallel \\ \quad \mathrm{O} \end{array}$$

Fats generally have saturated long chains of carbons in the acids from which they are formed. Oils are generally formed from acids with unsaturated bonds. The major difference is that fats are solids and oils are liquids.

21. Define saponification and write an equation for the saponification of tripalmitylglycerol; the triester of palmitic acid, $CH_3(CH_2)_{14}CO_2H$.

Solution

Saponification is the reaction of the triester in the presence of aqueous base to form glycerol and a salt of the fatty acid.

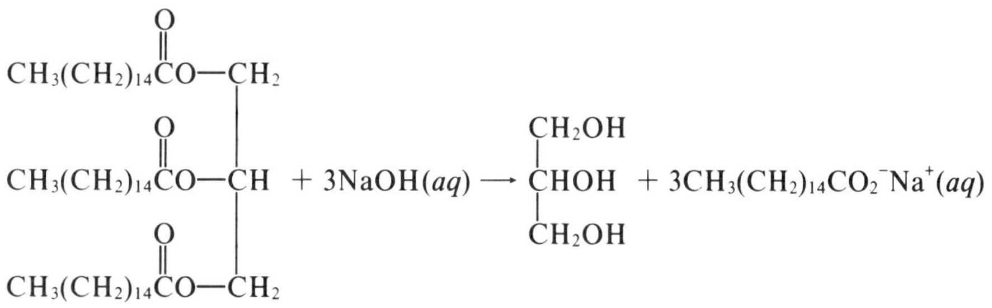

$$\begin{array}{c} CH_3(CH_2)_{14}\overset{O}{\overset{\|}{C}}O-CH_2 \\ CH_3(CH_2)_{14}\overset{O}{\overset{\|}{C}}O-CH \\ CH_3(CH_2)_{14}\overset{O}{\overset{\|}{C}}O-CH_2 \end{array} + 3NaOH(aq) \longrightarrow \begin{array}{c} CH_2OH \\ CHOH \\ CH_2OH \end{array} + 3CH_3(CH_2)_{14}CO_2^-Na^+(aq)$$

23. The molecular weight of an unknown triglyceride can be estimated by determining the mass of KOH required to saponify a known mass of lipid. What is the molecular weight of triglyceride if 3.12 mg of the triglyceride requires 0.590 mg of KOH for saponification?

Solution

lipid + 3KOH(aq) ⟶ glycerol + 3 salt

$$\text{mol. wt (lipid)} = 0.00312 \text{ g lipid} \times \frac{3 \text{ mol KOH}}{1 \text{ mol lipid}} \times \frac{56.1056 \text{ g KOH}}{\text{mol KOH}} \times \frac{1}{0.000590 \text{ g KOH}}$$

$$= 890 \text{ g/mol (tripalmitylglycerol)}$$

NUCLEIC ACIDS

25. Write the complete structure of the following segment of DNA, reading from the 5' end to the 3' end: G-T-C-A.

Solution

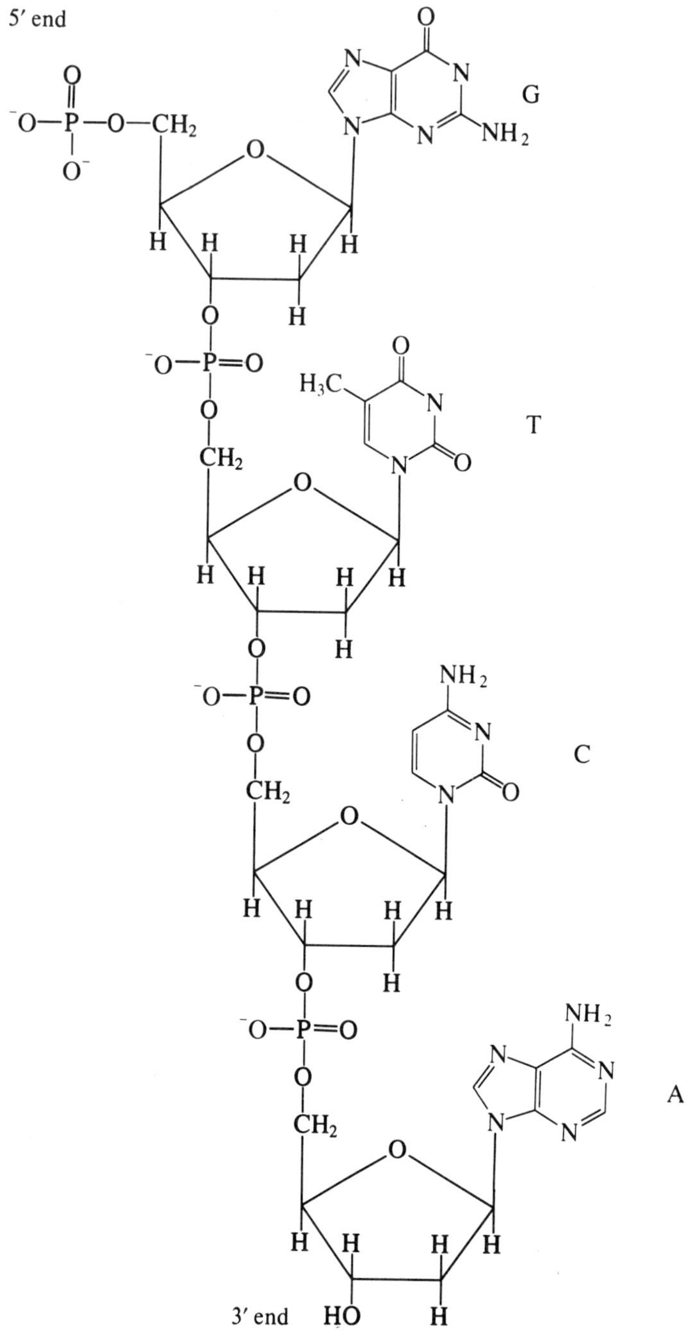

[Structure of a DNA tetranucleotide showing 5' end with phosphate-CH₂-sugar-G (guanine), linked via phosphodiester bonds through successive nucleotides T (thymine), C (cytosine), and A (adenine) at the 3' end with HO group]

5' end

G

T

C

A

3' end

27. What would be (a) the product of replication and (b) the product of transcription of the following fragment of a DNA molecule?

    5' G-T-C-A-A-T-G-G-A 3'

Solution

(a) 3' C-A-G-T-T-A-C-C-T 5'
(b) 5' U-C-C-A-U-U-G-A-C 3'

29. Complete the following table.

    Solution

	DNA	RNA
Sugar unit	2-deoxyribose	ribose
Purine bases	A G	A G
Pyrimidine bases	C T	C U

31. Explain how sickle-cell anemia occurs.

    Solution

    In sickle-cell anemia, the red blood cells contain an abnormal hemoglobin in which a negatively charged glutamic acid residue at the sixth position of the beta chain of normal adult hemoglobin is replaced by an uncharged valine residue. Because of this change, the hemoglobin is less soluble in water, causing the cells to "sickle" or clump, thus blocking the flow of blood.

33. How does the body build a defense against chronic infection?

    Solution

    Two processes defend against chronic infections: a process in which white blood cells overwhelm and destroy foreign matter; and an antibody immune response in which lymphocytes manufacture antibodies, which are large protein molecules that combine with the infecting substance called an antigen.

ADDITIONAL EXERCISES

35. How would you explain the observation that proteins can contain additional amino acids that are structurally related to the 20 genetically coded amino acids?

    Solution

    An amino acid in the protein sequence can be modified after it is incorporated into the protein, thus accounting for a nongenetically coded amino acid.

37. What evidence do we have that a twenty-first genetically coded amino acid will not be found?

    Solution

    All the triplet codons have been established as to which amino acid each one is coded for. There are no missing positions in the table nor is there any evidence that there may be other possibilities.

39. The synthesis of organic molecules in nature from carbon dioxide and water is an endothermic process. What is the source of the required energy?

    Solution

    Photosynthesis provides the energy for the conversion of CO_2 and water into oxygen and a carbohydrate. This occurs through a process in which the chloroplasts of green plants and algae convert light into chemical energy.

41. What is recombinant DNA technology, and what are some of its consequences for society? What ethical and moral concerns arise? Cite examples both of benefits and problems.

Solution

Recombinant DNA technology is based on the ideal that the DNA chain can be cleaved and a defective coding corrected, or perhaps even improved, for some supposed improved final result. Thus, not only could genetic diseases be cured but plants could be given resistance to disease and even frost. The implications for society are immense. However, no one knows the long-term effects of such changes on society or nature. Along with vast good looms potential disaster. Great caution must be exercised.

43. The degradation of both fatty acids, such as palmitic acid, and carbohydrates, such as glucose, leads to the formation of carbon dioxide and water. However, the degradation of palmitic acid produces $2\frac{1}{2}$ times as much energy as that of glucose, per gram of substance consumed. Explain this in terms of the average oxidation numbers of the carbon atoms in fatty acids, carbohydrates, and carbon dioxide.

Solution

In the carbohydrates the common structural feature is the unit

$$\begin{array}{c} C \\ | \\ H-C-O-H \\ | \\ C \end{array}$$

whereas in the long-chain hydrocarbons, the common feature is

$$\begin{array}{c} C \\ | \\ H-C-H \\ | \\ C \end{array}$$

Upon combustion any difference in energy must be related to the differences of the C—O—H group and the C—H group in their reactions with oxygen. From bond dissociation energies we have

BOND BREAKING
(ALL NUMERICAL VALUES IN kJ/MOL)

C—O—H + O=O	C—H + O=O
1 O—H + $\frac{1}{2}$O=O	1 C—H + O=O
464 + $\frac{1}{2}$(140)	415 + 140

BOND FORMING AS CO$_2$ AND H$_2$O

| $\frac{1}{2}$C=O + 1H—O | 1C=O + 1H—O |
| $\frac{1}{2}$(741) + 464 | 741 + 464 |

BOND FORMING − BOND BREAKING
370 + 464 − 464 − 70 = 300   741 + 464 − 415 − 140 = 650

The difference is 650 − 300 = 350 kV/mol. This is more than twice the heat released by the long-chain hydrocarbons compared to the carbohydrates. Taking into account the difference in mass per mole raises the difference to about $2\frac{1}{2}$ times the heat released by the long-chain hydrocarbons. As the oxidation of carbon increases from fatty acid (with no oxygens along the chain) to carbohydrates (with 1 oxygen along the chain or ring), to carbon dioxide (with the greatest oxidation — two oxygens), the energy that can be released is reduced.

Appendixes

Appendix C: General Physical Constants (1986 Values)

Avogadro's number	6.0221367×10^{23} mol^{-1}
Electron charge, e	$1.60217733 \times 10^{-19}$ coulomb (C)
Electron rest mass, m_e	9.109390×10^{-31} kg
Proton rest mass, m_p	$1.6726231 \times 10^{-27}$ kg
Neutron rest mass, m_n	$1.6749286 \times 10^{-27}$ kg
Charge-to-mass ratio for electron, e/m_e	$1.75881962 \times 10^{11}$ coulomb kg^{-1}
Faraday constant, F	9.6485309×10^4 coulomb/equivalent
Planck constant, h	$6.6260755 \times 10^{-34}$ J s
Boltzmann constant, k	1.380658×10^{-23} J K^{-1}
Gas constant, R	8.205784×10^{-2} L atm mol^{-1} K^{-1} $= 8.314510$ J mol^{-1} K^{-1}
Speed of light (in vacuum), c	2.99792458×10^8 m s^{-1}
Atomic mass unit ($= \frac{1}{12}$ the mass of an atom of the ^{12}C nuclide), amu	$1.6605402 \times 10^{-27}$ kg
Rydberg constant, R_∞	1.0973731534×10^7 m^{-1}

Appendix D: Solubility Products

Substance	K_{sp} at 25°C	Substance	K_{sp} at 25°C
Aluminum		Lead	
$Al(OH)_3$	1.9×10^{-33}	$Pb(OH)_2$	2.8×10^{-16}
Barium		PbF_2	3.7×10^{-8}
$BaCO_3$	8.1×10^{-9}	$PbCl_2$	1.7×10^{-5}
$BaC_2O_4 \cdot 2H_2O$	1.1×10^{-7}	$PbBr_2$	6.3×10^{-6}
$BaSO_4$	1.08×10^{-10}	PbI_2	8.7×10^{-9}
$BaCrO_4$	2×10^{-10}	$PbCO_3$	1.5×10^{-13}
BaF_2	1.7×10^{-6}	PbS	6.5×10^{-34}
$Ba(OH)_2 \cdot 8H_2O$	5.0×10^{-3}	$PbCrO_4$	1.8×10^{-14}
$Ba_3(PO_4)_2$	1.3×10^{-29}	$PbSO_4$	1.8×10^{-8}
$Ba_3(AsO_4)_2$	1.1×10^{-13}	$Pb_3(PO_4)_2$	3×10^{-44}
Bismuth		Magnesium	
$BiO(OH)$	1×10^{-12}	$Mg(OH)_2$	1.5×10^{-11}
$BiOCl$	7×10^{-9}	$MgCO_3 \cdot 3H_2O$	$ca\ 1 \times 10^{-5}$
Bi_2S_3	7.3×10^{-91}	$MgNH_4PO_4$	2.5×10^{-13}
Cadmium		MgF_2	6.4×10^{-9}
$Cd(OH)_2$	1.2×10^{-14}	MgC_2O_4	8.6×10^{-5}
CdS	2.8×10^{-35}	Manganese	
$CdCO_3$	2.5×10^{-14}	$Mn(OH)_2$	4.5×10^{-14}
Calcium		$MnCO_3$	8.8×10^{-11}
$Ca(OH)_2$	7.9×10^{-6}	MnS	4.3×10^{-22}
$CaCO_3$	4.8×10^{-9}	Mercury	
$CaSO_4 \cdot 2H_2O$	2.4×10^{-5}	$Hg_2O \cdot H_2O$	1.6×10^{-23}
$CaC_2O_4 \cdot H_2O$	2.27×10^{-9}	Hg_2Cl_2	1.1×10^{-18}
$Ca_3(PO_4)_2$	1×10^{-25}	Hg_2Br_2	1.26×10^{-22}
$CaHPO_4$	5×10^{-6}	Hg_2I_2	4.5×10^{-29}
CaF_2	3.9×10^{-11}	Hg_2CO_3	9×10^{-17}
Chromium		Hg_2SO_4	6.2×10^{-7}
$Cr(OH)_3$	6.7×10^{-31}	Hg_2S	8×10^{-52}
Cobalt		Hg_2CrO_4	2×10^{-9}
$Co(OH)_2$	2×10^{-16}	HgS	2×10^{-59}
$CoS(\alpha)$	4.5×10^{-27}	Nickel	
$CoS(\beta)$	6.7×10^{-29}	$Ni(OH)_2$	1.6×10^{-14}
$CoCO_3$	1.0×10^{-12}	$NiCO_3$	1.36×10^{-7}
$Co(OH)_3$	2.5×10^{-43}	$NiS(\alpha)$	2×10^{-27}
Copper		$NiS(\beta)$	8×10^{-33}
$CuCl$	1.85×10^{-7}	Potassium	
$CuBr$	5.3×10^{-9}	$KClO_4$	1.07×10^{-2}
CuI	5.1×10^{-12}	K_2PtCl_6	1.1×10^{-5}
$CuSCN$	4×10^{-14}	$KHC_4H_4O_6$	3×10^{-4}
Cu_2S	1.2×10^{-54}	Silver	
$Cu(OH)_2$	5.6×10^{-20}	$\tfrac{1}{2}Ag_2O\ (Ag^+ + OH^-)$	2×10^{-8}
CuS	6.7×10^{-42}	$AgCl$	1.8×10^{-10}
$CuCO_3$	1.37×10^{-10}	$AgBr$	3.3×10^{-13}
Iron		AgI	1.5×10^{-16}
$Fe(OH)_2$	7.9×10^{-15}	$AgCN$	1.2×10^{-16}
$FeCO_3$	2.11×10^{-11}	$AgSCN$	1.0×10^{-12}
FeS	8×10^{-26}	Ag_2S	8×10^{-58}
$Fe(OH)_3$	1.1×10^{-36}	Ag_2CO_3	8.2×10^{-12}

Solubility Products (continued)

Substance	K_{sp} at 25°C	Substance	K_{sp} at 25°C
Ag_2CrO_4	9×10^{-12}	TlSCN	5.8×10^{-4}
$Ag_4Fe(CN)_6$	1.55×10^{-41}	Tl_2S	9.2×10^{-31}
Ag_2SO_4	1.18×10^{-5}	$Tl(OH)_3$	1.5×10^{-44}
Ag_3PO_4	1.8×10^{-18}	Tin	
Strontium		$Sn(OH)_2$	5×10^{-26}
$Sr(OH)_2 \cdot 8H_2O$	3.2×10^{-4}	SnS	6×10^{-35}
$SrCO_3$	9.42×10^{-10}	$Sn(OH)_4$	1×10^{-56}
$SrCrO_4$	3.6×10^{-5}	Zinc	
$SrSO_4$	2.8×10^{-7}	$ZnCO_3$	6×10^{-11}
$SrC_2O_4 \cdot H_2O$	5.61×10^{-8}	$Zn(OH)_2$	4.5×10^{-17}
Thallium		ZnS	1×10^{-27}
TlCl	1.9×10^{-4}		

Appendix E:
Formation Constants for Complex Ions

Equilibrium	K_f
$Al^{3+} + 6F^- \rightleftharpoons [AlF_6]^{3-}$	5×10^{23}
$Cd^{2+} + 4NH_3 \rightleftharpoons [Cd(NH_3)_4]^{2+}$	4.0×10^6
$Cd^{2+} + 4CN^- \rightleftharpoons [Cd(CN)_4]^{2-}$	1.3×10^{17}
$Co^{2+} + 6NH_3 \rightleftharpoons [Co(NH_3)_6]^{2+}$	8.3×10^4
$Co^{3+} + 6NH_3 \rightleftharpoons [Co(NH_3)_6]^{3+}$	4.5×10^{33}
$Cu^+ + 2CN^- \rightleftharpoons [Cu(CN)_2]^-$	1×10^{16}
$Cu^{2+} + 4NH_3 \rightleftharpoons [Cu(NH_3)_4]^{2+}$	1.2×10^{12}
$Fe^{2+} + 6CN^- \rightleftharpoons [Fe(CN)_6]^{4-}$	1×10^{37}
$Fe^{3+} + 6CN^- \rightleftharpoons [Fe(CN)_6]^{3-}$	1×10^{44}
$Fe^{3+} + 6SCN^- \rightleftharpoons [Fe(NCS)_6]^{3-}$	3.2×10^3
$Hg^{2+} + 4Cl^- \rightleftharpoons [HgCl_4]^{2-}$	1.2×10^{15}
$Ni^{2+} + 6NH_3 \rightleftharpoons [Ni(NH_3)_6]^{2+}$	1.8×10^8
$Ag^+ + 2Cl^- \rightleftharpoons [AgCl_2]^-$	2.5×10^5
$Ag^+ + 2CN^- \rightleftharpoons [Ag(CN)_2]^-$	1×10^{20}
$Ag^+ + 2NH_3 \rightleftharpoons [Ag(NH_3)_2]^+$	1.6×10^7
$Zn^{2+} + 4CN^- \rightleftharpoons [Zn(CN)_4]^{2-}$	1×10^{19}
$Zn^{2+} + 4OH^- \rightleftharpoons [Zn(OH)_4]^{2-}$	2.9×10^{15}

Appendix F: Ionization Constants of Weak Acids

Acid	Formula	K_a at 25°C
Acetic	CH_3CO_2H	1.8×10^{-5}
Arsenic	H_3AsO_4	4.8×10^{-3}
	$H_2AsO_4^-$	1×10^{-7}
	$HAsO_4^{2-}$	1×10^{-13}
Arsenous	H_3AsO_3	5.8×10^{-10}
Boric	H_3BO_3	5.8×10^{-10}
Carbonic	H_2CO_3	4.3×10^{-7}
	HCO_3^-	7×10^{-11}
Cyanic	HCNO	3.46×10^{-4}
Formic	HCO_2H	1.8×10^{-4}
Hydrazoic	HN_3	1×10^{-4}
Hydrocyanic	HCN	4×10^{-10}
Hydrofluoric	HF	7.2×10^{-4}
Hydrogen peroxide	H_2O_2	2.4×10^{-12}
Hydrogen selenide	H_2Se	1.7×10^{-4}
	HSe^-	1×10^{-10}
Hydrogen sulfate ion	HSO_4^-	1.2×10^{-2}
Hydrogen sulfide	H_2S	1.0×10^{-7}
	HS^-	1.0×10^{-19}
Hydrogen telluride	H_2Te	2.3×10^{-3}
	HTe^-	1×10^{-5}
Hypobromous	HBrO	2×10^{-9}
Hypochlorous	HClO	3.5×10^{-8}
Nitrous	HNO_2	4.5×10^{-4}
Oxalic	$H_2C_2O_4$	5.9×10^{-2}
	$HC_2O_4^-$	6.4×10^{-5}
Phosphoric	H_3PO_4	7.5×10^{-3}
	$H_2PO_4^-$	6.3×10^{-8}
	HPO_4^{2-}	3.6×10^{-13}
Phosphorous	H_3PO_3	1.6×10^{-2}
	$H_2PO_3^-$	7×10^{-7}
Sulfurous	H_2SO_3	1.2×10^{-2}
	HSO_3^-	6.2×10^{-8}

Appendix G: Ionization Constants of Weak Bases

Base	Ionization equation	K_b at 25°C
Ammonia	$NH_3 + H_2O \rightleftharpoons NH_4^+ + OH^-$	1.8×10^{-5}
Dimethylamine	$(CH_3)_2NH + H_2O \rightleftharpoons (CH_3)_2NH_2^+ + OH^-$	7.4×10^{-4}
Methylamine	$CH_3NH_2 + H_2O \rightleftharpoons CH_3NH_3^+ + OH^-$	4.4×10^{-4}
Phenylamine (aniline)	$C_6H_5NH_2 + H_2O \rightleftharpoons C_6H_5NH_3^+ + OH^-$	4.6×10^{-10}
Trimethylamine	$(CH_3)_3N + H_2O \rightleftharpoons (CH_3)_3NH^+ + OH^-$	7.4×10^{-5}

Appendix H: Standard Electrode (Reduction) Potentials

Half-reaction	$E°$, V	Half-reaction	$E°$, V
$Li^+ + e^- \longrightarrow Li$	-3.09	$Zn^{2+} + 2e^- \longrightarrow Zn$	-0.763
$K^+ + e^- \longrightarrow K$	-2.925	$Cr^{3+} + 3e^- \longrightarrow Cr$	-0.74
$Rb^+ + e^- \longrightarrow Rb$	-2.925	$HgS + 2e^- \longrightarrow Hg + S^{2-}$	-0.72
$Ra^{2+} + 2e^- \longrightarrow Ra$	-2.92	$[Cd(NH_3)_4]^{2+} + 2e^- \longrightarrow Cd + 4NH_3$	-0.597
$Ba^{2+} + 2e^- \longrightarrow Ba$	-2.90	$Ga^{3+} + 3e^- \longrightarrow Ga$	-0.53
$Sr^{2+} + 2e^- \longrightarrow Sr$	-2.89	$S + 2e^- \longrightarrow S^{2-}$	-0.48
$Ca^{2+} + 2e^- \longrightarrow Ca$	-2.87	$[Ni(NH_3)_6]^{2+} + 2e^- \longrightarrow Ni + 6NH_3$	-0.47
$Na^+ + e^- \longrightarrow Na$	-2.714	$Fe^{2+} + 2e^- \longrightarrow Fe$	-0.440
$La^{3+} + 3e^- \longrightarrow La$	-2.52	$[Cu(CN)_2]^- + e^- \longrightarrow Cu + 2CN^-$	-0.43
$Ce^{3+} + 3e^- \longrightarrow Ce$	-2.48	$Cr^{3+} + e^- \longrightarrow Cr^{2+}$	-0.41
$Nd^{3+} + 3e^- \longrightarrow Nd$	-2.44	$Cd^{2+} + 2e^- \longrightarrow Cd$	-0.403
$Sm^{3+} + 3e^- \longrightarrow Sm$	-2.41	$Se + 2H^+ + 2e^- \longrightarrow H_2Se$	-0.40
$Gd^{3+} + 3e^- \longrightarrow Gd$	-2.40	$[Hg(CN)_4]^{2-} + 2e^- \longrightarrow Hg + 4CN^-$	-0.37
$Mg^{2+} + 2e^- \longrightarrow Mg$	-2.37	$ClO_4^- + H_2O + 2e^- \longrightarrow ClO_3^- + 2OH^-$	-0.36
$Y^{3+} + 3e^- \longrightarrow Y$	-2.37	$PbSO_4 + 2e^- \longrightarrow Pb + SO_4^{2-}$	-0.356
$Am^{3+} + 3e^- \longrightarrow Am$	-2.32	$In^{3+} + 3e^- \longrightarrow In$	-0.342
$Lu^{3+} + 3e^- \longrightarrow Lu$	-2.25	$[Ag(CN)_2]^- + e^- \longrightarrow Ag + 2CN^-$	-0.31
$\frac{1}{2}H_2 + e^- \longrightarrow H^-$	-2.25	$Co^{2+} + 2e^- \longrightarrow Co$	-0.277
$Sc^{3+} + 3e^- \longrightarrow Sc$	-2.08	$[SnF_6]^{2-} + 4e^- \longrightarrow Sn + 6F^-$	-0.25
$[AlF_6]^{3-} + 3e^- \longrightarrow Al + 6F^-$	-2.07	$Ni^{2+} + 2e^- \longrightarrow Ni$	-0.250
$Pu^{3+} + 3e^- \longrightarrow Pu$	-2.07	$Sn^{2+} + 2e^- \longrightarrow Sn$	-0.136
$Th^{4+} + 4e^- \longrightarrow Th$	-1.90	$CrO_4^{2-} + 4H_2O + 3e^- \longrightarrow Cr(OH)_3 + 5OH^-$	-0.13
$Np^{3+} + 3e^- \longrightarrow Np$	-1.86	$Pb^{2+} + 2e^- \longrightarrow Pb$	-0.126
$Be^{2+} + 2e^- \longrightarrow Be$	-1.85	$MnO_2 + 2H_2O + 2e^- \longrightarrow Mn(OH)_2 + 2OH^-$	-0.05
$U^{3+} + 3e^- \longrightarrow U$	-1.80	$[HgI_4]^{2-} + 2e^- \longrightarrow Hg + 4I^-$	-0.04
$Hf^{4+} + 4e^- \longrightarrow Hf$	-1.70	$2H^+ + 2e^- \longrightarrow H_2$	0.00
$SiO_3^{2-} + 3H_2O + 4e^- \longrightarrow Si + 6OH^-$	-1.70	$NO_3^- + H_2O + 2e^- \longrightarrow NO_2^- + 2OH^-$	$+0.01$
$Al^{3+} + 3e^- \longrightarrow Al$	-1.66	$[Ag(S_2O_3)_2]^{3-} + e^- \longrightarrow Ag^+ + 2S_2O_3^{2-}$	$+0.01$
$Ti^{2+} + 2e^- \longrightarrow Ti$	-1.63	$[Co(NH_3)_6]^{3+} + e^- \longrightarrow [Co(NH_3)_6]^{2+}$	$+0.1$
$Zr^{4+} + 4e^- \longrightarrow Zr$	-1.53	$S + 2H^+ + 2e^- \longrightarrow H_2S$	$+0.141$
$ZnS + 2e^- \longrightarrow Zn + S^{2-}$	-1.44	$Sn^{4+} + 2e^- \longrightarrow Sn^{2+}$	$+0.15$
$Cr(OH)_3 + 3e^- \longrightarrow Cr + 3OH^-$	-1.3	$Cu^{2+} + e^- \longrightarrow Cu^+$	$+0.153$
$[Zn(CN)_4]^{2-} + 2e^- \longrightarrow Zn + 4CN^-$	-1.26	$Co(OH)_3 + e^- \longrightarrow Co(OH)_2 + OH^-$	$+0.17$
$Zn(OH)_2 + 2e^- \longrightarrow Zn + 2OH^-$	-1.245	$[HgBr_4]^{2-} + 2e^- \longrightarrow Hg + 4Br^-$	$+0.21$
$[Zn(OH)_4]^{2-} + 2e^- \longrightarrow Zn + 4OH^-$	-1.216	$AgCl + e^- \longrightarrow Ag + Cl^-$	$+0.222$
$CdS + 2e^- \longrightarrow Cd + S^{2-}$	-1.21	$Hg_2Cl_2 + 2e^- \longrightarrow 2Hg + 2Cl^-$	$+0.27$
$[Cr(OH)_4]^- + 3e^- \longrightarrow Cr + 4OH^-$	-1.2	$ClO_3^- + H_2O + 2e^- \longrightarrow ClO_2^- + 2OH^-$	$+0.33$
$[SiF_6]^{2-} + 4e^- \longrightarrow Si + 6F^-$	-1.2	$Cu^{2+} + 2e^- \longrightarrow Cu$	$+0.337$
$V^{2+} + 2e^- \longrightarrow V$	$ca -1.18$	$[Fe(CN)_6]^{3-} + e^- \longrightarrow [Fe(CN)_6]^{4-}$	$+0.36$
$Mn^{2+} + 2e^- \longrightarrow Mn$	-1.18	$[Ag(NH_3)_2]^+ + e^- \longrightarrow Ag + 2NH_3$	$+0.373$
$[Cd(CN)_4]^{2-} + 2e^- \longrightarrow Cd + 4CN^-$	-1.03	$O_2 + 2H_2O + 4e^- \longrightarrow 4OH^-$	$+0.401$
$[Zn(NH_3)_4]^{2+} + 2e^- \longrightarrow Zn + 4NH_3$	-1.03	$[RhCl_6]^{3-} + 3e^- \longrightarrow Rh + 6Cl^-$	$+0.44$
$FeS + 2e^- \longrightarrow Fe + S^{2-}$	-1.01	$Ag_2CrO_4 + 2e^- \longrightarrow 2Ag + CrO_4^{2-}$	$+0.446$
$PbS + 2e^- \longrightarrow Pb + S^{2-}$	-0.95	$NiO_2 + 2H_2O + 2e^- \longrightarrow Ni(OH)_2 + 2OH^-$	$+0.49$
$SnS + 2e^- \longrightarrow Sn + S^{2-}$	-0.94	$Cu^+ + e^- \longrightarrow Cu$	$+0.521$
$Cr^{2+} + 2e^- \longrightarrow Cr$	-0.91	$TeO_2 + 4H^+ + 4e^- \longrightarrow Te + 2H_2O$	$+0.529$
$Fe(OH)_2 + 2e^- \longrightarrow Fe + 2OH^-$	-0.877	$I_2 + 2e^- \longrightarrow 2I^-$	$+0.5355$
$SiO_2 + 4H^+ + 4e^- \longrightarrow Si + 2H_2O$	-0.86	$[PtBr_4]^{2-} + 2e^- \longrightarrow Pt + 4Br^-$	$+0.58$
$NiS + 2e^- \longrightarrow Ni + S^{2-}$	-0.83	$MnO_4^- + 2H_2O + 3e^- \longrightarrow MnO_2 + 4OH^-$	$+0.588$
$2H_2O + 2e^- \longrightarrow H_2 + 2OH^-$	-0.828	$[PdCl_4]^{2-} + 2e^- \longrightarrow Pd + 4Cl^-$	$+0.62$

Standard Electrode (Reduction) Potentials (continued)

Half-reaction	$E°$, V	Half-reaction	$E°$, V
$ClO_2^- + H_2O + 2e^- \longrightarrow ClO^- + 2OH^-$	+0.66	$O_2 + 4H^+ + 4e^- \longrightarrow 2H_2O$	+1.23
$[PtCl_6]^{2-} + 2e^- \longrightarrow [PtCl_4]^{2-} + 2Cl^-$	+0.68	$MnO_2 + 4H^+ + 2e^- \longrightarrow Mn^{2+} + 2H_2O$	+1.23
$O_2 + 2H^+ + 2e^- \longrightarrow H_2O_2$	+0.682	$Cr_2O_7^{2-} + 14H^+ + 6e^- \longrightarrow 2Cr^{3+} + 7H_2O$	+1.33
$[PtCl_4]^{2-} + 2e^- \longrightarrow Pt + 4Cl^-$	+0.73	$Cl_2 + 2e^- \longrightarrow 2Cl^-$	+1.3595
$Fe^{3+} + e^- \longrightarrow Fe^{2+}$	+0.771	$HClO + H^+ + 2e^- \longrightarrow Cl^- + H_2O$	+1.49
$Hg_2^{2+} + 2e^- \longrightarrow 2Hg$	+0.789	$Au^{3+} + 3e^- \longrightarrow Au$	+1.50
$Ag^+ + e^- \longrightarrow Ag$	+0.7991	$MnO_4^- + 8H^+ + 5e^- \longrightarrow Mn^{2+} + 4H_2O$	+1.51
$Hg^{2+} + 2e^- \longrightarrow Hg$	+0.854	$Ce^{4+} + e^- \longrightarrow Ce^{3+}$	+1.61
$HO_2^- + H_2O + 2e^- \longrightarrow 3OH^-$	+0.88	$HClO + H^+ + e^- \longrightarrow \frac{1}{2}Cl_2 + H_2O$	+1.63
$ClO^- + H_2O + 2e^- \longrightarrow Cl^- + 2OH^-$	+0.89	$HClO_2 + 2H^+ + 2e^- \longrightarrow HClO + H_2O$	+1.64
$2Hg^{2+} + 2e^- \longrightarrow Hg_2^{2+}$	+0.920	$Au^+ + e^- \longrightarrow Au$	ca +1.68
$NO_3^- + 3H^+ + 2e^- \longrightarrow HNO_2 + H_2O$	+0.94	$NiO_2 + 4H^+ + 2e^- \longrightarrow Ni^{2+} + 2H_2O$	+1.68
$NO_3^- + 4H^+ + 3e^- \longrightarrow NO + H_2O$	+0.96	$PbO_2 + SO_4^{2-} + 4H^+ + 2e^- \longrightarrow PbSO_4 + 2H_2O$	+1.685
$Pd^{2+} + 2e^- \longrightarrow Pd$	+0.987		
$Br_2(l) + 2e^- \longrightarrow 2Br^-$	+1.0652	$H_2O_2 + 2H^+ + 2e^- \longrightarrow 2H_2O$	+1.77
$ClO_4^- + 2H^+ + 2e^- \longrightarrow ClO_3^- + H_2O$	+1.19	$Co^{3+} + e^- \longrightarrow Co^{2+}$	+1.82
$Pt^{2+} + 2e^- \longrightarrow Pt$	ca +1.2	$F_2 + 2e^- \longrightarrow 2F^-$	+2.87
$ClO_3^- + 3H^+ + 2e^- \longrightarrow HClO_2 + H_2O$	+1.21		

Appendix I: Standard Molar Enthalpies of Formation, Standard Molar Free Energies of Formation, and Absolute Standard Entropies [298.15 K (25°C), 1 atm]

Substance	$\Delta H_f°$, kJ mol^{-1}	$\Delta G_f°$, kJ mol^{-1}	$S_{298}°$, J K^{-1} mol^{-1}
Aluminum			
Al(s)	0	0	28.3
Al(g)	326	286	164.4
Al$_2$O$_3$(s)	−1676	−1582	50.92
AlF$_3$(s)	−1504	−1425	66.44
AlCl$_3$(s)	−704.2	−628.9	110.7
AlCl$_3 \cdot$ 6H$_2$O(s)	−2692	—	—
Al$_2$S$_3$(s)	−724	−492.4	—
Al$_2$(SO$_4$)$_3$(s)	−3440.8	−3100.1	239
Antimony			
Sb(s)	0	0	45.69
Sb(g)	262	222	180.2
Sb$_4$O$_6$(s)	−1441	−1268	221
SbCl$_3$(g)	−314	−301	337.7
SbCl$_5$(g)	−394.3	−334.3	401.8
Sb$_2$S$_3$(s)	−175	−174	182
SbCl$_3$(s)	−382.2	−323.7	184
SbOCl(s)	−374	—	—

Standard Molar Enthalpies of Formation, Standard Molar Free Energies of Formation, and Absolute Standard Entropies [298.15 K (25°C), 1 atm] (continued)

Substance	ΔH_f°, kJ mol^{-1}	ΔG_f°, kJ mol^{-1}	S_{298}°, J K^{-1} mol^{-1}
Arsenic			
As(s)	0	0	35
As(g)	303	261	174.1
As$_4$(g)	144	92.5	314
As$_4$O$_6$(s)	−1313.9	−1152.5	214
As$_2$O$_5$(s)	−924.87	−782.4	105
AsCl$_3$(g)	−258.6	−245.9	327.1
As$_2$S$_3$(s)	−169	−169	164
AsH$_3$(g)	66.44	68.91	222.7
H$_3$AsO$_4$(s)	−906.3	—	—
Barium			
Ba(s)	0	0	66.9
Ba(g)	175.6	144.8	170.3
BaO(s)	−558.1	−528.4	70.3
BaCl$_2$(s)	−860.06	−810.9	126
BaSO$_4$(s)	−1465	−1353	132
Beryllium			
Be(s)	0	0	9.54
Be(g)	320.6	282.8	136.17
BeO(s)	−610.9	−581.6	14.1
Bismuth			
Bi(s)	0	0	56.74
Bi(g)	207	168	186.90
Bi$_2$O$_3$(s)	−573.88	−493.7	151
BiCl$_3$(s)	−379	−315	177
Bi$_2$S$_3$(s)	−143	−141	200
Boron			
B(s)	0	0	5.86
B(g)	562.7	518.8	153.3
B$_2$O$_3$(s)	−1272.8	−1193.7	53.97
B$_2$H$_6$(g)	36	86.6	232.0
B(OH)$_3$(s)	−1094.3	−969.01	88.83
BF$_3$(g)	−1137.3	−1120.3	254.0
BCl$_3$(g)	−403.8	−388.7	290.0
B$_3$N$_3$H$_6$(l)	−541.0	−392.8	200
HBO$_2$(s)	−794.25	−723.4	40
Bromine			
Br$_2$(l)	0	0	152.23
Br$_2$(g)	30.91	3.142	245.35
Br(g)	111.88	82.429	174.91
BrF$_3$(g)	−255.6	−229.5	292.4
HBr(g)	−36.4	−53.43	198.59
Cadmium			
Cd(s)	0	0	51.76
Cd(g)	112.0	77.45	167.64
CdO(s)	−258	−228	54.8
CdCl$_2$(s)	−391.5	−344.0	115.3

Standard Molar Enthalpies of Formation, Standard Molar Free Energies of Formation, and Absolute Standard Entropies [298.15 K (25°C), 1 atm] (continued)

Substance	ΔH_f°, kJ mol^{-1}	ΔG_f°, kJ mol^{-1}	S_{298}°, J K^{-1} mol^{-1}
CdSO$_4$(s)	−933.28	−822.78	123.04
CdS(s)	−162	−156	64.9
Calcium			
Ca(s)	0	0	41.6
Ca(g)	192.6	158.9	154.78
CaO(s)	−635.5	−604.2	40
Ca(OH)$_2$(s)	−986.59	−896.76	76.1
CaSO$_4$(s)	−1432.7	−1320.3	107
CaSO$_4 \cdot$ 2H$_2$O(s)	−2021.1	−1795.7	194.0
CaCO$_3$(s) (calcite)	−1206.9	−1128.8	92.9
CaSO$_3 \cdot$ 2H$_2$O(s)	−1762	−1565	184
Carbon			
C(s) (graphite)	0	0	5.740
C(s) (diamond)	1.897	2.900	2.38
C(g)	716.681	671.289	157.987
CO(g)	−110.52	−137.15	197.56
CO$_2$(g)	−393.51	−394.36	213.6
CH$_4$(g)	−74.81	−50.75	186.15
CH$_3$OH(l)	−238.7	−166.4	127
CH$_3$OH(g)	−200.7	−162.0	239.7
CCl$_4$(l)	−135.4	−65.27	216.4
CCl$_4$(g)	−102.9	−60.63	309.7
CHCl$_3$(l)	−134.5	−73.72	202
CHCl$_3$(g)	−103.1	−70.37	295.6
CS$_2$(l)	89.70	65.27	151.3
CS$_2$(g)	117.4	67.15	237.7
C$_2$H$_2$(g)	226.7	209.2	200.8
C$_2$H$_4$(g)	52.26	68.12	219.5
C$_2$H$_6$(g)	−84.68	−32.9	229.5
CH$_3$COOH(l)	−484.5	−390	160
CH$_3$COOH(g)	−432.25	−374	282
C$_2$H$_5$OH(l)	−277.7	−174.9	161
C$_2$H$_5$OH(g)	−235.1	−168.6	282.6
C$_3$H$_8$(g)	−103.85	−23.49	269.9
C$_6$H$_6$(g)	82.927	129.66	269.2
C$_6$H$_6$(l)	49.028	124.50	172.8
CH$_2$Cl$_2$(l)	−121.5	−67.32	178
CH$_2$Cl$_2$(g)	−92.47	−65.90	270.1
CH$_3$Cl(g)	−80.83	−57.40	234.5
C$_2$H$_5$Cl(l)	−136.5	−59.41	190.8
C$_2$H$_5$Cl(g)	−112.2	−60.46	275.9
C$_2$N$_2$(g)	308.9	297.4	241.8
HCN(l)	108.9	124.9	112.8
HCN(g)	135	124.7	201.7
Chlorine			
Cl$_2$(g)	0	0	222.96
Cl(g)	121.68	105.70	165.09

Standard Molar Enthalpies of Formation, Standard Molar Free Energies of Formation, and Absolute Standard Entropies [298.15 K (25°C), 1 atm] (continued)

Substance	ΔH_f°, kJ mol^{-1}	ΔG_f°, kJ mol^{-1}	S_{298}°, J K^{-1} mol^{-1}
ClF(g)	−54.48	−55.94	217.8
ClF$_3$(g)	−163	−123	281.5
Cl$_2$O(g)	80.3	97.9	266.1
Cl$_2$O$_7$(l)	238	—	—
Cl$_2$O$_7$(g)	272	—	—
HCl(g)	−92.307	−95.299	186.80
HClO$_4$(l)	−40.6	—	—
Chromium			
Cr(s)	0	0	23.8
Cr(g)	397	352	174.4
Cr$_2$O$_3$(s)	−1140	−1058	81.2
CrO$_3$(s)	−589.5	—	—
(NH$_4$)$_2$Cr$_2$O$_7$(s)	−1807	—	—
Cobalt			
Co(s)	0	0	30.0
CoO(s)	−237.9	−214.2	52.97
Co$_3$O$_4$(s)	−891.2	−774.0	103
Co(NO$_3$)$_2$(s)	−420.5	—	—
Copper			
Cu(s)	0	0	33.15
Cu(g)	338.3	298.5	166.3
CuO(s)	−157	−130	42.63
Cu$_2$O(s)	−169	−146	93.14
CuS(s)	−53.1	−53.6	66.5
Cu$_2$S(s)	−79.5	−86.2	121
CuSO$_4$(s)	−771.36	−661.9	109
Cu(NO$_3$)$_2$(s)	−303	—	—
Fluorine			
F$_2$(g)	0	0	202.7
F(g)	78.99	61.92	158.64
F$_2$O(g)	−22	−4.6	247.3
HF(g)	−271	−273	173.67
Hydrogen			
H$_2$(g)	0	0	130.57
H(g)	217.97	203.26	114.60
H$_2$O(l)	−285.83	−237.18	69.91
H$_2$O(g)	−241.82	−228.59	188.71
H$_2$O$_2$(l)	−187.8	−120.4	110
H$_2$O$_2$(g)	−136.3	−105.6	233
HF(g)	−271	−273	173.67
HCl(g)	−92.307	−95.299	186.80
HBr(g)	−36.4	−53.43	198.59
HI(g)	26.5	1.7	206.48
H$_2$S(g)	−20.6	−33.6	205.7
H$_2$Se(g)	30	16	218.9
Iodine			
I$_2$(s)	0	0	116.14

Standard Molar Enthalpies of Formation, Standard Molar Free Energies of Formation, and Absolute Standard Entropies [298.15 K (25°C), 1 atm] (continued)

Substance	ΔH_f°, kJ mol^{-1}	ΔG_f°, kJ mol^{-1}	S_{298}°, J K^{-1} mol^{-1}
$I_2(g)$	62.438	19.36	260.6
$I(g)$	106.84	70.283	180.68
$IF(g)$	95.65	−118.5	236.1
$ICl(g)$	17.8	−5.44	247.44
$IBr(g)$	40.8	3.7	258.66
$IF_7(g)$	−943.9	−818.4	346
$HI(g)$	26.5	1.7	206.48
Iron			
$Fe(s)$	0	0	27.3
$Fe(g)$	416	371	180.38
$Fe_2O_3(s)$	−824.2	−742.2	87.40
$Fe_3O_4(s)$	−1118	−1015	146
$Fe(CO)_5(l)$	−774.0	−705.4	338
$Fe(CO)_5(g)$	−733.9	−697.26	445.2
$FeSeO_3(s)$	−1200	—	—
$FeO(s)$	−272	—	—
$FeAsS(s)$	−42	−50	120
$Fe(OH)_2(s)$	−569.0	−486.6	88
$Fe(OH)_3(s)$	−823.0	−696.6	107
$FeS(s)$	−100	−100	60.29
$Fe_3C(s)$	25	20	105
Lead			
$Pb(s)$	0	0	64.81
$Pb(g)$	195	162	175.26
$PbO(s)$ (yellow)	−217.3	−187.9	68.70
$PbO(s)$ (red)	−219.0	−188.9	66.5
$Pb(OH)_2(s)$	−515.9	—	—
$PbS(s)$	−100	−98.7	91.2
$Pb(NO_3)_2(s)$	−451.9	—	—
$PbO_2(s)$	−277	−217.4	68.6
$PbCl_2(s)$	−359.4	−314.1	136
Lithium			
$Li(s)$	0	0	28.0
$Li(g)$	155.1	122.1	138.67
$LiH(s)$	−90.42	−69.96	25
$Li(OH)(s)$	−487.23	−443.9	50.2
$LiF(s)$	−612.1	−584.1	35.9
$Li_2CO_3(s)$	−1215.6	−1132.4	90.4
Manganese			
$Mn(s)$	0	0	32.0
$Mn(g)$	281	238	173.6
$MnO(s)$	−385.2	−362.9	59.71
$MnO_2(s)$	−520.03	−465.18	53.05
$Mn_2O_3(s)$	−959.0	−881.2	110
$Mn_3O_4(s)$	−1388	−1283	156
Mercury			
$Hg(l)$	0	0	76.02

Standard Molar Enthalpies of Formation, Standard Molar Free Energies of Formation, and Absolute Standard Entropies [298.15 K (25°C), 1 atm] (continued)

Substance	ΔH_f°, kJ mol^{-1}	ΔG_f°, kJ mol^{-1}	S_{298}°, J K^{-1} mol^{-1}
$Hg(g)$	61.317	31.85	174.8
$HgO(s)$ (red)	−90.83	−58.555	70.29
$HgO(g)$ (yellow)	−90.46	−57.296	71.1
$HgCl_2(s)$	−224	−179	146
$Hg_2Cl_2(s)$	−265.2	−210.78	192
$HgS(s)$ (red)	−58.16	−50.6	82.4
$HgS(s)$ (black)	−53.6	−47.7	88.3
$HgSO_4(s)$	−707.5	—	—
Nitrogen			
$N_2(g)$	0	0	191.5
$N(g)$	472.704	455.579	153.19
$NO(g)$	90.25	86.57	210.65
$NO_2(g)$	33.2	51.30	239.9
$N_2O(g)$	82.05	104.2	219.7
$N_2O_3(g)$	83.72	139.4	312.2
$N_2O_4(g)$	9.16	97.82	304.2
$N_2O_5(g)$	11	115	356
$NH_3(g)$	−46.11	−16.5	192.3
$N_2H_4(l)$	50.63	149.2	121.2
$N_2H_4(g)$	95.4	159.3	238.4
$NH_4NO_3(s)$	−365.6	−184.0	151.1
$NH_4Cl(s)$	−314.4	−201.5	94.6
$NH_4Br(s)$	−270.8	−175	113
$NH_4I(s)$	−201.4	−113	117
$NH_4NO_2(s)$	−256	—	—
$HNO_3(l)$	−174.1	−80.79	155.6
$HNO_3(g)$	−135.1	−74.77	266.2
Oxygen			
$O_2(g)$	0	0	205.03
$O(g)$	249.17	231.75	160.95
$O_3(g)$	143	163	238.8
Phosphorus			
$P(s)$	0	0	41.1
$P(g)$	58.91	24.5	280.0
$P_4(g)$	314.6	278.3	163.08
$PH_3(g)$	5.4	13	210.1
$PCl_3(g)$	−287	−268	311.7
$PCl_5(g)$	−375	−305	364.5
$P_4O_6(s)$	−1640	—	—
$P_4O_{10}(s)$	−2984	−2698	228.9
$HPO_3(s)$	−948.5	—	—
$H_3PO_2(s)$	−604.6	—	—
$H_3PO_3(s)$	−964.4	—	—
$H_3PO_4(s)$	−1279	−1119	110.5
$H_3PO_4(l)$	−1267	—	—
$H_4P_2O_7(s)$	−2241	—	—
$POCl_3(l)$	−597.1	−520.9	222.5
$POCl_3(g)$	−558.48	−512.96	325.3

Standard Molar Enthalpies of Formation, Standard Molar Free Energies of Formation, and Absolute Standard Entropies [298.15 K (25°C), 1 atm] (continued)

Substance	ΔH_f°, kJ mol^{-1}	ΔG_f°, kJ mol^{-1}	S_{298}°, J K^{-1} mol^{-1}
Potassium			
K(s)	0	0	63.6
K(g)	90.00	61.17	160.23
KF(s)	−562.58	−533.12	66.57
KCl(s)	−435.868	−408.32	82.68
Silicon			
Si(s)	0	0	18.8
Si(g)	455.6	411	167.9
SiO$_2$(s)	−910.94	−856.67	41.84
SiH$_4$(g)	34	56.9	204.5
H$_2$SiO$_3$(s)	−1189	−1092	130
H$_4$SiO$_4$(s)	−1481	−1333	190
SiF$_4$(g)	−1614.9	−1572.7	282.4
SiCl$_4$(l)	−687.0	−619.90	240
SiCl$_4$(g)	−657.01	−617.01	330.6
SiC(s)	−65.3	−62.8	16.6
Silver			
Ag(s)	0	0	42.55
Ag(g)	284.6	245.7	172.89
Ag$_2$O(s)	−31.0	−11.2	121
AgCl(s)	−127.1	−109.8	96.2
Ag$_2$S(s)	−32.6	−40.7	144.0
Sodium			
Na(s)	0	0	51.0
Na(g)	108.7	78.11	153.62
Na$_2$O(s)	−415.9	−377	72.8
NaCl(s)	−411.00	−384.03	72.38
Sulfur			
S(s) (rhombic)	0	0	31.8
S(g)	278.80	238.27	167.75
SO$_2$(g)	−296.83	−300.19	248.1
SO$_3$(g)	−395.7	−371.1	256.6
H$_2$S(g)	−20.6	−33.6	205.7
H$_2$SO$_4$(l)	−813.989	690.101	156.90
H$_2$S$_2$O$_7$(s)	−1274	—	—
SF$_4$(g)	−774.9	−731.4	291.9
SF$_6$(g)	−1210	−1105	291.7
SCl$_2$(l)	−50	—	—
SCl$_2$(g)	−20	—	—
S$_2$Cl$_2$(l)	−59.4	—	—
S$_2$Cl$_2$(g)	−18	−32	331.4
SOCl$_2$(l)	−246	—	—
SOCl$_2$(g)	−213	−198	309.7
SO$_2$Cl$_2$(l)	−394	—	—
SO$_2$Cl$_2$(g)	−364	−320	311.8
Tin			
Sn(s)	0	0	51.55
Sn(g)	302	267	168.38

Standard Molar Enthalpies of Formation, Standard Molar Free Energies of Formation, and Absolute Standard Entropies [298.15 K (25°C), 1 atm] (continued)

Substance	ΔH_f°, kJ mol^{-1}	ΔG_f°, kJ mol^{-1}	S_{298}°, J K^{-1} mol^{-1}
SnO(s)	−286	−257	56.5
SnO$_2$(s)	−580.7	−519.7	52.3
SnCl$_4$(l)	−511.2	−440.2	259
SnCl$_4$(g)	−471.5	−432.2	366
Titanium			
Ti(s)	0	0	30.6
Ti(g)	469.9	425.1	180.19
TiO$_2$(s)	−944.7	−889.5	50.33
TiCl$_4$(l)	−804.2	−737.2	252.3
TiCl$_4$(g)	−763.2	−726.8	354.8
Tungsten			
W(s)	0	0	32.6
W(g)	849.4	807.1	173.84
WO$_3$(s)	−842.87	−764.08	75.90
Zinc			
Zn(s)	0	0	41.6
Zn(g)	130.73	95.178	160.87
ZnO(s)	−348.3	−318.3	43.64
ZnCl$_2$(s)	−415.1	−369.43	111.5
ZnS(s)	−206.0	−201.3	57.7
ZnSO$_4$(s)	−982.8	−874.5	120
ZnCO$_3$(s)	−812.78	−731.57	82.4
Complexes			
[Co(NH$_3$)$_4$(NO$_2$)$_2$]NO$_3$, cis	−898.7	—	—
[Co(NH$_3$)$_4$(NO$_2$)$_2$]NO$_3$, trans	−896.2	—	—
NH$_4$[Co(NH$_3$)$_2$(NO$_2$)$_4$]	−837.6	—	—
[Co(NH$_3$)$_6$][Co(NH$_3$)$_2$(NO$_2$)$_4$]$_3$	−2733	—	—
[Co(NH$_3$)$_4$Cl$_2$]Cl, cis	−997.0	—	—
[Co(NH$_3$)$_4$Cl$_2$]Cl, trans	−999.6	—	—
[Co(en)$_2$(NO$_2$)$_2$]NO$_3$, cis	−689.5	—	—
[Co(en)$_2$Cl$_2$]Cl, cis	−681.1	—	—
[Co(en)$_2$Cl$_2$]Cl, trans	−677.4	—	—
[Co(en)$_3$](ClO$_4$)$_3$	−762.7	—	—
[Co(en)$_3$]Br$_2$	−595.8	—	—
[Co(en)$_3$]I$_2$	−475.3	—	—
[Co(en)$_3$]I$_3$	−519.2	—	—
[Co(NH$_3$)$_6$](ClO$_4$)$_3$	−1035	−227	636
[Co(NH$_3$)$_5$NO$_2$](NO$_3$)$_2$	−1089	−418.4	350
[Co(NH$_3$)$_6$](NO$_3$)$_3$	−1282	−530.5	469
[Co(NH$_3$)$_5$Cl]Cl$_2$	−1017	−582.8	366
[Pt(NH$_3$)$_4$]Cl$_2$	−728.0	—	—
[Ni(NH$_3$)$_6$]Cl$_2$	−994.1	—	—
[Ni(NH$_3$)$_6$]Br$_2$	−923.8	—	—
[Ni(NH$_3$)$_6$]I$_2$	−808.3	—	—